Functional Nanometer-Sized Clusters of Transition Metals

Synthesis, Properties and Applications

RSC Smart Materials

Series Editors:
Hans-Jörg Schneider, *Saarland University, Germany*
Mohsen Shahinpoor, *University of Maine, USA*

Titles in this Series:

1: Janus Particle Synthesis, Self-Assembly and Applications
2: Smart Materials for Drug Delivery: Volume 1
3: Smart Materials for Drug Delivery: Volume 2
4: Materials Design Inspired by Nature
5: Responsive Photonic Nanostructures: Smart Nanoscale Optical Materials
6: Magnetorheology: Advances and Applications
7: Functional Nanometer-Sized Clusters of Transition Metals: Synthesis, Properties and Applications

How to obtain future titles on publication:
A standing order plan is available for this series. A standing order will bring delivery of each new volume immediately on publication.

For further information please contact:
Book Sales Department, Royal Society of Chemistry, Thomas Graham House, Science Park, Milton Road, Cambridge, CB4 0WF, UK
Telephone: +44 (0)1223 420066, Fax: +44 (0)1223 420247
Email: booksales@rsc.org
Visit our website at www.rsc.org/books

Functional Nanometer-Sized Clusters of Transition Metals
Synthesis, Properties and Applications

Edited by

Wei Chen
Institute of Applied Chemistry, Chinese Academy of Sciences,
Changchun, P. R. China
Email: weichen@ciac.jl.cn

and

Shaowei Chen
Department of Chemistry, University of California, USA
Email: shaowei@ucsc.edu

ROYAL SOCIETY
OF **CHEMISTRY**

THE QUEEN'S AWARDS
FOR ENTERPRISE:
INTERNATIONAL TRADE
2013

RSC Smart Materials No. 7

ISBN: 978-1-84973-824-8
PDF eISBN: 978-1-78262-851-4
ISSN: 2046-0066

Published by The Royal Society of Chemistry,
Thomas Graham House, Science Park, Milton Road,
Cambridge CB4 0WF, UK

Registered Charity Number 207890

For further information see our web site at www.rsc.org

Printed and bound by CPI Group (UK) Ltd, Croydon, CR0 4YY

Preface

Metal nanocrystals in the size range of 1 to 100 nm have attracted extensive attention in the past decades due to their unusual properties and potential applications in many areas. In particular, metal clusters with a core size smaller than 2 nm exhibit unique physical and chemical properties that are significantly different from those of the corresponding large nanoparticles and molecular compounds. Such a size range represents a bridge between atoms/molecules and nanoparticles, and thus represents a fascinating multidisciplinary research area. Yet, because of the ultrasmall dimensions, development of effective protocols for the size-controlled synthesis has been a significant challenge in the studies of metal nanoclusters. In recent years, various effective strategies have been described in the literature for the synthesis of metal nanoclusters with a specific number of metal atoms and surface protecting ligands. The successful preparation of composition-determined metal clusters renders it possible to study their size- and composition-related properties. Significantly, with a size comparable to the Fermi wavelength of electrons and consequently the formation of discrete electronic energy levels, metal nanoclusters exhibit unprecedented optical and electronic properties, including size-dependent energy level structures, photoluminescence, and catalytic properties. Therefore, metal nanoclusters have been found to show promising applications in nanoelectronics, catalysis, biological and chemical sensing, molecular imaging, biological labeling, biomedicine, and so on.

This book highlights some recent progress in the synthesis, characterization, interfacial engineering and applications of metal nanoclusters. The contributors to this book consist of leading experts in this field. For organothiol-stabilized metal nanoclusters, the mechanism of cluster formation is of critical importance to achieve structure-controlled synthesis. Tong and

RSC Smart Materials No. 7
Functional Nanometer-Sized Clusters of Transition Metals: Synthesis, Properties and Applications
Edited by Wei Chen and Shaowei Chen
© The Royal Society of Chemistry 2014
Published by the Royal Society of Chemistry, www.rsc.org

co-workers first review and discuss the relevant chemistry involved in the two-phase synthesis of alklychalcogenolate-stabilized metal nanoparticles. As an important research field of metal nanoclusters, González and López-Quintela summarize the recently developed strategies and synthetic routes for the preparation of photoluminescent atomic quantum clusters. Up to now, the research in the field of metal nanocluster is mainly concentrated on gold and silver metals. In this book, several chapters focus on Au and Ag nanoclusters. Bigioni *et al.* report the magic-number silver nanoclusters, Xu and Suslick summarize the work on water-soluble fluorescent silver nanoclusters, Yang and Wang highlight biomolecule-protected silver nanoclusters, Xie *et al.* present the newly-developed synthetic strategies, and Pradeep *et al.* discuss in detail the preparation and application of noble metal clusters in protein templates. As an important application of metal nanoclusters, López-Quintela *et al.*, Lu and Chen, and Tsukuda *et al.* review the catalytic properties of metal nanoclusters from different aspects. Wang and Ubaldo give a summary of the development of In Silico study of metal nanoclusters, which is helpful for the experimental syntheses and design of metal clusters. Zhou and Dong and Tanaka and Inouye highlight the biological applications of metal nanoclusters. Chen *et al.* review the recent advances in Janus nanoparticles through interfacial engineering. The editors express their deep appreciation to the authors for their support and contributions to the book.

This book will be a valuable reference for researchers in the general area of functional nanomaterials. It may also serve as a study guide for graduate students and senior undergraduate students who are interested in nanoscale materials chemistry and engineering.

Wei Chen
Shaowei Chen

Contents

Chapter 1   **Mechanistic Insights into the Brust–Schiffrin Synthesis of Organochalcogenolate-Stabilized Metal Nanoparticles**   1
Yuan Gao, Yangwei Liu, Ying Li, Oksana Zaluzhna and YuYe J. Tong

    1.1   Introduction   1
    1.2   Phase Transfer of Metal Ions: Formation of Inverse Micelle Encapsulated Water   2
         1.2.1   Proton NMR Evidence of Encapsulated Water   2
         1.2.2   Phase Transfer of Metal Ions: Formation of Metal Complex   5
    1.3   Addition of Ligand: Reduction of Metal Complex or Formation of Polymeric Species   8
         1.3.1   Alkylthiols RSH   8
         1.3.2   Dialkyl Diselenide RSe–SeR   11
         1.3.3   Dialkyl Ditelluride RTe–TeR   14
    1.4   Critical Role of Water   16
    1.5   Reduction by $NaBH_4$: Formation of Zero-Valence Metal Nanoparticles   19
         1.5.1   Normal Reduction Sequence   19
         1.5.2   Reversed Reduction Sequence   20
    1.6   Conclusions and Future Outlook   22
    Acknowledgements   23
    References   23

RSC Smart Materials No. 7
Functional Nanometer-Sized Clusters of Transition Metals: Synthesis, Properties and Applications
Edited by Wei Chen and Shaowei Chen
© The Royal Society of Chemistry 2014
Published by the Royal Society of Chemistry, www.rsc.org

**Chapter 2   New Strategies and Synthetic Routes to Synthesize
Fluorescent Atomic Quantum Clusters   25**
Beatriz Santiago González and M. Arturo López-Quintela

    2.1   Introduction   25
    2.2   Top-Down Approach   27
    2.3   Bottom-Up Approach   30
        2.3.1   Chemical Reduction   31
        2.3.2   Photoreduction   33
        2.3.3   Sonochemistry   34
        2.3.4   Electrochemical Synthesis   35
        2.3.5   Microemulsions   36
        2.3.6   Microwave-Assisted Synthesis   37
        2.3.7   Template-Assisted Synthesis   38
    2.4   Summary   43
    Acknowledgements   47
    References   47

Chapter 3   Silver Magic-Number Clusters and Their Properties   51
*Brian A. Ashenfelter, Anil Desireddy, Jingshu Guo,
Brian E. Conn, Wendell P. Griffith and Terry P. Bigioni*

    3.1   Introduction   51
    3.2   Synthesis   52
        3.2.1   Aqueous Synthesis   52
        3.2.2   Non-Aqueous Synthesis   55
        3.2.3   Solid-State Synthesis   57
    3.3   PAGE Separations   58
    3.4   Mass Spectrometry   61
    3.5   Optical Properties   65
        3.5.1   Electronic Structure Theory   66
        3.5.2   Absorption Spectroscopy   67
        3.5.3   Fluorescence Spectroscopy   70
        3.5.4   Spectroscopy Challenges   74
    3.6   Conclusions   75
    References   75

**Chapter 4   Synthesis and Applications of Water-Soluble Fluorescent
Silver Nanoclusters   80**
Hangxun Xu and Kenneth S. Suslick

    4.1   Introduction   80
    4.2   Synthesis of Water-Soluble Fluorescent Ag
        Nanoclusters   81
        4.2.1   Radiolytic Reduction Synthesis of Ag
            Nanoclusters   81

| | | 4.2.2 | Photochemical Reduction Synthesis of Ag Nanoclusters | 82 |

4.2.2 Photochemical Reduction Synthesis of Ag Nanoclusters 82

4.2.3 Sonochemical Preparation of Ag Nanoclusters 84

4.2.4 Microwave-Assisted Synthesis of Ag Nanoclusters 85

4.2.5 Chemical Reduction for Preparation of Ag Nanoclusters 86

4.3 Applications of Water-Soluble Fluorescent Ag Nanoclusters 92

4.3.1 Applications of Ag Nanoclusters in Chemical Sensing 92

4.3.2 Applications of Ag Nanoclusters in Biosensing 93

4.4 Conclusions 96

References 97

Chapter 5 Synthesis and Applications of Silver Nanoclusters Protected by Polymers, Protein, Peptide and Short Molecules 100
X. Yang and E. K. Wang

5.1 Introduction 100

5.2 Polymers Protected Silver Nanoclusters and Their Related Applications 102

5.3 Protein- and Peptide-Protected Silver Nanoclusters and Their Related Applications 108

5.3.1 Protein-Protected Silver Nanoclusters and Their Related Applications 108

5.3.2 Peptide-Protected Silver Nanoclusters and Their Related Applications 110

5.4 Short Molecule-Protected Silver Nanoclusters and Their Related Applications 113

5.5 Conclusion and Perspective 125

Acknowledgements 127

References 127

Chapter 6 Novel Synthetic Strategies for Thiolate-Protected Au and Ag Nanoclusters: Towards Atomic Precision and Strong Luminescence 131
Xun Yuan, Qiaofeng Yao, Yong Yu, Zhentao Luo and Jianping Xie

6.1 Introduction 131

6.2   General Synthetic Routes for
      Thiolate-Protected Au/Ag NCs: Brust and
      Brust-Like Methods   133
6.3   Reductive Decomposition of Au(ɪ)/Ag(ɪ)–SR
      Complexes   136
      6.3.1   Tailoring the Size of Au(ɪ)/Ag(ɪ)–SR
              Complexes   136
      6.3.2   Tailoring the Structure of Au(ɪ)/Ag(ɪ)–SR
              Complexes   139
      6.3.3   Tailoring the Reductive Decomposition
              Kinetics   141
6.4   Thiol Etching of Polydisperse Au/Ag NCs   146
      6.4.1   Size-Dependent Stability – the Driving Force
              for Size-Focusing   146
      6.4.2   Facile Size-Focusing Methods   150
      6.4.3   Tailoring the Thiol Etching Process to
              Synthesize Metastable Au NCs   151
      6.4.4   Versatile Size-Focusing Methods   156
6.5   Thiol Etching of Large Au/Ag NPs   156
      6.5.1   Delicate Selection of the Etchants   156
      6.5.2   Tailoring the Etching Environment   158
      6.5.3   Tailoring the NP Precursors   161
6.6   Conclusions   162
References   164

Chapter 7 **Noble Metal Clusters in Protein Templates**   169
           Thalappil Pradeep, Ananya Baksi and Paulrajpillai
           Lourdu Xavier

7.1   Introduction   169
      7.1.1   General Properties of Clusters   169
      7.1.2   Trends in the Choice of Ligands for Cluster
              Synthesis   171
      7.1.3   Protein Protected Metal
              Clusters—Conglomeration of Disciplines   173
7.2   Synthesis and Characterization   177
      7.2.1   General Synthetic Route and Separation of
              Protein Protected Clusters   177
      7.2.2   General Characterization   178
      7.2.3   Mass Spectrometry and Clusters   179
      7.2.4   Mass Spectrometric Studies on the Growth of
              Clusters in Protein Templates   184
      7.2.5   Conformational Changes in Proteins Upon
              Cluster Synthesis   188
      7.2.6   Peptide Protected Metal Clusters   191

7.3   Origin and Properties of Luminescence in Protein
       Protected Noble Metal Clusters   192
       7.3.1   Mechanism of Metal Ion Induced Quenching
               of Luminescence   196
7.4   Applications of NMQCs@Proteins   199
       7.4.1   Sensing   199
       7.4.2   Bio-imaging   202
       7.4.3   Molecular Imaging Guided Delivery of
               Therapeutics   207
       7.4.4   Other Applications   209
7.5   Gas Phase Clusters Derived From Protein Templates   209
7.6   Outlook   214
Acknowledgements   215
References   215

Chapter 8   Metal(0) Clusters in Catalysis   226
Noelia Vilar-Vidal, José Rivas and M. Arturo López-Quintela

8.1   Introduction   226
8.2   Metal Cluster Mediated Catalysis   228
       8.2.1   Homogeneous Catalysis   228
       8.2.2   Heterogeneous Catalysis   243
8.3   Conclusions and Remarks   254
Acknowledgements   255
References   255

**Chapter 9   Metal Nanoclusters: Size-Controlled Synthesis and
              Size-Dependent Catalytic Activity   261**
Yizhong Lu and Wei Chen

9.1   Introduction   261
9.2   Size-Controlled Synthesis of Metal Nanoclusters   262
       9.2.1 $Au_{25}(SR)_{18}$ Nanoclusters   264
       9.2.2 $Au_{38}(SR)_{24}$ Nanoclusters   266
       9.2.3 $Au_{102}(SR)_{44}$ Nanoclusters   268
       9.2.4 $Au_{144}(SR)_{60}$ Nanoclusters   270
       9.2.5   Other Atomic Monodisperse $Au_n(SR)_m$
               Nanoclusters   271
       9.2.6   Ag Nanoclusters   271
       9.2.7   Cu Nanoclusters   273
       9.2.8   Pt Nanoclusters   275
       9.2.9   Pd Nanoclusters   275
9.3   Size-Dependent Catalytic Activity of Metal
       Nanoclusters   276
       9.3.1   CO Oxidation   276
       9.3.2   Oxygen Reduction Reaction (ORR)   278

9.3.3   Aerobic Oxidation   282
9.3.4   Other Catalytic Applications   283
9.4   Conclusions and Future Outlook   285
Acknowledgements   285
References   286

Chapter 10 Metal Clusters in Catalysis   **291**
Seiji Yamazoe and Tatsuya Tsukuda

10.1   Introduction   291
        10.1.1   Why Metal Clusters?   291
        10.1.2   Classification   294
10.2   Stabilized/Protected Metal Cluster Catalysts   294
        10.2.1   Size-Controlled Synthesis   294
        10.2.2   Composition-Controlled Synthesis   299
        10.2.3   Catalytic Applications   301
10.3   Supported Metal Cluster Catalysts   306
        10.3.1   Size-Controlled Synthesis   306
        10.3.2   Composition-Controlled Synthesis   310
        10.3.3   Catalytic Applications   312
10.4   Summary and Prospects   316
References   317

Chapter 11 *In Silico* Studies of Functional Transition Metal
            Nanoclusters   **323**
Lichang Wang and Pamela C. Ubaldo

11.1   *In Silico* Synthesis and Characterization of
        Functional Transition Metal Nanoclusters   323
        11.1.1   Synthesis   324
        11.1.2   Characterization   328
11.2   Catalysis of Transition Metal Nanoclusters   330
        11.2.1   Sinter-Resistant Ir Nanoclusters   331
        11.2.2   Multicomponent Pt Alloy Nanoclusters for
                  O_2 Reduction   334
        11.2.3   Pt Nanoclusters for Activation of C–H
                  bonds in CH_4 338
11.3   Other Functionalities of Transition Metal
        Nanoclusters   340
        11.3.1   Pd Nanoclusters for Sensing CH_4 341
        11.3.2   Au Nanoclusters for Chiral
                  Recognition   344
        11.3.3   Ag Nanoclusters as Conductive Ink   346
11.4   Conclusions   348
References   349

Chapter 12 DNA-Templated Metal Nanoclusters and Their Applications 352

Zhixue Zhou and Shaojun Dong

12.1 Introduction 352
12.2 The Interactions Between DNA and Metal Ions 354
12.3 The Synthesis, Characterization and Unique Properties of DNA-Templated Metal NCs 355
   12.3.1 The Synthesis of DNA/Metal NCs 355
   12.3.2 Characterization 355
   12.3.3 The Unique Properties of DNA–Metal NCs 362
12.4 DNA-Templated Ag NCs and Applications 365
   12.4.1 DNA-Templated Ag NCs 365
   12.4.2 Application of Fluorescent DNA–Ag NCs 371
12.5 DNA-Templated Au NCs 383
12.6 Conclusions and Future Outlook 384
References 385

Chapter 13 Synthesis of Fluorescent Platinum Nanoclusters for Biomedical Imaging 391

Shin-ichi Tanaka and Yasushi Inouye

13.1 Introduction 391
13.2 Experimental 392
13.3 Blue-Emitting Platinum Nanoclusters 394
   13.3.1 Preparation of Blue-Emitting Platinum Nanoclusters 394
   13.3.2 Characterization 396
   13.3.3 Application to Bioimaging 396
13.4 Green-Emitting Platinum Nanoclusters 398
   13.4.1 Preparation of Green-Emitting Platinum Nanoclusters 398
   13.4.2 Characterization 400
   13.4.3 Application to Bioimaging 402
13.5 Conclusion 403
References 404

Chapter 14 Janus Nanoparticles by Interfacial Engineering 407

Yang Song, Xiaojun Liu and Shaowei Chen

14.1 Introduction 407
14.2 Polymer-Based Janus Structures 408
   14.2.1 Copolymerization 408
   14.2.2 Electrospinning 411
   14.2.3 Polymer–Inorganic Heterodimers 413

14.3   Metal Nanocrystals   414
        14.3.1   Solid Masks   414
        14.3.2   Soft Masks   418
        14.3.3   Controlled Phase Separation and Growth   420
        14.3.4   Bimetallic   423
14.4   Conclusion   426
Acknowledgements   427
References   427

Subject Index   **434**

CHAPTER 1

Mechanistic Insights into the Brust–Schiffrin Synthesis of Organochalcogenolate-Stabilized Metal Nanoparticles

YUAN GAO, YANGWEI LIU, YING LI, OKSANA ZALUZHNA AND YUYE J. TONG*

Department of Chemistry, Georgetown University, 37th & O Streets, NW, Washington, DC 20057, USA
*Email: yyt@georgetown.edu

1.1 Introduction

Metal nanoparticles (NPs) made of tens, hundreds, or thousands of atoms can have tunable chemical and physical properties as a function of NP size (number of atoms), elemental composition, and/or chemical environment (ligand-stabilized, matrix-embedded, or structurally-encaged). These NPs are artificial atoms[1–7] and novel building blocks for new materials that hold novel physicochemical properties as compared to the existing (atomic/molecular) materials. It is expected that these novel materials will enable widespread technological breakthroughs in the not too distant future, for instance in molecular and/or nano-electronics and clean energy generation.[8,9] Within this broad context, organoligands, particularly organothiolate-stabilized metal (mainly Au) NPs, have been subjected to intensive

RSC Smart Materials No. 7
Functional Nanometer-Sized Clusters of Transition Metals: Synthesis, Properties and Applications
Edited by Wei Chen and Shaowei Chen
© The Royal Society of Chemistry 2014
Published by the Royal Society of Chemistry, www.rsc.org

research over the last two decades due to their potential applications in nano-optics,[10] nano-electronics,[11] (bio)sensing[12] and medicinal science (theranostics).[13]

The first step towards any practical applications of metal NPs is the synthesis of these metal NPs, preferably air-stable and of homogeneous size distribution and known chemical composition. Among many synthetic methods, the *Brust–Schiffrin two-phase method* (BSM) synthesis worked out by Brust, Schiffrin, and company in 1994,[14] including its late variants, is definitively the most widely employed synthetic approach to make <5 nm organo-ligand-stabilized metal NPs. Briefly, a typical BSM consists of three steps: Step 1, metal ions are phase transferred (PT-ed) from an aqueous to an organic phase (usually toluene or benzene) with a PT reagent (usually tetraoctylammonium bromide (TOAB), *i.e.* R_4NBr, $R = C_8H_{17}$). Step 2, organochalcogen-containing ligand (usually RSH) is added to the separated organic phase during which Au^{III} cations can be reduced to Au^{I} cations. Step 3, metal ions residing in the separated organic phase are reduced into M^0 by a reducing reagent like $NaBH_4$ during which organochalcogenolate-protected metal NPs are formed.

Despite the prevailing use of the BSM in the synthesis of sub-5 nm metal (mainly Au) NPs (according to Thomson Reuters' Web of Knowledge, the original paper[14] has accumulated a current number of citations as high as 3755, and counting), mechanistic details of the BSM synthesis have been sketchy until very recently.[15–18] A long-held belief concerning the metal precursor in the synthesis of metal NPs, probably due to earlier papers by Whetten *et al.*,[19,20] has been that the metal-thiolate polymer, $[Au^ISR]_n$, is the metal ion precursor of metal NPs. However, a recent paper by Goulet and Lennox[18] has shown that the metal–TOA^+ complex, $[TOA][Au^IBr_2]$, can also be the major metal ion precursor. Our ensuing studies have not only confirmed the results of Goulet and Lennox, but also proposed that the BSM synthesis is an inverse micelle based approach based on their proton NMR results and showed *via* Raman spectroscopic study that the Au–S bond does not form until the formation of Au NPs.[15] In this chapter, we will review and discuss in various degrees of detail the relevant chemistry involved, particularly the role of encapsulated water, in the BSM synthesis of alkyl-chalcogenolate-stabilized metal NPs unravelled after the paper of Goulet and Lennox[18] and highlight the similarity and difference when ligands containing different chalcogen elements (S, Se, or Te) are used as the starting source of the NP-stabilizing agents.

1.2 Phase Transfer of Metal Ions: Formation of Inverse Micelle Encapsulated Water

1.2.1 Proton NMR Evidence of Encapsulated Water

The experimental evidence of possible encapsulated water by TOAB in an organic phase came first from the observation of a large down-field shift

(~2 ppm, due largely to the appearance of hydrogen bonding among the water molecules that strongly suggests the formation of water aggregates) of the water proton peak in C_6D_6 containing dissolved TOAB (0.03 mmol of TOAB in 0.8 mL C_6D_6) as compared to that of pure C_6D_6 after both being mixed with 0.105 mL Milli-Q water (18.2 MΩ) and the extra water layer being then removed, as shown in Figure 1.1. The water peak at 2.43 ppm was also reported in the Goulet and Lennox paper.[18]

More convincing and detailed evidence of the inverse micelle formation is shown in Figure 1.2, proton NMR spectra of a series of samples prepared by mixing various amounts of TOAB with 0.8 mL C_6D_6 and 0.210 mL Milli-Q water and then separating the undissolved water. The two clearly distinguishable regimes enable the critical micelle concentration (CMC) to be determined:[21] the intersection of the two dashed lines gives the CMC of TOAB in $C_6D_6 = 7.5$ mM, which is about 4 to 5 times smaller than the TOAB concentrations generally used in a typical BSM synthesis of metal NPs. That is, under the normal condition of the BSM synthesis, inverse micelles enclosed by TOAB are formed.

Unlike other well-known inverse micelle systems, such as sodium 2-ethylhexylsulfosuccinate (AOT), whose size can be readily varied by changing water/surfactant ratio,[22] our proton NMR data show that the inverse micelles of TOAB can be saturated even with a very small amount of water, as shown by the proton NMR spectra in Figure 1.3A and the corresponding normalized peak integrals in Figure 1.3B.[16] The samples for the spectra in Figure 1.3A were prepared with the same amount of TOAB (0.03 mmol) in

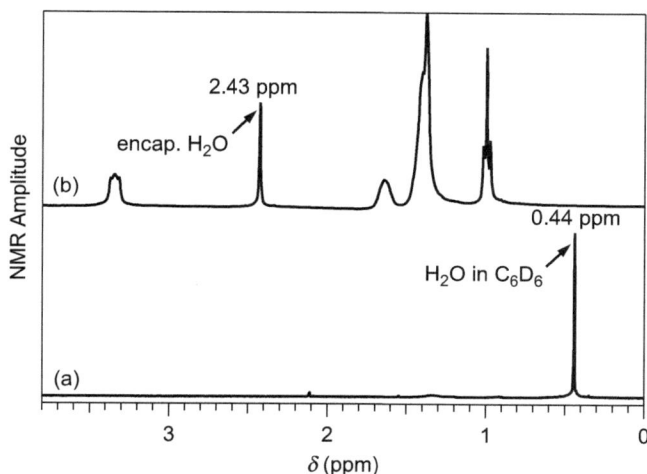

Figure 1.1 Proton NMR spectra of (a) water in pure C_6D_6 and (b) in C_6D_6 with dissolved TOAB. Both samples were prepared using the same volume of C_6D_6 (0.8 mL) and mixed with the same volume of water (0.105 mL) before being separated. The amount of TOAB was 0.03 mmol in (b). Modified from ref. 15. Copyright 2011, American Chemical Society.

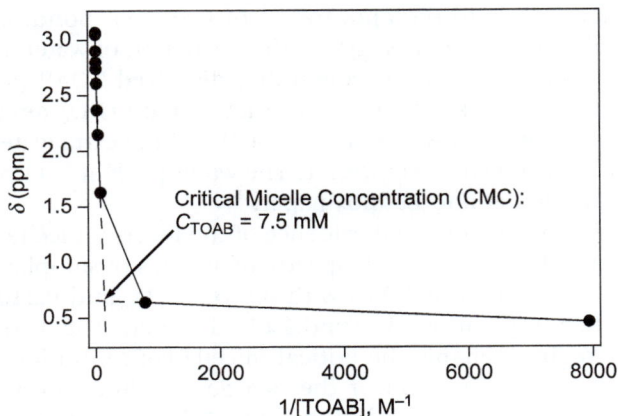

Figure 1.2   The proton peak shift of water in C_6D_6 solution of TOAB as a function of the TOAB concentration. The two clearly distinguishable variation regimes indicate the formation of inverse micelles.

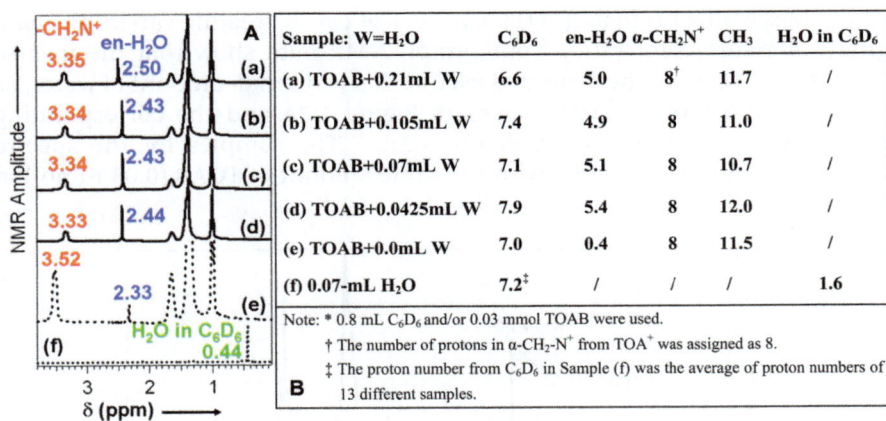

Sample: W=H₂O	C_6D_6	en-H₂O	a-CH₂N⁺	CH₃	H₂O in C_6D_6
(a) TOAB+0.21mL W	6.6	5.0	8†	11.7	/
(b) TOAB+0.105mL W	7.4	4.9	8	11.0	/
(c) TOAB+0.07mL W	7.1	5.1	8	10.7	/
(d) TOAB+0.0425mL W	7.9	5.4	8	12.0	/
(e) TOAB+0.0mL W	7.0	0.4	8	11.5	/
(f) 0.07-mL H₂O	7.2‡	/	/	/	1.6

Note: * 0.8 mL C_6D_6 and/or 0.03 mmol TOAB were used.
† The number of protons in a-CH₂-N⁺ from TOA⁺ was assigned as 8.
‡ The proton number from C_6D_6 in Sample (f) was the average of proton numbers of 13 different samples.

Figure 1.3   The proton NMR spectra (A) and the corresponding normalized peak integrals (B) of a series of samples whose preparation parameters are explained in (B).
Modified from ref. 16. Copyright 2011, American Chemical Society.

0.8 mL C_6D_6 but mixed with different amounts of water before the organic phase being separated for the NMR measurements. Interestingly, the amount of encapsulated water remained constant even if the water amount varied from 0.0425 mL to 0.21 mL, so did the peak positions of the encapsulated water and –CH₂N⁺.

With the information in Figure 1.3, we can estimate the average size of the inverse micelles. According to Figure 1.3B, the average number of protons from the H₂O in Samples (a–d) is 5.11 per TOAB. After subtracting the number of protons from the H₂O saturated in C_6D_6 (*i.e.* 1.63 per TOAB

(Sample (f)), the number of protons from the H_2O encapsulated in the inverse micelles formed by TOAB is 3.48, which means that the ratio of the encapsulated H_2O to TOAB is 1.74. Assuming that one molecule of TOAB is surrounded by one molecule of H_2O, 57% of the encapsulated H_2O might be located in the outer layer of the inverse micelle core. Together with the fact that the H_2O volume per molecule is 0.03 nm^3 at 20 °C (density of H_2O: 0.998 g mL^{-1}), the diameter of the inverse micelle core, *i.e.* the part of the H_2O encapsulated in the inverse micelle of TOAB, is ~2.5 nm, which is at the small end of known inverse micelles.[22]

1.2.2 Phase Transfer of Metal Ions: Formation of Metal Complex

The first step of the BSM synthesis is the phase transfer (PT) of metal ions from the initial aqueous solution to the organic phase using TOAB as PT agent. The proton spectra in Figure 1.4A were obtained after the PT of different amounts of Au ions into the organic phase.[16] The corresponding normalized proton peak integrals and sample preparation parameters are collected in Figure 1.4B. In addition to constant C_6D_6 volume (0.8 mL) and amount of TOAB (0.03 mmol), the volumes of the aqueous solutions were kept the same as used for Figure 1.3. Thus, the corresponding TOAB:Au ratios were 1:1, 2:1, 3:1, and 5:1 for spectra (g) to (j) respectively. For comparison, the ^{1}H NMR spectra of the synthesized [TOA][AuBr$_4$] complex (0.0125 mmol) in 0.8 mL of C_6D_6 with (0.09 mL) and without H_2O mixing are shown in spectra (k) and (l) respectively.

As can be seen in Figure 1.4A, the introduction of metal species not only caused clear variation of peak positions of the encapsulated water and α-

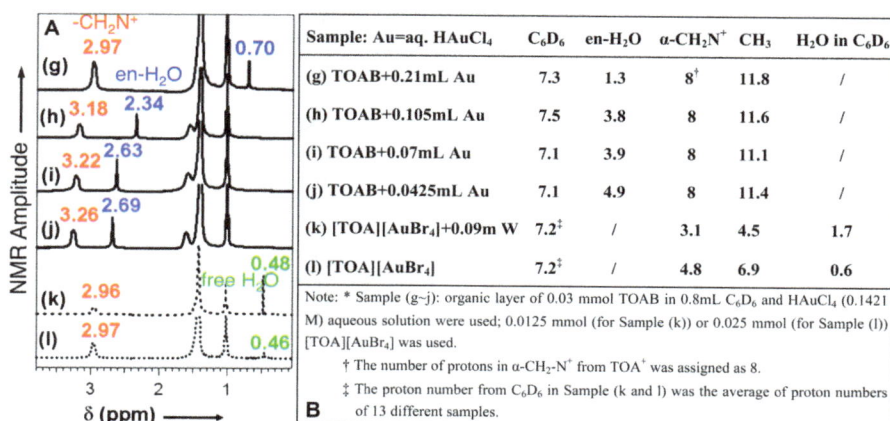

Sample: Au=aq. HAuCl$_4$	C_6D_6	en-H_2O	α-CH$_2$N$^+$	CH$_3$	H_2O in C_6D_6
(g) TOAB+0.21mL Au	7.3	1.3	8†	11.8	/
(h) TOAB+0.105mL Au	7.5	3.8	8	11.6	/
(i) TOAB+0.07mL Au	7.1	3.9	8	11.1	/
(j) TOAB+0.0425mL Au	7.1	4.9	8	11.4	/
(k) [TOA][AuBr$_4$]+0.09m W	7.2‡	/	3.1	4.5	1.7
(l) [TOA][AuBr$_4$]	7.2‡	/	4.8	6.9	0.6

Note: * Sample (g~j): organic layer of 0.03 mmol TOAB in 0.8mL C_6D_6 and HAuCl$_4$ (0.1421 M) aqueous solution were used; 0.0125 mmol (for Sample (k)) or 0.025 mmol (for Sample (l)) [TOA][AuBr$_4$] was used.

† The number of protons in α-CH$_2$-N$^+$ from TOA$^+$ was assigned as 8.

‡ The proton number from C_6D_6 in Sample (k and l) was the average of proton numbers of 13 different samples.

Figure 1.4 The proton NMR spectra (A) and the corresponding normalized proton peak integrals and sample preparation parameters (B). Modified from ref. 16. Copyright 2011, American Chemical Society.

proton in $-CH_2N^+$, but also the amount of encapsulated water. The pair $(H_2O/-CH_2N^+)$ peak values are now down-field shifted as the Au content decreases: 0.70/2.97, 2.34/3.18, 2.63/3.22, and 2.69/3.26 ppm for TOAB : Au ratios of 1:1, 2:1, 3:1, and 5:1, respectively. Notice that the α-^{1}H peak position of $-CH_2N^+$ in the TOAB : Au = 1:1 case [Figure 1.4(g)] is the same as that of the synthesized [TOA][AuBr$_4$] complex [Figure 1.4(k), (l)]. This indicates strongly that one [TOA]$^+$ cation was associated with one [Au^{III}X$_4$]$^-$ anion with X = Br$^-$ or/and Cl$^-$. As the TOAB : Au ratio increases, which corresponds to a decrease of Au content because a constant TOAB amount was used, more and more [TOA]$^+$ cations can be associated with halogen anions (*e.g.*, Br$^-$ or Cl$^-$) other than the Au^{III} complex. However, the fact that only one α-^{1}H peak of $-CH_2N^+$ was observed in all cases is a sign of fast exchange between the Au^{III} complex units and the halogen anions in the inverse micelles.

Now taking 2.97 ppm [Figure 1.4(g), (k), (l)] and 3.34 ppm [Figure 1.3(a) to (d)] as the α-^{1}H peak positions of the $-CH_2N^+$ in two extremities, *i.e.*, all Au^{III} units *vs.* all halogen anions, we calculated the expected peak position using the fast-exchange model $\delta(\alpha\text{-}^1H) = 2.97y + 3.34(1-y)$ (in ppm) where y is the fraction of the Au^{III} units among the total anions (*i.e.*, [AuX$_4$]$^-$ and X$^-$). We found 3.16, 3.22, and 3.27 ppm for TOAB : Au ratios of 2:1, 3:1, and 5:1, respectively. These should be compared with the experimentally observed values: 3.18, 3.22, and 3.26 ppm [Figure 1.4(h) to (j)]. That these two sets of values are in excellent agreement indicates strongly that the partitioning of the Au^{III} units and the halogen anions in a given inverse micelle follows the nominal stoichiometry.

Figure 1.5 presents the ^{1}H NMR spectra of samples containing Ag ions in the organic phase after the PT. The samples were prepared with 0.8 mL C$_6$D$_6$,

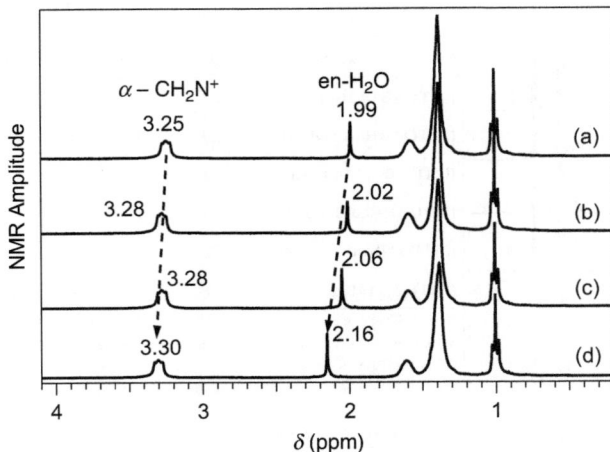

Figure 1.5   Proton NMR spectra of reaction intermediate C$_6$D$_6$ solutions in the 1st step of BSM synthesis of Ag NPs as a function of TOAB : Ag ratio: (a) 3:1, (b) 3.5:1, (c) 4:1, and (d) 5:1.

0.03 mmol of TOAB, 0.4 mL aqueous solution of $AgNO_3$ of different concentrations. After the PT of Ag ions, the organic phase was collected and used for NMR measurements. As can be seen from the figure, the peak positions of water are very similar to those in Figure 1.4A, indicating the formation of inverse micelles. Moreover, the same trend in down-field shift and increase of water amount were observed as the Ag content decreases in the order TOAB : Ag = 3 : 1, 3.5 : 1, 4 : 1, and 5 : 1. The peak integrals for the encapsulated water (normalized by setting that of α-^1H of $-CH_2N^+$ to 8) are 2.7, 2.9, 3.4, and 3.7 respectively, which are also close to those found in the Au case (see Figure 1.4B).

We show in Figure 1.6 the ^1H NMR spectra of the organic phase after PT of Pd^{II} (from $PdCl_2$) or Pt^{VI} (from H_2PtCl_6). The samples were again prepared with constant C_6D_6 volume (0.8 mL) and constant TOAB amount (0.03 mmol) but various metal content: TOAB : Pd or Pt = 1 : 1, 2 : 1, 3 : 1, and 5 : 1 for spectra (a) to (d) respectively. Interestingly, as the metal content decreases, we observe again the down-field shift and amplitude increase of the water peak for both Pd^{II} and Pt^{VI} samples, as those observed for the Au and Ag samples in Figures 1.4A and 1.5, respectively, also suggesting the formation of inverse micelles. However, the trend of variation for the α-^1H peak of $-CH_2N^+$ in the Pd^{II} case (left panel in Figure 1.6) is opposite to those of the three other metals, for which the chemical reason is still unclear but probably has to do with the difference in the type of complex structure formed with TOA^+. Notwithstanding such difference, the water aggregates behave in a remarkably similar fashion that is related directly to the formation of inverse micelles, as clearly alluded to by the results shown in Figure 1.6.

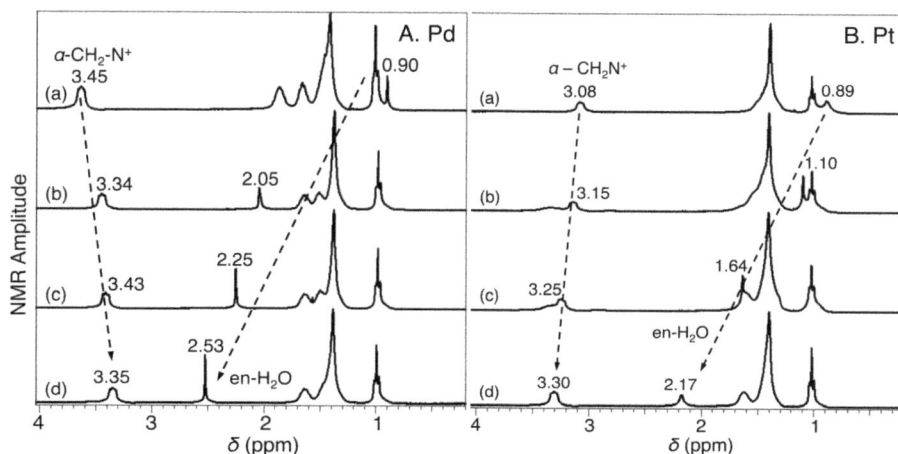

Figure 1.6   Proton NMR spectra of reaction intermediate C_6D_6 solutions in the 1st step of the BSM synthesis of Pd (A) and Pt (B) NPs as a function of TOAB : Pd(Pt) ratio: (a) 1 : 1, (b) 2 : 1, (c) 3 : 1, and (d) 4 : 1.

1.3 Addition of Ligand: Reduction of Metal Complex or Formation of Polymeric Species

1.3.1 Alkylthiols RSH

The second step in a typical BSM synthesis is to add ligand to the organic phase that contains PT-ed metal ions. This is *the* step where divergence exists in the literature as to what precursor species become the PT-ed Au^{III} ions. Earlier work by Whetten asserted that the Au^{III} ions were reduced to Au^{I} and formed [Au^{I}RS]$_n$ polymeric species.[19,20] This assertion has been widely accepted, cited, or/and assumed as the Au precursor species in the BSM synthesis until the Goulet and Lennox paper[18] in which [TOA][Au^{I}X$_4$] and [TOA][Au^{I}X$_2$] complexes were shown to be the relevant metal ion precursors, which also applies to Ag and Cu. Our work confirms this important discovery by showing that no metal–sulfur bond is formed during this step, as presented in Figure 1.7.[15]

It is expected that a Au–S bond will form when RSH self-assembles on a Au surface or polymeric [AuSR]$_n$ species are formed. This is indeed what we have observed, as shown in Figure 1.7 by the presence of a Au–SR vibrational band at 327 cm^{-1} in the Raman spectra (c) and (d) for C$_{12}$SH self-assembling on a rough Au surface and for the synthesized [AuSR]$_n$-like polymer respectively.

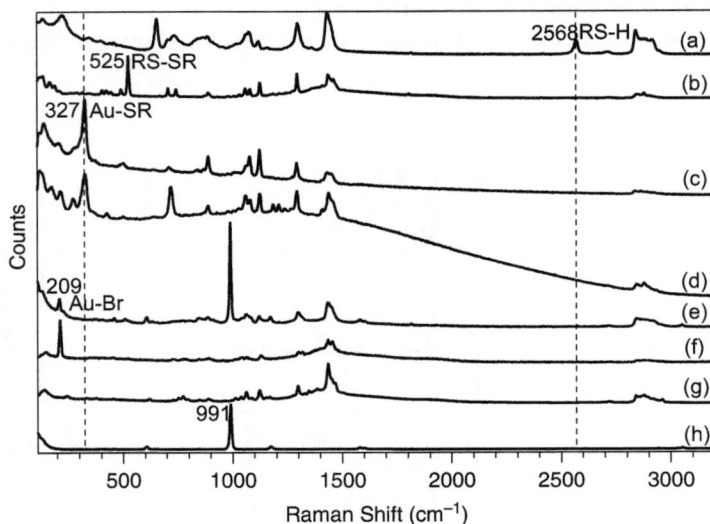

Figure 1.7   Raman spectra of (a) dodecanethiol, (b) didodecyl disulfide, (c) dodecanethiol self-assembled on a rough Au electrode, (d) synthesized [AuSR]$_n$-like polymer, (e) the concentrated C$_6$H$_6$ layer of HAuCl$_4$ and 3 equiv. of TOAB after the addition of 2 equiv. of dodecanethiol, (f) synthesized [TOA][AuBr$_2$] complex, (g) TOAB, and (h) C$_6$H$_6$.
Modified from ref. 15. Copyright 2011, American Chemical Society.

However, no such vibrational band was observed when $C_{12}SH$ was added to an organic phase of TOAB that contained PT-ed Au ions obtained in the first step of a BSM synthesis (see spectrum (e) in Figure 1.7), indicating no formation of the Au–S bond that is expected for the presence of polymeric $[AuSR]_n$ species. The appearance of a Au–Br band at 209 cm^{-1} in spectrum (e) is direct experimental evidence confirming the exchange of Br$^-$ from TOAB with Cl$^-$ in the original $(AuCl_4)^-$.

Figure 1.8 shows the Raman spectra of species formed in the 2nd step of the BSM synthesis of Ag (A) and Cu (B) NPs and of some reference materials.[15] Again, no Ag–S or Cu–S vibrational band was observed among the species formed in the 2nd step of the BSM synthesis: spectrum (c) in Figure 1.8A and B. No RS–SR but RS–H observed in the Ag case indicates that no reduction of Ag^I took place. On the other hand, a strong RS–SR band was observed in the Cu case, illustrating that the added $C_{12}SH$ reduced Cu^{II} to Cu^I without forming a Cu–S bond.

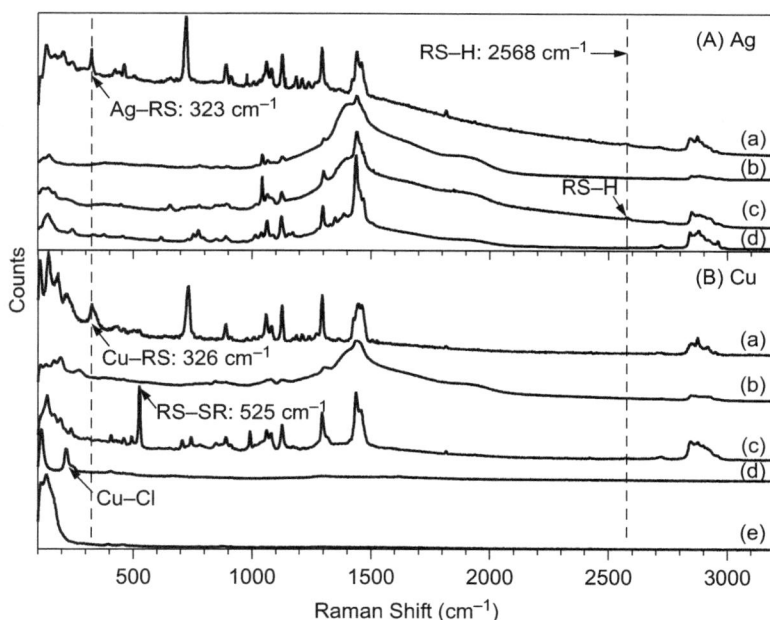

Figure 1.8   Raman spectra of species formed in the 2nd step of the BSM synthesis of Ag (A) and Cu (B) NPs. (A): (a) synthesized $[AgSR]_n$ polymer, (b) concentrated organic phase of TOAB containing PT-ed Ag^I ions, (c) concentrated organic phase after adding dodecanethiol to the pre-concentrated (b) sample, and (d) TOAB. (B): (a) synthesized $[Cu^ISR]_n$ polymer, (b) synthesized $[TOA]_2[CuX_4]$, (c) concentrated toluene solution of $[TOA]_2[CuX_4]$ plus 3 equiv. of $C_{12}SH$, (d) CuCl$_2 \cdot$ H$_2$O, and (e) CuX$_2$. (X: Cl or Br).
Modified from ref. 15. Copyright 2011, American Chemical Society.

A (left panel)

en-H$_3$O$^+$ 4.16 3.12 (m)
5.56 3.23 (n)
4.69 3.27 (o)
3.96 3.30 (p)
-S-S-CH$_2$- 3.10 (q)
HS-CH$_2$- 3.09 2.58 (r)
3.11 2.17 (s)

δ (ppm)

Sample: C$_{12}$=C$_{12}$H$_{25}$	C$_6$D$_6$	en-H$_3$O$^+$	α-CH$_2$N$^+$	CH$_2$SH	(-CH$_2$S)$_2$
(m) Sample (g) + C$_{12}$SH	7.1	1.7	8†	0.3	3.4
(n) Sample (h) + C$_{12}$SH	7.0	2.7	8	0.3	1.7
(o) Sample (i) + C$_{12}$SH	7.3	4.0	8	0.2	1.2
(p) Sample (j) + C$_{12}$SH	7.3	4.4	8	0.2	0.7
(q) Sample (k) + C$_{12}$SH	7.2‡	/	3.9	1.0	2.1
(r) [TOA][AuBr$_2$] + (C$_{12}$S)$_2$	7.2‡	/	2.5	/	2.6
(s) [TOA][AuBr$_2$] + C$_{12}$SH	7.2‡	/	4.2	1.6	/

Note: * Sample (m~p): the ratio of S to Au was 2: 1; Sample (q~s): the ratio of S to Au was 3: 1; 0.011 mmol (for Sample (r)) or 0.025 mmol (for Sample (s)) [TOA][AuBr$_2$] was used.

† The number of protons in α-CH$_2$-N$^+$ from TOA$^+$ was assigned as 8.

‡ The proton number from C$_6$D$_6$ in Sample (q-s) was the average of proton numbers of 13 different samples.

B

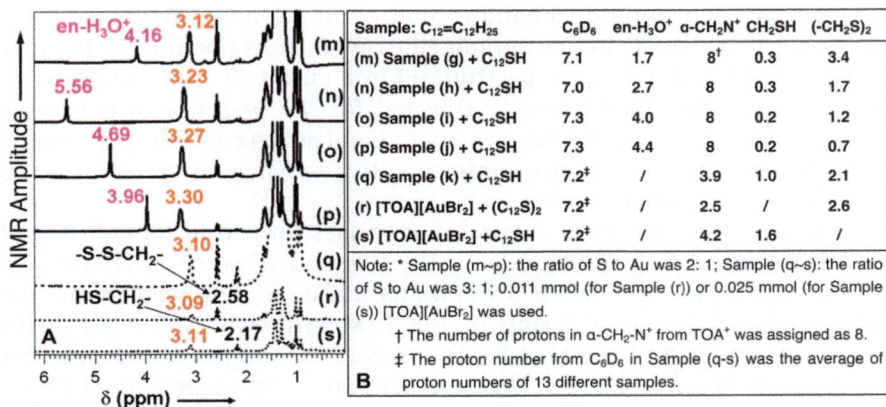

Figure 1.9　(A) Proton NMR spectra of samples (g) to (k) shown in Figure 1.4A after the addition of C$_{12}$SH and of reference disulfide (r) and thiol (s). (B) The corresponding normalized proton peak integrals and sample preparation parameters.
Modified from ref. 16. Copyright 2011, American Chemical Society.

Figure 1.9A presents the proton NMR spectra of the organic phase obtained in the 2nd step of the BSM synthesis for samples (g) to (k) shown in Figure 1.4A, together with those of reference disulfide (spectrum (r)) and thiol (spectrum (s)).[16] The formation of disulfide (peak at 2.58 ppm) in all samples indicates that the added thiol reduced Au^{III} to Au^{I} ions, which was also observed by Goulet and Lennox previously.[18] The appearance of even more down-field shifted water peaks evidences the acidification of the encapsulated water by the reaction

$$[TOA][Au^{III}X_4] + 2RSH \rightarrow [TOA][Au^{I}X_2] + RSSR + 2HX \tag{1.1}$$

in which acidic protons were generated. That is, the existing inverse micelle encapsulated water or/and organic-solvent-dissolved water provided a hydrophilic receiving medium for the reaction-generated protons, enabling Reaction (1.1) to proceed forward readily.

Figure 1.10 compares the proton NMR spectra of the reaction intermediate C$_6$D$_6$ solutions in the 2nd step of the BSM synthesis of Pd (a) and Pt (b) NPs. As can be seen, the added C$_{12}$SH did not react with PT-ed Pd^{II} ions but reduced PT-ed Pt^{IV} ions. As in the case of Au, the latter led to the formation of disulfide:

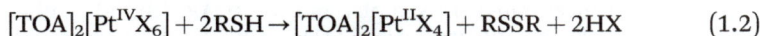

$$[TOA]_2[Pt^{IV}X_6] + 2RSH \rightarrow [TOA]_2[Pt^{II}X_4] + RSSR + 2HX \tag{1.2}$$

The large down-field shift of the encapsulated water peak (5 ppm) was the result of Reaction (1.2).

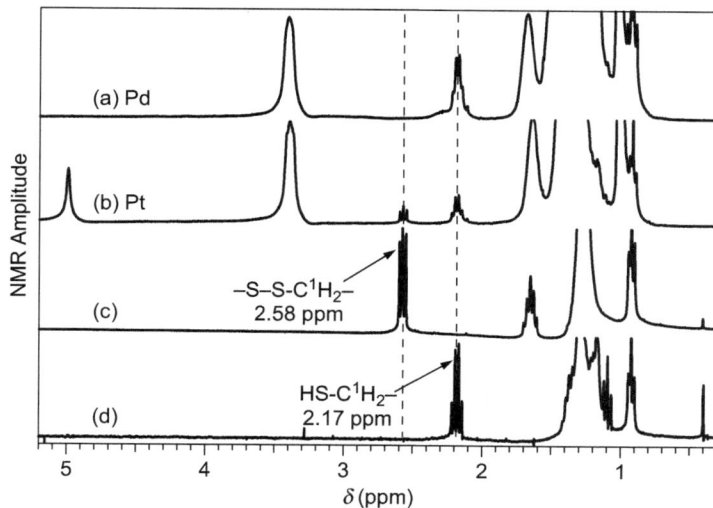

Figure 1.10 Proton NMR spectra of the reaction intermediate C_6D_6 solutions of (a) Pd ($PdCl_2 + 2TOAB + 3C_{12}SH$) and (b) Pt ($H_2PtCl_6 + 5TOAB + 3C_{12}SH$) samples. (c) and (d) are reference spectra for $(C_{12}S)_2$ and $C_{12}SH$ respectively.

1.3.2 Dialkyl Diselenide RSe–SeR

When using Se- and Te- (*vide infra*) containing organo-ligands as metal NP-stabilizers, one has to use air-stable diselenides or ditellurides rather than air-instable selenols or tellurols. Earlier work by Ulman *et al.*[23] showed that Au NPs synthesized with one-phase (1p) BSM had better quality than those synthesized by two-phase BSM. Our research has shown that this is highly likely related to the fact that only one dominant single metal precursor species exists for the 1p BSM synthesis while multiple metal precursor species co-exist in the 2p BSM synthesis.

Figure 1.11 presents the ^1H and ^{13}C NMR spectra of the reaction intermediate solution (a)/(c) and the starting dioctyl-diselenide $(C_8Se)_2$ (b)/(d) of the 1p BSM synthesis in which the THF solution of $(C_8Se)_2$ and aqueous solution of $HAuCl_4$ were directly mixed and no PT agent was used.[24] The ^1H peak of $^1H_2CSe-$ at 2.75 ppm in the starting ligand [Figure 1.11(b)] disappeared in the 1p intermediate [Figure 1.11(a)], which indicates that the Se–Se bond of the starting ligand was broken, *i.e.*, the reaction between $(C_8Se)_2$ and $HAuCl_4$ occurred. This also corroborates with the large ^{13}C shift for C_1/C_2 of the reaction intermediate solution [Figure 1.11(c)] *vs.* those of the starting ligand [Figure 1.11(d)]. Most importantly, the eight ^{13}C peaks in Figure 1.11(c) imply that the 1p intermediate consists dominantly of a single type of metal complex.

Figure 1.12 shows the corresponding Raman spectra of the starting ligand (a), the reaction intermediate solution (b), and aqueous solution of

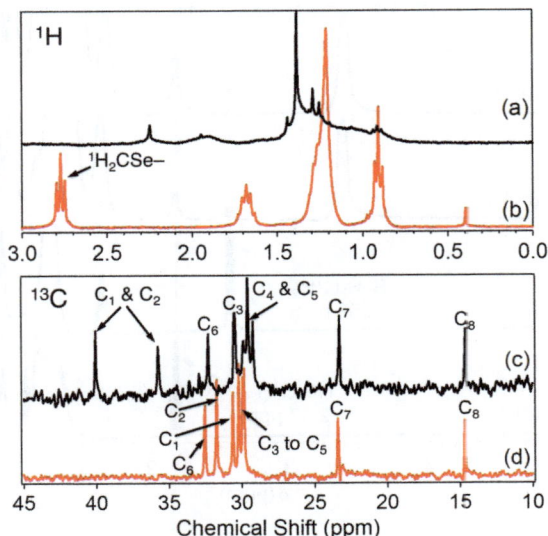

Figure 1.11 ^1H (a and b) and ^{13}C NMR (c and d) spectra of the 1p intermediate (a and c) and the pure $(C_8Se)_2$ (b and d) respectively. The arrows indicate the assignments of the respective peaks.
Modified from ref. 24. Copyright 2012, The Royal Society of Chemistry.

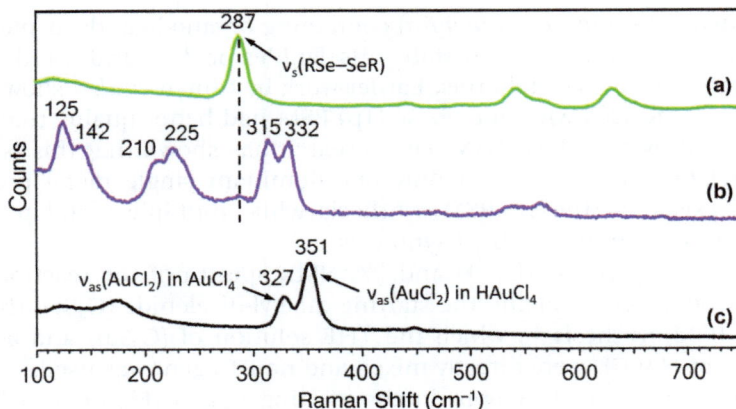

Figure 1.12   Raman spectra of (a) dioctyl diselenide, (b) the concentrated 1p intermediate after removing THF, *i.e.*, 1 equiv. of HAuCl$_4$ and 0.5 equiv. of dioctyl diselenide, and (c) aqueous HAuCl$_4$ solution.
From ref. 24. Copyright 2012, The Royal Society of Chemistry.

HAuCl$_4$.[24] The RSe–SeR vibrational band at 287 cm^{-1} observed in the starting ligand [spectrum (a)] largely disappeared in spectrum (b) of the reaction intermediate solution, indicating the majority of Se–Se bonds were broken, in agreement with the NMR results discussed above. This is further corroborated by the disappearance of the characteristic bands of starting $HAuCl_4$. Moreover, aided by DFT calculations,[24] the band 142 cm^{-1} was assigned to Au–Se vibration. Additionally, XPS measurements showed that the oxidation state of Se in the reaction intermediate solution was 2+ instead of the starting −1 and that of Au was −2. Thus, by combining the above NMR, Raman, and XPS results altogether, it was concluded that the reaction between HAuCl and (C_8Se) did occur and produced dominantly one type of metal complex, highly likely in the form of $Cl_2AuSe(C_8)Cl_2$.

^{1}H NMR (A) and Raman (B) spectra of the reaction intermediate solutions obtained in the 2nd step of the 2p BSM synthesis with different TOAB : Au ratios but the same Au : Se (1 : 1) ratio are presented in Figure 1.13.[24] Both ^{1}H NMR and Raman of the simply mixed C_6D_6 solution of TOAB and $(C_8Se)_2$ (spectrum (a) in Figure 1.13A and B) proved that no reaction took place between the two species, although the slight blue-shift of the RSe–SeR band[25] (from 287 to 292 cm^{-1}) does indicate a certain degree of interaction between them. However, reaction occurred when the Au^{III} ions were added, but was not as fast as observed in the 1p BSM synthesis, as indicated by the

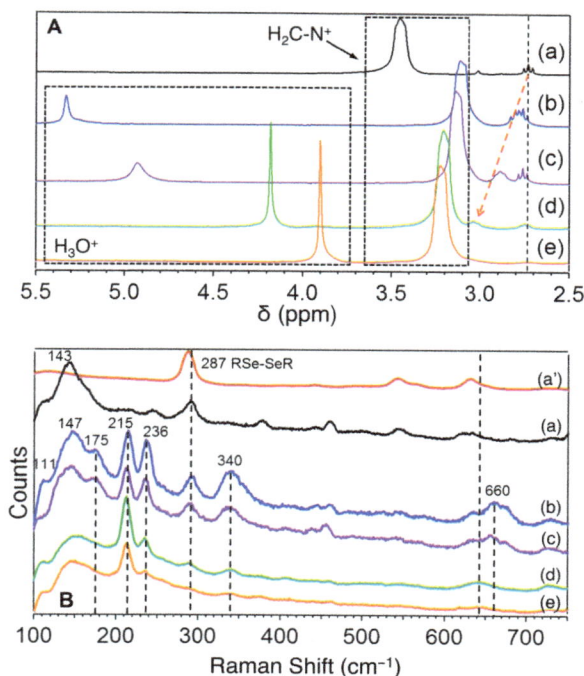

Figure 1.13 (A) ^{1}H NMR spectra of (a) $3TOAB + 0.5(C_8Se)_2$, (b) $1TOAB + 1HAuCl_4 + 0.5(C_8Se)_2$, (c) $1.25TOAB + 1HAuCl_4 + 0.5(C_8Se)_2$, (d) $2TOAB + 1HAuCl_4 + 0.5(C_8Se)_2$, and (e) $3TOAB + 1HAuCl_4 + 0.5(C_8Se)_2$ in C_6D_6. (B) Raman spectra of (a') $(C_8Se)_2$ and of (a)–(e) as those in A. Modified from ref. 24. Copyright 2012, The Royal Society of Chemistry.

persisting proton peak of $-C^1H_2N^+$ at 2.73 ppm in Figure 1.13A and the RSe–SeR vibrational band at 292 cm^{-1} in Figure 1.13B. As TOAB : Au ratio increased, spectral features (2.73 ppm peak in (A) and 287 cm^{-1} band in (B)) related to the RSe–SeR bond decreased and eventually disappeared completely (from (a) to (e)), implying that the majority of RSe–SeR bonds were broken.

The observation of the hydronium ions (H_3O^+), as indicated in Figure 1.13A, is intriguing because a simple phase transfer of Au^{III} ions does not generate observable hydronium ions. Although the subsequent addition of thiol does, where the hydrogen in the –SH group acts like a reductant, $(C_8Se)_2$ does not have such an obvious proton source. Thus, hydrolysis of encapsulated water, such as Se–Se $+ 2H_2O \rightarrow 2SeOH + H^+$, might have happened. We speculate that the band at 660 cm^{-1} is related to a hydrogen-bonding-like Se–H_2O interaction[26] before the Se–Se bond breaking and the band at 642 cm^{-1} to a Se–OH like bond after the Se–Se bond breaking. On the other hand, the appearance of the further down-field shifted water whose amount increases as the TOAB : Au ratio decreases suggests the existence of inverse micelles and that the degree of the hydrolysis reaction depended on the amount of encapsulated water: the larger the latter is, the higher the former will be. The available XPS data for the 3TOAB : 1Au (e) and 2TOAB : 1 Au (d) samples indicate that the oxidation states of Au in these two 2p samples (84.2 eV and 84.3 eV respectively) were basically the same as that of Au in the 1p synthesis (84.25 eV) but those of the Se (57.98 eV and 58.05 eV) were different from the 1p synthesis (58.4 eV).[27] That is, the smaller binding energies in the former indicate that the charge screening of Se was higher than that in the latter, which is consistent with the proposed Se–OH bonding.

Both the ^{1}H NMR (Figure 1.13A) and Raman (Figure 1.13B) data strongly suggest that more than one species exist as the 2p intermediates. The appearance of encapsulated H_3O^+ as shown in Figure 1.13A suggests that hydrolysis of water had happened and likely led to species such as $X_2AuSe(C8)(OH)X$ (X = Br or Cl) as one of the intermediates whose exact compositions still need to be further studied. One important difference between the 1p and 2p BSM syntheses, however, is that the results shown in Figures 1.11 and 1.12 were highly reproducible but not those in Figure 1.13 which appeared to be highly dependent on many hard-to-control factors.

1.3.3 Dialkyl Ditelluride RTe–TeR

Among organochalcogenolate-stabilized metal NPs, organotellurolate-metal NPs are the least investigated system. Figure 1.14 presents the proton NMR spectra of (a) the intermediate reaction C_6D_6 solution obtained in the 2nd step of the BSM synthesis of Au NPs using $(C_{12}Te)_2$, (b) that obtained in the 1st step, (c) pure $(C_{12}Te)_2$ solution, and (d) pure $C_{12}SH$ solution. The first observation is that, in great contrast to using thiol (c), there was no acidic H^+ formation when $(C_{12}Te)_2$ was used (a): the peak position of the

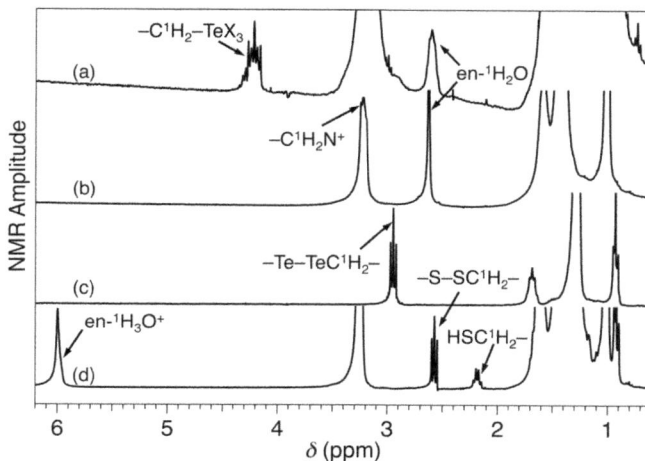

Figure 1.14 Proton NMR spectra (in C_6D_6) of (a) the intermediate solution of the Au of $(3TOAB + HAuCl_4 + 0.33(C_{12}Te)_2)$, (b) the organic phase after PT $(3TOAB + HAuCl_4)$, (c) $(C_{12}Te)_2$, and (d) the solution formed by addition of 2 equiv. of $C_{12}SH$ to the solution in (b). The concentration of each chemical was kept the same in all samples. Modified from ref. 28. Copyright 2012, American Chemical Society.

encapsulated water remained almost unchanged as compared to the organic phase right after PT (b), although broadened significantly. Instead, a peak assignable to $RTeX_3$ appeared (a). Also, little unreacted $(C_{12}Te)_2$ is left.[28]

Raman measurements[28] corroborate well with the NMR ones. Figure 1.15 compares the Raman spectrum (a) of the sample (a) in Figure 1.14 with that of (b) pure $(C_{12}Te)_2$, (c) the synthesized $[TOA][Au^ITeBr_2]$ complex, and (d) TOAB. Several observations can be made here. First, disappearance of the RTe–TeR band at 194 cm^{-1} in the reaction intermediate indicates that the reaction between the added $(C_{12}Te)_2$ and the PT-ed $[TOA][Au^{III}X_4]$ complex broke the Te–Te bond. Second, the appearance of the 209 cm^{-1} band[15,29] in (a) that can be reasonably assigned to the Au–Br$_2$ band in the synthesized $[TOA][Au^IBr_2]$ complex (c) suggests that the initial Au^{III} ions were reduced to Au^I ions. Third, that no Au–Te vibrational band (~ 190 cm^{-1}) was observed implies that no Au–Te bond was formed during the reaction. Fourth, assisted by the DFT calculations, the 280 cm^{-1} band can be assigned to RTe–X$_3$.

Moreover, the available XPS data showed that the oxidation state of Au in the reaction intermediate [sample a in Figure 1.14(a) and 1.15(a)] was +1 and that of Te was +2. Therefore, summarizing the above discussion, we propose the following stoichiometric reaction for the reduction of Au^{III} to Au^I by the ditelluride, whose oxidation state changes accordingly from −1 to +2:

$$[TOA][Au^{III}X_4] + 3(RTe)_2 \rightarrow [TOA][Au^IX_2] + 3RTeX_3 \qquad (1.3)$$

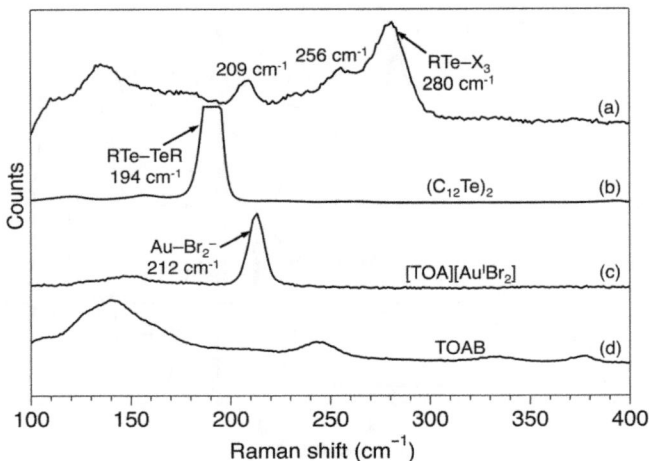

Figure 1.15   Raman spectra of (a) dried intermediate of the Au system $(3TOAB + HAuCl_4 + 0.33(C_{12}Te)_2)$, (b) $(C_{12}Te)_2$, (c) $[TOA][AuBr_2]$, and (d) TOAB.
Modified from ref. 28. Copyright 2012, American Chemical Society.

On the other hand, the ditelluride does not react with PT-ed Ag complex if added to the separated Ag-containing organic phase.

1.4   Critical Role of Water

As alluded to in the previous discussion, the inverse micelle encapsulated water offers a hydrophilic micro-environment as a proton accepting reaction medium, enabling Reaction (1) to proceed. Such an important role was further confirmed by using the synthesized, anhydrous metal complex as the starting material. We have observed that the reduction of the synthesized $[TOA][AuBr_4]$ complex to $[TOA][AuBr_2]$ by thiols in an anhydrous organic (toluene or benzene) solvent was extremely slow but happened almost instantaneously in the presence of H_2O.[16]

It has been observed that if the inverse micelle encapsulated water is the main form of water existing in the BSM synthesis, then addition of thiol does not lead to the formation of Au–S bonds, $i.e.$, formation of polymeric $[AuSR]_n$ species. However, if more water beyond saturating the inverse micelles exists, then addition of thiol can lead to the formation of polymeric $[AuSR]_n$ species providing that the thiol : Au ratio is larger than 2. For instance, if the water layer of the original aqueous solution of Au salt was not discarded after the PT of Au ions to the organic phase and to which thiol was added, then stirring would lead to the formation of a white cloudy material that sits between the aqueous and organic layers. Such a white cloudy material gave solution Raman spectrum (c) in Figure 1.16 in which those of the

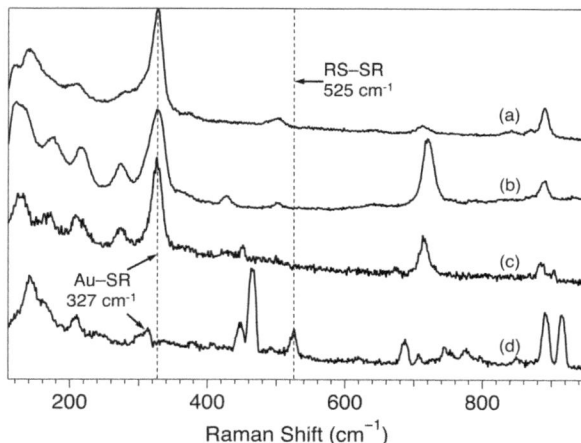

Figure 1.16 Raman spectra of (a) self-assembled $C_{12}SH$ on a rough Au surface, (b) synthesized $[AuSR]_n$, (c) white cloudy material formed after $C_{12}SH$ was added to a Au ion-PT-ed solution that retained the original aqueous layer of Au salt solution, and (d) separated organic phase after addition of $C_{12}SH$ and being stirred for 24 h. (a) and (b) are the same spectra as Figure 1.7(c) and (d) but are reproduced here to facilitate the comparison and discussion.
Modified from ref. 16. Copyright 2011, American Chemical Society.

self-assembled thiol on a rough Au surface (a) and the synthesized polymeric $[AuSR]_n$ species (b) are also presented for comparison.[16]

That Figure 1.16(c) is almost identical to Figure 1.16(b) suggests strongly that the white cloudy material was the polymeric $[AuSR]_n$ species. On the other hand, stirring the organic phase obtained in the 2nd step of a BSM synthesis in the air long enough can also lead to the formation of a small amount of polymeric $[AuSR]_n$ species in addition to forming disulfide, as indicated by Figure 1.16d. Summarily, Figure 1.17 presents a general metal NP formation mechanism[15] in a BSM synthesis (A) and reaction conditions[16] that lead to different, complex *vs.* polymer, reaction routes.

It has also been discovered that the encapsulated water is also essential for S–S or Se–Se bond breaking in a BSM synthesis if the former is used as the source of stabilizing ligand. This is illustrated by the proton spectra of the reaction intermediate solutions with and without encapsulated water presented in Figure 1.18.[30] For both $(C_{12}S)_2$ and $(C_{12}Se)_2$, the pre-existence of encapsulated water enabled S–S and Se–Se bond breaking to happen, as indicated by the appearance of the acidified water peaks in Figure 1.18(a) and (b) for disulfide and in Figure 1.18(d) and (e) for diselenide with TOAB : Au ratio $= 3:1$ and $2:1$ respectively. On the other hand, the presence of remaining disulfide as suggested by the residual peak at 2.57 ppm suggests that diselenide was more reactive than disulfide. Without the presence of

A

B

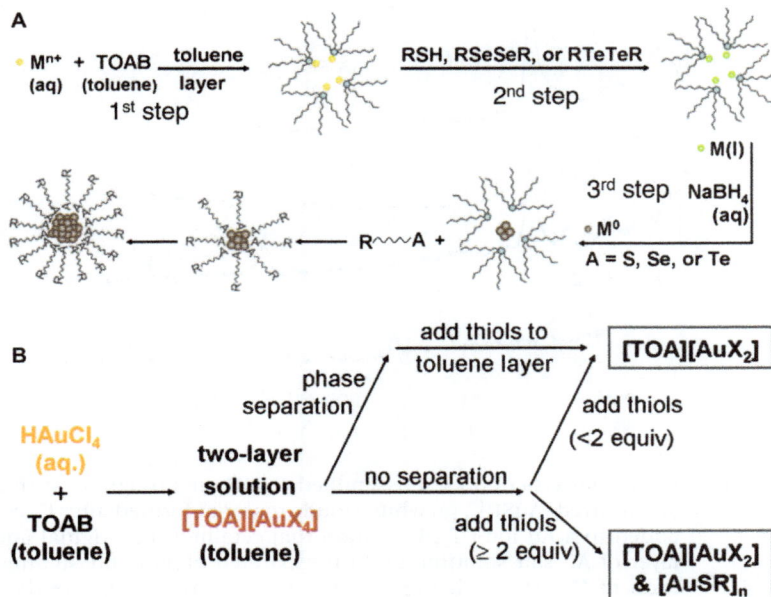

Figure 1.17   (A) A general Au NP formation mechanism, and (B) complex *vs.* polymeric Au ion intermediates in a BSM synthesis.
Modified from ref. 15 and 16. Copyright 2011, American Chemical Society.

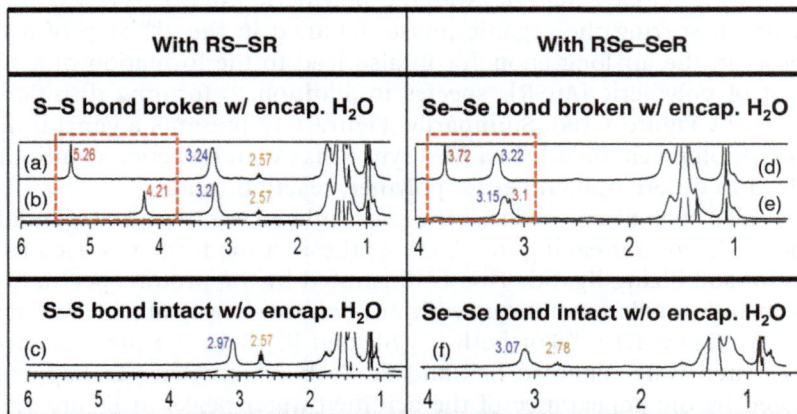

Figure 1.18   Proton NMR spectra of reaction intermediate solutions with and without encapsulated water for didodecane disulfide and didodecane diselenide.
Modified from ref. 16 and 30. Copyright 2012, American Chemical Society.

inverse micelle encapsulated water, no S–S and Se–Se bond breaking took place, as confirmed by spectra in Figure 1.18(c) and (f) where no acidified water peaks but those of disulfide (2.57 ppm) and diselenide (2.78 ppm) are observed.

Since no viable proton sources other than water existed in the reaction solutions and no vibrational evidence of Au–S[15,31] and Au–Se[24] bond formation was observed, the appearance of acidic protons as described above strongly implies that the overall reaction may involve hydrolysis, most likely in the form:

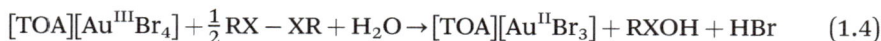

$$[\text{TOA}][\text{Au}^{\text{III}}\text{Br}_4] + \tfrac{1}{2}\text{RX} - \text{XR} + \text{H}_2\text{O} \rightarrow [\text{TOA}][\text{Au}^{\text{II}}\text{Br}_3] + \text{RXOH} + \text{HBr} \qquad (1.4)$$

where X stands for either S or Se. The presence of paramagnetic Au^{II} ions was alluded to, although not definitively, by the preliminary electrochemistry and UV characterizations that at least ascertained that the species in the reaction solutions were neither Au^{III} nor Au^{I}.

1.5 Reduction by NaBH₄: Formation of Zero-Valence Metal Nanoparticles

1.5.1 Normal Reduction Sequence

The last step of a normal BSM synthesis is to reduce the metal ions to the zero valence state, usually by NaBH_4, and form metal NPs.[14] The exact chemical state of the metal ion precursors and the composition of the reaction intermediate solutions before this final stage are critically important in determining how much control, and therefore how good the quality of the metal NPs, can be achieved. Figure 1.19 compares the TEM images of three Au NP samples made with the same starting (0.025 mmol) $[\text{TOA}][\text{Au}^{\text{I}}\text{Br}_2]$ complex in a mixture of 10 mL toluene and 0.21 mL water but slightly different ligand compositions before the addition of aqueous solution of NaBH_4: (a) $3\text{C}_{12}\text{SH}$, (b) a mixture of $1\text{C}_{12}\text{SH}$ and $1(\text{C}_{12}\text{S})_2$, and (c) $1.5(\text{C}_{12}\text{S})_2$.[17]

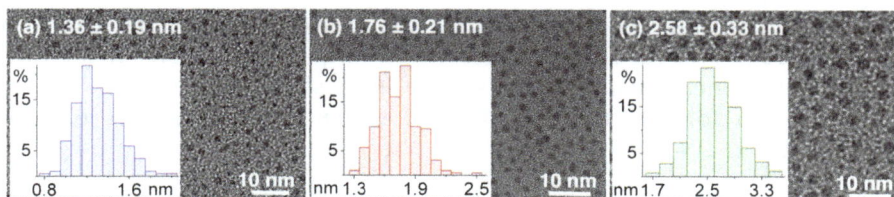

Figure 1.19 TEM images with corresponding size distribution histograms of the Au NPs formed from a mixture of $[\text{TOA}][\text{Au}^{\text{I}}\text{Br}_2]$ toluene solution and water with (a) 3 equiv. of C_{12}SH, (b) a mixture of 1 equiv. of C_{12}SH and 1 equiv. of $(\text{C}_{12}\text{S})_2$, (c) 1.5 equiv. of $(\text{C}_{12}\text{S})_2$. All three samples had the same S : Au ratio = 3 : 1.
Modified from ref. 17. Copyright 2011, The Royal Society of Chemistry.

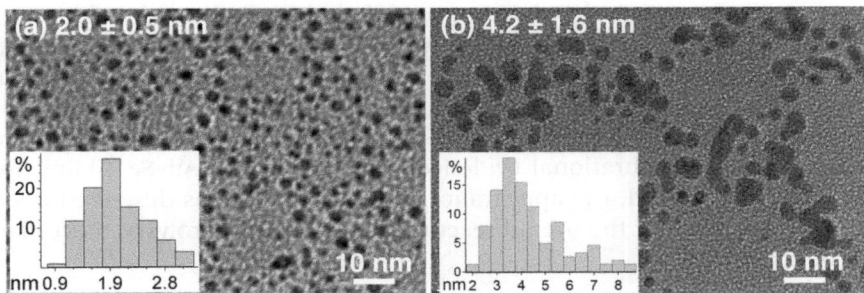

Figure 1.20   TEM images of octaneselenolate-stabilized Au NPs synthesized by (a)
one-phase and (b) two-phase BSM syntheses.
Modified from ref. 24. Copyright 2012, The Royal Society of
Chemistry.

Since Au^I was the starting metal ion, the addition of thiol would not lead to
Reaction (1) and neither would it form thiolate polymer. As can be seen, a
mixture of different ligands, which is usually the case in a normal BSM
synthesis of thiolate-stabilized Au NPs, degrades the quality of the Au NPs
formed. If a single type of ligand is used, thiol would be a better choice than
disulfide.

   In Section 1.3.2, we discussed the chemical state of the species existing in
the reaction intermediate solutions of the 1p *vs.* 2p BSM synthesis of orga-
noselenolate-stabilized Au NPs. It was concluded that a single dominant
metal precursor species existed in the reaction intermediate solution of the
1p BSM synthesis but multiple species co-existed in that of the 2p BSM
synthesis. The TEM images of the final Au NPs made by the 1p (a) and 2p (b)
BSM synthesis are presented in Figure 1.20.[24] As can be clearly seen, the
former generated much better quality Au NPs than the latter did.

   In Section 1.4, we discussed the enabling effect of the inverse micelle
encapsulated water to break the S–S or Se–Se bond that led to different
chemical states of the reaction solutions before the addition of $NaBH_4$.
Figure 1.21[30] presents the TEM images of the Au NPs formed after adding
$NaBH_4$ to reaction solutions that gave the proton NMR spectra shown in
Figure 1.18 (a), (c), (d) and (g). The contrast in quality of Au NPs made with
and without encapsulated water is quite impressive: from 1.9 ± 0.3 nm to
3.6 ± 1.6 nm for disulfide and from 2.4 ± 0.8 nm to 5.8 ± 2.1 nm for
diselenide.

1.5.2   Reversed Reduction Sequence

In addition to the identification of $[TOA][Au^{III}Br_4]$ and $[TOA][Au^IBr_2]$ complexes as the major intermediate species in the BSM synthesis by Goulet and
Lennox,[18] we have demonstrated, as discussed in the previous sections, that
the Au–S bond, and metal–ligand bond in general, does not form until the
metal ions are reduced to zero valence atoms and forming nascent metal

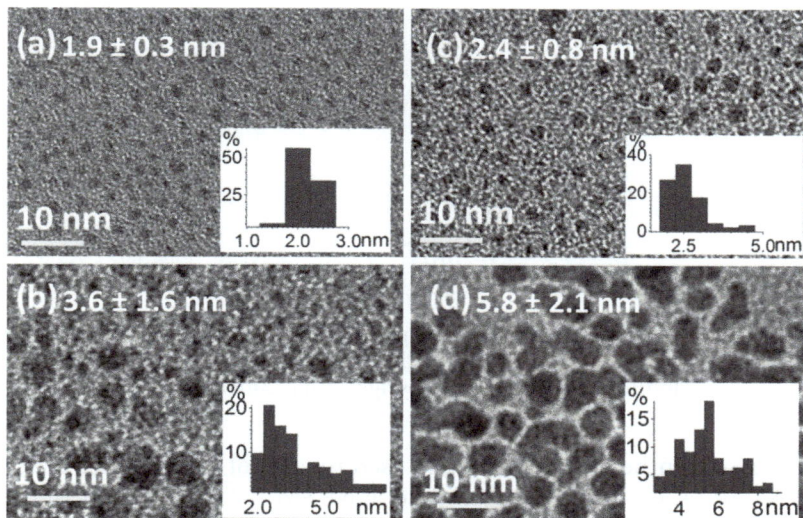

Figure 1.21 TEM images and corresponding size distribution histograms of Au NPs obtained by adding aqueous solution of 10 equiv. of $NaBH_4$ to the reaction intermediate solutions that gave the proton NMR spectra shown in Figure 1.18 (a), (c), (d), and (e) for image (a), (b), (c), and (d) respectively.

Modified from ref. 30. Copyright 2012, American Chemical Society.

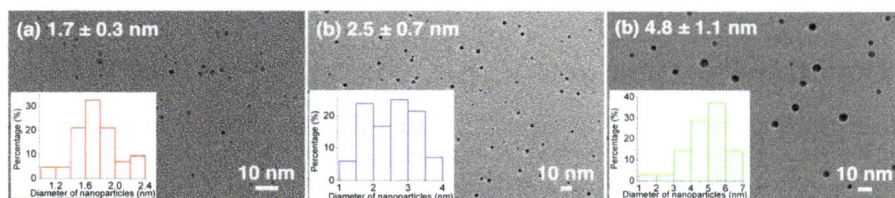

Figure 1.22 TEM images with size distribution histograms of the dodecanethiolate-protected Au NPs obtained through the reversed BSM synthesis with stirring time interval between adding reductant ($NaBH_4$) and the ligand ($C_{12}SH$) of (a) 10 s, (b) 10 min, and (c) 24 h. The same S : Au ratio $= 3 : 1$ was used in all three syntheses.

Modified from ref. 15. Copyright 2011, American Chemical Society.

clusters, as illustrated in Figure 1.17A. This discovery led us to experiment reversing the order of adding stabilizing ligand and reductant $NaBH_4$ to the reaction solution,[15] *i.e.*, adding $NaBH_4$ first then followed by adding the ligand. Indeed, narrowly distributed thiolate-stabilized Au NPs could be made by using the reversed BSM synthesis and the size could be controlled by the time interval between the addition of the reductant and the ligand even with the same S : Au ratio, as shown in Figure 1.22.[15] Notice that in the

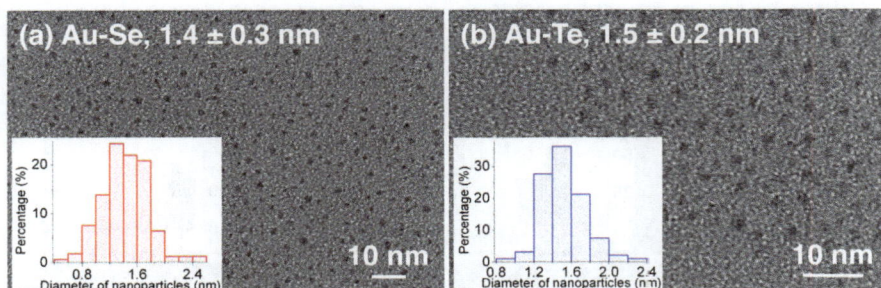

Figure 1.23   TEM images and the corresponding size distribution histograms of (a)
                   octylselenolate- and (b) dodecyltellurolate-stabilized Au NPs.
                   Modified from ref. 15. Copyright 2011, American Chemical Society.

normal BSM synthesis, S : Au ratio is used to control the NP size. Thus, the
same S : Au ratio would lead to the same NP size.

One unique feature of the reversed BSM synthesis is that it enables sub-2
nm selenolate- or tellurolate-stabilized Au NPs to be synthesized as demonstrated by Figure 1.23,[15] which had never been achieved previously. In the
syntheses that led to the formation of those Au NPs, the starting source for
Au ions was the synthesized [TOA][Au^{I}Br$_2$] complex and dioctyl diselenide
or didodecyl ditelluride as stabilizing ligand. 10 s lapsed between adding
the reductant and the ligand. It is also worth noting that the reversed BSM
applies equally well to synthesizing small Ag and Cu NPs.

1.6   Conclusions and Future Outlook

In this chapter, we have discussed in rather detailed fashion metal NP formation chemistry, particularly the chemical state of the intermediate species
leading to the final formation of metal NPs, in light of the recent advanced
mechanistic understanding of the popular BSM synthesis. The most salient
point of the latter is that the BSM synthesis is basically an inverse micelle
process in which encapsulated water plays a critical role in determining the
details of metal NP formation chemistry. For instance, it provides a hydrophilic proton accepting reaction-facilitating medium if thiol is used as the
source of stabilizing ligand (Reaction (1.1) and (1.2)). In the case of disulfide
or diselenide, it also acts as a bond breaking enabler (Reaction (1.4)). On the
other hand, what critical role the encapsulated water would play in the case
of ditellurite (Reaction (1.3)) is still not clear.

It is important to point out that the inverse micelles formed by TOAB may
behave differently from the traditionally well-studied inverse micelles such
as AOT in that the wall of TOAB inverse micelles may not be as closely
packed, therefore as rigid, as that of AOT-like inverse micelles because of the
four equal-length, therefore space-occupying branches, particularly for
shorter ones. Moreover, it has been observed that the TOAB inverse micelles

do not have an easily measurable size range that can be varied because they only need a tiny amount of water to get saturated. Still, their physico-chemical properties warrant further investigations due to the prevailing use of the BSM synthesis in making sub-5 nm metal NPs. Overall, it is expected that the chemistry detailed in this chapter will be helpful to any further attempts at improving synthesis of metal NPs of better quality and with better control of properties.

Acknowledgements

The research in the Tong Lab at Georgetown University is financially supported by NSF (CHE-0923910 and CHE-0456848) and by DOE-BES (DE-FG02–07ER15895).

References

1. R. C. Ashoori, *Nature*, 1996, **379**, 413.
2. P. L. McEuen, *Science*, 1997, **278**, 1729.
3. D. F. Perepichka and F. Rosei, *Angew. Chem., Int. Ed.*, 2007, **46**, 6006.
4. A. J. Quinn and G. Redmond, *Prog. Solid State Chem.*, 2005, **33**, 263.
5. H. Wang, D. W. Brandl, P. Nordlander and N. J. Halas, *Acc. Chem. Res.*, 2007, **40**, 53.
6. M. A. Tofanelli and C. J. Ackerson, *J. Am. Chem. Soc.*, 2012, **134**, 16937.
7. M. Walter, J. Akola, O. Lopez-Acevedo, P. D. Jadzinsky, G. Calero, C. J. Ackerson, R. L. Whetten, H. Groenbeck and H. Hakkinen, *Proc. Natl. Acad. Sci. U. S. A.*, 2008, **105**, 9157.
8. C. P. Collier, T. Vossmeyer and J. R. Heath, *Annu. Rev. Phys. Chem.*, 1998, **49**, 371.
9. *Nanoparticles: Building Blocks for Nanotechnology*, ed. V. M. Rotello, Kluwer Academic/Plenum Publishers, New York, 2004.
10. A. Kyrychenko, G. V. Karpushina, D. Svechkarev, D. Kolodezny, S. I. Bogatyrenko, A. P. Kryshtal and A. O. Doroshenko, *J. Phys. Chem. C*, 2012, **116**, 21059.
11. M. Manheller, S. Karthaeuser, R. Waser, K. Blech and U. Simon, *J. Phys. Chem. C*, 2012, **116**, 20657.
12. F. P. Zamborini, L. Bao and R. Dasari, *Anal. Chem.*, 2011, **84**, 541.
13. S. E. Lohse and C. J. Murphy, *J. Am. Chem. Soc.*, 2012, **134**, 15607.
14. M. Brust, M. Walker, D. Bethell, D. J. Schiffrin and R. Whyman, *J. Chem. Soc., Chem. Commun.*, 1994, 801.
15. Y. Li, O. Zaluzhna, B. Xu, Y. Gao, J. M. Modest and Y. J. Tong, *J. Am. Chem. Soc.*, 2011, **133**, 2092.
16. Y. Li, O. Zaluzhna and Y. Y. J. Tong, *Langmuir*, 2011, 27, 7366.
17. Y. Li, O. Zaluzhna and Y. Y. J. Tong, *Chem. Commun.*, 2011, **47**, 6033.
18. P. J. G. Goulet and R. B. Lennox, *J. Am. Chem. Soc.*, 2010, **132**, 9582.

19. T. G. Schaaff, M. N. Shafigullin, J. T. Khoury, I. Vezmar, R. L. Whetten, W. G. Cullen, P. N. First, C. GutierrezWing, J. Ascensio and M. J. JoseYacaman, *J. Phys. Chem. B*, 1997, **101**, 7885.
20. M. M. Alvarez, J. T. Khoury, T. G. Schaaff, M. Shafigullin, I. Vezmar and R. L. Whetten, *Chem. Phys. Lett.*, 1997, **266**, 91.
21. L. u. Klíčová, P. Šebej, P. Štacko, S. K. Filippov, A. Bogomolova, M. Padilla and P. Klán, *Langmuir*, 2012, **28**, 15185.
22. M. P. Pileni, *J. Phys. Chem.*, 1993, **97**, 6961.
23. C. K. Yee, R. Jordan, A. Ulman, H. White, A. King, M. Rafailovich and J. Sokolov, *Langmuir*, 1999, **15**, 3486.
24. O. Zaluzhna, Y. Li, C. Zangmeister, T. C. Allison and Y. J. Tong, *Chem. Commun.*, 2012, **48**, 362.
25. S. W. Han and K. Kim, *J. Colloid Interface Sci.*, 2001, **240**, 492.
26. F. K. Huang, R. C. Horton, D. C. Myles and R. L. Garrell, *Langmuir*, 1998, **14**, 4802.
27. C. D. Wagner, L. H. Gale and R. H. Raymond, *Anal. Chem.*, 1979, **51**, 466.
28. Y. Li, O. Zaluzhna, C. D. Zangmeister, T. C. Allison and Y. J. Tong, *J. Am. Chem. Soc.*, 2012, **134**, 1990.
29. P. Braunstein and R. J. H. Clark, *J. Chem. Soc., Dalton Trans.*, 1973, 1845.
30. O. Zaluzhna, Y. Li, T. C. Allison and Y. J. Tong, *J. Am. Chem. Soc.*, 2012, **134**, 17991.
31. P. G. Lustemberg, C. Vericat, G. A. Benitez, M. E. Vela, N. Tognalli, A. Fainstein, M. L. Martiarena and R. C. Salvarezza, *J. Phys. Chem. C*, 2008, **112**, 11394.

CHAPTER 2

New Strategies and Synthetic Routes to Synthesize Fluorescent Atomic Quantum Clusters

BEATRIZ SANTIAGO GONZÁLEZ AND M. ARTURO LÓPEZ-QUINTELA*

Dipartimento di Scienza dei Materiali, Università degli Studi di Milano-Bicocca, via Cozzi 55, I-20125 Milano, Italy
*Email: malopez.quintela@usc.es

2.1 Introduction

Metal nanoparticles with dimensions from ~ 2 to 100 nm show smoothly size-tunable electronic properties, being well-studied and widely-employed for sensing, bioassays, and nanomedicine.[1–3] But, when the particle size becomes comparable to the Fermi wavelength of an electron (~ 0.52 nm for gold and silver)—here named as atomic quantum clusters (AQCs)—new optical, electrical and chemical properties appear, which are dramatically different from what is expected. Due to their very small sizes, the density of states in AQCs is not high enough to show the typical surface plasmon resonance (SPR) features of nanoparticles. Instead, they have discrete electronic energy levels and display "molecule-like" optical characteristics, providing the bridge between the larger metal nanoparticles and atoms.

RSC Smart Materials No. 7
Functional Nanometer-Sized Clusters of Transition Metals: Synthesis, Properties and Applications
Edited by Wei Chen and Shaowei Chen
© The Royal Society of Chemistry 2014
Published by the Royal Society of Chemistry, www.rsc.org

For large AQCs (*i.e.* those formed by a number of atoms, $N > \approx 10$–15) the UV-Vis spectra consist of a continuous increase in absorbance with decreasing wavelength, whereas small AQCs ($N < \approx 10$–15) display molecule-like bands (Figure 2.1).[4,5]

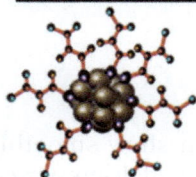

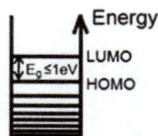

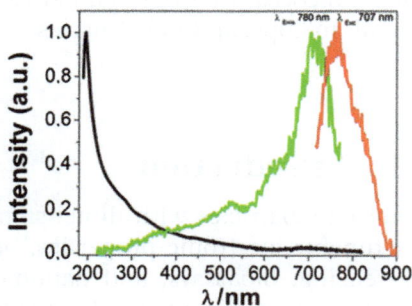

Another important feature of AQCs is their strong luminescent properties, offering high quantum yields (QY), large molar extinctions, good photostability and, in some cases, large Stokes shifts. Hence, comparing with luminescent materials developed until now—quantum dots (QDs),[6] dye-doped nanoparticles,[7] up-converting lanthanide-doped nanoparticles,[8] carbon nanodots[9]—metal clusters display numerous advantages thanks to

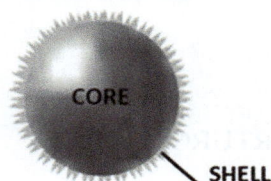

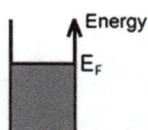

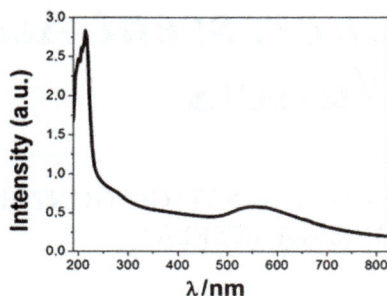

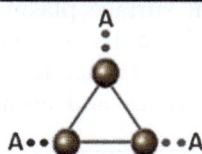

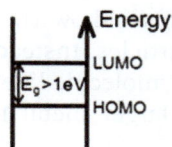

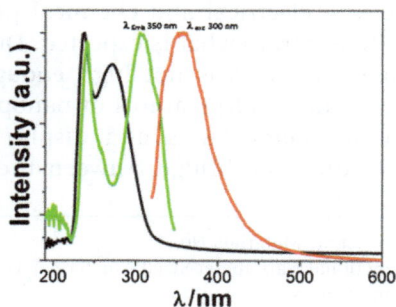

Figure 2.1   Schematic properties of nanoparticles and clusters.
                (Adapted from ref. 5.)

their low hydrodynamic sizes and inert nature, but especially due to their biocompatibility and photostability. For these reasons, they emerge as a very interesting alternative to apply in different areas like biosensors,[10] molecular imaging,[11–13] optoelectronic,[14] nanomedicine[15] and environmental analysis.[16,17]

During past decades, multiple methods have been developed to synthesize metal clusters, allowing the preparation of clusters with different sizes protected with different capping agents (such as small molecules, polymers, biomolecules, proteins...), exhibiting different emissions. The great efforts made in the study of the preparation of AQCs, mainly during the last decade, resulted in a wide variety of strategies, many of them being one pot routes where the clusters can be synthesized in a matter of minutes. Generally, they can be classified as top-down routes or bottom-up approaches.

It was observed that, similar to QDs, by controlling the cluster size it is possible to obtain size tunable luminescent metal clusters. This size-dependence phenomenon is also reflected in Figure 2.1, where, as an example, the luminescent properties for small and large clusters are shown.

The most studied and widely synthesized are gold clusters, but silver analogues are also very common nowadays thanks to their high luminescence quantum yield (QY). More recently, copper and platinum are also being extensively explored, as well as some bimetallic/alloy AQCs.

2.2 Top-Down Approach

In general, the *top-down* approach consists of obtaining the clusters from their bigger precursors, either nanoparticles or bulk metal. It was found that some strong ligands such as thiols,[18] as well as some biomolecules[19,20] or multivalent polymers,[21] had a capacity to produce a core reduction by removing atoms from the surface of the nanoparticle or breaking the nanoparticle into smaller pieces leading to stable clusters. This etching based strategy has been established as a common route for the preparation of AQCs.

Glutathione, (GSH) a natural tripeptide consisting of glutamic acid, cysteine and glycine, is widely used as a stabilizer for synthesizing metal clusters, as well as to etch nanoparticles. For example, Muhammed *et al.* have obtained Au_8 (λ excitation (λ_{exc}) 370 and λ emission (λ_{emis}) at 465 nm) and Au_{25} (λ_{exc} at 535 nm and λ_{emis} at 700 nm) from mercaptosuccinic acid-protected gold nanoparticles by etching with excess glutathione at pH ~ 3 and pH 7–8 respectively.[18] The QYs calculated for Au_8 and Au_{25} were 15% and 0.19%, respectively.

In a similar approach, bright-red-emitting Au_{23} AQCs were synthesized by an interfacial etching procedure using $Au_{25}SG_{18}$ as starting material. Au_{23} is highly water-soluble and brightly fluorescent. With QY of 1.3% Au_{23} clusters exhibited quenching of fluorescence in the presence of Cu^{2+} ions and thus can be used as a Cu^{2+}-detection sensor.[22]

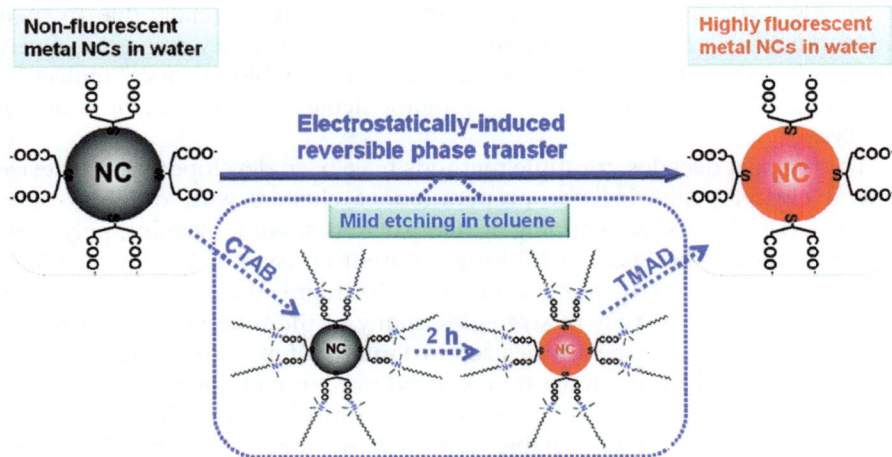

Figure 2.2   Schematic illustration of the process to generate highly fluorescent metal clusters by a phase transfer cycle (ref. 23). Copyright © 2011, American Chemical Society.

Yuan *et al.*[23] proposed a new method for generating highly fluorescent Ag, Au, Pt, and Cu clusters using a mild etching by a phase transfer *via* electrostatic interactions. They demonstrated, using Ag as a model, that a simple and fast phase transfer cycle can turn polydisperse nonfluorescent, and unstable silver clusters (in aqueous solution) into monodisperse, highly fluorescent, and stable Ag clusters in aqueous solution protected by the same original thiol ligand (glutathione) (Figure 2.2). The synthetic protocol was then extended to fabricate highly fluorescent silver clusters protected by custom-designed peptides with different functionalities (*e.g.*, carboxyl, hydroxyl, and amine). Ag, Au, Pt and Cu clusters show emissions at 660, 700, 435, and 438 nm, respectively, exciting at 480, 540, 380, and 378 nm, respectively, measured in toluene.

More recently, Pt clusters have also been obtained through this top-down approach by Le Guével *et al.*[24] also using glutathione for ligand etching of platinum nanoparticles. Platinum clusters show excitation at 490 nm and emission at 570 nm with a calculated QY of 17% (Figure 2.3).

Rao and Pradeep[25] obtained a crude mixture of red and blue–green-emitting clusters $Ag_8(H_2MSA)_8$ and $Ag_7(H_2MSA)_7$, (H_2MSA, mercaptosuccinic acid), by an interfacial etching reaction from H_2MSA-protected silver nanoparticles as a precursor. Both clusters are luminescent, Ag_7 λ_{exc} 350 and λ_{emis} 440 nm, and Ag_8 λ_{exc} 550 and λ_{emis} 650 nm, and show QYs of 9% and 0.3%, respectively.

More recently, Guo *et al.*[26] have obtained fluorescent Au_8 clusters (λ_{exc} 360 nm and λ_{emis} 436 nm), generated from heterophase ligand-exchange induced etching of DDA (dodecylamine) capped gold nanoparticles using

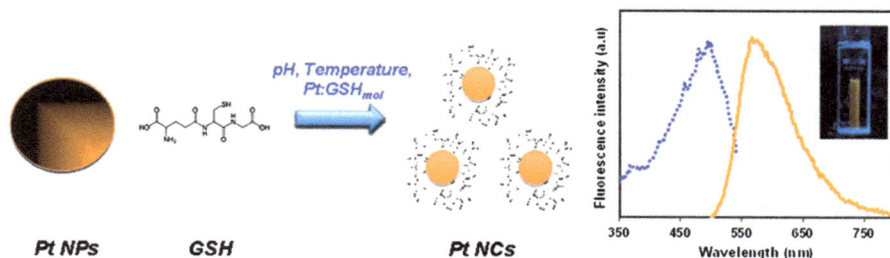

Figure 2.3   Yellow-emitting platinum clusters obtained by etching (ref. 24). Copyright © 2012, American Chemical Society.

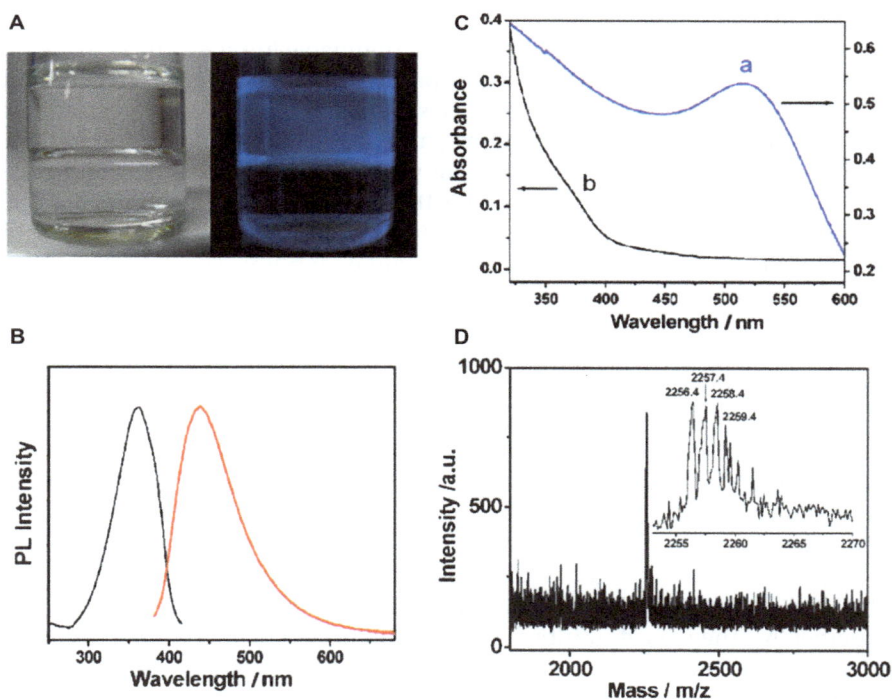

Figure 2.4   (A) Photographs of gold AQCs solutions under room light (left) and UV-lamp irradiation (365 nm, right). (B) Excitation (left) and emission (right) spectra of synthesized fluorescent gold AQCs. (C) UV-Vis absorbance spectra of DDA capped gold nanoparticle precursors (a, blue line) and the obtained fluorescent gold AQCs (b, black line). (D) MALDI-TOF mass spectrum of the obtained gold AQCs (positive spectrum, reflector mode, with 6-aza-2-thiothymine as assisted matrix). Inset: the isotopic pattern of the peak at m/z 2256.4 (reproduced from ref. 26).

one kind of thiol ligand (2-(dimethylamino)ethanethiol, DMAET). They have also shown that the synthesized clusters, with a QY of $19 \pm 2\%$, display electrochemiluminescence (Figure 2.4).

Huang and coworkers[27] created a series of water-soluble alkenethiol-gold clusters with tunable fluorescence through modification of the chain length of the alkenethiols. The synthetic procedure consists of adding the alkenethiol to the preformed gold nanoparticles stabilized by THCP (Tetrakis (hydroxymethyl)phosphonium chloride). They found a decrease in cluster size upon increasing the size of the thiol capping agent used. Clusters displayed fluorescence emission between 500 and 620 nm, with the highest QY obtained using 11-mercaptoundecanoic acid. These clusters can be used as a system for sensing Hg(II) ions based on fluorescence quenching.

Following the same strategy,[28] wavelength-tunable luminescent Au clusters and Au/Ag clusters have been obtained controlling the molar ratios of THPC to Au ions and of Ag ions.

They used 11-mercaptoundecanoic acid (11-MUA) to react with the as-prepared nanoparticles to prepare wavelength-tunable luminescent 11-MUA–Au clusters and 11-MUA–Au/Ag clusters, respectively. Au and Ag/Ag clusters exhibit a number of attractive optical properties: tunable luminescence wavelengths (456–640 nm), long lifetimes (>250 ns), and large Stokes shifts (>100 nm).

Lin et al.[11] reported the synthesis of water-soluble fluorescent gold clusters capped with dihydrolipoic acid (DHLA) using the etching approach. The resulting clusters have a QY of $\approx$ 1–3%, low photobleaching compared to organic fluorophores, and very good colloidal stability. They can be conjugated with EDC (carbodiimide) chemistry to biologically relevant molecules such as PEG (polyethylene glycol) BSA (serum albumin), avidin and streptavidin, having great potential in biological labeling applications. Zhou et al.[19] have introduced an etching approach to prepare water-soluble blue fluorescent, sub-nanometer-sized gold clusters from gold nanocrystals (nanoparticles and nanorods) with the assistance of biomolecules (amino acids, peptide, protein, and DNA) under sonication. The obtained clusters (Au_8) showed high and stable photoluminescence in water as well as solvent-dependent photoluminescence properties.

Highly fluorescent and water-soluble Au_8 metal clusters were also obtained through a ligand induced etching process by Duan and Nie. Specifically, they used multivalent coordinating polymers such as polyethylenimine (PEI) which are able to "etch" preformed colloidal gold nanocrystals, leading to atomic gold clusters that are highly fluorescent at 445 and 505 nm upon UV light excitation.[21]

Large clusters were obtained by Muhammed et al., also employing an etching route. Luminescent Au_{38} clusters with a QY of 4% were obtained via core etching of gold nanoparticles by BSA (serum albumin). Their luminescent properties were exploited as a "turn-off" sensor for the detection of Cu^{2+} ions and a "turn-on" for glutathione detection.[29]

2.3   Bottom-Up Approach

Using this strategy, clusters are built from smaller structures at atomic level such as single atoms or ions. Here, the most commonly used techniques differ depending on the size of the clusters to be obtained. Normally, wet synthetic protocols include the reduction of metal salts using strong or weak reducing agents, or by other approaches like electrochemistry, photo-chemistry, sonochemistry, microwave-assisted methods, *etc.*, in the presence of an appropriate stabilizer. Some of the stabilizers consist of templates with predefined structures acting as nanoreactors where the clusters are obtained. In this case, the strategy is usually named as template-assisted synthesis. Among the capping molecules used for the synthesis one can find polymers, dendrimers, and biomolecules such as oligonucleotides, peptides, proteins, *etc.* Therefore, choosing the suitable stabilizer, together with a kinetic control of the process are the key points to achieve relatively monodisperse fluorescent clusters.

2.3.1 Chemical Reduction

Traditional wet chemical reduction consists of the reduction of the corresponding precursor (normally the corresponding metal salt) in the presence of small molecules, *e.g.* thiols, phosphines, tetraalkylammonium salts, *etc.* The most commonly used reducing agent is sodium borohydride, but some other mild reducing agents, such as THCP[30] and ascorbic acid[31] are also used, as well as some stabilizers which can act as reducing agents like *e.g.* thiols,[32] proteins[33] and amino acids.[34] Through this strategy different experimental parameters can be adjusted in order to control the cluster size: metal to ligand ratio, the type and concentration of the reducing agent, synthesis temperature, pH, *etc.*

For example, DHLA (dihydrolipoic acid)–Au clusters have been obtained by reducing both LA (lipoic acid) and gold salt using NaBH$_4$ as reducing agent in aqueous solution. The obtained Au clusters possess near-infrared fluorescence emission (λ_{exc} 550 nm λ_{emis} 684 nm) with a QY $\sim 0.6\%$, showing good biocompatibility verified by a cytotoxicity test, and average fluorescence lifetime two orders of magnitude longer than the lifetime of cellular autofluorescence, which make them promising candidates as novel fluorescent probes in biological applications.[35]

Au$_{11}$ clusters were successfully synthesized *via* a modified Brust two-phase method in the presence of carboxylic acid-terminated thiolated molecules. Through the ion-pair interaction, molecules of tetraoctylammonium cation (TOA$^+$) are combined with the COO$^-$ groups of the carboxylic acid-terminated thiolated molecules that are absorbed on the surface of the gold clusters by self-assembly to form the high-density ligands (SCH$_2$CH$_2$COO$^-$)$_7$ (TOA$^+$)$_7$ (Figure 2.5). Au$_{11}$ clusters exhibit blue luminescence with a QY of 8.6%.[36]

Shang *et al.*[30] reported a simple one-pot strategy for preparing small fluorescent Au clusters using a mild reductant, THPC in the presence of D-penicillamine (DPA). DPA capped Au clusters show excitation and emission bands at 400 nm and 610 nm respectively, with a QY of 1.3%. Adding borohydride to the as-prepared clusters and analyzing the samples by XPS they conclude that the luminescence of the prepared Au clusters is closely related

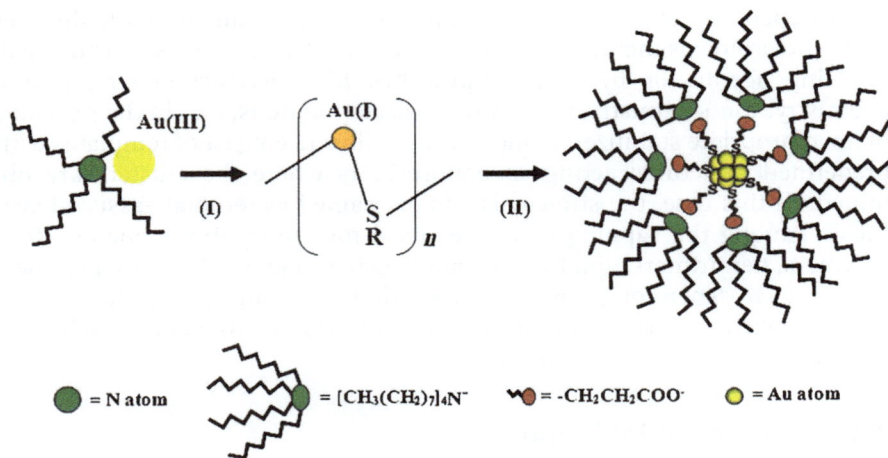

Figure 2.5 Schematic illustration of the formation of $Au_{11}(SCH_2CH_2COO^-)_7$ $(TOA^+)_7$ clusters (ref. 36).
Copyright © 2009 WILEY-VCH Verlag GmbH & Co. KGaA, Weinheim.

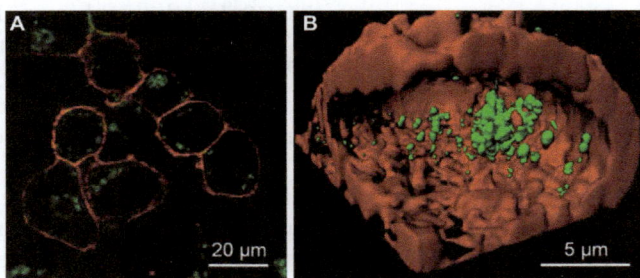

Figure 2.6 (A) Confocal image of HeLa cells after incubation with DPA–Au clusters for 2 h. (B) Cross-section of a 3D image reconstruction, showing internalized DPA–Au clusters. Membranes were stained with the red dye DiD. Images were taken by 2-photon excitation at 810 nm (reproduced from ref. 30).

to the presence of Au(I) on their surfaces. In a first optical imaging application, they studied the internalization of the Au clusters by live HeLa cells using confocal microscopy with two-photon excitation (Figure 2.6). Cell viability assay also revealed the good biocompatibility of these Au clusters demonstrating a great potential of DPA-stabilized Au clusters as fluorescent nanoprobes in bioimaging and related applications.

More recently, Sun *et al.* described an 11-MUA-directed, one-pot method for the synthesis of water-soluble fluorescent Au AQCs. The as-prepared water-soluble Au AQCs with average diameters of 1.8–0.4 nm exhibit a unique fluorescence excitation at 285 nm, a maximum emission at 608 nm, and a QY of 2.4%. The fluorescence of 11-MUA-Au AQCs can be quenched by several metal ions but selectively by Cr^{3+} ions when using EDTA as the masking agent for other metal ions.[32]

Phosphines have also been used in metal cluster synthesis. Thus, for example Shichibu *et al.*[37] obtained Au_{13} clusters using the capability of hydrochloric acid to promote the convergence of the nuclearity in polydisperse phosphine-coordinated gold clusters, providing a new route to obtain mainly geometrically magic-number Au_{13} clusters. The as-prepared clusters show strong near-infrared photoluminescence emission (λ_{emis} 766 nm).

2.3.2 Photoreduction

Photoreduction processes have also recently received great attention. Compared with other approaches, photoreduction avoids the use of additional reducing agents, being a clean, non-toxic and, in many cases, less time-consuming approach.[38]

Generally, polymers with different functional groups are employed for the preparation of luminescent clusters using photoreduction. Thus, fluorescent silver clusters were prepared in polymer microgels of poly(N-iso-propylacrylamide-acrylic acid-2-hydroxyethylacrylate) by UV irradiating the mixture of polymer microgel and Ag^+.[39] Shen *et al.* have also employed a multiarm star poly(acrylic acid) "molecular hydrogel" as template with the assistance of UV exposure to obtain fluorescent silver clusters with various emission bands in the region 450–750 nm (Figure 2.7).[40]

Díez *et al.* have systematically investigated the optical properties of Ag clusters in solution and their response to the environment. They have employed poly(methacrylic acid) (PMAA) and visible light to initiate the reaction. Silver clusters show emission properties and also electrochemiluminescence and solvatochromic properties with QYs around 5–6%.[41]

Water-soluble fluorescent copper, silver and gold AQCs with QYs of 2.2, 6.8 and 5.3%, respectively, have been prepared by irradiating the solution of their precursors with an ultraviolet light source (8 W, wavelength = 365 nm)

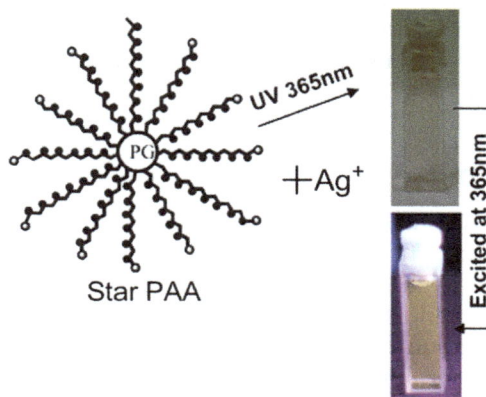

Figure 2.7 Water-soluble fluorescent Ag clusters prepared using star-branched polyglycerol-*block*-poly(acrylic acid) (PG-*b*-PAA) templates (ref. 40). Copyright © 2007 WILEY-VCH Verlag GmbH & Co. KGaA, Weinheim.

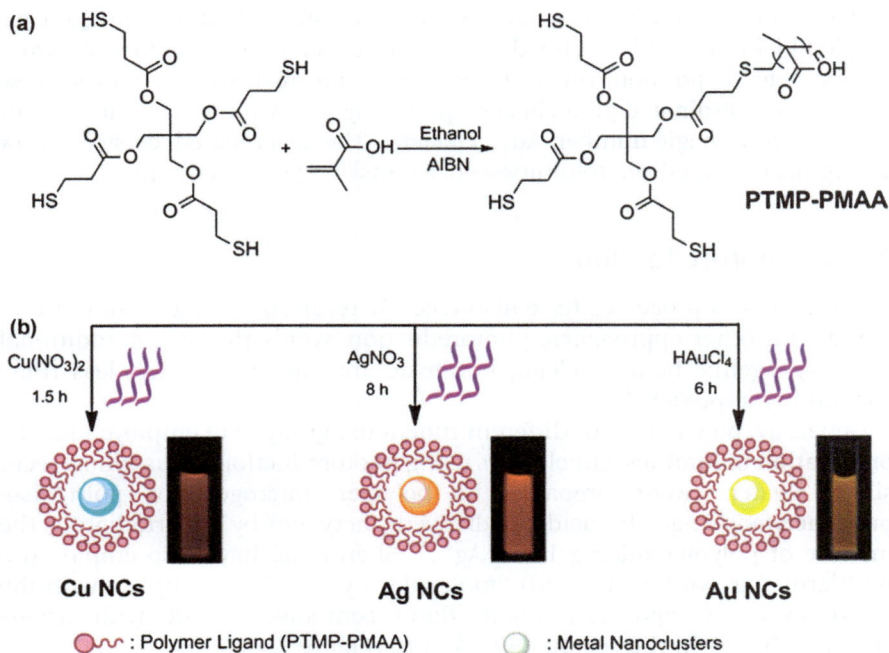

Figure 2.8 Reaction scheme for (a) synthesis of the thioether polymer ligand PTMP–PMAA and (b) preparation of AQCs stabilized by polymer ligands (reproduced from ref. 38).

in the presence of poly(methacrylic acid) functionalized with pentaerythritol tetrakis 3-mercaptopropionate (PTMP–PMAA) (Figure 2.8).[38]

Huang *et al.* have used a multifunctional polymer ligand containing thiol, thioether, and ester functional groups, to prepare intense blue fluorescent (λ_{emis} 430 nm, λ_{exc} 365 nm) gold AQCs with a diameter of ~ 1.2 nm and a high QY of 24.3% in organic media without using additional phase-transfer reagents. This strategy is also suitable to obtain fluorescent Ag and Pt AQCs, but with lower QYs.[42]

2.3.3 Sonochemistry

Another alternative, environmentally friendly, method that avoids the typical chemical reduction using sodium borohydride is sonochemistry. This method has been well-established for the synthesis of different nanoparticles and nanostructured materials, but has received less attention for the preparation of AQCs.[43] The chemical effects of high intensity ultrasound derive primarily from acoustic cavitation: the formation, growth, and implosive collapse of bubbles in a liquid irradiated with ultrasound. Localized hot spots with temperatures of 5000 K and pressures of hundreds of bars can be generated during ultrasonic irradiation of

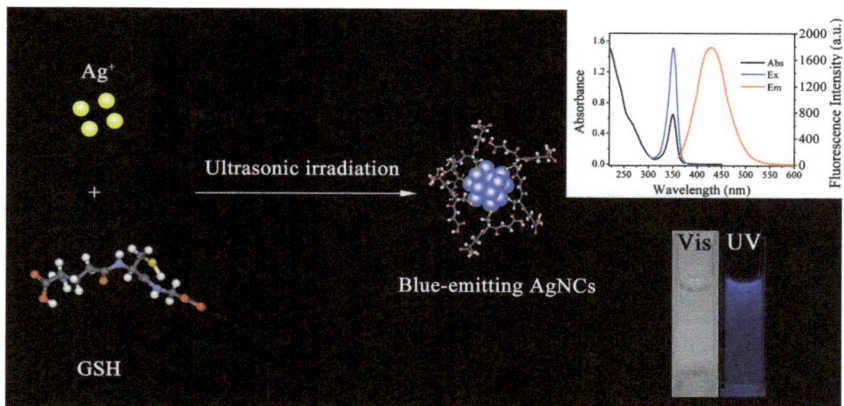

Figure 2.9 Schematic representation of a one-pot sonochemical approach to prepare highly blue-emitting Ag clusters using glutathione as a stabilizing agent in aqueous solution (reproduced from ref. 46).

water. As a consequence, highly reactive species, including $HO_2^{\bullet}$, $H^{\bullet}$, $OH^{\bullet}$ and perhaps e^-_{aq} are formed during aqueous sonolysis, similar to the effects of γ-ray or UV irradiation of aqueous solutions. The advantages of sonochemistry include nonhazardous, quick reaction rate, controllable reaction conditions and the ability to form nanoparticles with uniform shapes, narrow size-distribution and high purity. Thus, for example, sonochemistry has been successfully used by Liu *et al.* to synthesize luminescent Au and alloy Au@Ag AQCs using BSA as capping agent. Au AQCs show a QY of 8%, with emission at 670 nm and excitation at 365 nm while Au@Ag AQCs show a higher QY of 9.6%, and a blue-shifted emission at 565 nm. These clusters can be applied to sense Cu(II).[44]

Xu and Suslick reported the sonochemical synthesis of highly fluorescent Ag AQCs using a polyelectrolyte, polymethylacrylic acid (PMAA), as template. This method resulted in a distribution of Ag AQCs sizes and fluorescent emission properties in the region 500–800 nm at different excitation wavelengths. These properties can be easily controlled varying the synthetic conditions.[45]

Zhou *et al.* also employed sonochemical synthesis to obtain fluorescent glutathione-stabilized silver clusters. Silver clusters show a QY of 1.9% emitting at 430 nm when excited at 350 nm (Figure 2.9).[46]

2.3.4 Electrochemical Synthesis

Electrochemical synthesis, first introduced by Reetz and Helbig[47] in 1994 for the preparation of nanoparticles, is a very simple technique, allowing the preparation of monodisperse AQC samples in just a matter of minutes. This strategy consists of the anodic dissolution of an electrode in the presence of a stabilizer and the subsequent electroreduction of the preformed cations.

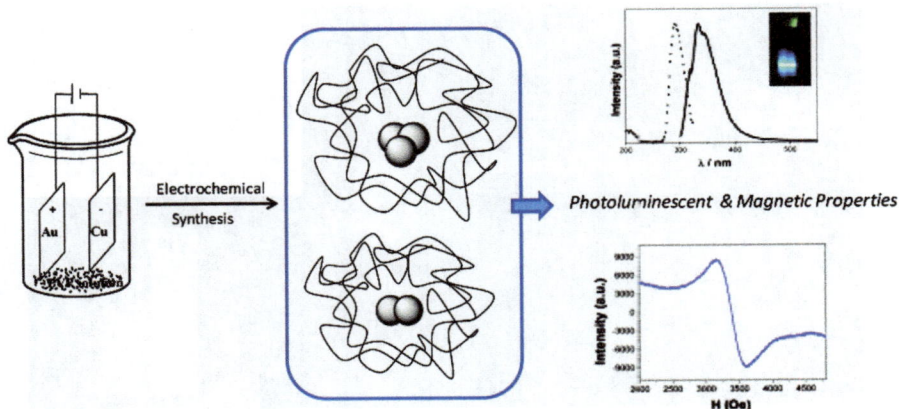

Figure 2.10 Synthesis of Au AQCs containing only two and three atoms in the presence of PVP *via* a simple electrochemical technique. These clusters show stable photoluminescent and magnetic properties (ref. 49).
Copyright © 2010, American Chemical Society.

This technique allows very precise control of the kinetic reaction by adjusting some synthetic parameters, such as the density current, temperature, *etc.*[48] In this way, luminescent gold, silver and copper AQCs have been prepared by this technique.

Blue luminescent poly(*N*-vinylpyrrolidone), PVP-protected gold AQCs with only 2 or 3 atoms and high QYs (12.5%) have been obtained at 25 °C and with a time of synthesis of only 300 s, observing that higher temperatures and higher synthesis time lead to bigger size distributions or even nanoparticles.[49] These results reveal the versatility of this route, which is also appropriate to synthesize silver and copper analogues (Figure 2.10).[50,51]

Small fluorescent atomic copper AQCs, stabilized by tetrabutylammonium nitrate, have also been synthesized by this electrochemical technique. These small clusters (Cu_N, $N < \approx 14$) show photoluminescence in the visible range with QYs (13%) and are very stable in both apolar and polar solvents.[50]

More recently, the same technique was applied to obtain hydrophilic/hydrophobic Ag_5 and Ag_6 blue luminescent clusters (QY 1.2%). In this case the use of dodecanethiol (DDT) and tetrabutylammonium (TBA) as capping agents allows silver clusters to be soluble in both water and organic solvents.[51]

2.3.5 Microemulsions

Microemulsions, with either water dispersed in oil or oil dispersed in water, and with dimensions of the liquid cores at the nanoscale level (nanodroplets), can also be used as chemical nanoreactors. One of the advantages of this method is that the size of the nanoreactor can be easily tuned by adjusting the composition of the microemulsions through the ratio of the liquid used as nanodroplet to the surfactant. By an appropriate choice of the nanodroplet sizes and the reactant concentrations the reaction kinetics can be controlled to synthesize clusters. For example, Ag_n clusters ($n \leq 10$) were prepared using a water-in-oil microemulsion consisting of a mixture of sodium bis(2-ethylhexyl) sulfosuccinate, isooctane, and water. Silver clusters show very large bandgaps (≈ 2.3 eV), fluorescence and molecular-like paramagnetic properties.[52] More recently, Vázquez-Vázquez *et al.*[53] reported the synthesis of small atomic copper clusters in microemulsions, Cu_N ($N \leq 13$), with different core sizes prepared using a microemulsion consisting of sodium dodecyl sulfate (SDS, as surfactant), isopentanol (as cosurfactant), cyclohexane (as the oily phase), and Cu(II) sulfate solution (as the aqueous phase). They proved that the core size can be easily controlled by adjusting the amount of the reducing agent, obtaining photoluminescent copper clusters using very low percentages of reducing agent.

2.3.6 Microwave-Assisted Synthesis

Microwave irradiation is also widely used nowadays for material synthesis due to fast response, low energy consumption and ease of operation. This method provides rapid and homogeneous heating helping to control the kinetics of reactions and the morphology of the products, and hence can be applied to prepare nanocrystals with high quality and narrow size distribution.

Shang *et al.* prepared dihydrolipoic acid (DHLA) capped fluorescent gold clusters by irradiating the solution with microwaves during the synthesis. They observed that by using microwave irradiation the QY of gold clusters was enhanced about five-fold and the reaction time was shortened from several hours to several minutes. The as-synthesized DHLA–Au clusters possessed bright near-infrared fluorescence (QY: 2.9%), and they can be used to sense Hg^{2+} in HeLa cells using spinning disc confocal microscopy (Figure 2.11).[54]

Au_{16}@BSA clusters were also obtained by a one-step microwave-assisted method by Yue *et al.*[55] The luminescent properties of such clusters make them appropriate as a luminescent enhanced sensor for detection of silver (I) ions with high selectivity and sensitivity.

Microwave irradiation was also used for the synthesis of water-soluble fluorescent Ag AQCs.[56] In a typical synthesis poly(methacrylic acid) (PMAA) and silver nitrate at specific pH were used as reactants. As-prepared Ag AQCs, with an average size of 1.6 nm, show red fluorescence emission around 608 nm and a characteristic absorption peak at about 508 nm.

Cu clusters[57] can also be obtained using a microwave-assisted polyol method without using additional protective and reducing agents. The as-prepared copper clusters of 2 nm show emission at 475 nm, exciting at 355 nm, with a QY in ethylene glycol of 0.65% (Figure 2.12).

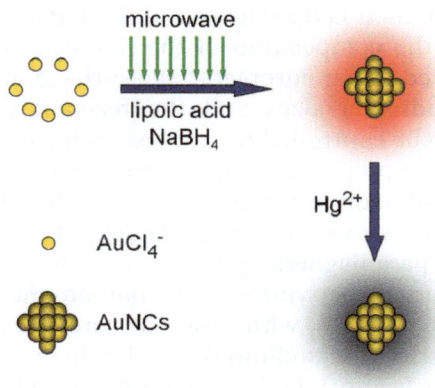

Figure 2.11 Schematic illustration of microwave-assisted synthesis of fluorescent Au clusters and their fluorescence quenching upon interaction with Hg^{2+} ions (reproduced from ref. 54).

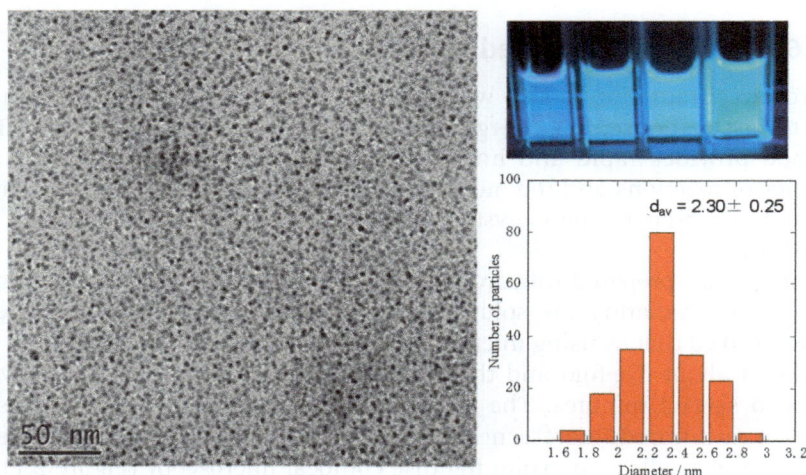

Figure 2.12 Cu clusters prepared by the microwave-assisted polyol method (reproduced from ref. 57).

2.3.7 Template-Assisted Synthesis

Many fluorescent clusters have been prepared following this scheme process, being one of the most widespread nowadays. It is very simple, and many templates can be used: polymers, dendrimers, and biomolecules such as oligonucleotides, peptides and proteins. These templates act as stabilizers and they not only provide different configurations and spaces where metal clusters are obtained, but they also make it possible to predefine a specific morphology and tune the core size, allowing in many cases the achievement of monodisperse cluster sizes.

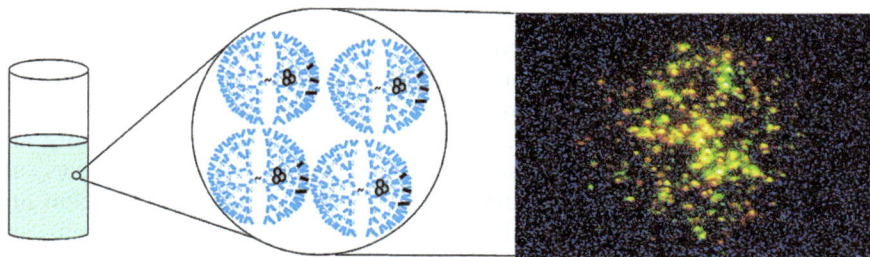

Figure 2.13 Highly fluorescent and photostable silver AQCs prepared photo-chemically within poly(amidoamine) dendrimer hosts in aqueous solutions (ref. 58).
Copyright © 2002, American Chemical Society.

2.3.7.1 Dendrimers

Dendrimers have a high capacity for sequestering ions, trapping them and forming AQCs. One example is the fourth-generation polyamidoamine dendrimer (PAMAM (G4-OH)), which has been used as a molecular template in the synthesis of fluorescent metal AQCs. Zheng and Dickson reported the synthesis of luminescent silver AQCs in PAMAM (poly(amidoamine)) through a direct photoreduction method (Figure 2.13). Silver AQCs showed five emission peaks ranging from 530 nm to 650 nm.[58]

The same group also obtained Au_8 AQCs with a high QY of 42%,[59] and they proved that by controlling the relative PAMAM/Au concentrations it is possible to obtain gold AQCs with emission colors ranging from the UV to near-infrared region.[60] Jao *et al.* reported the use of dendrimers with terminal amine and hydroxyl groups as a microcavity-template upon microwave irradiation for obtaining gold clusters with very high QYs (20% to 62%).[61] Bao *et al.*[31] developed a method for synthesizing fluorescent gold clusters using physiological temperature (37 °C) and ascorbic acid (vitamin C), a mild biologically derived reductant, in the presence of PAMAM dendrimer templates. Gold clusters produced by this method show emission at 460 nm, exciting at 387 nm. Gold clusters with multiple color emission can also be obtained by this method introducing a minor variation.

More recently, Jin *et al.*[62] prepared Pt_5 AQCs by reducing H_2PtCl_6 with $NaBH_4$ in the presence of PAMAM (G4-OH). Then, mercaptoacetic acid (MAA) was added to replace the PAMAM (G4-OH) ligand. The emission wavelength and lifetime of the Pt_5 AQCs (470 nm, 8.8 ± 0.5 ns) were different from those of oxidized PAMAM (G4-OH) (450 nm, 6.3 ± 0.5 ns) showing a QY of 18% in water.

2.3.7.2 Nucleotides

This strategy is very simple. It consists of the reduction of metal ions in the presence of appropriate quantities of nucleotides, generally using sodium borohydride as reducing agent.

Deoxyribonucleic acids, especially the cytosine nucleobase, exhibit strong affinity for silver ions. Furthermore, the flexibility, facile synthesis, small size, and sequence tunability enable DNA molecules to be good protection groups for very small, few-atom luminescent silver clusters. DNA–Ag AQCs are robust fluorescence materials with good QY, high photostability and small size. Interestingly, the use of DNA-templated silver AQCs allows tunable emission spectra, which depend on the DNA sequence. Dickson and coworkers[63] provided a detailed description of this fluorescent system, and this motivated continued research in the preparation and characterization of DNA–Ag AQCs. In general, silver AQCs are obtained by reduction of silver nitrate in the presence of the ligand by sodium borohydride. The mass spectrometry analysis revealed cluster sizes ranging from Ag_1 to Ag_4/oligonucleotide. Subsequent optimizations of the synthetic procedure led to improvements in the monodispersity of the samples. Thus, for example, a near-IR emitter (λ_{exc} 750 nm/λ_{emis} 810 nm) Ag_{10}-DNA encapsulated with excellent photostability, and 30% fluorescence QY[64] was reported. Using DNA microarrays to optimize the scaffold sequences, Richards *et al.* created five highly photostable Ag AQC fluorophores with tunable emission from the blue to the near-IR and excellent photophysical properties.[65] Vosch *et al.* also reported strongly emissive individual DNA-encapsulated Ag AQCs as single-molecule fluorophores,[66] and Sharma *et al.* reported Ag AQC species with an excitation maximum at 650 nm and emission maximum at 700 nm, a fluorescence QY of 17% and 2.6 ns lifetime (Figure 2.14).[67]

O'Neill *et al.*[68] presented the loop-dependent synthesis of fluorescent Ag AQCs on DNA hairpins with 3–12 cytosines in the loop. They showed a loop-dependent fluorescence of Ag AQCs, and QY $\approx$ 3%.

Jia *et al.* reported for the first time the synthesis of DNA-hosted copper AQCs with excitation and emission maxima at 344 nm and 593 nm, respectively and a decay time near 4.5 μs (Figure 2.15).[69] They described a sensitive fluorescent diagnostic of the mismatch type in a specific

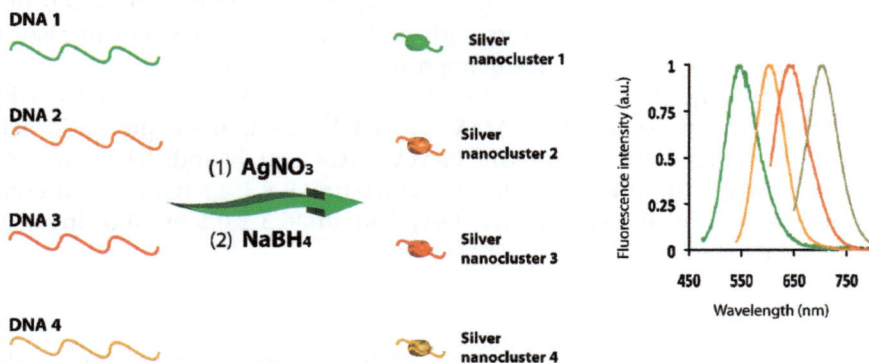

Figure 2.14   Palette of fluorescent silver AQCs, templated and functionalized by DNA, produced by Sharma *et al.* (reproduced from ref. 67).

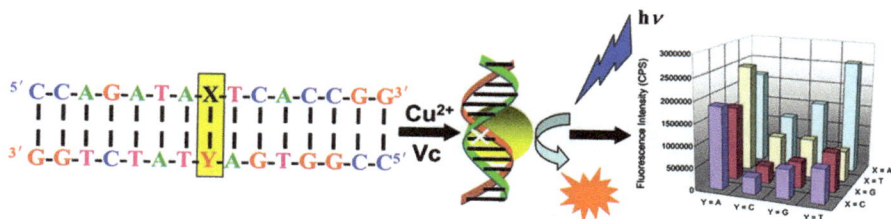

Figure 2.15 DNA-hosted copper AQCs for fluorescent identification of single nucleotide polymorphisms (ref. 69).
Copyright © 2012, American Chemical Society.

DNA sequence, providing some light on the luminescent mechanism of Cu AQCs.

More recently, the amino acid histidine (L-3,4 dihydroxyphenylalanine, L-DOPA) was used as both reductant and stabilizer to produce fluorescent Au clusters in a simple reaction.[34] Within a short reaction time of 15 min, this strategy allows the fabrication of homogeneous Au clusters having the capability to sense ferric ions. The as-prepared Au clusters exhibited a fluorescence emission at 525 nm and a QY of 1.7%. On the basis of an aggregation-induced fluorescence quenching mechanism, these fluorescent Au clusters offer acceptable sensitivity, high selectivity, and a limit of detection of 3.5 μM for the determination of Fe^{3+} ions, which is lower than the maximum level (0.3 mgL^{-1}, equivalent to 5.4 μM) of Fe^{3+} permitted in drinking water by the U.S. Environmental Protection Agency.

2.3.7.3 Proteins

Size control can be achieved using proteins as templates,[70] with the advantage of having metal clusters embedded in a functional macromolecule matrix. Increasing research interest in these types of systems indicates an emerging trend for such synthetic methods. Multiple inorganic nanomaterials and nanostructures have been formed by biomineralization or biomimetic mineralization in natural and synthetic proteins.[71] Bio-medically important functional proteins such as *e.g.* lactotransferrin[72] or insulin[73] have successfully been used to obtain metal clusters.

Chaudhari *et al.* studied the time-dependent biomineralization of Au^{3+} by native lactoferrin (NLf) and bovine serum albumin (BSA) resulting in near-infrared (NIR) luminescent gold quantum clusters which occur through a protein-bound Au^{1+} intermediate and subsequent emergence of free protein. They demonstrated the importance of obtaining monodispersed protein-protected clusters for the cluster crystallization, which is necessary to undertake a systematic study of its characterization and to determine the type of macromolecule–metal cluster interaction. Such studies could be the basis for creating tailor-made genetically engineered proteins to form clusters with desired properties.[74]

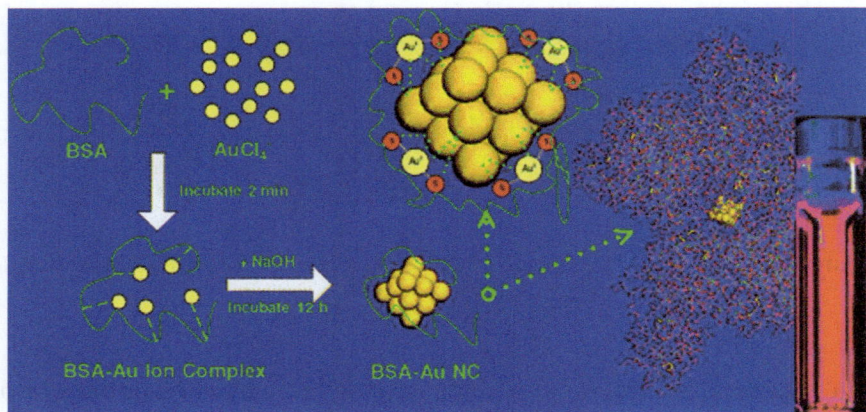

Figure 2.16 Simple, one-pot, "green" synthetic route, based on the "biomineralization" capability of bovine serum albumin (BSA) for the preparation of highly stable Au AQCs with red emission and high QY (ref. 70). Copyright © 2009, American Chemical Society.

Gold AQCs with red emissions can be directly obtained using a common protein to sequester and reduce Au precursors *in situ* (Figure 2.16).[70] In line with this, as-prepared BSA-Au AQCs, consisting of 25 gold atoms with excitation and emission bands located at 480 nm and 640 nm, respectively, and a QY of 6% were obtained.

AuAg alloy AQCs can also be prepared using proteins as templates. Mixing as-synthesized protein-protected Au and Ag clusters resulted in the formation of alloy clusters in bovine serum albumin (BSA). They showed emission at 707 nm when excited at 370 nm, with a calculated QY of 2.8%. Time resolved measurements revealed a tri-exponential decay in the nanosecond range: 0.12 ns (71%), 1.20 ns (22%) and 11.8 ns (7%).[75]

Human serum albumin (HSA) can also be used to obtain fluorescent AQCs similarly to bovine serum albumin (BSA). Accordingly, highly fluorescent Ag_9/HSA and Ag_{14}/HSA have been obtained using HSA as protein template at physiological temperature (37 °C). The important features of Ag_9/HSA are (i) a one-pot, green synthesis having blue emission centered at 460 nm, (ii) high QYs (16%) and fast fluorescence lifetimes (7.51 ns), and (iii) further reduction of such clusters gives red AQCs (emission at 690 nm), which can further be reoxidized to form Ag_9/HSA with almost 100% recovery. The important features of Ag_{14}/HSA are (i) high QYs (11%) and fast fluorescence lifetimes (2.72 ns) and (ii) oxidation gives Ag_9/HSA (emission at 460 nm), which can further be reduced to red-emitting AQCs. The novelty of this study is that it is possible to interconvert both Ag/HSA AQCs, depending on the experimental conditions, besides the fact that they show two-photon excitation properties that enable them to be used in several important applications, like bioimaging (Figure 2.17).[76]

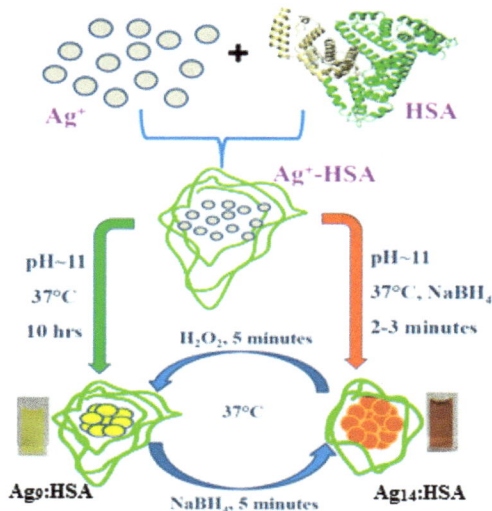

Figure 2.17 Schematic representation of the formation and interconversion of Ag_9/HSA and Ag_{14}/HSA (ref. 76).
Copyright © 2012, American Chemical Society.

2.4 Summary

Table 2.1 shows a summary of the photophysical parameters of metal clusters obtained by different approaches.

One can try to rationalize the results observed in Table 2.1. For this purpose we will try to apply the spherical Jellium model to obtain an approximate description of the observed dependence of the emission energy on the number of atoms, N, in metal clusters. According to this simple model, the band gap of a cluster with N atoms will be given by the simple relation, $E_g = E_{Fermi}/N^{1/3}$, (where $E_{Fermi} =$ Fermi energy of the bulk metal, and $E_g =$ band gap energy of the cluster). If we assume that the band gap energy is proportional to the fluorescence emission or excitation energy, then we can predict a linear relationship between such energies and $N^{-1/3}$. Figures 2.18 and 2.19 show that, indeed, there is a relatively good linear relationship for all clusters displayed in Table 2.1. The fitted values of the slopes were: 10.0 ± 0.05 eV (excitation energy) and 8.0 ± 0.7 eV (emission energy) for copper clusters, and 6.6 ± 0.3 eV (excitation energy) and 5.0 ± 0.2 eV (emission energy) for the rest of the clusters (mainly gold and silver). It can be seen that the value obtained from the emission energy matches reasonably well with the E_{Fermi} value for bulk gold and silver (5.3 eV), and for bulk copper (7.0 eV). Moreover, from the fittings it can be predicted that $\Delta E = E_{exc} - E_{emis} = 1.6N^{-1/3}$ for Au and Ag, and $2.0N^{-1/3}$ for Cu. This indicates that this simple Jellium model can be used to obtain an approximate value for the band gap energy of clusters from their emission energies. At the

Table 2.1 Photophysical parameters of some AQCs obtained using different approaches.

Synthesis	Exc/Emis (nm)	QY (%)	Lifetime (ns)	Capping agent	Size	Ref.
Chemical reduction	355/417	8.6	—	MPA (Mercaptopropionic acid) Tetraoctyl ammonium (TOA$^+$)	Au$_{11}$	36
Chemical reduction	360/766	—	—	Dppe (1,2-bis(diphenylphosphino) ethane)	Au$_{13}$	37
Electrochemical	240,270/315,335,350	12.5	1.12, 4.11, 10.6	Poly(N-vinylpyrrolidone) (PVP)	Au$_2$ (a), Au$_3$ (b)	49
Electrochemical	325/400 350/420	1.2	0.476, 1.64, 5.11 (Emis 400 nm) 0.565, 1.85, 5.75	Dodecanethiol (DDT)/Tetrabutyl ammonium (TBA)	Ag$_5$ (a) and Ag$_6$ (b)	51
Electrochemical	300/410	13	—	Tetrabutylammonium (TBA) nitrate	Cu$_N$ $< \approx 14$	50
Etching	353/445	10–20	—	PolyethylenImine (PEI)	Au$_8$	21
Etching	360/436	19	—	DMAET (2-(dimethylamino ethanethiol)	Au$_8$	26
Etching	370/475 535/700	15, 0.19	0.03, 1.6, 5.3 (Au$_8$)	Glutathione	Au$_8$ (a), Au$_{25}$ (b)	18
Etching	532/685	1.3	0.04, 2.4, and 68.5	Glutathione	Au$_{23}$	22
Etching	350/440 550/650	9, 0.3	0.035, 37.2, 37.2, 5.68 (Ag$_7$) 0.012, 0.396, 2.10, 8.31 (Ag$_7$)	H$_2$MSA (mercaptosuccinic acid)	Ag$_7$ (a), Ag$_8$ (b)	25

Method	Ex/Em	QY	MW / ratio	Template	Cluster	Ref
Microemulsions	270,290/350	—	—	—	$Ag_{N<10}$	52
Microemulsions	290/333	—	—	—	Cu_N ($N \leq 13$),	53
Microwave	350/604	—	—	Serum albumin (BSA)	Au_{16}	55
Photoreduction	360/610	5.3	—	Pentaerythritol tetrakis 3-mercaptopropionate	Au_5	38
Photoreduction	365/430	24.3	—	Thioether polymer	Au_7	42
Photoreduction	525/625	5-6	0.96, 2.4-3.1, 148 000-164 000, 20 000-45 000	Poly-(methacrylic acid) (PMAA)	Ag_2, Ag_3 and Ag_5	41
Templating	384/450	42	7.5 and 2800	(Poly)aminoamide (PAMAM)	Au_8	59
Templating	330/385	70	3.5	(Poly)aminoamide PAMAM	Au_5 (a)	60
	385/456	42	7.5		Au_8 (b)	
	433/510	25	5.2		Au_{13} (c)	
	670/751	15	3.6		Au_{23} (d)	
	765/879	10	—		Au_{31} (e)	
Templating	480/640	6		Serum albumin (BSA)	Au_{25}	33
Templating	370/707	2.8	0.12, 1.20 and 11.8	Serum albumin (BSA)	$(AuAg)_{38}$ (a)	75
	370/660	6.8	0.14, 1.30 and 32.9		Au_{38} (b)	
	370/670	1.9	0.16, 0.96 and3.9		Ag_{38} (c)	
Templating	380/460	16	7.51	Human serum albumin (HSA)	Ag_9	76
	480/690	11	2.72		Ag_{14}	
Templating	750/810	30	1.8	DNA	Ag_{10}	64
Templating	380/470	18	8.8	(Poly)aminoamide (PAMAM)	Pt_5	62

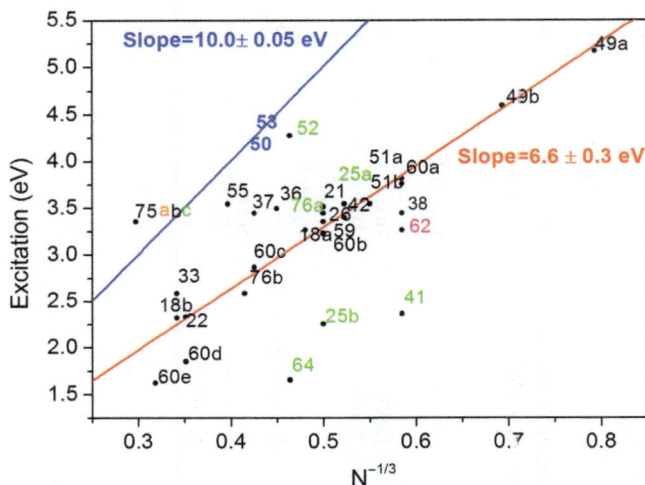

Figure 2.18   Linear fitting of the excitation energies to the Jellium model for gold (black), silver (green), copper (blue), platinum (pink) and alloy AuAg (orange) clusters displayed in Table 2.1.

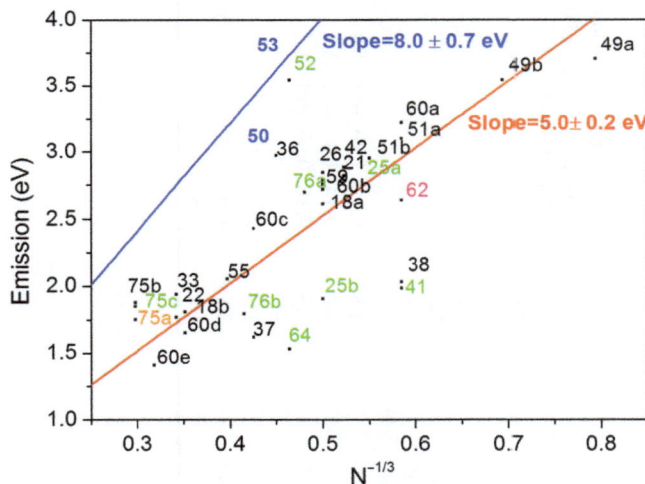

Figure 2.19   Linear fitting of the emission energies to the Jellium model for gold (black), silver (green), copper (blue), platinum (pink) and alloy AuAg (orange) clusters displayed in Table 2.1.

same time, these results show that an approximate estimation of the number of atoms of the cluster can also be made from this simple model, a result already pointed out by some authors.[49–51,77] In the figures one can see that there are some deviations from the fitted curves, indicating that there are, as expected, other factors—different from the number of atoms—contributing to the band gap energy of clusters. From inspection one can see that the major deviations come from cases where large Stokes shifts were reported, which can be related to some charge-transfer processes (between the clusters and the ligands), which cannot be accounted for by the very simple Jellium model because it does not take into account the influence of the ligands.

In summary, it can be said that remarkable progress has been made during the last decade in the aqueous synthesis of fluorescent metal clusters. Multiple methods have been developed to synthesize metal clusters, allowing clusters to be obtained with different sizes and protected with different capping agents (such as biomolecules, proteins, *etc.*), providing different luminescent properties.

Clusters with high QYs have been obtained; generally observing the highest values for clusters obtained in dendrimers and DNA templates. These capping agents seem to preserve better the inherent fluorescent properties of metal clusters protecting them against cluster–cluster interactions or possible interactions with the solvent.

It is also important to point out the generally great photostability obtained for many metal clusters. This important aspect, together with their tiny sizes, their biocompatibility and their relatively high QYs (some of them reaching values close to those of QDs), make them a very attractive alternative to traditional fluorophores, with significant potential applications as biological labels or optoelectronic devices.

Acknowledgements

We want to acknowledge the financial support of the MCI, Spain (MAT2010-20442), Xunta de Galicia (Grupos Ref.Comp.2010/41) and Obra Social Fundación La Caixa, Spain.

References

1. J. Li, X. Li, H. J. Zhai and L. S. Wang, *Science*, 2003, **299**, 864.
2. C. M. Cobley, E. Katz and I. Willner, *Angew. Chem., Int. Ed.*, 2004, **43**, 6042.
3. J. Y. Chen, E. C. Cho, L. V. Wang and Y. N. Xia, *Chem. Soc. Rev.*, 2011, **40**, 44.
4. G. H. Woehrle, M. G. Warner and J. E. Hutchison, *J. Phys. Chem. B*, 2002, **106**, 9979.
5. J. Calvo, J. Rivas and M. A. López-Quintela, Synthesis of Subnanometric Nanoparticles, in *Encyclopedia of Nanotechnology*, ed. B. Bhushan, Springer–Verlag, New York, 2012, 2639–2648.
6. X. Michalet, F. Pinaud, L. Bentolila, J. Tsay, S. Doose, J. Li, G. Sundaresan, A. Wu, S. Gambhir and S. Weiss, *Science*, 2005, **307**, 538.
7. J. Yan, M. C. Estévez, J. E. Smith, K. Wang, X. He, L. Wang and W. Tan, *Nano Today*, 2007, **2**, 44.

8. D. K. Chatterjee, M. K. Gnanasammandhan and Y. Zhang, *Small*, 2010, **6**, 2781.
9. S. N. Baker and S. A. Baker, *Angew. Chem., Int. Ed.*, 2010, **49**, 6726.
10. L. Shang and S. Dong, *J. Mater. Chem.*, 2008, **18**, 4636.
11. C.-A. Lin, T.-Y. Yang, C.-H. Lee, S. H. Huang, R. A. Sperling, M. Zanella, J. K. Li, J.-L. Shen, H.-H. Wang, H.-I. Yeh, W. J. Parak and H. Chang, *ACS Nano*, 2009, **3**, 395.
12. C. Wang, Y. Wang, L. Xu, D. Zhang, M. Liu, X. Li, H. Sun, Q. Lin and B. Yang, *Small*, 2012, **8**, 3137.
13. J. Zhang, Y. Fu, C. V. Conroy, Z. Tang, G. Li, R. Y. Zhao and G. Wang, *J. Phys. Chem. C*, 2012, **116**, 26561.
14. C. Lin, C. Lee, J. Hsieh, H. Wang, J. Li, J. Shen, W. Chan, H. Yeh and W. Chang, *J. Med. Biol. Eng.*, 2009, **29**, 276.
15. Z. F. Liu, L. M. Tong and T. Zhu, *Chem. Soc. Rev.*, 2011, **40**, 1296.
16. W. Chen and S. W. Chen, *Angew. Chem., Int. Ed.*, 2009, **48**, 4386.
17. J. A. Ho, H.-C. Chang and W.-T. Su, *Anal. Chem.*, 2012, **84**, 3246.
18. M. A. H. Muhammed, S. Ramesh, S. Sinha, S. Pal and T. Pradeep, *Nano Res.*, 2008, **1**, 333.
19. R. Zhou, M. Shi, X. Chen, M. Wang and H. Chen, *Chem. – Eur. J.*, 2009, **15**, 4944.
20. X. Yuan, M. I. Setyawati, A. S. Tan, C. N. Ong, D. T. Leong and J. Xie, *NPG Asia Mater.*, 2013, **5**, e39, DOI: 10.1038/am.2013.3.
21. H. Duan and S. Nie, *J. Am. Chem. Soc.*, 2007, **129**, 2412.
22. M. Muhammed, P. Verma, S. Pal, R. Kumar, S. Paul, R. Omkumar and T. Pradeep, *Chem. Eur. J.*, 2009, **15**, 10110.
23. X. Yuan, Z. Luo, Q. Zhang, X. Zhang, Y. Zheng, J. Y. Lee and J. Xie, *ACS Nano*, 2011, **5**, 8800.
24. X. Le Guével, V. Trouillet, C. Spies, G. Jung and M. Schneider, *J. Phys. Chem. C*, 2012, **116**, 6047.
25. T. U. B. Rao and T. Pradeep, *Angew. Chem., Int. Ed.*, 2010, **49**, 3925.
26. W. Guo, J. Yuanab and E. Wang, *Chem. Commun.*, 2012, **48**, 3076.
27. C.-C. Huang, Z. Yang, K.-H. Lee and H.-T. Chang, *Angew. Chem., Int. Ed.*, 2007, **46**, 6824.
28. C.-C. Huang, H.-Y. Liao, Y.-C. Shiang, Z.-H. Lin, Z. Yang and H.-T. Chang, *J. Mater. Chem.*, 2009, **19**, 755.
29. M. Muhammed, P. Verma, S. Pal, A. Retnakumari, M. Koyakutty, S. Nair and T. Pradeep, *Chem. Eur. J.*, 2010, **16**, 10103.
30. L. Shang, R. M. Dörlich, S. Brandholt, R. Schneider, V. Trouillet, M. Bruns, D. Gerthsen and G. U. Nienhaus, *Nanoscale*, 2011, **3**, 2009.
31. Y. P. Bao, C. Zhong, D. M. Vu, J. P. Temirov, R. B. Dyer and J. S. Martinez, *J. Phys. Chem. C*, 2007, **111**, 12194.
32. J. Sun, J. Zhang and Y. Jin, *J. Mater. Chem. C*, 2013, **1**, 138.
33. C. Guo and J. Irudayaraj, *Anal. Chem.*, 2011, **83**, 2883.
34. X. Yang, M. Shi, R. Zhou, X. Chen and H. Chen, *Nanoscale*, 2011, **3**, 2596.
35. L. Shang, N. Azadfar, F. Stockmar, W. Send, V. Trouillet, M. Bruns, D. Gerthsen and G. U. Nienhaus, *Small*, 2011, **18**, 2614.

36. Z. Wang, W. Cai and J. Sui, *ChemPhysChem*, 2009, **10**, 2012.
37. Y. Shichibu and K. Konishi, *Small*, 2010, **6**, 1216.
38. H. Zhang, X. Huang, L. Li, G. Zhang, I. Hussain, Z. Li and B. Tan, *Chem. Commun.*, 2012, **48**, 567.
39. J. G. Zhang, S. Q. Xu and E. Kumacheva, *Adv. Mater.*, 2005, **17**, 2336.
40. Z. Shen, H. W. Duan and H. Frey, *Adv. Mater.*, 2007, **19**, 349.
41. I. Díez, M. Pusa, S. Kulmala, H. Jiang, A. Walther, A. S. Goldmann, A. H. E. Muller, O. Ikkala and R. H. A. Ras, *Angew. Chem., Int. Ed.*, 2009, **48**, 2122.
42. X. Huang, B. Li, L. Li, H. Zhang, I. Majeed, I. Hussain and B. Tan, *J. Phys. Chem. C*, 2012, **116**, 448.
43. H. Xu, B. W. Zeiger and K. S. Suslick, *Chem. Soc. Rev.*, 2013, **42**, 2555.
44. H. Liu, X. Zhang, X. Wu, L. Jiang, C. Burda and J.-J. Zhu, *Chem. Commun.*, 2011, **47**, 4237.
45. H. Xu and K. S. Suslick, *ACS Nano*, 2010, **4**, 3209.
46. T. Zhou, M. Rong, Z. Cai, C. J. Yanga and X. Chen, *Nanoscale*, 2012, **4**, 4103.
47. M. T. Reetz and W. Helbig, *J. Am. Chem. Soc.*, 1994, **116**, 1401.
48. M. L. Rodríguez-Sánchez, M. J. Rodríguez, M. C. Blanco, J. Rivas and M. A. López-Quintela, *J. Phys. Chem. B*, 2005, **109**, 1183.
49. B. Santiago González, M. J. Rodríguez, C. Blanco, J. Rivas, M. A. López-Quintela and J. M. G. Martinho, *Nano Lett.*, 2010, **10**, 4217.
50. N. Vilar-Vidal, M. C. Blanco, M. A. López-Quintela, J. Rivas and C. Serra, *J. Phys. Chem. C*, 2010, **114**, 15924.
51. B. Santiago González, C. Blanco and M. A. López-Quintela, *Nanoscale*, 2012, **4**, 7632.
52. A. Ledo-Suárez, J. Rivas, C. F. Rodríguez-Abreu, M. J. Rodríguez, E. Pastor, A. Hernández-Creus, S. B. Oseroff and M. A. López-Quintela, *Angew. Chem., Int. Ed.*, 2007, **46**, 8823.
53. C. Vázquez-Vázquez, M. Bañobre-López, A. Mitra, M. A. López-Quintela and J. Rivas, *Langmuir*, 2009, **25**, 8208.
54. L. Shang, L. Yang, F. Stockmar, R. Popescu, V. Trouillet, M. Bruns, D. Gerthsen and G. U. Nienhaus, *Nanoscale*, 2012, **4**, 4155.
55. Y. Yue, T.-Y. Liu, H.-W. Li, Z. Liu and Y. Wu, *Nanoscale*, 2012, **4**, 2251.
56. R. Li, C. Wang, F. Bo, Z. Wang, H. Shao, S. Xu and Y. Cui, *ChemPhysChem*, 2012, **13**, 2097.
57. H. Kawasaki, Y. Kosaka, Y. Myoujin, T. Narushima, T. Yonezawa and R. Arakawa, *Chem. Commun.*, 2011, **47**, 7740.
58. J. Zheng and R. M. Dickson, *J. Am. Chem. Soc.*, 2002, **124**, 13982.
59. J. Zheng, J. T. Petty and R. M. Dickson, *J. Am. Chem. Soc.*, 2003, **125**, 7780.
60. J. Zheng, C. Zhang and R. M. Dickson, *Phys. Rev. Lett.*, 2004, **93**, 077402.
61. Y.-C. Jao, M.-K. Chen and S.-Y. Lin, *Chem. Commun.*, 2010, **46**, 2626.
62. T. Jin, S. I. Tanaka, J. Miyazaki, D. K. Tiwari and Y. Inouye, *Angew. Chem., Int. Ed.*, 2011, **50**, 431.
63. J. T. Petty, J. Zheng, N. V. Hud and R. M. Dickson, *J. Am. Chem. Soc.*, 2004, **126**, 5207.

64. J. T. Petty, C. Fan, S. P. Story, B. Sengupta, A. St. John Iyer, Z. Prudowsky and R. M. Dickson, *J. Phys. Chem. Lett.*, 2010, **1**, 2524.
65. C. I. Richards, S. Choi, J. C. Hsiang, Y. Antoku, T. Vosch, Y. L. Tzeng and R. M. Dickson, *J. Am. Chem. Soc.*, 2008, **130**, 5038.
66. T. Vosch, Y. Antoku, J.-C. Hsiang, C. I. Richards, J. I. Gonzalez and R. M. Dickson, *Proc. Natl. Acad. Sci. U. S. A.*, 2007, **104**, 12616.
67. J. Sharma, H. C. Yeh, H. Yoo, J. H. Werner and J. S. Martinez, *Chem. Commun.*, 2010, **46**, 3280.
68. P. R. O'Neill, L. R. Velazquez, D. G. Dunn, E. G. Gwinn and D. K. Fygenson, *J. Phys. Chem. C*, 2009, **113**, 4229.
69. X. Jia, J. Li, L. Han, J. Ren, X. Yang and E. Wang, *ACS Nano*, 2012, **6**, 3311.
70. J. Xie, Y. Zheng and J. Y. Ying, *J. Am. Chem. Soc.*, 2009, **131**, 888.
71. P. L. Xavier, K. Chaudhari, A. Baksi and T. Pradeep, *Nano Rev.*, 2012, **3**, 14767.
72. P. L. Xavier, K. Chaudhari, P. K. Verma, S. K. Pal and T. Pradeep, *Nanoscale*, 2010, **2**, 2769.
73. C. L. Liu, H. T. Wu, Y. H. Hsiao, C. W. Lai, C. W. Shih, Y. K. Peng, K. C. Tang, H. W. Chang, Y. C. Chien, J. K. Hsiao, J. T. Cheng and P. T. Chou, *Angew. Chem., Int. Ed.*, 2011, **50**, 7056.
74. K. Chaudhari, P. L. Xavier and T. Pradeep, *ACS Nano*, 2011, **5**, 8816.
75. J. S. Mohanty, P. L. Xavier, K. Kamalesh Chaudhari, M. S. Bootharaju, N. Goswami, S. K. Palc and T. Pradeep, *Nanoscale*, 2012, **4**, 4255.
76. U. Anand, S. Ghosh and S. Mukherjee, *J. Phys. Chem. Lett.*, 2012, **3**, 3605.
77. J. Oliver-Meseguer, J. R. Cabrero-Antonino, I. Domínguez, A. Leyva-Pérez and A. Corma, *Science*, 2012, **338**, 1452.

CHAPTER 3

Silver Magic-Number Clusters and Their Properties

BRIAN A. ASHENFELTER, ANIL DESIREDDY, JINGSHU GUO,
BRIAN E. CONN, WENDELL P. GRIFFITH AND
TERRY P. BIGIONI*

Department of Chemistry, University of Toledo, 2801 W. Bancroft St.,
Toledo, OH 43606, USA
*Email: Terry.Bigioni@utoledo.edu

3.1   Introduction

Small metal clusters offer an opportunity to study a wide range of physical and chemical phenomena, including the origins of nanostructure stability, catalysis, optical absorption and fluorescence, as well as therapeutic applications. While a tremendous amount of work has already been done for gold, comparatively little has been done for silver despite its technological importance. The study of silver often provides information that is complementary to that for gold. Further, it is sometimes the case that silver cluster properties are superior to those of gold, such as in the case of antibacterial and optical properties, particularly plasmons and fluorescence. It is for these reasons that we focus our attention here on small magic-number silver clusters and review recent progress in the field. This review covers different synthetic strategies and electrophoretic separations, as well as characterization by mass spectrometry and optical absorption and fluorescence spectroscopy.

RSC Smart Materials No. 7
Functional Nanometer-Sized Clusters of Transition Metals: Synthesis, Properties and Applications
Edited by Wei Chen and Shaowei Chen
© The Royal Society of Chemistry 2014
Published by the Royal Society of Chemistry, www.rsc.org

3.2   Synthesis

It is the ligand shell that imparts solubility to nanoparticles. In general, there are two options: water-soluble or organic-soluble. For noble metal nanoparticles, the metal salts and reducing agents are in general water-soluble, which is ideal for synthesizing water-soluble particles. For organic-soluble particles, however, a compromise of some kind needs to be made since some of the reagents require water as a solvent and the final product requires an organic solvent. In general this is done with an ionic surfactant, which facilitates phase transfer. Recently, solid-state reactions have also been developed, which could be considered a subset of these classifications depending on the ligand and the final solvent used.

Noble metal cluster synthesis can then be classified as aqueous or non-aqueous. Aqueous synthesis is preferable as it is straightforward, homogeneous and more green. There are also well-developed separations that greatly aid in the subsequent analysis of these materials, including gel electrophoresis. Mass spectrometry of aqueous noble metal clusters is more challenging, however, since it is more difficult to remove the solvent. Organic syntheses are still quite popular, for both historical reasons[1] and because ligands such as phenylethanethiol have been quite successfully employed in single crystal growth and X-ray structure determination. Separations of organic-soluble clusters are more challenging, and generally involve size-selective precipitation, columns, and centrifugation.

The first clearly magic-numbered silver clusters that were produced were actually synthesized as single-sized products. The first report appeared in 2009, wherein a water-based synthesis was developed using captopril, glutathione, and cysteine as the protecting ligands.[2,3] In this case, the cluster formulae have yet to be positively identified, however the spectrum of the glutathione cluster matches that of $Ag_{32}(SG)_{19}$.[4,5] Based on the spectra, it is quite possible that the captopril and cysteine clusters share the same formula. That same year,[6] silver clusters protected with various phenyl containing ligands were synthesized in DMF. These were later identified as $Ag_{44}(SPh)_{30}^{4-}$ clusters,[7] where the 4– charge is carried by the core such that the electron count is 18.[8]

The first family of silver magic-numbered clusters was reported in 2010.[4] The clusters were synthesized in water with glutathione as the protecting ligand. Electrophoretic separations showed that the mass distribution was composed of discrete clusters, each forming a distinct band with a variety of different colours observed. Variations in synthetic conditions produced the same family of silver clusters where the differences were exhibited only in the mass distribution, as expected for magic-numbered clusters.

3.2.1   Aqueous Synthesis

Silver nitrate is a popular metal source and sodium borohydride is the most common reducing agent for silver nanoparticle syntheses, each of which is water-soluble. Generally, aqueous-based syntheses are done by reducing the silver salt in the presence of excess ligands. If the ligands are also water-soluble, then such a reaction would be expected to occur in a single homogeneous phase. Adding a water-soluble thiolate ligand to a solution containing silver cations typically produces an insoluble precursor, however, such that the reaction is not truly a one-phase reaction. This precursor might be a silver thiolate, in analogy to the gold thiolate produced during gold cluster syntheses, although it has yet to be definitely identified. In some cases the precursor can be solubilized by changing the reaction conditions, such as pH, however the role of homogeneous *versus* inhomogeneous reaction mixtures in producing the final product is still not well understood. Reduction of the silver precursor produces water-soluble silver clusters.

Although there were several reports of very small Ag_n clusters (where $n < 10$), Kitaev's group was the first to report the synthesis of a truly magic-number silver cluster, in particular in a size range similar to gold.[2] They first synthesized a mixture of silver cluster sizes in water with cysteine, glutathione, and captopril ligands. Once synthesized, the clusters underwent cyclic oxidation and reduction processes, using hydrogen peroxide to oxidize and sodium borohydride to reduce. These processes were repeated until the spectrum evolved to that of a single-sized species. In this case, the cyclic oxidation and reduction destroyed all but the most stable cluster size. Electrophoresis showed that the captopril and glutathione products were single sized, although a second band was present in the cysteine product. This was a very nice demonstration of the molecular nature of magic-number clusters and showed the potential for the synthesis of single-sized molecular products. The same group reported the synthesis of single-sized clusters using a mixture of ligands, namely captopril and glutathione, at higher pH.[3] Adjusting the pH after mixing the precursors with KOH produced a homogeneous solution without insoluble thiolate polymers. It was also found that using a mixture of ligands slowed down the reduction steps and improved the stability of the clusters.

The first family of silver magic-numbered clusters was prepared by the reduction of silver nitrate with sodium borohydride in the presence of excess glutathione as a protecting ligand.[4] A typical synthesis involved mixing silver nitrate and glutathione in water in a 1 to 4 molar ratio followed by reduction with as much as 10 molar equivalents of sodium borohydride. Mixing glutathione with silver nitrate formed a white solid precursor, similar to the Au(I)SR polymer commonly formed during Au cluster syntheses. Upon dropwise addition of the reducing agent, the solution changed in colour from yellow to orange initially and then finally to a rich dark brown over the course of an hour, indicating the formation of silver clusters.

The as-prepared clusters were separated by gel electrophoresis into a family of magic-number clusters containing more than 21 different bands. This discrete size distribution is in contrast to the continuous size distributions of typical nanoparticle preparations. The different colours of the bands indicated that each band contained chemically distinct species. It is important to point out that when synthesis conditions were changed, such as the reducing agent addition rate, metal to ligand ratio, solvent, and temperature, the same colour bands appeared in the same order in the gel, but with different concentrations. In other words, changing synthesis conditions did not change the identity of the clusters produced but only their relative abundance. This is a hallmark of magic-number clusters.

Based on this work, earlier work can now be reinterpreted. For example, the cluster denoted as band 6 in reference 4 has since been identified as $Ag_{32}(SG)_{19}$.[5] Spectra that resemble the $Ag_{32}(SG)_{19}$ cluster can be found in several earlier papers, including those of Cathcart and coworkers[2,3] and Wu et al.,[9] leading one to opine that the 32 atom silver cluster might share the trait of the 25 atom gold cluster in that the identity of the ligand is not of great importance in determining the atomic structure and therefore the electronic structure and optical spectra of the clusters. It remains to be seen if these silver clusters are indeed isostructural, however, or if the absorption profile has a deeper origin.

Various other methods of synthesizing small Ag clusters have also been reported, but without good evidence for whether or not their sizes follow any sort of magic number rules. Most famously, Dickson's group reported the first fluorescent silver by photoreducing Ag_2O to small Ag_n fluorescent clusters (where $n = 2$ to 8).[10,11] Zheng and Dickson also reported a photosynthetic method to prepare fluorescent water-soluble silver nanoclusters encapsulated by a dendrimer.[12] Rao and Pradeep also reported a synthetic route for red and blue–green fluorescent clusters through interfacial etching.[13] In this route, plasmonic silver particles capped with mercaptosuccinic acid were prepared in a water–methanol mixture by reducing silver nitrate with aqueous sodium borohydride in the presence of the ligand. This material was then used as a precursor for the interfacial etching process and the production of small fluorescent clusters. Both Kumacheva and coworkers and Adhikari and Banerjee reported the synthesis of water-soluble fluorescent silver nanoclusters by using UV light as a reducing source and using microgels as a template.[14,15] Shang and Dong also produced water-soluble Ag clusters by photoreduction, but instead used a polyelectrolyte template.[16] In general, these sorts of templated syntheses are sensitive to the pH since deprotonation of the acid groups provides a site for silver ions to coordinate during nanoparticle growth.

There is still a significant amount of confusion in the literature regarding size determination of small silver clusters, however, primarily due to the great challenges involved with obtaining accurate mass spectrometry measurements of silver clusters and the ineffectiveness of electron microscopy for very small metal clusters. For example, Adhikari and Banerjee synthesized water-soluble fluorescent clusters for Hg(II) sensing using lipoic acid as a protecting ligand.[17] The authors stated that the average diameter of their Ag particles was 2.44 nm, yet they also rely heavily on mass spectrometry and state that only Ag_4 and Ag_5 clusters were found in their study. Clearly one or both of these conclusions is incorrect. Mass spectrometry likely measured only fragments, as discussed below in section 3.4, while there is a tendency for larger particles to contribute more to TEM and XRD measurements than smaller clusters. Similarly, since small clusters are difficult to image but are more fluorescent than large nanoparticles they can, if present, make it appear as if larger plasmonic nanoparticles are fluorescent. Evidently, more care must be taken with silver cluster research than with gold.

3.2.2   Non-Aqueous Synthesis

Two general strategies are used to synthesize silver nanoparticles that are soluble in non-aqueous solvents. If a water-soluble metal salt and reducing agent are used, then a two-phase synthesis is a common approach to satisfying the solvation requirements of both reagents and the final product. This was first demonstrated by Brust *et al.* using two phases of roughly equal volume, each containing some of the starting materials.[18] A phase transfer agent was used to transport the gold ions into the non-aqueous phase before the reducing agent was added, such that the gold could be reduced at the phase boundary and immediately passivated by the ligands. While the result was a major advance in nanoparticle synthesis, the size distribution was broad and the average size was relatively large.

Although the two-phase approach was the first employed for ligating metal nanoparticles,[18] it is seldom employed in the synthesis of magic-number clusters. This might be due to the difficulty of synthesizing very small nanoparticles and controlled size distribution. For example, Murray *et al.* synthesized toluene-soluble 4-*tert*-butylbenzyl mercaptan-capped silver particles by a two-phase method.[19] In this synthesis, tetraoctylammonium bromide was used to phase transfer silver nitrate from water into toluene, which contained the ligands, before reduction with ice-cold aqueous sodium borohydride. This produced small silver nanoparticles, however most of the nanoparticles fell in a size range from 1–3 nm. Formally, two-phase syntheses also include the so-called inverse micelle approach, in which a microemulsion of aqueous reagents is made in a non-polar solvent like hexane or toluene, which contains the ligands.[20] It is difficult to imagine that the growth conditions in two-phase syntheses could be very uniform, and indeed the product tends to have a broad size distribution.

One-phase syntheses attempt to create more uniform reaction conditions by choosing reagents and solvents that are compatible, which can sometimes lead to very narrow size distributions.[6] The first one-phase synthesis of monolayer-protected silver nanoparticles was done in an ethanol solvent and produced dodecanethiol-passivated nanoparticles that were 3.1 ± 0.6 nm in diameter.[21] Although it is unlikely that the product contained a significant amount of magic-number clusters, this was the first demonstration of ligand-passivated silver nanoparticle synthesis. Unfortunately, while ethanol was a good solvent for the reagents it was not a good solvent for the final product, which led to low yields. It is clear from this result that a one-phase synthesis would require better choices of metal source and reducing agent for the given solvent and final product.

In 2009, Bakr *et al.* reported a synthesis of organic-soluble silver clusters, which were referred to as intensely and broadly absorbing nanoparticles (IBANs).[6] In this one-phase synthesis, the silver salt (AgNO$_3$), ligand (various substituted aryl-thiols) and reducing agent (NaBH$_4$) were dissolved in the same polar organic solvent (DMF). Although plasmonic particles initially formed, after 4 hours the plasmonic colour disappeared. It was speculated that the plasmonic particles were digested to either smaller seeds, silver ions, or even layered compounds like thiolates that could act as an intermediate precursor for synthesis, although there was no direct evidence for this. Next, a small quantity of aqueous NaBH$_4$ was added, since water increases the strength of the reducing agent, and the solution quickly darkened. After storing this solution in the freezer for several days, a final pink-coloured solution of IBANs was obtained.

It was argued that the IBAN synthesis produced single-sized products, using results from ultracentrifugation, gel electrophoresis, and optical absorption spectroscopy.[6] For example, the highly structured absorption spectra are unlikely to be those of mixtures since a superposition of spectra for different species would tend to average over features in such a way as to lose detail. Subsequent studies of silver clusters ligated with 4-fluor-othiophenol (4FTP) and 2-naphthalenethiol (2NPT) appear to confirm these findings.[7] Further, both 4FTP and 2NPT silver clusters were found to have the same stoichiometry, namely $[Ag_{44}(SR)_{30}]^{4-}$ clusters. This molecular formula is significant since it has a closed electronic shell with 18 electrons, as discussed below. It is also significant that this appears to be a single-sized synthesis of a relatively high-nuclearity cluster, albeit in low yield.

Also in 2009, Jin *et al.* published a synthetic route for producing thiolate-protected Ag$_7$ clusters in high yield.[9] The particles were synthesized in ethanol, a poor solvent, to limit the growth of the clusters. The silver salt and ligand were also cooled down to slow down the cluster growth rate. The ligands used were 2,3-dimercaptosuccinic acid (DMSA) such that the clusters had a molecular formula of Ag$_7$(DMSA)$_4$, as determined by mass spectrometry. Once synthesized, the clusters could then be redispersed in water at basic pH. In contrast to the work above, this is significant since it is a high-yield synthesis of an apparently single-sized product, however it is an extremely small cluster and it is not clear if it was an especially stable fragment that was observed in the mass spectrum given the extreme fragility of silver clusters and the early date of this work (see section 3.4).

In 2011, Tsukuda *et al.* reported a synthesis similar to the Bakr synthesis using 4-(*tert*-butyl)benzyl as a protecting ligand.[52] This synthesis was done in THF at 0 °C and finished in a few hours. In this synthesis, an ice-cold aqueous reducing agent was added to slow down the reduction. Clusters were identified with mass spectrometry as containing 280 silver atoms and 120 ligand molecules.

In 2013, the first crystal structure of silver clusters was reported for Ag$_{14}$ fluorescent clusters synthesized by Zheng *et al.*, which had a 6-atom octahedral Ag core and a cube-shaped protective shell.[22] The remaining 8 silver atoms are on the corners of the cube with bridging sulfur atoms from the thiol ligands roughly defining the edges of the cube. The outermost 8 silver atoms required protection from bulky coordinating phosphine ligands, however, such that a mixture of thiol and phosphine ligands was required to stabilize the cluster. In this synthesis, a mixture of triphenyl-phosphine and 3,4-difluoro-benzenethiol ligands was used. This work has been followed up by two additional crystal structures, containing 16 and 32 Ag atoms, each of which with rather irregular cores and mixed thiol–phosphine ligand layers.[23] These mixed ligand results are the only reported silver cluster crystal structures to date.

3.2.3 Solid-State Synthesis

Aqueous and non-aqueous synthesis categorizations describe the vast majority of syntheses, however Pradeep *et al.* recently reported a solid-state synthetic route.[24] In this case, the solid reagents were first mixed and ground together to initiate the reaction before being collected with a solvent and non-solvent. A $1:5$ molar ratio of silver nitrate and mercaptosuccinic acid were ground into a homogeneous powdered solid before solid sodium bor-ohydride was added. Over the course of grinding, the colour of the solid material changed from white to orange and finally to brown, indication a solid-state reaction had occurred at the interface of the grains. Water was then added to the solid mixture to obtain the final fluorescent clusters. The clusters were then purified by precipitating and washing with alcohol. The components of the synthesis were then separating by gel electrophoresis, which produced only two visible bands. This suggests that the synthesis produced only two cluster sizes as the major product. The most abundant clusters were analyzed by electrospray ionization mass spectrometry and assigned the formula $Ag_9(SR)_7$. It is quite likely that this is a fragment of the original larger cluster, however, as it is now known that silver clusters are quite fragile and easily fragment even using gentle ionization methods.[5] It is also expected that such a small cluster would have an absorption edge considerably higher in energy.[4]

It is reasonable to think that this solid-state-only reaction might not go to completion, since it is difficult to imagine grains small enough to be completely consumed over the course of the as-described reaction. The solid reaction mixture was ultimately dissolved in water, however, where any unreacted material could be converted to the final product. While the reaction might not have been entirely in the solid state, it is conceivable that given enough time it could be run to completion without solvents. It is important to point out, however, that this represents a new class of nano-particle synthesis that could be quite useful given the gram scale of this demonstration reaction. Further, it shows how robust magic-number cluster syntheses could be with regard to concentrations, which also gives impetus to scaling up to industrially-relevant syntheses.

3.3   PAGE Separations

Size separations of organic-soluble clusters are typically done using column chromatography or size selective precipitation. While these methods have been used successfully to isolate different sizes of clusters, they remain somewhat challenging methods. Water-soluble clusters often carry a charge such that electrophoretic separation and purification are possible. Poly-acrylamide gel electrophoresis (PAGE) is a simple, fast and effective tech-nique to separate aqueous clusters and has a long and successful history of use in biochemistry.

In a typical PAGE experiment, clusters are driven through the porous poly-accrylamide gel by an applied electric field, which acts on the surface charge of the clusters. The applied voltage is normally 200–300 V, which generates a significant amount of Joule heating such that water cooling is needed. Gels must be run with an electrolyte to support electrophoresis. Often the standard glycine and tris base buffers are used, with pH 8.8, although clusters that are not sensitive to pH may be run with any compatible electrolyte.

The small size of the pores in the gel increases the effective viscosity of the medium in a way that depends on the hydrodynamic radius of the clusters, such that larger clusters are retarded more so than smaller clusters as they move through the polymer. This is the basis for size separation. The mixture separates into discrete bands of material within the gel, each containing a different cluster size, as shown in Figure 3.1.

For proteins, an ionic surfactant is used to denature the protein such that the mobility is related to the length and therefore the mass of the protein. Sodium dodecyl sulfate (SDS) is most often used for protein PAGE separ-ations, commonly known as SDS-PAGE. The ionic surfactant also ensures that the surface charge density of different proteins is the same, *i.e.* it is independent of the sequence, such that they are subjected to the same electromotive forces. This allows distance traveled to be interpreted in terms of protein mass.

Metal clusters cannot denature, however, nor do they have a sequence. The ligands that carry the charge on the clusters are all identical such that the surface charge density from cluster to cluster should not depend on size. The use of ionic surfactants is therefore unnecessary, so SDS is not used for metal cluster PAGE separations. Also, since metal clusters are much more compact than denatured proteins, the gel density must be much higher for clusters than proteins. Gel densities of 30% are typically used to achieve good separations of metal clusters, although densities as low as 20% are also effective. Gel densities higher than 30% can increase the resolution of the separations but also increase the gel running time. As a result, gel densities higher than 30% are rarely used since the extremely small pore sizes make the required running time disadvantageous.

The PAGE gel consists of two layers of different density. The high-density (30%) gel layer is known as the resolving gel, as its purpose is to separate the mixture and resolve its components. For protein separations, density

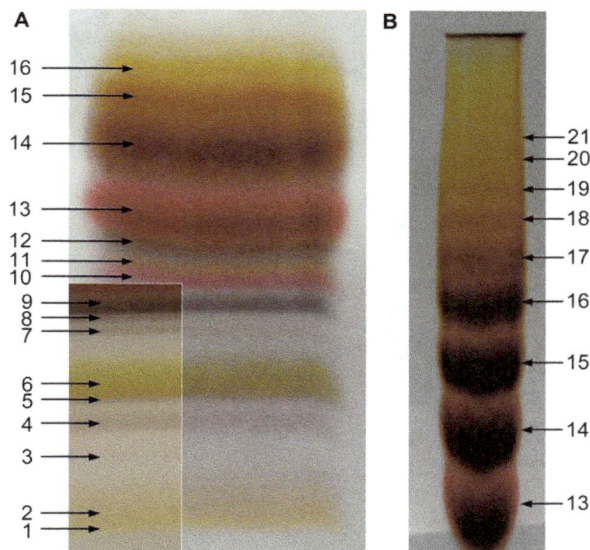

Figure 3.1 PAGE results for Ag:SG clusters. (A) Bands of discrete Ag:SG clusters. The smallest clusters are at the bottom and have the lowest index. Inset: finer bands revealed at a higher concentration. (B) Discrete higher-mass clusters, synthesized with less glutathione.
(Reprinted with permission from Kumar, S.; Bolan, M. D.; Bigioni, T. P., Glutathione-Stabilized Magic-Number Silver Cluster Compounds. *J. Am. Chem. Soc.* **2010**, *132* (38), 13141–13143. Copyright 2010 American Chemical Society.)

gradients are often employed to obtain better separations. For nanoparticle separations, however, the resolving gel density is generally uniform. This layer makes up most of the gel thickness. Above the resolving gel is a relatively thin layer of low-density ($\sim 5\%$) gel that is called the stacking gel, as its purpose is to spatially focus the clusters at the top of the resolving gel. This "stacking" event allows the clusters to enter the resolving gel at approximately the same time, minimizing broadening of the resultant bands.

Polyacrylamide gels are made by polymerizing the acrylamide monomer with the bis-acrylamide cross linker, as shown in Figure 3.2. Stock solutions of 40% T and 4% C acrylamide and bis-acrylmide solutions are mixed with 1.5 M tris HCl (pH 8.8) to achieve the desired gel density, where % T (total monomer concentration) and % C (weight percentage of cross linker) are defined as:

$$\% \, T = \frac{g \, acrylamide + g \, bis\text{-}acrylamide}{total \, volume} \times 100$$

$$\% \, C = \frac{g \, bis\text{-}acrylamide}{g \, acrylamide + g \, bis\text{-}acrylamide} \times 100$$

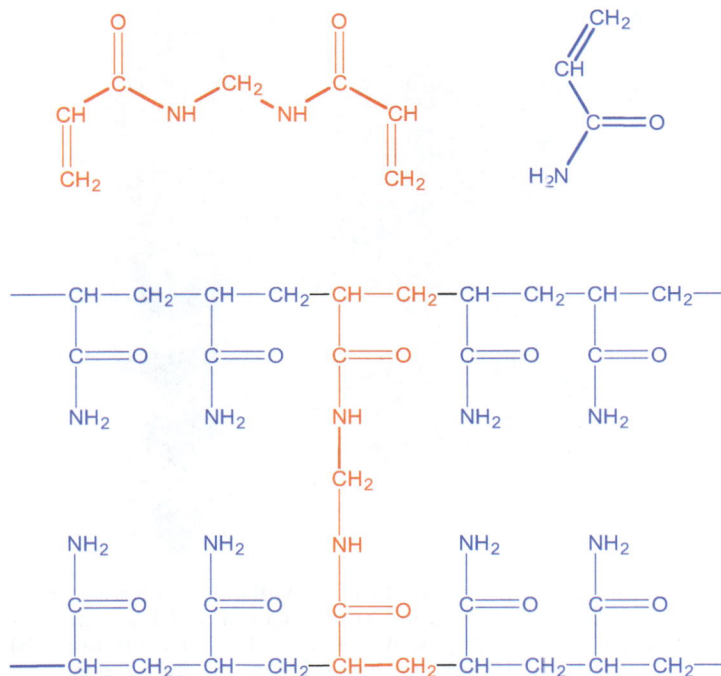

Figure 3.2 Schematic of polymerization of acrylamide (blue) monomers with crosslinker (red). The monomers are shown at the top and the polymer at the bottom. The concentration of the cross-linker determines the gel density and pore size.

The average pore size of the gel is controlled by % T and % C. Higher % T gels contain more total monomer and are therefore higher in density, resulting in smaller pore sizes for better separation and resolution.

Casting of the gel is rather straightforward. Once the acrylamide/bis-acrylamide monomer solution is prepared, ammonium persulfate (APS) is added to initiate the polymerization reaction and tetramethylethylenediamine (TEMED) is added to catalyze the reaction. The rate of polymerization depends on the concentration of the catalyst and monomer as well as the temperature. Once the solution is cast between the glass plates, *tert*-butyl alcohol is poured on the top of the gel to ensure a smooth and level surface. After ~45 minutes, the alcohol is removed, the gel is rinsed with DI water, and the stacking gel is cast (typically a 5% gel). Sample wells are defined by inserting a comb into the stacking gel. The comb is removed after polymerization, leaving wells that serve to define the lanes. Unreacted monomers can be minimized by allowing the gel to polymerize overnight and then running the gel for an hour before loading samples into the wells.

Samples are prepared in aqueous solutions containing 5–10% glycerol. The glycerol increases the density of the solution such that it may settle to the bottom of each well. While proteins must be stained by a non-specifically binding dye in order to visualize their position and determine their mass, this is not required for metal clusters as they are stronger light absorbers than typical organic dyes. As a result, the native analyte can be easily visualized and tracked as it passes through the gel and separates, allowing one to evaluate the optimal stopping time for achieving the best separations. Once the optimal separation of bands is achieved, each band can then be excised such that the separated clusters can be extracted from the gel and studied further in their purified form.

3.4 Mass Spectrometry

The use of mass spectrometry (MS) in the analysis of noble metal magic-number clusters has grown by leaps and bounds over the last decade. The main attraction of this method of analysis is that apart from single-crystal X-ray diffraction (SC-XRD), mass spectrometry is the only technique that is capable of providing full characterization of the three defining properties of these clusters: core size, nature of the protecting ligand, and metal–ligand stoichiometry (molecular formula). SC-XRD tends to be extremely time-consuming, difficult and requires a large quantity of sample (of the order of tens of milligrams). The advantage of MS for noble metal magic-number cluster analysis is that only a small amount of sample (<1 milligram) is required for complete analysis, simultaneously providing information on core size, ligand, and in many cases molecular formula. The type of information that can be derived from MS analysis, though, greatly depends on the type of mass spectrometer and the nature of the sample.

The typical mass spectrometer is composed of four basic parts: the ionization source, the mass analyzer, the detector, and the vacuum system. The nature of the ionization source has the most marked effect on the result. The ionization sources that have been applied to noble metal magic-number clusters have been laser desorption/ionization (LDI) and more recently matrix-assisted laser desorption/ionization (MALDI) and electrospray ionization (ESI). These ion sources have been successfully applied to the analysis of gold magic-number clusters. In the matrix-less LDI, it has been shown for the Au system in most cases that the generation of cluster ions occurs only when the fluence of the laser is high enough to photolytically cleave the C–S bond, thereby resulting in the detection of only metal core–sulfur clusters. Due to the poor resolution obtained by LDI, the resultant information derived has been limited to approximate core mass and therefore size. Regardless, early LDI experiments by Whetten *et al.*[1] were instrumental in the development of the techniques that would eventually lead to the identification and formula assignment of the Au cluster series.[25,26] A decade later, Dass *et al.* demonstrated the utility of adding an organic matrix to protect the cluster during ionization and their ability to detect intact clusters.[27] Using the matrix *trans*-2-[3-(4-*tert*-butylphenyl)-2-methyl-2-propenylidene] malononitrile (DCTB) they were able to detect intact

$Au_{25}(SCH_2CH_2Ph)_{18}$ with no loss of ligand, although significant loss of $Au_4(SC_8H_9)_4$ did occur. Kumara and Dass also recently demonstrated the use of MALDI-TOF MS to characterize mixed-metal alloy nanomolecules with 38 metal atoms.[28]

Although the incorporation of the matrix was an important improvement to the LDI technique and necessary to eliminate the problem of ligand loss through photolytic cleavage, it was not until the use of ESI that the first unfragmented mass spectra were obtained by Tsukuda *et al.* for a series of glutathione-protected Au clusters[25] and their chemical formulae corrected shortly thereafter.[26] Since then a number of researchers have utilized ESI-MS to characterize a range of gold magic-numbered clusters.[29–49]

In great contrast to the Au magic-numbered clusters, one major limitation of MS-based analysis of Ag clusters is their relative lower stability. The presence of 2 naturally occurring abundant Ag isotopes (^{107}Ag and ^{109}Ag), which significantly broadens the mass spectral peaks for clusters containing several silver atoms, has proven to be another significant complicating factor. The effects of these limitations are clearly evidenced by the fact that almost all of the published reports on Ag clusters before 2012 have been for only very small clusters or fragments with molecular weights <2000 Da[3,9,13,24,50] and that precise mass assignments for larger Ag clusters had not yet appeared in scientific literature.[6,19,51,52] Wu and coworkers used ESI-MS to characterize their $Ag_7L_4^-$ cluster with *meso*-2,3-dimercaptosuccinic acid ligands.[9] This work highlighted the utility of tandem mass spectrometry in nanocluster analysis. In these experiments the $Ag_7S_4^-$ fragment confirmed that the charge of the cluster came from the core and that only one thiol of the DMSA ligand binds to the Ag_7 core. Using the matrix *trans*-2-[3-(4-*tert*-butylphenyl)-2-methyl-2-propenyldidene]malonitrile (DCTB) and MALDI-MS, Rao and Pradeep assigned chemical formulae to their $Ag_7L_7^-$ and $Ag_8L_8^-$ clusters protected with mercaptosuccinic acid (H_2MSA).[13] Due to significant sodium adduction, LDI-MS and ESI-MS were also used to confirm the assigned formulae. In this case, one possible source of the sodium adducts in the MALDI-MS measurements could have been contamination of the matrix. Often MALDI matrices are recrystallized to remove cations (like Na^+ and K^+), which adversely affect the quality of MS data. Rao and coworkers used negative ionization ESI-MS to characterize their $Ag_9S_7^{2-}$ cluster with H_2MSA ligands.[24]

Guo *et al.* only recently demonstrated the utility of MS in the analysis of large intact Ag magic-numbered clusters in their characterization of $Ag_{32}(SG)_{19}$ (SG: glutathione).[5] This was the first published report for MS detection of a thiol-protected Ag cluster containing >8 Ag atoms. Very soon after that Harkness and coworkers published their results for two intensely- and broadly-absorbing nanoparticles with molecular formula $Ag_{44}L_{30}^{4-}$ where L was 4-fluorothiophenol or 2-naphthalenethiol.[7] A summary of all published reports where mass spectrometry has been used to characterize thiol-protected Ag clusters is provided in Table 3.1.

Table 3.1 List of thiol-protected Ag clusters analyzed by mass spectrometry.

Chemical Formula	Ligand (L)	Ionization Source	Polarity	Ref.	Additional Comments
Fragments, Ag < 5	Captopril	ESI	–ve	3	
$Ag_7L_4^-$	*meso*-2,3-Dimercaptosuccinic acid (DMSA)	ESI	–ve	9	Utilized tandem mass spectrometry; $Ag_7S_4^-$ fragment confirmed charge on core and that only one thiol of DMSA binds to Ag_7 core
$Ag_7L_7^-$ $Ag_8L_8^-$	Mercaptosuccinic acid (H_2MSA)	MALDI LDI ESI	–ve	13	MALDI with matrix *trans*-2-[3-(4-*tert*-butylphenyl)-2-methyl-2-propenylididene]malononitrile (DCTB)
$Ag_9L_7^{2-}$	Mercaptosuccinic acid (H_2MSA)	MALDI LDI ESI	–ve	24	
$Ag_{32}L_{19}^{-}$ $Ag_{44}L_{30}^{4-}$	Glutathione (SG) *para*-Mercaptobenzoic acid (pMBA) 2-Naphthalenethiol (2-NPT) 4-Fluorothiophenol (4-FTP)	ESI ESI ESI	–ve –ve –ve	5 54 7	
$Ag_{75}L_{40}$	Glutathione (SG) Phenylethanethiol (PET)	MALDI	+ve	55	ESI and MALDI MS with glutathione ligands unsuccessful; success with MALDI using DCTB matrix after phase transfer to toluene and consequent ligand exchange to PET
$Ag_{\sim140}L_{\sim53}$ $Ag_{152}L_{60}$ $Ag_{\sim223}L_{\sim108}$ $Ag_{\sim280}L_{\sim120}$	4-*tert*-Butylbenzyl mercaptan (BBT) Phenylethanethiol (PET) 4-(*tert*-Butyl)benzyl mercaptan (SBB)	LDI MALDI ESI LDI	+ve +ve +ve	19 56 52	Estimated the masses of the entire cluster and the core using ESI MS and LDI MS, respectively. Difference between ESI and LDI MS determined masses provided approximate for mass of ligands alone.
$Ag_{\sim923}L_{\sim351}$	Phenylethanethiol (PET)	MALDI	+ve	57	

The success of the pioneering work by Guo and coworkers lies in their realization of commonalities between the extremely fragile and labile Ag magic-numbered clusters and noncovalent protein complexes.[5] In fact, significant progress has been made to improve the MS detection of fragile noncovalent complexes over the last decade. These demonstrated that optimization of the ESI interface conditions, making them as gentle as possible, was essential for reduction of dissociation and reliable detection of complexes while still achieving sufficient desolvation. In the optimization of ESI and other instrumental parameters, the most important to the ionization, transmission and eventual detection of intact Ag magic-numbered clusters were size of the initial spray droplet, source temperature, collisional cooling, and collisional energy. Indeed, using nanospray (instead of the conventional ESI) results in much smaller initial droplet sizes, which require fewer and less energetic collisions for complete ion desolvation and hence result in a much gentler ionization process and more reliable detection of labile complexes.[53] Reduction of the source temperature reduces the possibility of thermal decomposition of the labile complexes. Potential energy of the cluster ions, which if too high could result in gas-phase decomposition, can be reduced by collisional cooling through gentle collision with bath gas in the fore-region of the mass spectrometer.[54] Guo and coworkers demonstrated that the trap/transfer collision energies (CE) had the most marked effect on reducing fragmentation of the $Ag_{32}(SG)_{19}$ cluster.[5] By decreasing the trap CE relative to the transfer CE and consequently collision-induced dissociation, they showed that the fragmentation of the cluster could be minimized while still allowing for only low-energy collisions, which are necessary to remove weakly bound species such as solvent or adducts. The largest improvement in detection of intact cluster species was demonstrated to be a combination of increased collisional cooling and decreased trap/transfer collision energies. Ion source temperature was shown to the smallest effect.[5]

Work by Guo *et al.* also highlighted that along with instrumental parameters, solvent plays an important role in the stabilization of Ag magic-numbered clusters.[5] Complete reduction of fragmentation as evidenced by peaks with *m/z* 527 and 820 corresponding to small fragments $Ag_2(SG)$ and $Ag_2(SG)_2$, was only possible with the addition of ammonium acetate and the reduction of the solution pH to 5 (see Figure 3.3). The importance of solution optimization is also more importantly evidenced by the narrowing of the ion distributions for each individual charge state, and the increase in the peak intensity of the intact species at *m/z* 2320 to become the most dominant species in the mass spectrum. Overnight aging of cluster solutions under these conditions also resulted in cleaner mass spectra with improved signal-to-noise ratios. The positive effect of ammonium acetate and decreased pH on the detection of $Ag_{32}(SG)_{19}$ could have been due to the increased ionic strength of the solution.

Using ESI-MS, Desireddy *et al.* compared the fragments found in fresh and aged solutions of $Ag_{32}(SG)_{19}$ and showed that the aged sample contained a

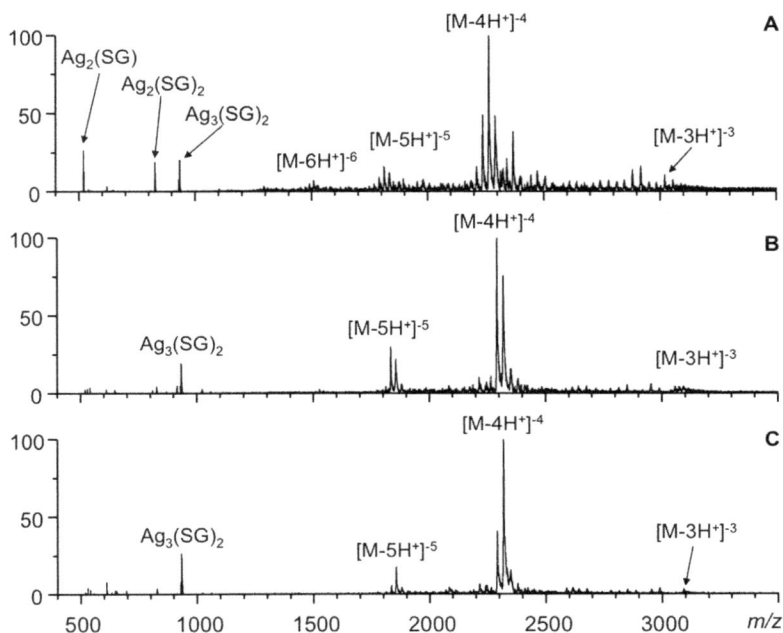

Figure 3.3   Negative nanoESI mass spectra of solutions of $Ag_{32}(SG)_{19}$ in (A) 50% methanol in water, (B) 50% methanol containing 5 mM ammonium acetate pH 7, and (C) 50% methanol containing 5 mM ammonium acetate pH 5. Fragments identified are all singly deprotonated species. (Reprinted with permission from Guo, J.; Kumar, S.; Bolan, M.; Desireddy, A.; Bigioni, T. P.; Griffith, W. P., Mass Spectrometric Identification of Large Silver Cluster Compounds: the case of $Ag_{32}(SG)_{19}$. *Anal. Chem.* **2012**, *84* (12), 5304–5308. Copyright 2012 American Chemical Society.)

significantly higher abundance of glutathionate species. This work highlighted the utility of mass spectrometry in monitoring the stability, solution phase and gas phase decomposition of nanoclusters. They showed that although gas phase dissociation of the clusters can occur during mass analysis, the differences between the fresh and aged samples do suggest that the anionic silver glutathionate species played a role in the solution phase decay mechanism of $Ag_{32}(SG)_{19}$.[55]

3.5   Optical Properties

The electronic structure of small metal particles can deviate markedly from that of the bulk, which can lead to distinctly different optical properties. Kubo was the first to address the electronic structure for fine metallic particles, in 1962.[56] His statistical approach was based on the Drude free-electron model[57] and on the fact that as fewer atoms were included in the metal particle, the density of states would decrease in a corresponding fashion. Kubo calculated the statistical average energy gap between states to be $<\delta> \propto E_F/\nu N$, where E_F is the position of the Fermi level with respect to the bottom of the conduction band, N is the number of atoms and ν is the valence for the metal. Here, he only considered irregularly shaped metal particles to avoid symmetry-induced degeneracies and ensure purely statistical electronic level spacings. Today, with the availability of atomic coordinates from single-crystal X-ray diffraction from Au clusters and their input into advanced computation tools,[58] a much more sophisticated picture of metal nanoparticle electronic structure is emerging. For silver clusters, however, the level of understanding lags behind that of gold.

To date, only one total structure for a Ag cluster has been reported using single crystal X-ray diffraction,[59] albeit for a mixed ligand shell containing both thiolates and phosphines. In the absence of total structure determination, understanding the electronic structure of small silver clusters remains a challenge. The optical properties of these metal clusters can help shed some light on their electronic structure, however. Absorption and fluorescence can indicate electronic transitions within the clusters, subject to selection rules. Together with theoretical models, this can help provide some insight into the location of discrete electronic states until total structure determination is possible.

3.5.1   Electronic Structure Theory

The stability of small Au "magic-number" clusters has been attributed to electronic shell closings[8] since large HOMO–LUMO gaps make it more difficult for chemical reactions to occur. This is in analogy to noble gases. The smallest clusters are known to be most stable owing, at least in part, to their very large HOMO–LUMO gaps. How these clusters evolve from the ligand-field splitting superatomic electronic structures to plasmonic nanoparticles remains an open question. However, a basic atom-like picture of the electronic structure of small Au clusters has emerged. While there are indications that the structure of Au and Ag magic-number clusters might be different,[4] there are also indications that the electron configuration rules might describe both metals equally well.

Metal clusters can be treated somewhat like atoms in that the free electrons (*e.g.* 5s electrons for Ag, 6s electrons for Au) of the metal core are localized in space much like electrons are localized in space by the nucleus of an atom. In fact, the electronic structure of clusters with approximately spherical cores can be described in much the same way as that of atoms. While the radial part of their respective potentials is quite different, spherically symmetric clusters will have electronic eigenstates with the same angular harmonics as an atom.[60]

This "superatom" model can be used to predict the stability of magic-number clusters utilizing the Aufbau principle and the superatomic orbitals.

Using the angular momentum quantum number and radial nodes analogous to that of atomic systems, the shell filling is as follows:

$$1S^2 | 1P^6 | 1D^{10} | 2S^2 \ 1F^{14} | 2P^6 \ 1G^{18} | \ldots$$

where the vertical bars indicate the largest HOMO–LUMO gaps and therefore the most important shell closings.[61] Using this model, one would predict significant shell closings for $n = 2, 8, 18, 34, 58, \ldots$, where n is the number of free electrons in the metal core. The electron count must also account for ligand interactions with the core, as they could contribute or withdraw electrons to or from the core. For clusters of the form $Au_N X_M L_S^z$ (X = one-electron withdrawing ligand, *e.g.* thiolate; L = Lewis base type ligand, *e.g.* ammonia), where z is the total charge on the core, the electron count can be determined as follows:[61,62]

$$n = N\nu - M - z \qquad (3.1)$$

where ν is the valence of the metal.

Clusters containing "magic-numbers" of electrons have been known for many years in the gas phase,[60] but have since been identified in the condensed phase.[1,26] For example, $Au_{25}(SR)_{18}^-$ has an 8 electron core,[58,63] $Ag_{44}(SR)_{30}^{4-}$ has an 18 electron core,[7] and $Au_{102}(SR)_{44}$ has a 58 electron core.[64] In contrast, $Au_{38}(SR)_{24}$ has a 14 electron core.[65] This can be explained by the symmetry of the bi-icosahedral cluster core. The non-spherical symmetry splits the d orbitals such that 14 electrons fill a sub shell of the cluster. Based on these previous results, one would predict that the recently reported $Ag_{32}(SR)_{19}$ cluster[5] should have a prolate symmetry and a similar electron count to that of $Au_{38}(SR)_{24}$.

3.5.2 Absorption Spectroscopy

Noble metal nanoparticles have long been known to be strong absorbers of light. Most notably, the surface plasmon resonance in colloidal Au and Ag particles has been studied extensively.[66] It is known that as the particle size decreases, the plasmon resonance is damped and eventually vanishes for particles below ~ 2 nm in diameter.[67] The UV-Vis absorption spectra of these few atom metal clusters exhibit features starkly different than those of larger particles.[67] The characteristic absorption features from these clusters resemble molecular-like single-electron transitions, rather than the canonical collective plasmon resonances of larger nanoparticles. Absorption spectroscopy is therefore an important way to probe and better understand the electronic structure of small metal clusters, particularly in the absence of solved crystal structures.

A simulated absorption spectrum for a hypothetical 25 atom Ag cluster was recently calculated by the Aikens group using time-dependent density functional theory, using the coordinates of a $Au_{25}(SR)_{18}^-$ cluster, in support of experimental efforts to synthesize *p*-MBA passivated silver clusters known as intensely and broadly absorbing nanoparticles (IBANs).[6] The various

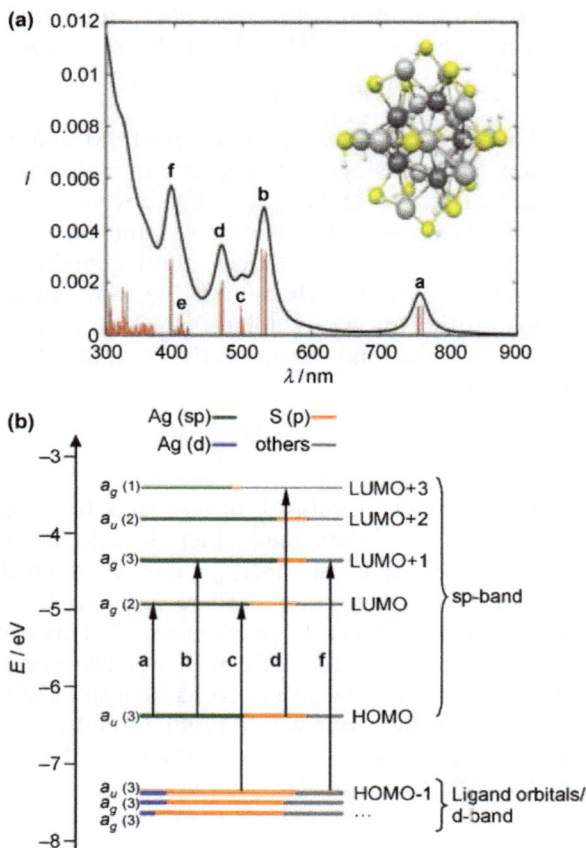

Figure 3.4   (a) theoretical absorption spectrum for $[Ag_{25}(SH)_{18}]^-$ (b) the theoretical orbital diagram for the same model compound.
(Reprinted with permission from Bakr, O. M.; Amendola, V.; Aikens, C. M.; Wenseleers, W.; Li, R.; Dal Negro, L.; Schatz, G. C.; Stellacci, F., Silver Nanoparticles with Broad Multiband Linear Optical Absorption. *Angew. Chem. Int. Ed.* **2009**, *48* (32), 5921–5926. Copyright 2009 Wiley and Sons, inc.)

transitions between superatom states have been described in detail.[6] As depicted in Figure 3.4(b), low energy transitions (a, b, and c) have been attributed to excitation across the HOMO–LUMO gap between superatomic orbitals, while higher energy transitions have been assigned to transitions between ligand states and higher lying superatomic orbitals.[6]

Although it was clear at the time that the Ag:*p*-MEA cluster was likely not $Ag_{25}(SR)_{18}^-$, this early work represented an important step in understanding the basic nature of silver cluster electronic structure. Since then, other efforts have been made by the Landman group to calculate the electronic structure of a large silver cluster, reported to be $Ag_{152}(SR)_{60}$.[68] This has not only provided an alternate perspective but also a first look at high angular momentum states for Ag. The cluster was modeled as $Ag_{152}(SR)_{60}{}^{2+}$ with a 90 electron closed shell configuration using spin density-functional theory.[69] In this case, the HOMO were the 1H superatomic orbitals, split by the ligands and straddling the 2D orbitals, and the LUMO were the 1I orbitals.

3.5.2.1 Size Effects on Absorption

The size of metal clusters can have a dramatic effect on their absorption characteristics. As particles grow from a few atom clusters to above several nanometers their absorption evolves from molecular-like in nature to plasmonic. This was nicely demonstrated by Bakr *et al.* by heating a solution of Ag IBAN clusters to 90 °C while monitoring the absorbance.[6] A clear evolution from a highly structured molecular-like spectrum into a single plasmonic peak was observed over the course of two hours, as shown in Figure 3.5. The emergence of the plasmon was attributed to the growth of the clusters as heating proceeded, therefore the time evolution of the spectra in Figure 3.5 could be interpreted as the size evolution from molecular clusters to plasmonic nanoparticles.

3.5.2.2 Ligand Effects on Absorption

Theoretical results have indicated that the superatomic orbitals involved in the observed optical transitions in IBANs each have a significant

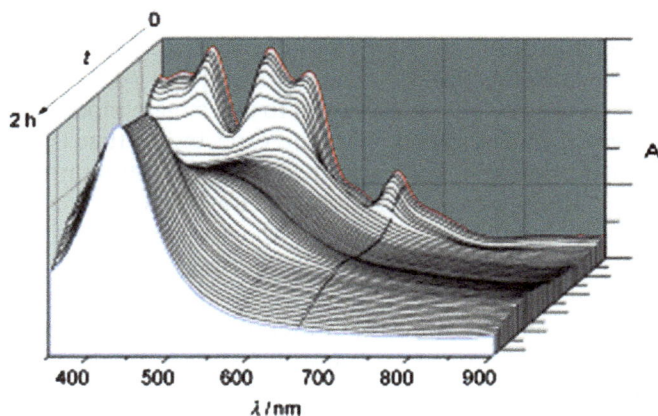

Figure 3.5 UV-Vis-NIR spectra of a solution of IBANs heated to 90 °C in DMF. Initially particles have molecule-like absorbance peaks (red) but transform into a single plasmonic peak (blue) after 2 hr.
(Reprinted with permission from Bakr, O. M.; Amendola, V.; Aikens, C. M.; Wenseleers, W.; Li, R.; Dal Negro, L.; Schatz, G. C.; Stellacci, F., Silver Nanoparticles with Broad Multiband Linear Optical Absorption. *Angew. Chem. Int. Ed.* **2009**, *48* (32), 5921–5926. Copyright 2009 Wiley and Sons, inc.)

contribution from sulfur.[6] Based on this, it was proposed that changes to the charge density of the sulfur atom on the aryl thiolates could result in perturbations to the orbitals, which could manifest in the structure of the optical spectra. This was illustrated by using multiple aryl thiolate ligands to protect the clusters, each with different substituents at different positions on the ring and each having a different effect on the charge density around the sulfur atom.

By comparing the spectra of six different ligands, the effect of the substituents can immediately be seen (see Figure 3.6).[6] The absorption features were influenced most strongly by the addition of a second thiol group in the *meta* position (recall the thiol is in the *para* position). In this case, the peaks collapse into only three peaks that resemble the $Ag_{32}(SG)_{19}$ cluster[4] or the presumably related Ag captopril cluster.[3] The spectrum is also strongly affected by the substitution of a methanol group at the *meta* position, however the resulting spectrum more resembled that of the fluorinated ligands.

It is important to consider two other ligand effects. First, the IBAN spectra (and presumably the IBAN clusters themselves) are unique to phenyl thiol ligands. While the band 13 cluster in the Ag:glutathionate system strongly resembles the IBANs in colour and size, their spectra are quite different (see Figure 3.6). This seems to indicate that the nature of the ligand, phenyl *versus* alkyl, plays a determinant role in the structure of the Ag core and therefore the electronic structure of the clusters. Further, in the above case of IBAN ligand substitution, it was only when substituents were located close to the metal surface that the spectra were strongly modified.[6] It is quite possible then that those substituents could have modified the atomic structure of the clusters, rather than simply modifying the local charge density, ultimately leading to changes in the spectra.

If modifications to the ligands near the metal can so strongly affect the optical spectra of the clusters, it is natural to investigate modifications to the ligands far away from the metal, with the expectation that there would be little effect on the spectra. This is indeed the case. Muhammed and coworkers analyzed the optical absorption spectra of clusters containing different functional end groups on a lipoic acid-based ligand.[70] Gel electrophoresis indicated shifts in mobility of the clusters, which confirmed the presence of the different functional groups on the corresponding particles. Spectra in Figure 3.7 show that differently ligated clusters can indeed produce the same characteristic absorption spectrum when it is the end group functionality that is modified. This is an important result as it provides excellent insight for developing more sophisticated cluster-based applications.

3.5.3   Fluorescence Spectroscopy

Normally metal has a very low photoluminescence quantum yield, due to the myriad of non-radiative decay channels available. For example, the measured quantum yield of bulk Au and Cu metal is in the order of 10^{-10}.[71]

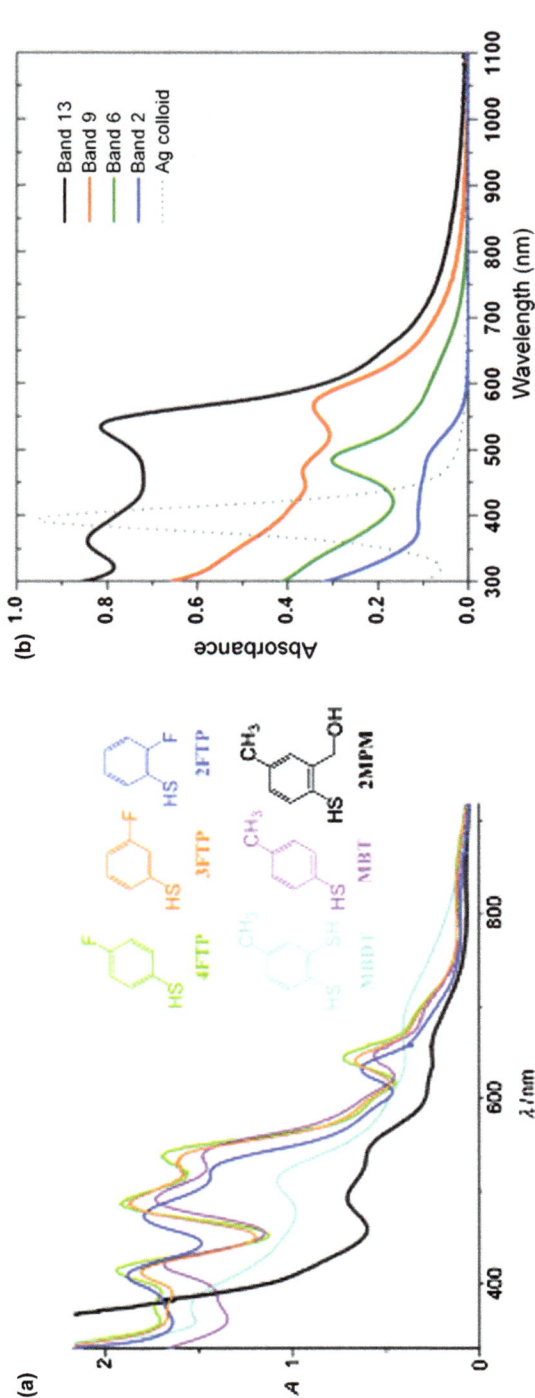

Figure 3.6 Left: Absorption spectra of IBANs using different aryl thiol ligands in DMF. (Reprinted with permission from Bakr, O. M.; Amendola, V.; Aikens, C. M.; Wenseleers, W.; Li, R.; Dal Negro, L.; Schatz, G. C.; Stellacci, F., Silver Nanoparticles with Broad Multiband Linear Optical Absorption. *Angew. Chem. Int. Ed.* **2009**, *48* (32), 5921–5926. Copyright 2009 Wiley and Sons, inc.) Right: Absorption spectra of different sizes of silver:glutathionate clusters in water. The cluster labeled Band 6 is the $Ag_{32}(SG)_{19}$ cluster. Colloidal Ag is shown for comparison. (Reprinted with permission from Kumar, S.; Bolan, M. D.; Bigioni, T. P., Glutathione-Stabilized Magic-Number Silver Cluster Compounds. *J. Am. Chem. Soc.* **2010**, *132* (38), 13141–13143. Copyright 2010 American Chemical Society.)

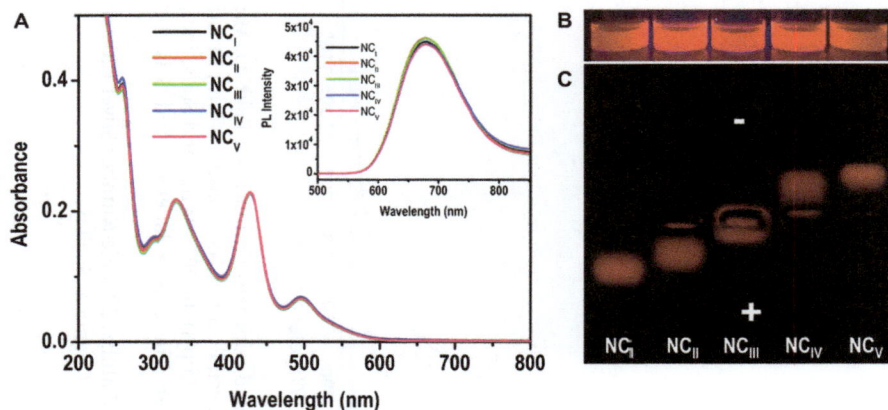

Figure 3.7 (a) Optical absorption of nanocrystals with ligands of different func-
tionality: 100% LA-PEG-COOH (NC$_{SI}$), 50% LA-PEG-COOH and 50%
LA-PEG-OCH$_3$ (NC$_{SII}$), 100% LA-PEG-OCH$_3$ (NC$_{SIII}$), a mixture of 50%
LA-PEG-NH$_2$ and 50% LA-PEG-OCH$_3$ (NC$_{SIV}$), 100% TA-PEG-NH$_2$ (NC$_{SV}$),
and the parent NP$_5$. (b) Photograph of the aqueous dispersion of the
samples upon UV light irradiation and (c) gel electrophoresis image.
(Reprinted with permission from Muhammed, M. A. H.; Aldeek, F.;
Palui, G.; Trapiella-Alfanso, L.; Mattoussi, H., Growth of *In Situ* Func-
tionalized Luminescent Silver Nanoclusters by Direct Reduction and
Size Focusing. *ACS Nano* **2012**, *6* (10), 8950–8961. Copyright 2012
American Chemical Society.)

These non-radiative decay channels can be suppressed in small particles,
however. If large enough energy gaps emerge within the electronic structure
of small metal nanoparticles, radiative decay can become more probable
than non-radiative decay since it can become easier to emit a photon than a
phonon. As a result, the quantum yields of small metal nanoparticles can be
many orders of magnitude larger than those of the bulk (10^{-4}–10^{-5}).[72,73]

For magic-number clusters, photon emission is the result of radiative
recombination across the HOMO–LUMO gap. The quality of luminescence
has been shown to be strongly dependent on the cluster size, capping ligand,
and local chemical environment. Analysis of this emission is helpful in
developing the superatomic orbital picture, therefore we will limit the
following discussion to ligand-protected clusters despite a significant
amount of excellent work being done with highly-fluorescent biotemplated
Ag clusters.

3.5.3.1 Size Effects on Fluorescence

The size, and hence number of atoms, in the metal clusters plays an important role in their emission. Ag clusters with 7 to 44 metal atoms have
been reported with emissions that shift from the visible to the near infrared
with increasing size.[17,59,74–82] A summary of these clusters, their emission
and quantum yield (QY) appear in Table 3.2. In general smaller clusters have

Table 3.2 Luminescent properties of Ag nanoclusters of varying size.

Cluster size (# of Ag atoms)	Emission (nm)	Excitation (nm)	QY (%)	Ref.
7	440	350	5	13
8	650	550	0.3	13
8	685	490	1.46	80
9	495	489	2.6	86
14	536	360	0.1	22
16	647	489	4.2	84
44	680		5	6,81
44	1375		0.01	6,81

larger HOMO–LUMO gaps,[4] therefore it is not surprising that emission from electronic transitions appear at higher energy for those structures. For example, IBAN clusters prepared by Bakr *et al.* produced broad emission centered around 1375 nm.[6] This is in contrast to Ag_7 particles reported by Sun *et al.* that produced emission at 440 nm.[83]

3.5.3.2 Effect of Chemical Environment on Fluorescence

The local environment of Ag nanoclusters is of importance if they are to be used for biological sensing and chemical applications.[84] Fluorescence from common fluorophores is affected by environmental factors such as their interaction with solvent molecules. Temperature, pH, solvent polarity, and fluorophore concentration can also play large roles in emission peak position and intensity. Colour tunable luminescence has been reported for polymethacrylic acid stabilized Ag clusters by adjusting the solvent polarity.[85] A red shift in emission has been reported for clusters stabilized in solvents of decreasing polarity such as solutions of water–methanol and water–ethanol as well as several organic solvents like DMF, THF and acetone.[84,85] Figure 8(a) shows how the emission of Ag clusters is enhanced by increasing the amount of ethanol in water, while Figure 8(b) shows how the fluorescence is also enhanced by dissolving the clusters in solvents that are mixtures of water and organic solvents with decreasing polarity. A similar shift is observed in the fluorescence of molecules with n–π* transitions.

The ligand–solvent interaction can also have an important effect on metal cluster luminescence. Diez *et al.* reported a red shift in emission for increasing ratios of Ag/ligand in clusters synthesized with methacrylic acid (MAA).[85] At higher ratios there are fewer ligands available to passivate the surface of the metal core, which increases the interactions between the solvent and core, causing a shift to lower energies.

Outside species can contribute to significant quenching of fluorescence. For example, changes to the structure and chemical environment can lead to static quenching through formation of non-fluorescent species. The presence of metal ions and molecules such as cysteine has been reported to quench the fluorescence of clusters.[13,17,86] For example, mercury (II) ions

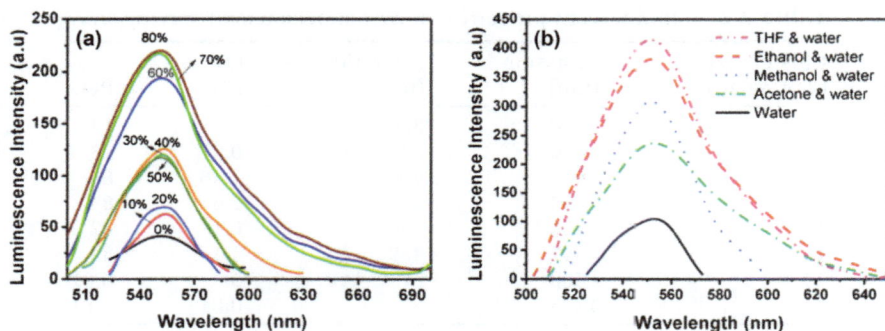

Figure 3.8   (a) Emission spectra of Ag clusters in water with increased amounts of ethanol. (b) Emission spectra from clusters in mixtures of water, water/acetone, water/methanol, water/ethanol, and water/THF. Decreasing polarity causes an increase in emission intensity.
(Reproduced from ref. 84 with permission from the PCCP Owner Societies)

appear to cause the greatest quenching effect of metal ions because the Ag core donates electron density to the Hg cation, thus becoming partially oxidized. The formation of a Ag amalgam completely quenches the cluster fluorescence, however.[17] Introduction of cysteine or other sulfur containing compounds also quench the fluorescence if they induce aggregation of the particles.[86]

3.5.4   Spectroscopy Challenges

Challenges remain regarding the identification of basic optical properties of silver magic-number clusters. For example, attributing fluorescence from any one cluster size is very difficult as a small amount of a strongly fluorescent impurity can dominate measurements of less fluorescent species. To that end, Desireddy et al. reported on the temporal stability and decay of glutathione-protected silver clusters, wherein aged clusters were separated by gel electrophoresis in order to identify the products of aging.[87] All cluster sizes were found to decay into smaller sizes, with a general trend of smaller clusters being more stable than larger clusters. Two particular clusters, namely band 6 ($Ag_{32}(SG)_{19}$) and band 2, were the most stable (see Figure 3.1 for band indices).[4] It is this 32 atom cluster that might be the most commonly observed due to its inherent stability[2,3] and its distinct spectrum.[4] While the band 2 cluster might be more stable, the absorption spectrum is less distinct than the band 6 cluster. As a result, a mixture of the two clusters might be expected to produce a spectrum that might not reveal the presence of the band 2 cluster, but instead might appear to be the pure band 6 cluster. Further, the band 2 cluster is more fluorescent than the band 6 cluster, therefore the fluorescence spectrum of the mixture ought to appear to be that of the smaller cluster only.

Based on the work of Desireddy *et al.*, samples of band 6 clusters will decay to produce band 2 clusters, meaning that all but the most freshly separated clusters are expected to be mixtures. Such a mixture could be dominated by the absorption spectrum of band 6 and dominated by fluorescence from band 2 clusters. For example, recent reports of "highly fluorescent" silver clusters contain absorption spectra that resemble the larger band 6 cluster but could very well contain both band 6 and band 2 clusters.[16,50,86] The absorption spectra seem to contain the signature peak of $Ag_{32}(SG)_{19}$ that appears near 490 nm, but the emission spectra show luminescence that reflect a much higher intensity than one would expect from the large $Ag_{32}(SG)_{19}$ cluster. As a result, this high fluorescence is likely emission due to the presence of a second smaller species, one with a larger HOMO–LUMO gap and a less distinct absorption fingerprint, such as the band 2 cluster. Nevertheless, a more comprehensive look at the emission of families of clusters, carried out with the proper care, will be essential to preventing the misidentification of true properties from a convolution of mixed species resulting from the rapid decay of larger clusters.

3.6 Conclusions

Small metal clusters possess many unique optical properties. The characteristic absorption and fluorescence of these materials provide an interesting view into their electronic behavior. "Superatomic" theory provides some possible solutions to their structural details and their interesting "magic-numbered" stability. Size, ligand, and chemical environment have all been shown to play critical roles in the absorption and emission quality of Ag nanoclusters. Smaller clusters give rise to greater stability and larger HOMO–LUMO gaps. The bright fluorescent emission produced by these clusters could have possibility for many uses in many unique applications such as biolabeling, microscopy, light harvesting, chemical sensing, *etc*. The more that is learned about their chemical structure and properties, the more which applications different clusters are suited for will be determined.

References

1. R. L. Whetten, J. T. Khoury, M. M. Alvarez, S. Murthy, I. Vezmar, Z. L. Wang, P. W. Stephens, C. L. Cleveland, W. D. Luedtke and U. Landman, *Adv. Mater.*, 1996, **8**, 428–433.
2. N. Cathcart, P. Mistry, C. Makra, B. Pietrobon, N. Coombs, M. Jelokhani-Niaraki and V. Kitaev, *Langmuir*, 2009, **25**, 5840–5846.
3. N. Cathcart and V. Kitaev, *J. Phys. Chem. C*, 2010, **114**, 16010–16017.
4. S. Kumar, M. D. Bolan and T. P. Bigioni, *J. Am. Chem. Soc.*, 2010, **132**, 13141–13143.
5. J. Guo, S. Kumar, M. Bolan, A. Desireddy, T. P. Bigioni and W. P. Griffith, *Anal. Chem.*, 2012, **84**, 5304–5308.

6. O. M. Bakr, V. Amendola, C. M. Aikens, W. Wenseleers, R. Li, L. Dal Negro, G. C. Schatz and F. Stellacci, *Angew. Chem. Int. Ed.*, 2009, **48**, 5921–5926.
7. K. M. Harkness, Y. Tang, A. Dass, J. Pan, N. Kothalawala, V. J. Reddy, D. E. Cliffel, B. Demeler, F. Stellacci, O. M. Bakr and J. A. McLean, *Nanoscale*, 2012, **4**, 4269–4274.
8. M. Walter, J. Akola, O. Lopez-Acevedo, P. D. Jadzinsky, G. Calero, C. J. Ackerson, R. L. Whetten, H. Gronbeck and H. Hakkinen, *Proc. Natl. Acad. Sci. U.S.A.*, 2008, **105**, 9157–9162.
9. Z. Wu, E. Lanni, W. Chen, M. E. Bier, D. Ly and R. Jin, *J. Am. Chem. Soc.*, 2009, **131**, 16672–16674.
10. L. A. Peyser, A. E. Vinson, A. P. Bartko and R. M. Dickson, *Science*, 2001, **291**, 103–106.
11. L. A. Peyser, T.-H. Lee and R. M. Dickson, *J. Phys. Chem. B*, 2002, **106**, 7725–7728.
12. J. Zheng and R. M. Dickson, *J. Am. Chem. Soc.*, 2002, **124**, 13982–13983.
13. T. U. B. Rao and T. Pradeep, *Angew. Chem. Int. Ed.*, 2010, **49**, 3925–3929.
14. J. Zhang, S. Xu and E. Kumacheva, *Adv. Mat.*, 2005, **17**, 2336–2340.
15. B. Adhikari and A. Banerjee, *Chem. Eur. J.*, 2010, **16**, 13698–13705.
16. L. Shang and S. Dong, *Chem. Comm.*, 2008, **2008**, 1088–1090.
17. B. Adhikari and A. Banerjee, *Chem. Mater.*, 2010, **22**, 4364–4371.
18. M. Brust, M. Walker, D. Bethell, D. J. Schiffrin and R. Whyman, *J. Chem. Soc., Chem. Commun.*, 1994, **1994**, 801–802.
19. M. R. Branham, A. D. Douglas, A. J. Mills, J. B. Tracy, P. S. White and R. W. Murray, *Langmuir*, 2006, **22**, 11376–11383.
20. J. P. Wilcoxon, R. L. Williamson and R. Baughman, *J. Chem. Phys.*, 1993, **98**, 9933–9950.
21. S. Murthy, T. P. Bigioni, Z. L. Wang, J. T. Khoury and R. L. Whetten, *Mater. Lett.*, 1997, **30**, 321–325.
22. H. Yang, J. Lei, B. Wu, Y. Wang, M. Zhou, A. Xia, L. Zheng and N. Zheng, *Chem. Comm*, 2013, **2013**, 300–302.
23. H. Yang, Y. Wang and N. Zheng, *Nanoscale*, 2013, **5**, 2674–2677.
24. T. U. B. Rao, B. Nataraju and T. Pradeep, *J. Am. Chem. Soc.*, 2010, **132**, 16304–16307.
25. Y. Negishi, Y. Takasugi, S. Sato, H. Yao, K. Kimura and T. Tsukuda, *J. Am. Chem. Soc.*, 2004, **126**, 6518–6519.
26. Y. Negishi, K. Nobusada and T. Tsukuda, *J. Am. Chem. Soc.*, 2005, **127**, 5261–5270.
27. A. Dass, A. Stevenson, G. R. Dubay, J. B. Tracy and R. W. Murray, *J. Am. Chem. Soc.*, 2008, **130**, 5940–5946.
28. C. Kumara and A. Dass, *Nanoscale*, 2012, **4**, 4084–4086.
29. Y. Shichibu, Y. Negishi, T. Tsukuda and T. Teranishi, *J. Am. Chem. Soc.*, 2005, **127**, 13464–13465.
30. Y. Negishi, Y. Takasugi, S. Sato, H. Yao, K. Kimura and T. Tsukuda, *J. Phys. Chem. B*, 2006, **110**, 12218–12221.

31. A. P. Gies, D. M. Hercules, A. E. Gerdon and D. E. Cliffel, *J. Am. Chem. Soc.*, 2007, **129**, 1095–1104.
32. Y. Negishi, N. K. Chaki, Y. Shichibu, R. L. Whetten and T. Tsukuda, *J. Am. Chem. Soc.*, 2007, **129**, 11322–11323.
33. J. B. Tracy, G. Kalyuzhny, M. C. Crowe, R. Balasubramanian, J. P. Choi and R. W. Murray, *J. Am. Chem. Soc.*, 2007, **129**, 6706–6707.
34. J. B. Tracy, M. C. Crowe, J. F. Parker, O. Hampe, C. A. Fields-Zinna, A. Dass and R. W. Murray, *J. Am. Chem. Soc.*, 2007, **129**, 16209–16215.
35. N. K. Chaki, Y. Negishi, H. Tsunoyama, Y. Shichibu and T. Tsukuda, *J. Am. Chem. Soc.*, 2008, **130**, 8608–8610.
36. C. A. Fields-Zinna, M. C. Crowe, A. Dass, J. E. Weaver and R. W. Murray, *Langmuir*, 2009, **25**, 7704–7710.
37. C. A. Fields-Zinna, J. S. Sampson, M. C. Crowe, J. B. Tracy, J. F. Parker, A. M. deNey, D. C. Muddiman and R. W. Murray, *J. Am. Chem. Soc.*, 2009, **131**, 13844–13851.
38. H. Qian, Y. Zhu and R. Jin, *ACS Nano*, 2009, **3**, 3795–3803.
39. J. P. Choi, C. A. Fields-Zinna, R. L. Stiles, R. Balasubramanian, A. D. Douglas, M. C. Crowe and R. W. Murray, *J. Phys. Chem. C*, 2010, **114**, 15890–15896.
40. L. A. Angel, L. T. Majors, A. C. Dharmaratne and A. Dass, *ACS Nano*, 2010, **4**, 4691–4700.
41. J. F. Parker, C. A. Fields-Zinna and R. W. Murray, *Acc. Chem. Res.*, 2010, **43**, 1289–1296.
42. M. Z. Zhu, H. F. Qian and R. Jin, *J. Phys. Chem. Lett.*, 2010, **1**, 1003–1007.
43. H. Qian and R. Jin, *Chem. Commun.*, 2011, **47**, 11462–11464.
44. X. Yang, M. Shi, R. Zhou, X. Chen and H. Chen, *Nanoscale*, 2011, **3**, 2596–2601.
45. A. Ghosh, T. Udayabhaskararao and T. Pradeep, *J. Phys. Chem. Lett.*, 2012, **3**, 1997–2002.
46. C. Lavenn, F. Albrieux, G. Bergeret, R. Chiriac, P. Delichere, A. Tuel and A. Demessence, *Nanoscale*, 2012, **4**, 7334–7337.
47. S. Xie, M. C. Paau, Y. Zhang, S. Shuang, W. Chan and M. M. Choi, *Nanoscale*, 2012, **4**, 5325–5332.
48. P. R. Nimmala, B. Yoon, R. L. Whetten, U. Landman and A. Dass, *J. Phys. Chem. A*, 2013, **117**, 504–517.
49. Y. Yu, X. Chen, Q. F. Yao, Y. Yu, N. Yan and J. P. Xie, *Chem. Mater.*, 2013, **25**, 946–952.
50. X. Yuan, Z. Luo, Q. Zhang, X. Zhang, Y. Zheng, J. Y. Lee and J. Xie, *ACS Nano*, 2011, **5**, 8800–8808.
51. S. Kumar, M. D. Bolan and T. P. Bigioni, *J. Am. Chem. Soc.*, 2010, **132**, 13141–13143.
52. Y. Negishi, R. Arai, Y. Niihoria and T. Tsukuda, *Chem. Commun.*, 2011, **47**, 5693–5695.
53. J. L. Benesch, B. T. Ruotolo, D. A. Simmons and C. V. Robinson, *Chem. Rev.*, 2007, **107**, 3544–3567.

54. K. M. Harkness, B. C. Hixson, L. S. Fenn, B. N. Turner, A. C. Rape, C. A. Simpson, B. J. Huffman, T. C. Okoli, J. A. McLean and D. E. Cliffel, *Anal. Chem.*, 2010, **82**, 9268–9274.
55. A. Desireddy, S. Kumar, J. S. Guo, M. D. Bolan, W. P. Griffith and T. P. Bigioni, *Nanoscale*, 2013, **5**, 2036–2044.
56. R. Kubo, *J. Phys. Soc. Japan*, 1962, **21**, 976.
57. N. W. Ashcroft and N. D. Mermin, *Solid State Physics*, W.B. Saunders Company, 1976.
58. M. Zhu, C. M. Aikens, F. J. Hollander, G. C. Schatz and R. Jin, *J. Am. Chem. Soc.*, 2008, **130**, 5883–5885.
59. H. Yang, J. Lei, B. Wu, Y. Wang, M. Zhou, A. Xia, L. Zheng and N. Zheng, *ChemComm*, 2013, **49**, 300–302.
60. W. A. de Heer, *Rev. Mod. Phys.*, 1993, **65**, 611–676.
61. M. Walter, J. Akola, O. Lopez-Acevedo, P. D. Jadzinsky, G. Calero, C. J. Ackerson, R. Whetten, H. Gronbeck and H. Hakkinen, *Proceedings of the National Academy of Science*, 2008, **105**, 9157–9162.
62. C. Aikens, *Journal of Physical Chemistry Letters*, 2011, **2**, 99–104.
63. M. W. Heaven, A. Dass, P. S. White, K. M. Holt and R. W. Murray, *J. Am. Chem. Soc.*, 2008, **130**, 3754–3755.
64. P. D. Jadzinsky, G. Calero, C. J. Ackerson, D. A. Bushnell and R. D. Kornberg, *Science*, 2007, **318**, 430–433.
65. H. Qian, W. T. Eckenhoff, Y. Zhu, T. Pintauer and R. Jin, *J. Am. Chem. Soc.*, 2010, **132**, 8280–8281.
66. U. Kreibig and M. Vollmer, *Optical Properties of Metal Clusters*, Springer-Verlag, Berlin, 1995.
67. R. B. Wyrwas, M. M. Alvarez, J. T. Khoury, R. C. Price, T. G. Schaaff and R. L. Whetten, *Eur. Phys. J. D*, 2007, **43**, 91–95.
68. I. Chakraborty, A. Govindarajan, J. Erusappan, A. Ghosh, T. Pradeep, B. Yoon, R. L. Whetten and U. Landman, *Nano Lett.*, 2012, **12**, 5861–5866.
69. R. N. Barnett and U. Landman, *Phys. Rev. B*, 1993, **48**, 2801–2097.
70. M. A. H. Muhammed, F. Aldeek, G. Palui, L. Trapiella-Alfanso and H. Mattoussi, *ACS Nano*, 2012, **6**, 8950–8961.
71. A. Mooradian, *Phys. Rev. Lett.*, 1969, **22**, 185–187.
72. J. P. Wilcoxon, J. E. Martin, F. Parsapour, B. Wiedenman and D. F. Kelley, *J. Chem. Phys.*, 1998, **108**, 9137.
73. T. P. Bigioni, R. L. Whetten and O. Dag, *J. Phys. Chem. B*, 2000, **104**, 6983–6986.
74. O. M. Bakr, V. Amendola, C. M. Aikens, W. Wenseleers, R. Li, L. D. Negro, G. C. Schatz and F. Stellacci, *Angewandte*, 2009, **48**, 5921–5926.
75. T. U. B. Rao and T. Pradeep, *Angewandte*, 2010, **49**, 3925–3929.
76. X. L. Guevel, B. Hotzer, G. Jung, K. Hollemeyer, V. Trouillet and M. Schneider, *Journal of Physical Chemistry C*, 2011, **115**, 10955–10963.
77. M. Pelton, Y. Tang, O. M. Bakr and F. Stellacci, *J. Am. Chem. Soc.*, 2012, **134**, 11856–11859.

78. M. A. H. Muhammed, F. Aldeek, G. Palui, L. Trapiella-Alfanso and H. Mattoussi, *ACS Nano*, 2012, **10**, 8950–8961.
79. S. Huang, C. Pfeiffer, J. Hollmann, S. Friede, J. J.-C. Chen, A. Beyer, B. Haas, K. Volz, W. Heimbrodt, J. M. M. Martos, W. Chang and W. J. Parak, *Langmuir*, 2012, **28**, 8915–8919.
80. X. Yuan, Y. Tay, X. Dou, Z. Luo, D. T. Leong and J. Xie, *Analytical Chemistry*, 2013, **85**, 1913–1919.
81. P. T. K. Chin, M. V. D. Linden, E. J. V. harten, A. barendregt, M. T. M. Rood, A. J. Koster, F. W. B. V. Leeuwen, C. D. M. Donega, A. J. R. Heck and A. Meijerink, *Nanotechnology*, 2013, **24**, 1–7.
82. X. Yuan, M. I. Setyawati, A. S. Tan, C. N. Ong, D. T. Leong and J. Xie, *Asia Materials*, 2013, 5.
83. Y. Sun, K. Balasubramanian, T. U. B. Rao and T. Pradeep, *J. Phys. Chem. C*, 2011, **115**, 20380–20387.
84. Y. Li, X. Wang, S. Xu and W. Xu, *Phys. Chem. Chem. Phys.*, 2013, **15**, 2665–2668.
85. I. Díez, M. Pusa, S. Kulmala, H. Jiang, A. Walther, A. S. Goldmann, A. H. E. Müller, O. Ikkala and R. H. A. Ras, *Angew. Chem. Int. Ed.*, 2009, **48**, 2122–2125.
86. X. Yuan, Y. Tay, X. Dou, Z. Luo, D. T. Leong and J. Xie, *Anal. Chem.*, 2013, **85**, 1913–1919.
87. A. Desireddy, S. Kumar, J. Guo, M. D. Bolan, W. P. Griffith and T. P. Bigioni, *Nanoscale*, 2013, **5**, 2036–2044.

CHAPTER 4

Synthesis and Applications of Water-Soluble Fluorescent Silver Nanoclusters

HANGXUN XU*[a] AND KENNETH S. SUSLICK*[b]

[a] CAS Key Laboratory of Soft Matter Chemistry, Department of Polymer Science and Engineering, University of Science and Technology of China, Hefei, Anhui, 230026, P. R. China; [b] Department of Chemistry, University of Illinois at Urbana-Champaign, Urbana, IL 61801 USA
*Email: hxu@ustc.edu.cn; ksuslick@illinois.edu

4.1 Introduction

Ag nanoclusters consisting of several to a hundred atoms with diameters less than 2 nm show molecule-like properties, including discrete electronic transitions and strong fluorescence.[1-4] They have received considerable research interest due to their unique optical, electrical and chemical properties and potential applications in chemical- and bio- sensing.[5-8] Ag nanoclusters exhibit excellent photostability, large Stokes shifts and high quantum yields, and therefore, they have become an important class of fluorophores for applications in imaging.[6,7] Since most applications involve aqueous conditions,[9] the synthesis of aqueous-stable and water-soluble Ag nanoclusters is critical. The synthesis of water-soluble Ag nanoclusters, however, is challenging because few-atom silver nanoclusters are highly reactive and will quickly aggregate to form large nanocrystals thus reducing their surface energy. Recently, many different synthetic approaches including radiolytic, chemical, sonochemical,

RSC Smart Materials No. 7
Functional Nanometer-Sized Clusters of Transition Metals: Synthesis, Properties and Applications
Edited by Wei Chen and Shaowei Chen
© The Royal Society of Chemistry 2014
Published by the Royal Society of Chemistry, www.rsc.org

photochemical, and microwave methods have been developed to prepare water-soluble fluorescent Ag nanoclusters.[9] This chapter will highlight recent advances in the synthesis of water-soluble fluorescent Ag nanoclusters and briefly introduce their applications in chemical- and bio- sensing.

Historically, well before the development of modern analytical techniques, photographers were the first to generate and utilize Ag nanoclusters.[10,11] Photographers used papers impregnated with silver halides to make images even in the early nineteenth century. When a photographic emulsion containing silver halides was exposed to light, silver halides were decomposed to form Ag ions, bromine, and free electrons followed by formation of Ag crystals. Associated with this process, species like Ag_2^+, Ag_2^0, Ag_3^0, Ag_3^+, Ag_4^0, and Ag_4^+ were also generated.[11] This was believed to be the first chemical process to make reactive Ag nanoclusters, although photographers did not understand that Ag nanoclusters were formed nor did they realize that Ag nanoclusters consisting of few atoms were fluorescent. In the 1970s, discrete absorption and fluorescence from Ag nanoclusters were spectroscopically detected in cryogenic noble gas matrix.[12] During the condensation of Ag with Ar gas, excited Ag_2^* and Ag_3^* were formed and chemiluminescence was observed.[13] Ag nanoclusters formed under these extreme conditions, however, are not amenable to real world applications.

4.2 Synthesis of Water-Soluble Fluorescent Ag Nanoclusters

4.2.1 Radiolytic Reduction Synthesis of Ag Nanoclusters

Radiolytic reduction of Ag ions in aqueous solutions with common polyelectrolytes as stabilizers usually leads to the formation of large Ag nanoparticles with typical surface plasmon resonance bands around 380–400 nm. During short γ-ray irradiation, Henglein and co-workers observed the formation of Ag nanoclusters consisting of a few Ag atoms in the presence of polyphosphate or polyacrylate.[14–17] Ag nanoclusters obtained in this way exhibit absorption bands characteristic of Ag nanoclusters with a few Ag atoms at 275 nm, 300 nm, 330 nm, and 345 nm.[14] Unfortunately, Ag nanoclusters prepared *via* this approach were short lived and formed large Ag nanoparticles after several hours. Pulse radiolysis experiments indicated that Ag nanoclusters also form in the solution without stabilizers,[18] but the lifetimes of bare Ag nanoclusters were too short to be detected during the synthesis of Ag nanoparticles using the γ-ray irradiation method, and therefore the existence of Ag nanoclusters had been overlooked.

The formation of Ag nanoclusters in aqueous solutions under γ-ray irradiation is thought to proceed *via* the following process:

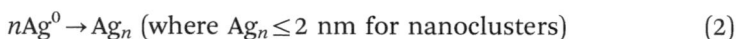

$$Ag^+ + e_{aq}^- \text{ or } R^\bullet \rightarrow Ag^0 \tag{1}$$

$$nAg^0 \rightarrow Ag_n \text{ (where } Ag_n \leq 2 \text{ nm for nanoclusters)} \tag{2}$$

Highly reactive species like radicals or solvated electrons can reduce Ag^+ to form Ag atoms in the solution. Meanwhile polyelectrolytes in the solution prevented the aggregation of Ag nanoclusters to form large Ag nanoparticles. Even so, these Ag nanoclusters were not stable and were very sensitive to UV light, being easily converted to large metallic Ag particles under UV irradiation.[15] In addition, Ag nanoclusters formed in this way were highly reactive due to the fact that the sizes of Ag nanoclusters were very small and essentially all Ag atoms were "surface atoms".[16] If a nucleophilic reagent such as NH_3, SH^- or CN^- is added to the Ag nanocluster solution, a strong absorption peak corresponding to Ag nanoparticles will quickly appear and all the characteristic absorption peaks of Ag nanoclusters disappear due to the conversion of Ag nanoclusters to Ag nanoparticles.[19]

Radiolytic reduction was able to fabricate water-soluble Ag nanoclusters conveniently, but Henglein and co-workers did not report the observation of fluorescence from Ag nanoclusters in aqueous solutions. A detailed study of the fluorescence properties of water-soluble Ag nanoclusters produced by radiolytic irradiation of Ag^+ was only recently published in 2005 by Treguer *et al.*[20]

4.2.2   Photochemical Reduction Synthesis of Ag Nanoclusters

The fluorescence of Ag nanoclusters in the solid state was first observed by photochemical reduction of AgO films by Dickson and co-workers in 2001.[21] In this case, fluorescent Ag nanoclusters were formed by light-induced decomposition of AgO. They also first reported the preparation of water-soluble fluorescent Ag nanoclusters *via* a photochemical reduction method using dendrimers as stabilizers.[22] Ag nanoclusters prepared in this way exhibited much narrower and more stable emission spectra than those individual Ag nanoclusters observed in AgO films.

For the formation of Ag nanoclusters *via* UV irradiation it is crucial that there be polymeric capping agents in solution. A variety of polyelectrolytes with simple structures can be used to prepare water-soluble fluorescent Ag nanoclusters *via* a photochemical approach; for example, highly fluorescent Ag nanoclusters were synthesized in aqueous solutions containing polyelectrolytes such as poly(methacrylic acid) (PMAA) and poly(acrylic acid) (PAA).[23] Ag nanoclusters obtained in this way exhibited excitation-dependent emission, relatively high quantum yield, and good photostability. Interestingly, these Ag nanoclusters also showed solvato-fluorochromic properties and were electrochemiluminescent (Figure 4.1).[24] The optical absorption and fluorescence emission properties of Ag nanoclusters prepared in this way can be tuned by choice of solvent. When the solvent was changed from water to methanol, a ~ 70 nm red shift of the absorption peak was observed. In addition, these Ag nanoclusters also showed cathodic hot electron-induced electrochemiluminescence. This phenomenon was believed to proceed *via* a redox excitation pathway, where the Ag nanoclusters were oxidized by the cathodically produced oxidizing radicals in the solution. In a

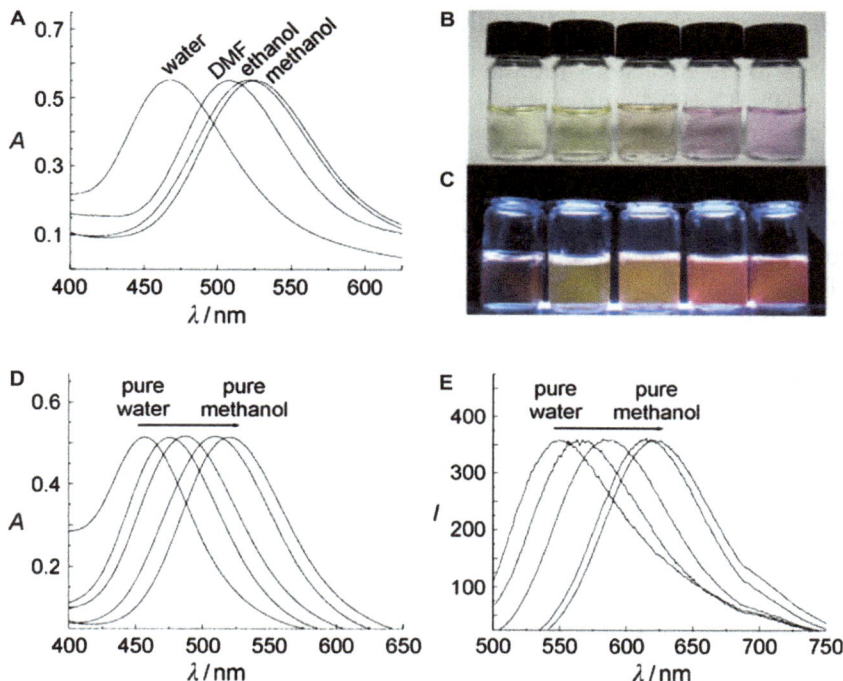

Figure 4.1   (A) UV-Vis absorption spectra of Ag nanoclusters in different solvents. Ag nanoclusters in water–methanol mixtures ranging from pure water (left) to pure methanol (right) under (B) visible light and (C) UV light illumination. (D) Absorption and (E) emission spectra of Ag nanoclusters shown in (B).
Reproduced with permission from ref. 24. Copyright 2009 John Wiley & Sons.

similar approach, poly(methacrylic acid) functionalized with pentaerythritol tetrakis-3-mercaptopropionate (PTMP–PMAA) was successfully used to prepare water-soluble Ag nanoclusters *via* a photochemical reduction method.[25] This polymer can also be used to make other water-soluble fluorescent metal nanoclusters like Cu nanoclusters and Au nanoclusters.

Polymer microgels (poly(N-isopropylacrylamide-acrylic acid-2-hydroxyethyl acrylate)) and multiarm star polyglycerol-b-polyacrylic acid (PG-b-PAA) copolymers were also used as templates to synthesize water-soluble fluorescent Ag nanoclusters using photochemical reduction.[26,27] The optical properties and photoluminescence of Ag nanoclusters prepared in the interior of macromolecular templates can be finely controlled using appropriate UV-irradiation time. Because both templates contain pH or temperature sensitive polymers, the fluorescent properties of Ag nanoclusters can be additionally tuned using external stimuli (*i.e.*, pH and temperature). Photochemical reduction is a clean synthetic approach and avoids the addition of external reducing agents.

4.2.3   Sonochemical Preparation of Ag Nanoclusters

Similar to the effects of γ-ray and deep-UV irradiation, highly reactive species like $OH^\bullet$, $H^\bullet$, $HO_2^\bullet$, and perhaps solvated electrons, e_{aq}^- are formed during ultrasonic irradiation of aqueous solutions.[28–30] High intensity ultrasound has found many important applications in the synthesis of nanostructured materials.[31,32] Extreme conditions inside collapsing bubbles with temperatures up to 20 000 K and pressures as high as 4000 bar have been spectroscopically determined during single-bubble cavitation.[33] The implosive collapse of cavitating bubbles can generate a number of reactive species that are chemically reductive and can be used to reduce Ag^+ in the solution.

Sonication of aqueous $AgNO_3$ solutions with dissolved PMAA easily forms fluorescent Ag nanoclusters.[34] The optical and fluorescence properties of Ag nanoclusters can be controlled by varying the duration of sonication (Figure 4.2). Prolonged sonication resulted in the formation of large Ag nanoparticles with a surface plasmon resonance band at 390 nm and decrease of the fluorescence intensity. TEM images indicate that all the sonochemically prepared Ag nanoclusters were smaller than 2 nm in diameter (Figure 4.2) and that the sizes of Ag nanoclusters increased as the sonication time increased. Sonochemically synthesized Ag nanoclusters exhibit excitation-dependent fluorescence. Varying the stoichiometry of the carboxylate groups relative to Ag^+ and the polymer molecular weight can also change the absorption and fluorescence of Ag nanoclusters prepared in this way. The quantum yield is measured to be ∼11%. Thus, sonochemistry provides a uniquely simple approach to make water-soluble Ag nanoclusters with various optical and fluorescence properties.

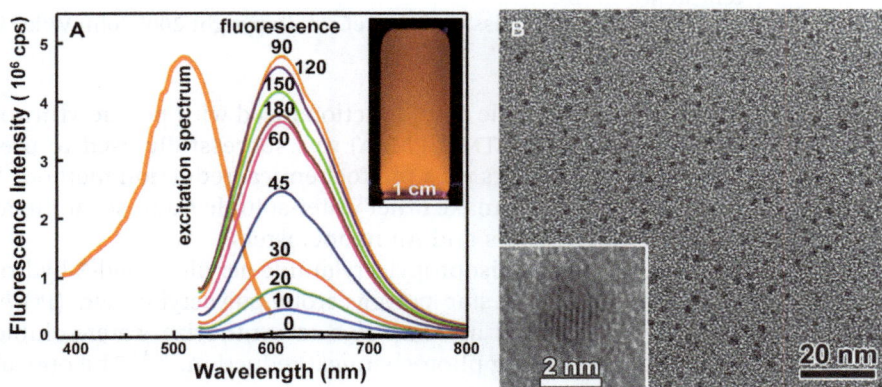

Figure 4.2   (A) Fluorescence emission spectra of sonochemically prepared Ag nanoclusters with sonication time varying from 0 to 180 min (inset: vial containing sonochemically prepared Ag nanoclusters illuminated by a 365 nm UV lamp). (B) TEM images of as-prepared Ag nanoclusters after 90 min sonication (inset shows a single magnified Ag nanocluster). Reproduced with permission from ref. 34. Copyright 2010 American Chemical Society.

4.2.4   Microwave-Assisted Synthesis of Ag Nanoclusters

Microwave irradiation is becoming an important tool in daily life and even in chemical reactions. The first use of microwaves as a heating method in chemical modifications can be traced back to the 1950s.[35] Microwave-assisted synthesis has gained wide acceptance since the report of using microwave irradiation in organic synthesis in 1986 and results primarily from rapid superheating of solvents.[36] The use of microwave heating for organic synthesis and chemical preparation of nanostructured materials has been extensively investigated in recent decades. Recently, microwave irradiation has also been applied to the synthesis of water-soluble fluorescent Ag nanoclusters.

Microwave irradiation of a PMAA solution with Ag^+ provided a rapid synthetic approach to prepare water-soluble fluorescent Ag nanoclusters (fluorescence appeared in 30 s).[37,38] The as-prepared Ag nanoclusters exhibited strong fluorescence emission at 575 nm with excitation at 510 nm (Figure 4.3A).[37] TEM imaging indicated that Ag nanoclusters prepared in this way were monodisperse and highly uniform with a diameter around 2 nm (Figure 4.3B).[37] The quantum yield was measured to be ∼6%. The fluorescence of Ag nanoclusters prepared using the microwave irradiation method varied with different irradiation time. The fluorescence intensity gradually increased up to 70 s of irradiation and then gradually decreased until it completely disappeared after 140 s. In addition to the duration of microwave irradiation, the solution pH also affected the fluorescence properties of Ag nanoclusters. Maximum intensity was achieved when the initial pH was ∼7. Water-soluble Ag nanoclusters prepared using microwave irradiation showed high selectivity for sensing Cr^{3+} ion in aqueous solutions.[37] Microwave-assisted synthesis offers a unique means for rapid and

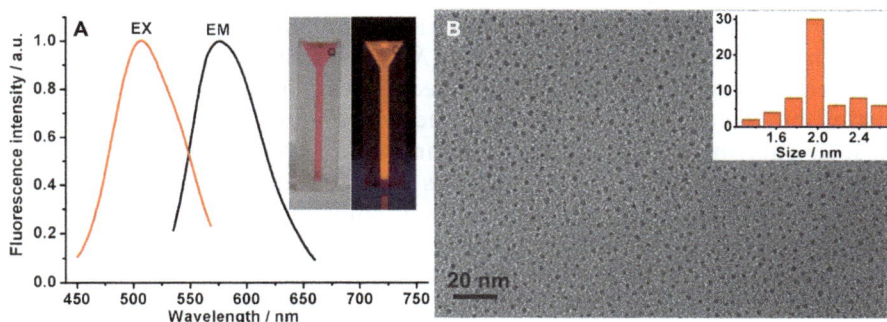

Figure 4.3   (A) Excitation and emission spectra of Ag nanoclusters synthesized *via* microwave irradiation (inset shows photographs of a solution of Ag nanoclusters under room light and 365 nm UV light illumination). (B) A typical TEM image of highly fluorescent Ag nanoclusters (inset shows the diameter histogram of Ag nanoclusters).
Reproduced with permission from ref. 37. Copyright 2011 Royal Society of Chemistry.

uniform heating of solutions; therefore, it can generate more homogeneous nucleation and shorter crystallization time for nanoclusters.

4.2.5 Chemical Reduction for Preparation of Ag Nanoclusters

Water-soluble Ag nanoclusters can be obtained through chemical reduction of Ag^+ using $NaBH_4$ in aqueous solutions or by bubbling an alkaline Ag^+ solution with H_2 and CO in the presence of polyphosphate.[16] Such chemical reduction methods, however, simultaneously produced large metallic Ag nanoparticles in the solution as confirmed by a strong absorption band at 380 nm corresponding to Ag nanoparticles. The reason for the formation of metallic Ag nanoparticles is probably the rapid aggregation of synthesized Ag nanoclusters due to the lack of appropriate stabilizers.

To avoid the formation of Ag nanoparticles in this process, the presence of stabilizers with high affinity for Ag^+ in the solution is necessary. DNA, peptides, proteins, and polythiol- or polyamine- appended molecules are commonly used stabilizers when chemical reduction is used to make water-soluble fluorescent Ag nanoclusters. DNA was the first used by Dickson *et al.* as a template to produce stable, water-soluble, nanoparticle-free, and fluorescent Ag nanoclusters with $NaBH_4$ as the reducing agent.[39–41] Ag^+ ions have strong interactions with oligonucleotide bases (N3 in the pyrimidines and N7 in the purines) but not with the negatively charged phosphates of the backbone. Upon adding $NaBH_4$ to the mixture of DNA and silver nitrate, small Ag nanoclusters were formed and no nanoparticles were observed. Mass spectral analysis indicated that this approach produced a maximum of 4 Ag atoms in the DNA template. ^1H NMR spectra revealed that peaks corresponding to cytosine H5 and H6 were significantly shifted upfield, indicating that Ag had a high affinity for cytosine bases on single-stranded DNA. Circular dichroism (CD) results also proved that Ag nanoclusters had high affinity with nucleobases, especially with cytosine. As a result, oligonucleotide strands used in the preparation of Ag nanoclusters usually contained a high percentage of cytosines in their sequences. Single-stranded DNA consisting of 12 cytosine bases were found to generate high quantum yield, near-IR emitting Ag nanoclusters (Figure 4.4).[40] The quantum yield of Ag nanoclusters prepared in this way was 17%. Chemical reduction preparations of Ag nanoclusters using DNA as a template exhibit excellent photostability and show essentially no photoblinking on experimentally relevant time scales (0.1 to >1000 ms).

Due to the fact that there were a number of different Ag nanoclusters formed with various numbers of Ag atoms, multiple emitting (red and blue–green) species were detected when the cytosine-rich oligonucleotide was used as a template. When 3 different oligonucleotides (dT_{12}, $dT_4C_4T_4$ and $dC_4T_4C_4$, where T = thymine and C = cytosine) were employed to make fluorescent Ag nanoclusters, both absorption and fluorescence spectral results indicated that thymine-rich oligonucleotides directed the formation of Ag nanoclusters that showed only blue–green emission, whereas

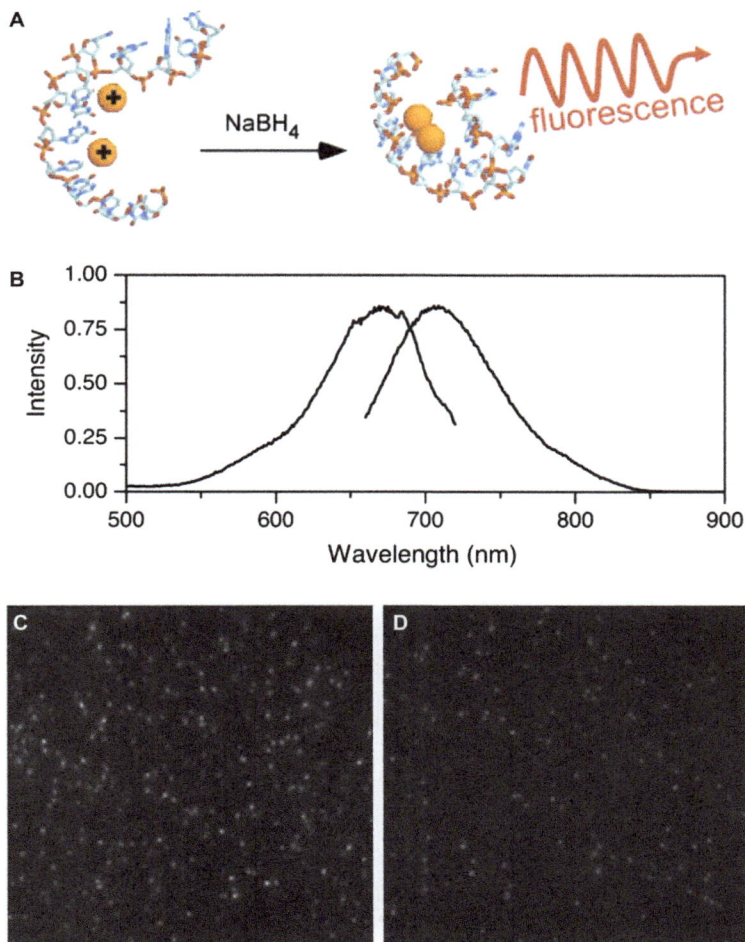

Figure 4.4 (A) Schematic illustration of the formation of Ag nanoclusters. After complexation of Ag^+ to single strand DNA containing 12 cytosines, the mixture was reduced with $NaBH_4$ and the fluorescent Ag nanoclusters formed. (B) Normalized excitation and emission spectra of the Ag nanoclusters. (C) Images of single Ag nanoclusters in a poly(vinyl alcohol) (PVA) film. (D) For comparison, the fluorescent image of single Cy5.29 molecules (a common fluorophore) in a PVA film. The image dimensions are 40 ìm×40 μm, and imaging conditions of (C) and (D) are identical.
Reproduced from ref. 40. Copyright 2007 American Association for the Advancement of Science.

cytosine-rich oligonucleotides led to the formation of both red- and blue–green emitting Ag nanoclusters.[42] This study indicated that base sequence influenced the fluorescence of Ag nanoclusters. Sequence-dependent fluorescence from Ag nanoclusters with different sequences and secondary

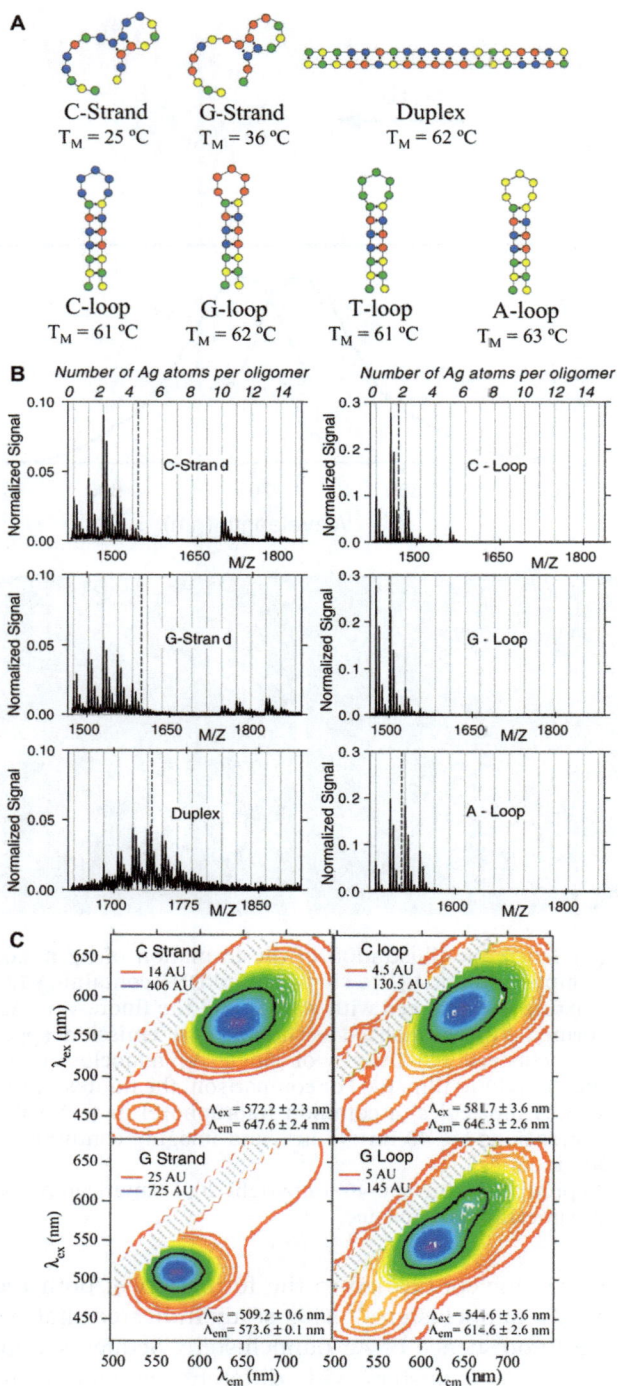

A

C-Strand
$T_M = 25 °C$

G-Strand
$T_M = 36 °C$

Duplex
$T_M = 62 °C$

C-loop
$T_M = 61 °C$

G-loop
$T_M = 62 °C$

T-loop
$T_M = 61 °C$

A-loop
$T_M = 63 °C$

B

Number of Ag atoms per oligomer

Number of Ag atoms per oligomer

C-Strand

C - Loop

G-Strand

G - Loop

Duplex

A - Loop

C

C Strand
— 14 AU
— 406 AU
$\lambda_{ex} = 572.2 ± 2.3$ nm
$\lambda_{em} = 647.6 ± 2.4$ nm

C loop
— 4.5 AU
— 130.5 AU
$\lambda_{ex} = 581.7 ± 3.6$ nm
$\lambda_{em} = 646.3 ± 2.6$ nm

G Strand
— 25 AU
— 725 AU
$\lambda_{ex} = 509.2 ± 0.6$ nm
$\lambda_{em} = 573.6 ± 0.1$ nm

G Loop
— 5 AU
— 145 AU
$\lambda_{ex} = 544.6 ± 3.6$ nm
$\lambda_{em} = 614.6 ± 2.6$ nm

structures of the bases that comprised the DNA strand was further investigated by Gwinn *et al.* using six 19-base DNA oligomers (Figure 4.5).[43] The fluorescence intensities and maximum emission peaks can be finely controlled by the DNA sequences and their secondary structures. Only one emission peak was observed when Ag nanoclusters were synthesized using the smallest hairpin. Other sequences, however, generated multiple emission peaks. Four different types of fluorescent Ag nanoclusters possibly containing different numbers of Ag atoms can be distinguished based on the fluorescence wavelengths and chemical stability. Gwinn *et al.* found that red and green emissions derived from Ag nanoclusters with 13 and 11 atoms respectively when DNA hairpin with 9 cytosines in the loop was used.[43] Microarrays with different DNA sequences were applied to identify the emission properties of Ag nanoclusters generated *in situ*. It was found that sequences indeed can induce the generation of Ag nanoclusters with different emission peaks from visible to near-IR. Therefore, the precise control of the fluorescent properties of Ag nanoclusters may become possible through choice of specifically designed DNA sequences and structures.

Ag nanoclusters were also used to make a fluorescence logic gate with the G-quadruplex in combination with a hairpin and i-motif.[44] The DNA used can fold into a hairpin with a C-loop and it also had several guanines in the stem. Therefore, this DNA can form a G-quadruplex upon the addition of K^+ in the solution. In addition, the DNA strand can generate an i-motif due to the presence of cytosines when the solution pH is lowered. Ag nanoclusters prepared at the hairpin state showed fluorescence peaks at 570 nm and 640 nm. When K^+ was added, only the 570 nm peak was strong. Lowering the solution pH completely quenched the fluorescence, and addition of K^+ and acid led to an emission peak at 601 nm. This ion-tuned fluorescence logic gate based on Ag nanoclusters can be illustrated in Scheme 4.1.[44] The fluorescence properties of Ag nanoclusters were also used to prepare logic devices based on programmable DNA-regulated Ag nanocluster signal transducers.[45] Using the DNA-encoding strategy, aqueous Ag nanoclusters were used as signal transducers to convert DNA inputs into fluorescence outputs for the construction of various DNA-based logic gates (AND, OR, INHIBIT, XOR, NOR, NAND, *etc.*) (Scheme 4.1). Because Ag nanoclusters can interact with many other biomolecules such as peptides, proteins, and RNA and are responsive to external stimuli, this kind of logic device may lead to a new generation of signal transducers and could be used to fabricate a variety of biomolecular logic systems.

Figure 4.5   (A) Schematic representations of the 19-base DNA oligomers used by Gwinn *et al.* Blue = cytosine (C), green = thymine (T), red = guanine (G), and yellow = adenine (A). (B) Mass spectra of the DNA–Ag solutions, for DNA sequences as labeled. (C) Contour maps of fluorescence emission *vs.* excitation from the DNA–Ag solutions. The black contour lies at the half-maximum intensity.
Reproduced from ref. 43. Copyright 2008 John Wiley & Sons.

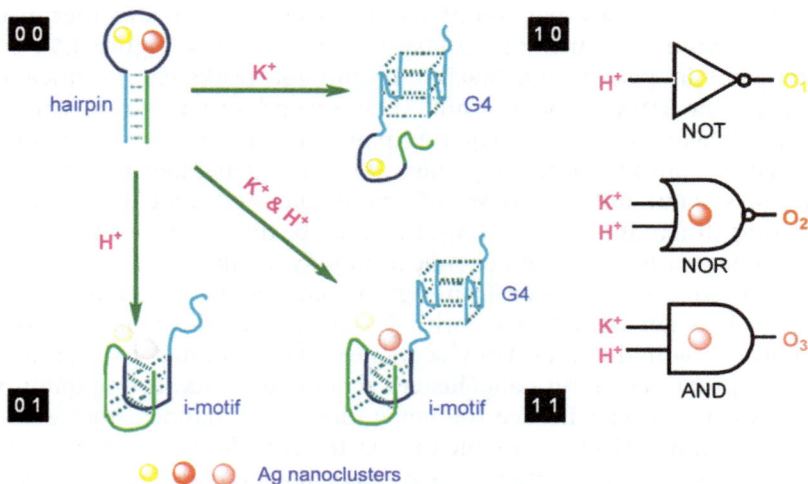

HP26: 5'-GGGTTAGGGTCCCCCCACCCTTACCC-3'

Scheme 4.1   (Left) Schematic chemical diagram of logic operations based on HP26-tuned fluorescent Ag nanoclusters. (Right) Corresponding symbols of logic gates. K^+ and H^+ serve as two independent inputs to trigger the allosterism of HP26 and modulate the fluorescence output. Ag nanoclusters are shown as spheres.
Reproduced with permission from ref. 44. Copyright 2011 American Chemical Society.

Lipoic acid and polyethylene glycol (PEG) modified lipoic acids are another class of versatile ligands for the synthesis of water-soluble fluorescent Ag nanoclusters.[46–48] Lipoic acid is not soluble in aqueous solutions. To make it soluble and stabilize aqueous Ag nanoclusters, $NaBH_4$ was required to reduce lipoic acid to form dihydrolipoic acid, which is soluble in water and can adsorb on the surface of Ag nanoclusters. High-resolution mass spectral study indicated that Ag nanoclusters prepared in this way primarily consisted of Ag_4 and Ag_5. PEG modified lipoid acids showed the capability to control the size of Ag nanoclusters *via* varying the stoichiometry of the Ag^+ to the ligands. Nanoparticles were formed for Ag to ligand ratios between 1000 and 5, while Ag nanoclusters were obtained for Ag to ligand ratios between 1 and 0.1 (Figure 4.6).[48] The unique advantage of using PEG modified ligands was the ability to control the surface functionalities of Ag nanoclusters *via* using different functionalized PEG ligands. For example, carboxylate and amine terminated PEGs can be used during this one phase growth reaction. Surface functionality of Ag nanoclusters is very important because this allows further fine tuning of the reactivity of Ag nanoclusters with target molecules like proteins, peptides, and DNA. One might expand this procedure to potential applications of Ag nanoclusters in biological sciences.

Size focusing or core etching methods using PEG modified lipoic acids as stabilizing agents have also been developed.[48] Unlike other synthetic

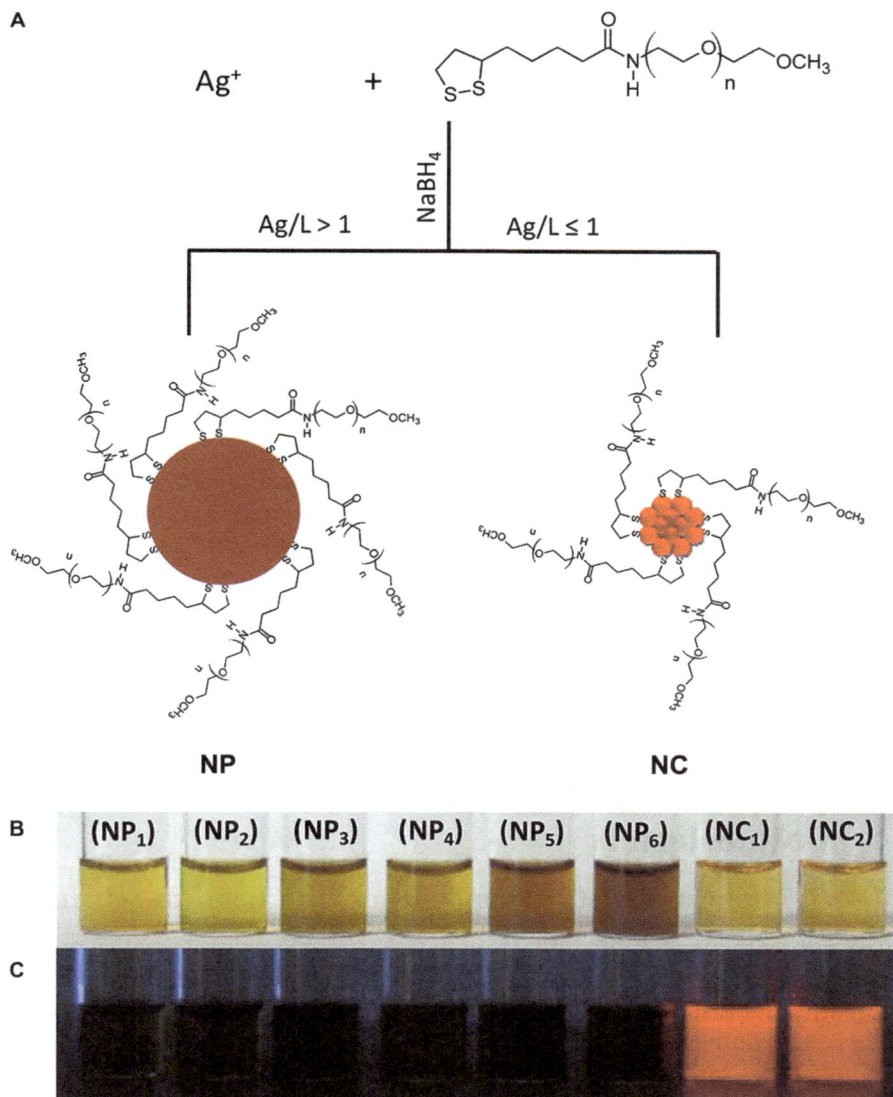

Figure 4.6 (A) Synthetic strategy of Ag nanoparticles and nanoclusters with different core sizes. The Ag to ligand ratios are varied to control the formation of either Ag nanoparticles or nanoclusters. (B) and (C) Photographs of aqueous solutions of Ag nanoparticles and nanoclusters synthesized at various Ag to ligand ratios under room light and UV irradiation, respectively.

Reproduced with permission from ref. 48. Copyright 2012 American Chemical Society.

approaches that all use Ag^+ as the nanocluster precursors, the size focusing method uses pre-formed Ag nanoparticles as precursors. Relatively large amount of ligands were required in this top-down route. TEM images clearly demonstrated that Ag nanoparticles with a rather broad size range (~ 2–7 nm) were reduced in size down to Ag nanoclusters with an average diameter of ~ 1.3 nm. The exact mechanism for the conversion of Ag nanoparticles to Ag nanoclusters was not clear but this process is probably due to etching of Ag nanoparticles with excess free thiols in the solution leading to reduction in size and polydispersity.[48]

Other organic ligands have also been used to synthesize water-soluble fluorescent nanoclusters *via* chemical reduction. For example, a multistage "cyclic reduction under oxidative conditions" (with the dubious acronym, CROC) approach has been applied to prepare Ag nanoclusters with water-soluble chiral thiols (captopril ((2*S*)-1-[(2*S*)-2-methyl-3-sulfanylpropa-noyl]pyrrolidine-2-carboxylic acid), glutathione (GSH = γ-Glu-Cys-Gly), and cysteine) as strong protecting and stabilizing ligands.[49] In this synthetic approach, $NaBH_4$ was used as the reducing agent and hydrogen peroxide was used as an oxidizing agent. Mass spectral analysis indicated that Ag nanoclusters prepared in this way primarily consisted of 22–28 Ag atoms. A protein stabilizer, bovine pancreatic α-chymotrypsin (CHT), was also developed for preparing fluorescent Ag nanoclusters of 1 nm average diameter.[50] In a kinetically controlled manner, Ag nanoclusters with less than 10 Ag atoms were obtained in microemulsions using $NaH_2PO_2 \cdot H_2O$ as a mild reducing agent.[51]

4.3 Applications of Water-Soluble Fluorescent Ag Nanoclusters

4.3.1 Applications of Ag Nanoclusters in Chemical Sensing

The fluorescence of water-soluble Ag nanoclusters is highly sensitive to local changes in their microenvironment, and can be employed to prepare highly sensitive fluorescent chemical and biological sensors in aqueous conditions. There are numerous recent reports on the applications of fluorescent Ag nanoclusters for sensing inorganic ions and biomolecules.

Cu^{2+} is an environmental pollutant and also a key trace element in biological systems. Photochemically produced Ag nanoclusters stabilized by PMAA were first used to detect Cu^{2+}.[52] The limit of detection (LOD, $s/n = 3$) was 8 nM, which is significantly lower than the EPA limit for Cu^{2+} in drinking water. The quenching of fluorescence was due to the complex formation between Cu^{2+} and carboxylic groups around fluorescent Ag nanoclusters. DNA-templated Ag nanoclusters were also developed for quantitative determination of Cu^{2+} in a label free format.[53] Such DNA–Ag nanoclusters showed a maximum fluorescence emission peak at 624 nm that quickly decreases upon the addition of Cu^{2+}. Further experiments indicated that the dominant factor causing the quenching of fluorescence was mainly the metal–metal interaction between Ag and Cu. The detection limit of this method is 10 nM. Cu^{2+} can also be detected using DNA-templated Ag nanoclusters *via* a "turn-on" approach.[54] The introduction of Cu^{2+} resulted in the formation of DNA–Cu/Ag nanoclusters with enhanced fluorescence and much higher quantum yield. Although 3-mercaptopropionic acid (MPA) can quench the fluorescence of Ag nanoclusters, the MPA induced fluorescence quenching of DNA–Cu/Ag nanoclusters was suppressed by Cu^{2+} *via* reduction of thiols to form disulfide which was not able to interact with Ag nanoclusters to quench their fluorescence. This method allowed "turn-on" detection of Cu^{2+} at concentrations as low as 2.7 nM.

Water-soluble fluorescent Ag nanoclusters can also be used as Hg^{2+} sensors due to the strong fluorescence quenching capability of Hg^{2+}.[55,56] The LOD of lipoic acid capped Ag nanoclusters was measured to be 0.1 nM which is substantially lower than the maximum permissible concentration limit of Hg^{2+} in drinking water (10 nM), set by US EPA.[46] The fluorescence quenching of Ag nanoclusters may occur through Hg^{2+} mediated interparticle aggregation. A "turn-on" approach to Hg^{2+} detection has also been developed based on specifically designed DNA making use of the T–T mismatch coordinating capability of Hg^{2+}.[57] A single strand DNA with an inner C-loop was first hybridized to another single strand sequence of DNA with a few T mismatches around the loop. The Ag nanoclusters synthesized using this designed DNA only exhibited weak fluorescence. Upon addition of Hg^{2+}, however, a $T–Hg^{2+}–T$ coordination was formed which increased the duplex stability and changed the environment around the C-loop. This structure change caused dramatic enhancement of fluorescence intensity thereby allowing highly sensitive detection of Hg^{2+} in aqueous solutions.[57]

Selective sensing of Cr^{3+} in aqueous solutions has been reported[37] with Ag nanoclusters prepared using a microwave-assisted method with PMAA as stabilizer. Cr^{3+} readily quenches fluorescent Ag nanoclusters at relatively low concentrations and the detection limit was found to be 28 nM. Anions such as S^{2-} can also be detected using fluorescent DNA-templated Au/Ag nanoclusters with an LOD for sulfide of 0.83 nM.[58]

Most of the sensing applications involving fluorescent Ag nanoclusters have been carried out in aqueous solutions. Test paper based sensing (analogous to pH paper and blood glucose test strips) may be more convenient for practical field applications. PMAA stabilized Ag nanoclusters synthesized by photochemical reduction have been shown to respond to Cu^{2+} in aqueous solutions.[52] Ag nanoclusters can also be immobilized on cellulose filter paper for label-free detection of Cu^{2+}.[59] After binding to cellulose papers, Ag nanoclusters still maintained their fluorescence toward Cu^{2+} and the decrease of fluorescent signals can be easily observed under UV irradiation (Figure 4.7).[59] Such a paper-based sensing platform provides a new route for using water-soluble fluorescent Ag nanoclusters as optical reporters.

4.3.2 Applications of Ag Nanoclusters in Biosensing

Cysteine (Cys), homocysteine (Hcy), and glutathione (GSH) are important small thiol-containing biomolecules. The thiol groups not only can stabilize

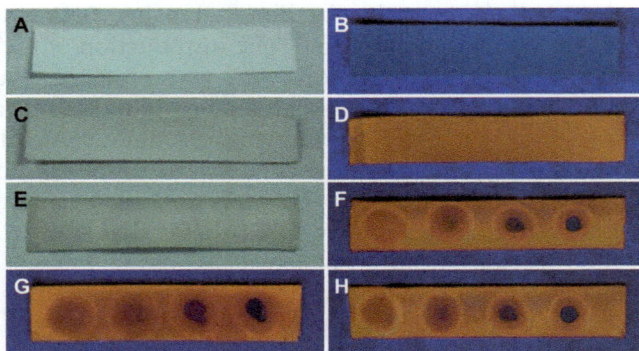

Figure 4.7 Ag nanoclusters on a paper platform for detection of Cu^{2+}: comparisons of the paper strips before (A and B) and after (C and D) immobilization of Ag nanoclusters. The photographs of the test papers after adding different concentrations of Cu^{2+} in (E and F) deionized water, (G) river water, and (H) barreled drinking water, from left to right: 0, 2×10^{-5}, 10^{-4}, 10^{-3} M. (A, C and E) and (B, D and F) are the same strips under white light and UV light illumination, respectively.
Reproduced with permission from ref. 59. Copyright 2012 Royal Society of Chemistry.

Ag nanoclusters but can also induce the oxidation of Ag nanoclusters, which leads to a decrease of fluorescence intensity.[60] A simple fluorescence quenching method was developed for detection of cysteine based on PMAA stabilized Ag nanoclusters.[61] The binding of Cys with Ag nanoclusters can readily induce the quenching of Ag nanoclusters while other α-amino acids do not. This method allowed Ag nanoclusters to selectively detect Cys with a LOD of 20 nM.[61] Due to steric hindrance, large thiol-containing compounds like bovine serum albumin (BSA) and GSH cannot efficiently quench the fluorescence of Ag nanoclusters.

Although the fluorescence of Ag nanoclusters can be efficiently quenched by thiol-containing compounds, in some cases thiols can be used to enhance the fluorescence of Ag nanoclusters. A "turn-on" assay was introduced to selectively detect thiol-containing compounds.[62] For example, the fluorescence intensity of Ag nanoclusters prepared with single strand DNA consisting of 12 cytosines can be enhanced in the presence of thiol-containing compounds. This "turn-on" approach was able to detect GSH with a LOD at 6.2 nM.[62]

The sequence of DNA can greatly influence the fluorescence of Ag nanoclusters, which permits their broad use for detection of specific DNA sequences, even to a single mutated nucleotide in a double stranded DNA. For example, if a nucleotide mismatch occurred two bases away from the nanocluster formation site (C_6 loop in the duplex), the fluorescence of Ag nanoclusters is quenched.[63] The location of the C_6 loop can be used to tune the fluorescence of Ag nanoclusters: locating the C_6-loop one or two bases away from the mutation site produced the most intense fluorescence, whereas moving the C_6-loop further from the mutation site, the fluorescence intensity became much weaker.

The fluorescence of DNA-templated Ag nanoclusters can be enhanced in the presence of guanine-rich DNA sequences.[64] This interesting phenomenon can also be exploited for DNA sensing (Figure 4.8).[64] The remaining segment of strand 1 bearing dark (non-fluorescent) Ag nanoclusters was used as a hybridization sequence to pair another single strand. After a guanine-rich complement tail hybridized with strand 1, an intense red emission fluorescence was observed, with emission intensity enhanced up to 500-fold and the enhancement exponentially increased with increasing number of guanine bases in proximity to the Ag nanoclusters. This type of Ag nanocluster can detect target DNA sequences with an exceptionally high signal-to-noise ratio of 175.

Ag nanoclusters have also been developed as fluorescent protein sensors. Aptamer-templated Ag nanoclusters have been synthesized and used for

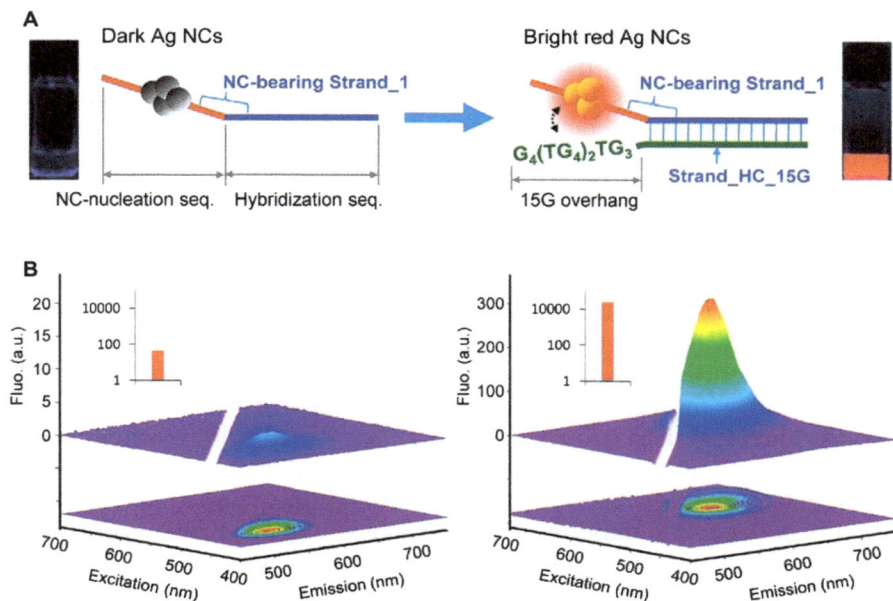

Figure 4.8   (A) Schematic showing the red fluorescence enhancement of DNA–Ag nanoclusters through proximity with a G-rich overhang, $3'$-G4(TG4)TG3, caused by DNA hybridization and photographs of the resulting emission under 366 nm UV irradiation. (B) 3D- and 2D-contour plots of excitation/emission spectra of the Ag nanoclusters before (left) and after (right) hybridizing nanocluster-bearing Strand 1 with Strand HC–15G. Inset: Integrated red fluorescence emission subtracted with buffer fluorescence in arbitrary units, which indicates an enhancement of $\sim$500-fold after duplex formation.

Reproduced with permission from ref. 64. Copyright 2010 American Chemical Society.

selective detection of thrombin.[65] A specific DAN scaffold containing a cytosine-rich DNA sequence (12-mer) was used to stabilize fluorescent Ag nanoclusters, and a thrombin-binding aptamer sequence (29-mer) was added to that to recognize thrombin. The fluorescence of the resulting Ag nanoclusters can be quenched *via* target binding of thrombin to the DNA scaffold. The detection limit of thrombin in this method was measured to be 1 nM. DNA-templated highly fluorescent (quantum yield >50%) Ag nanoclusters can be used to detect single-strand DNA binding proteins (SSBs).[66] When SSBs interacted with DNA-templated Ag nanoclusters, the DNA scaffold altered their configurations leading to the quenching of fluorescence, with a resulting LOD of 0.2 nM. This method allowed selective detection of SSBs over other tested proteins, including trypsin, lysozyme, myoglobin, BSA, thrombin, *etc.*

4.4   Conclusions

Water-soluble fluorescent Ag nanoclusters have now emerged as an important class of luminescent nanomaterials. Recent advances in the synthesis of fluorescent Ag nanoclusters *via* various synthetic methodologies have provided insights into the origins of their excellent photophysical properties as well as the development of applications in chemical- and bio-sensing, bio-imaging, and single molecule microscopy. Ag nanoclusters are among the strongest fluorescent emitters, exhibiting high quantum yields, high emission rates, large Stokes shifts, high molecular extinction coefficients, and excellent photostability. Their fluorescent properties are controlled through the details of their synthesis and stabilizing surface bound polyelectrolytes and are sensitive to the surrounding conditions.

It would be highly desirable to correlate the size of Ag nanoclusters with their corresponding optical properties. Current synthetic approaches cannot precisely allow production of Ag nanoclusters with specific numbers of Ag atoms. Therefore, it is still challenging to develop efficient synthetic routes to prepare truly monodispersed, highly emissive Ag nanoclusters. Although DNA, polyelectrolytes, proteins, and thiol-containing compounds have been used to protect and stabilize Ag nanoclusters, a general understanding of how such surface modifications affect the chemical and physical properties of Ag nanoclusters is still incomplete. Much work remains to be done before we will be able to achieve control over size and surface chemistry of each individual Ag nanocluster.

Another area where Ag nanoclusters may find useful future applications is catalysis. Ag nanoparticles have already been used to catalyze a variety of oxidation reactions[67–69] and Ag nanoclusters deposited on alumina have been demonstrated to catalyze the reduction of various nitro aromatic compounds with good recyclability.[70] Further applications of water-soluble Ag nanoclusters as catalysts in aqueous solutions are likely to continue and expand, and it will be interesting to see if the fluorescent properties of Ag nanoclusters prove useful as mechanistic probes during such catalysis.

References

1. J. Zheng, P. R. Nicovich and R. M. Dickson, *Annu. Rev. Phys. Chem.*, 2007, **58**, 409.
2. J. P. Wilcoxon and B. L. Abrams, *Chem. Soc. Rev.*, 2006, **35**, 1162.
3. I. Diez and R. H. A. Ras, *Nanoscale*, 2011, **3**, 1963.
4. Y. Z. Lu and W. Chen, *Chem. Soc. Rev.*, 2012, **41**, 3594.
5. L. Shang, S. J. Dong and G. U. Nienhaus, *Nano Today*, 2011, **6**, 401.
6. S. Choi, R. M. Dickson and J. H. Yu, *Chem. Soc. Rev.*, 2012, **41**, 1867.
7. Y. C. Shiang, C. C. Huang, W. Y. Chen, P. C. Chen and H. T. Chang, *J. Mater. Chem.*, 2012, **22**, 12972.
8. A. Latorre and A. Somoza, *ChemBioChem*, 2012, **13**, 951.
9. H. X. Xu and K. S. Suslick, *Adv. Mater.*, 2010, **22**, 1078.
10. D. N. Rogers, *Chemistry of Photography*, The Royal Society of Chemistry, Cambridge, 2006.
11. http://www.cheresources.com/content/articles/other-topics/chemistry-of-photography.
12. S. Fedrigo, W. Harbich and J. Buttet, *J. Chem. Phys.*, 1993, **99**, 5712.
13. W. Schulze, I. Rabin and G. Ertl, *ChemPhysChem*, 2004, **5**, 403.
14. A. Henglein, *Chem. Phys. Lett.*, 1989, **154**, 473.
15. T. Linnert, P. Mulvaney, A. Henglein and H. Weller, *J. Am. Chem. Soc.*, 1990, **112**, 4657.
16. A. Henglein, *J. Phys. Chem.*, 1993, **97**, 5457.
17. B. G. Ershov and A. Henglein, *J. Phys. Chem. B*, 1998, **102**, 10663.
18. P. Mulvaney and A. Henglein, *Chem. Phys. Lett.*, 1990, **168**, 391.
19. A. Henglein, P. Mulvaney and T. Linnert, *Faraday Discuss.*, 1991, **92**, 31.
20. M. Treguer, F. Rocco, G. Lelong, A. Le Nestour, T. Cardinal, A. Maali and B. Lounis, *Solid State Sci.*, 2005, 7, 812.
21. L. A. Peyser, A. E. Vinson, A. P. Bartko and R. M. Dickson, *Science*, 2001, **291**, 103.
22. J. Zheng and R. M. Dickson, *J. Am. Chem. Soc.*, 2002, **124**, 13982.
23. L. Shang and S. J. Dong, *Chem. Commun.*, 2008, 1088.
24. I. Diez, M. Pusa, S. Kulmala, H. Jiang, A. Walther, A. S. Goldmann, A. H. E. Muller, O. Ikkala and R. H. A. Ras, *Angew. Chem., Int. Ed.*, 2009, **48**, 2122.
25. H. Zhang, X. Huang, L. Li, G. W. Zhang, I. Hussain, Z. Li and B. Tian, *Chem. Commun.*, 2012, **48**, 567.
26. J. G. Zhang, S. Q. Xu and E. Kumacheva, *Adv. Mater.*, 2005, **17**, 2336.
27. Z. Shen, H. W. Duan and H. Frey, *Adv. Mater.*, 2007, **19**, 349.
28. P. Riesz and T. Kondo, *Free Radical Biol. Med.*, 1992, **13**, 247.
29. K. S. Suslick, Y. Didenko, M. M. Fang, T. Hyeon, K. J. Kolbeck, W. B. McNamara, M. M. Mdleleni and M. Wong, *Philos. Trans. R. Soc., A*, 1999, **357**, 335.
30. E. Ciawi, J. Rae, M. Ashokkumar and F. Grieser, *J. Phys. Chem. B*, 2006, **110**, 13656.
31. J. H. Bang and K. S. Suslick, *Adv. Mater.*, 2010, **22**, 1039.

32. H. X. Xu, B. W. Zeiger and K. S. Suslick, *Chem. Soc. Rev.*, 2013, **42**, 2555.
33. K. S. Suslick and D. J. Flannigan, *Annu. Rev. Phys. Chem.*, 2008, **59**, 659.
34. H. X. Xu and K. S. Suslick, *ACS Nano*, 2010, **4**, 3209.
35. A. Hoz, A. Díaz-Ortiz and A. Moreno, *Chem. Soc. Rev.*, 2005, **34**, 164.
36. R. Gedye, F. Smith, K. Westaway, H. Ali, L. Baldisera, L. Laberge and J. Rousell, *Tetrahedron Lett.*, 1986, **27**, 279.
37. S. H. Liu, F. Lu and J. J. Zhu, *Chem. Commun.*, 2011, **47**, 2661.
38. R. Q. Li, C. L. Wang, F. Bo, Z. Y. Wang, H. B. Shao, S. H. Xu and Y. P. Cui, *ChemPhysChem*, 2012, **13**, 2097.
39. J. T. Petty, J. Zheng, N. V. Hud and R. M. Dickson, *J. Am. Chem. Soc.*, 2004, **126**, 5207.
40. T. Vosch, Y. Antoku, J. C. Hsiang, C. I. Richards, J. I. Gonzalez and R. M. Dickson, *Proc. Natl. Acad. Sci. U. S. A.*, 2007, **104**, 12616.
41. C. M. Ritchie, K. R. Johnsen, J. R. Kiser, Y. Antoku, R. M. Dickson and J. T. Petty, *J. Phys. Chem. C*, 2007, **111**, 175.
42. B. Sengupta, C. M. Ritchie, J. G. Buckman, K. R. Johnsen, P. M. Goodwin and J. T. Petty, *J. Phys. Chem. C*, 2008, **112**, 18776.
43. E. G. Gwinn, P. O'Neill, A. J. Guerrero, D. Bouwmeester and D. K. Fygenson, *Adv. Mater.*, 2008, **20**, 279.
44. T. Li, L. Zhang, J. Ai, S. Dong and E. Wang, *ACS Nano*, 2011, **5**, 6334.
45. Z. Z. Huang, Y. Tao, F. Pu, J. S. Ren and X. G. Qu, *Chem. – Eur. J*, 2012, **18**, 6663.
46. B. Adhikari and A. Banerjee, *Chem. Mater.*, 2010, **22**, 4364.
47. P. T. K. Chin, M. van der Linden, E. J. Harten, A. Barendregt, M. T. M. Rood, A. J. Koster, F. W. B. Leeuwen, C. M. Donega, A. J. R. Heck and A. Meijerink, *Nanotechnology*, 2013, **24**, 075703.
48. M. A. H. Muhanmmed, F. Aldeek, G. Palui, L. Trapiella-Alfonso and H. Mattoussi, *ACS Nano*, 2012, **6**, 8950.
49. N. Cathcart, P. Mistry, C. Makra, B. Pietrobon, N. Coombs, M. Kelokhani-Niaraki and V. Kitaev, *Langmuir*, 2009, **25**, 5840.
50. S. S. Narayanan and S. K. Pal, *J. Phys. Chem. C*, 2008, **112**, 4874.
51. A. Ledo-Suarez, J. Rivas, C. F. Rodriguez-Abreu, M. J. Rodriguez, E. Pastor, A. Hernandez-Creus, S. B. Oseroff and M. A. Lopez-Quintela, *Angew. Chem., Int. Ed.*, 2007, **46**, 8823.
52. L. Shang and S. J. Dong, *J. Mater. Chem.*, 2008, **18**, 4636.
53. M. Zhang and B. C. Ye, *Analyst*, 2011, **136**, 5139.
54. Y. T. Su, G. Y. Lan, W. Y. Chen and H. T. Chang, *Anal. Chem.*, 2010, **82**, 8566.
55. W. Guo, J. Yuan and E. Wang, *Chem. Commun.*, 2009, 3395.
56. G. Y. Lan, W. Y. Chen and H. T. Chan, *RSC Adv*, 2011, **1**, 802.
57. L. Deng, Z. Zhou, J. Li, T. Li and S. Dong, *Chem. Commun.*, 2011, **47**, 11065.
58. W. Y. Chen, G. Y. Lan and H. T. Chan, *Anal. Chem.*, 2011, **83**, 9450.
59. X. J. Liu, C. H. Zong and L. H. Lu, *Analyst*, 2012, **137**, 2406.
60. B. Y. Han and E. Wang, *Biosens. Bioelectron.*, 2011, **26**, 2585.
61. L. Shang and S. J. Dong, *Biosens. Bioelectron.*, 2009, **24**, 1569.

62. Z. Z. Huang, F. Pu, Y. Lin, J. S. Ren and X. Qu, *Chem. Commun.*, 2011, **47**, 3487.
63. W. Guo, J. Yuan, Q. Dong and E. Wang, *J. Am. Chem. Soc.*, 2010, **132**, 932.
64. H. C. Yeh, J. Sharma, J. J. Han, J. S. Martinez and J. H. Werner, *Nano Lett.*, 2010, **10**, 3106.
65. J. Sharma, H. C. Yeh, H. Yoo, J. H. Werner and J. S. Martinez, *Chem. Commun.*, 2011, **47**, 2294.
66. G. Y. Lan, W. Y. Chen and H. T. Chang, *Analyst*, 2011, **136**, 3623.
67. K. Watanabe, D. Menzel, N. Nilius and H. J. Freund, *Chem. Rev.*, 2006, **106**, 4301.
68. Y. Y. Chen, C. Wang, H. Y. Liu, J. S. Qiu and X. H. Bao, *Chem. Commun.*, 2005, 5298.
69. T. Mitsudome, S. Arita, H. Mori, T. Mizugaki, K. Jitsukawa and K. Kaneda, *Angew. Chem., Int. Ed.*, 2008, **47**, 138.
70. A. Leelavathi, T. U. B. Rao and T. Pradeep, *Nanoscale Res. Lett.*, 2011, **6**, 1.

CHAPTER 5

Synthesis and Applications of Silver Nanoclusters Protected by Polymers, Protein, Peptide and Short Molecules

X. YANG[a,b] AND E. K. WANG[*a]

[a] State Key Laboratory of Electroanalytical Chemistry, Changchun Institute of Applied Chemistry, Chinese Academy of Sciences, Changchun, 130022, Jilin, P. R. China; [b] University of Chinese Academy of Sciences, Beijing, 100039, P. R. China
*Email: ekwang@ciac.jl.cn

5.1   Introduction

The physical and chemical properties of silver are highly dependent on their size, in particular in the nanometer range. Bulk silver is electrically conducting and a good optical reflector due to the sea of freely moving delocalized electrons in the conduction band. On the other hand, silver nanoparticles display various colours due to surface plasmon resonance. Silver nanoclusters (Ag NCs) have attracted much research interest because of their unique size-dependent optical, electronic, magnetic, and catalytic properties that bridge the gap between small molecules (*e.g.* organometallic compounds) and bulk crystals (diameter typically >2 nm). As the size of silver nanoparticles decreases to the Fermi wavelength of an electron,

RSC Smart Materials No. 7
Functional Nanometer-Sized Clusters of Transition Metals: Synthesis, Properties and Applications
Edited by Wei Chen and Shaowei Chen
© The Royal Society of Chemistry 2014
Published by the Royal Society of Chemistry, www.rsc.org

discrete energy levels begin to form, and interactions with light lead to electronic transitions between different energy levels and strong photoluminescence. In fact, various studies have shown luminescent Ag NCs, which also demonstrate unique catalytic activity that varies strongly with the size of the nanoparticles. For instance, recent studies have demonstrated that Ag NCs might combine with oxygen to form a Ag NC and oxygen complex due to spin accommodation, which may be exploited for the catalysis of oxygen reduction reactions.[1]

Despite having excellent optical, electronic, magnetic, and catalytic properties, Ag NCs without stabilization would strongly interact with each other and aggregate irreversibly so as to reduce their surface energy. Therefore, a proper stabilizing scaffold is indispensable. Over the years, various scaffolds have been utilized for the stabilization of Ag NCs, such as cryogenic noble gas matrices, zeolites, polymers, DNA, peptides and short molecules. Cryogenic noble gas matrices were used for detailed spectroscopic study of Ag NCs in the 1970s and zeolites have been used as scaffolds for the synthesis of Ag NCs for decades. Recently, different organic scaffolds have been developed for the synthesis of luminescent Ag NCs since the first report by Dickson's group in 2002.[2] Due to the well adjusted interaction between ligands and Ag NCs and the big potential of organic scaffolds in Ag NC synthesis, numerous organic scaffolds and approaches have been utilized for the synthesis of Ag NCs such as polymers, DNA, proteins, and even simple molecules.

Various applications of Ag NCs have been developed based on their size-dependent optical, electronic, magnetic, and catalytic properties. As the size of Ag NCs is small enough, discrete energy levels begin to form, whereby interactions with light lead to electronic transitions between different energy levels and hence strong photoluminescence. Ag NCs have been utilized as substitutes for quantum dots for bioimaging and nanosensing for the detection of different targets such as metal ions, small biomolecules, proteins, nucleic acids, *etc.* Also, the excellent catalytic properties of Ag NCs allow their potential application as catalysts for various chemistry and electrochemistry reactions.

In this chapter, we will first introduce the synthesis and properties of Ag NCs protected by polymers, and highlight recent advances in their related applications in nanosensing, cell imaging and catalysis. The synthesis approaches, properties and related applications of Ag NCs protected by proteins and peptides will be summarized subsequently. As polymers, proteins and peptides are all constructed of short molecules, Ag NCs protected by short molecules are very important in the detailed study of polymer, protein and peptide-protected Ag NCs. Thus, we will focus on the synthesis approaches, properties and related applications of Ag NCs protected by short molecules afterwards. In the final section, we will give a brief conclusion and discuss some of the challenges that researchers currently face in their research on Ag NCs.

5.2 Polymers Protected Silver Nanoclusters and Their Related Applications

Zheng and Dickson reported the first stable aqueous solution of fluorescent Ag NCs in 2002.[2] OH-terminated dendrimers PAMAM G4 and G2 (fourth- and second-generation) were used as scaffolds to host, stabilize and solubilize Ag NCs in both aerated and deaerated solutions (Figure 5.1). By dissolving 0.5 μmol of PAMAM G4 and 1.5 μmol of AgNO$_3$ into 1 mL of distilled water (18 MΩ) and adjusting to neutrality with acetic acid, silver ions readily interacted with the dendrimer. To create dendrimer-encapsulated Ag NCs, no reducing agents were added to the reactions. The detailed structures of thus formed Ag NCs were further determined by electrospray ionization mass spectroscopy (ESI-MS), and the ESI-MS results showed that the Ag NCs were mainly constructed of four species such as [PAMAM + Ag], [PAMAM + Ag$_2$], [PAMAM + Ag$_3$], and [PAMAM + Ag$_4$]. The Ag NCs in the dendrimers exhibited distinct fluorescence peaks ranging from 533 to 648 nm with high water solubility and photostability. Therefore, the Ag NCs protected by PAMAM could be suitable labels to study chemical and biological systems.[2] Balogh and coworkers also utilized PAMAM G5 to prepare aqueous silver(I)-dendrimer complexes (with a molar ratio of 25 Ag$^+$ per dendrimer) at a

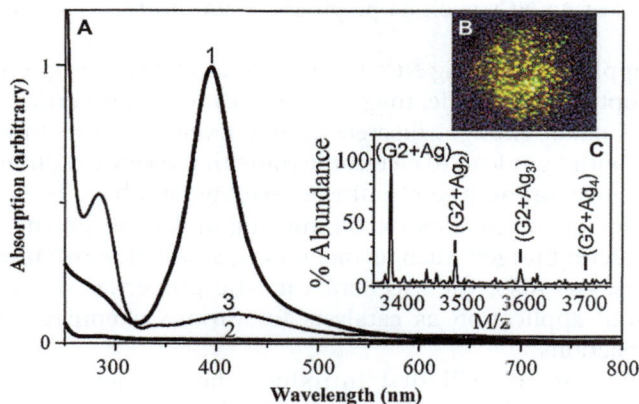

Figure 5.1 (A) UV-vis spectra of aqueous Ag–dendrimer solutions. (1) Strong plasmon absorption (398 nm) characteristic of large, nonfluorescent dendrimer-encapsulated silver nanoparticles prepared through NaBH$_4$ reduction of silver ions in the dendrimer host (1 : 12 dendrimer:Ag), (2) absorption spectrum of nonfluorescent 1 : 3 (dendrimer : Ag) solution before photoactivation, and (3) the sample solution after photoacivation to yield highly fluorescent silver nanodots. (B) Fluorescence image from solution 3 (250×, 1 s exposure on Kodak DCS620X colour digital camera, 476 nm excitation). (C) Electrospray ionization mass spectrum of photoactivated G2-OH PAMAM (MW: 3272 amu)–AgNO$_3$ solution. Ag$_n$ nanodot peaks are spaced by the Ag atomic mass (107.9 amu) and only appear in the fluorescent, photoactivated nanodot solutions (reprinted with permission from ref. 2).

biologic pH of 7.4. Conversion of silver(I)-dendrimer complexes into dendrimer templated Ag NCs was achieved by irradiating the solutions with UV light to reduce the bound Ag^+ cations to zerovalent Ag^0 atoms, which were simultaneously trapped in the dendrimer network, resulting in the formation of $\{(Ag^0)_{25}$-PAMAM_E5.NH$_2\}$, $\{(Ag^0)_{25}$-PAMAM_E5.NGly$\}$, and $\{(Ag^0)_{25}$-PAMAM_E5.NSAH$\}$ dendrimer nanocomposites, respectively. The cytotoxicity of dendrimers and related Ag NCs was evaluated using an XTT colorimetric assay of cellular viability and the cellular uptake of Ag NCs was determined by transmission electron and confocal microscopy. The results indicated that $\{(Ag^0)_{25}$-PAMAM_E5.NH$_2\}$, $\{(Ag^0)_{25}$-PAMAM_E5.NGly$\}$, and $\{(Ag^0)_{25}$-PAMAM_E5.NSAH$\}$ formed primarily single particles with diameters between 3 and 7 nm. The dendrimer-protected Ag NCs were fluorescent, and their surface charge, cellular internalization, toxicity, and cell labelling capabilities were determined by the surface functionalities of dendrimer templates which allowed potential application in cell biomarkers.[3]

In addition to dendrimers, there were several other groups devoted to the synthesis of metal nanoparticles in polymer templates in the absence of typical chemical reducing agents. Kumacheva *et al.* used poly(*N*-isopropylacrylamide-acrylic acid-2-hydroxyethyl acrylate) microgel particles as a scaffold to synthesize luminescent Ag NCs (Figure 5.2). The pH of the microgel dispersion was adjusted to the required value by adding 0.1 M solution of NaOH or HCl. A freshly prepared aqueous 0.2 M $AgNO_3$ solution was then added to the microgel dispersion in a molar ratio $[Ag^+]$: $[COOH]$ of 1 : 1. The system was mixed for *ca.* 30 min, dialyzed against deionized water, and subjected to UV-irradiation at 365 nm for various time intervals. The luminescence emissions of Ag NCs could be easily adjusted from 530 nm to 600 nm by changing the pH value of the microgel dispersion. Since the microgel served as the template for the synthesis of Ag NCs allowing for the controlled nucleation and growth of very small NCs and protecting these Ag NCs from interactions with photoluminescence quenchers in the bulk dispersion, such luminescent microgels held great potential for the typical polymer microgels such as drug delivery, fabrications of photonic crystals and chemical and biosensing.[4]

Shang and Dong reported the synthesis of luminescent and water soluble Ag NCs using a simple anionic polyelectrolyte such as poly(methacrylic acid) (PMAA) as the template. The common polyelectrolyte PMAA could function as an ideal template to generate highly luminescent Ag NCs; and compared with previous reports, PMAA offered several crucial advantages as the template. PMAA carried carboxylic acid groups capable of coordinating with Ag^+ ions; PMAA chains at low ionization exhibited the necessary hydrophobicity due to the presence of methyl groups in the side chain, and hydrophobic regions facilitated the formation of Ag NCs; PMAA responded reversibly to the variations of environmental pH, ionic strength and temperature, which could extend the application of the emissive nanoclusters into areas such as biomedicine. The luminescence quantum yield of such PMAA-protected Ag NCs was about 18.6%. The luminescence of PMAA templated Ag NCs was

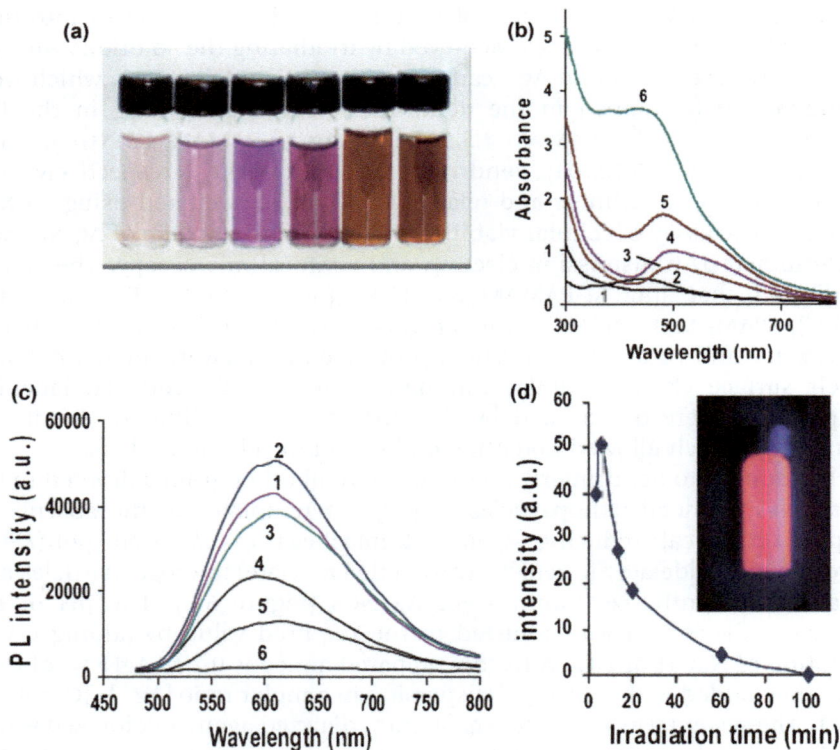

Figure 5.2 Typical appearance (a) and UV-vis spectra (b) of poly(NIPAM-AA-HEA) microgel dispersions after mixing with Ag$^+$ ions and UV-irradiating them for different time intervals, in minutes: (1) 3; (2) 6; (3) 13; (4) 20; (5) 40; (6) 100. In (a), dispersions from left to right correspond to spectra (1)–(6); in (b) $\lambda_{irr} = 365$ nm, pH 8.14. (c) Evolution of PL spectra and (d) PL intensity of Ag nanoclusters photogenerated in microgels. UV-irradiation time in minutes: (1) 3, (2) 6, (3) 13, (4) 20, (5) 40, (6) 100; pH 8.14, $\lambda_{irr} = 365$ nm, $\lambda_{ex} = 450$ nm. The inset shows a photoluminescent hybrid microgel obtained after 10 min UV irradiation (reprinted with permission from ref. 4).

found to be quenched effectively by Cu^{2+}, but not by other common metal ions.[5] By virtue of the specific response toward the analytes, Shang and Dong developed a novel and simple luminescent method for the detection of Cu^{2+} with a low detection limit of 8 nM. Such water-soluble Ag NCs were also utilized for the detection of cysteine based on their novel photoluminescence properties. The luminescence of PMAA templated Ag NCs was found to be quenched effectively by cysteine, but not when the other alpha-amino acids were present. By virtue of the specific response, a new, simple and sensitive luminescent assay for the detection of cysteine was developed with a detection limit of 20 nM.[6] Ras *et al.* also used PMAA as scaffolds to synthesize luminescent Ag NCs through a photoreduction approach.

Visible light, instead of ultraviolet or γ irradiation, was used to initiate the reduction of silver forming Ag NCs in the solutions. The luminescence emissions could be adjusted by changing the molar ratio of Ag^+ and MAA, and in different water–methanol mixtures. The detailed structures of PMAA-protected Ag NCs were further determined by using matrix-assisted laser desorption time of flight mass spectroscopy (MALDI-TOF-MS). The MALDI-TOF MS results showed that such synthesized Ag NCs mostly consisted of by Ag_2 and Ag_3. Most importantly, this was the first report of metal nanoclusters that exhibited electrochemiluminescence (ECL). The large tuneable shift in the optical characteristics, the high photoluminescence quantum yield and the ECL might confer remarkable advantages to the Ag NCs over large nanoparticles in applications such molecular sensing.[7] The absorption properties of different hybrid systems consisting of Ag_1, Ag_2, or Ag_3 atomic clusters and PMAA were calculated using time-dependent density-functional theory (TDDFT) by the Hancock group. The polymer was found to have an extensive structural-dependency on the spectral patterns of the hybrid systems relative to the bare clusters. The absorption spectrum could be "tuned" to the visible range for hybrid systems with an odd number of electrons per silver cluster, whereas for hybrid systems comprising an even number of electrons per silver cluster, the leading absorption edge could be shifted up to 4.5 eV. The results gave theoretical support to the experimental observations on the absorption in the visible range in metal cluster–polymer hybrid structures.[8] Xu and Suslick reported a convenient sonochemical approach to the synthesis of stable, water-soluble, luminescent Ag NCs using PMAA as the capping agent. The Ag NCs displayed a maximum luminescence emission at 610 nm upon excitation at 510 nm, and the optical and luminescence properties of the Ag NCs could be easily controlled by varying the synthetic conditions, such as sonication time, stoichiometry of the carboxylate groups to Ag^+, and polymer molecular weight.[9]

Xu and coworkers developed a new solid template with a hydrophilic surface and a hydrophobic core to load Ag^+ ions. The core/shell structure as the template was similar to PAMAM dendrimers and could result in a "cage effect", which provided enhanced protection capability to Ag NCs and prevented the continued growth of nanoclusters to nanoparticles. The detailed structures of Ag NCs were determined by laser desorption time of flight mass spectroscopies (LDI-TOF-MS) and the LDI-TOF-MS results showed that the Ag NCs mainly consisted of Ag_1, Ag_2, Ag_3, Ag_4, Ag_5 and Ag_6. The Ag NCs displayed a luminescence emission at about 560 nm with a quantum yield of 6% and lifetime of 0.460 ns. Because of the protective structure of the three-dimensional network of the polymer template, the Ag NCs had such high stability that there were little changes in the fluorescence spectra of Ag NCs even after nine-month storage in the dark. And this highly stable fluorescent Ag NC had successfully been applied for bioimaging.[10] Zhu *et al.* synthesized highly luminescent Ag NCs with a microwave-assisted method. In the study, a mixed aqueous solution of PMAA sodium salt was kept under microwave irradiation for 70 s. The water-soluble Ag NCs obtained through the rapid and green microwave-assisted process exhibited strong luminescence, and could be used as a novel luminescence probe for the detection of Cr^{3+} ions with high sensitivity and selectivity.[11]

Mattoussi and coworkers recently developed a one phase growth reaction to prepare a series of silver nanoparticles and luminescent Ag NCs using $NaBH_4$ reduction of silver nitrate in the presence of molecular scale ligands made of polyethylene glycol (PEG) appended with lipoic acid (LA) groups at one end and reactive (-COOH/-NH$_2$) or insert (-OCH$_3$) functional groups at the other. The PEG segment in the ligand promoted solubility in a variety of solvents including water, while LAs provided multidentate coordinating groups that promoted Ag-ligand complex formation and strong anchoring onto the nanoparticle and nanocluster surface. The size and properties of Ag NCs were primarily controlled by varying the Ag-to-ligand (Ag : L) molar ratios and the molar amount of $NaBH_4$ used. Luminescent Ag NCs formed at lower molar ratios of Ag : L, while nonluminescent silver nanoparticles were produced at higher molar ratios. Interestingly, nonluminscent silver nano-particles could be converted into luminescent Ag NCs, *via* a process referred to as "size focusing", in the presence of added excess ligands and reducing agent. Such synthesized Ag NCs emitted in the far red region at about 680 nm with a quantum yield of *ca.* 12% and could be redispersed in a number of solvents with varying polarity while maintaining their optical and spectroscopic properties.[12]

Another type of polyelectrolyte polyethyleneimine (PEI) was utilized as scaffold for the synthesis of luminescent Ag NCs. Different from previous polymer templated Ag NCs, PEI capped Ag NCs showed high stability which made them stable at room temperature for at least 30 days due to the strong interaction between N atoms and Ag atoms. The PEI-protected Ag NCs showed a blue emission at 455 nm. Luo *et al.* utilized PEI capped Ag NCs to develop a fluorometric method for the determination of hydrogen peroxide and glucose at high sensitivity. The luminescence of the PEI templated Ag NCs could be particularly quenched by H_2O_2. The oxidization of glucose by glucose oxidase coupled with the luminescence quenching of PEI-protected Ag NCs by H_2O_2 can be used to detect glucose. The method has been used for the detection of glucose in human serum samples with satisfactory results.[13]

Recently, Ershov and Henglein found that Ag NCs containing several atoms could be obtained in the presence of polyphosphate or polyacrylate in aqueous solution during γ-ray irradiation. The synthesis process was very simple. Typically, an aqueous solution of a silver salt (*e.g.*, $AgClO_4$) and a polymer (sodium polyacrylate) were deaerated by bubbling the solution with argon. The mixed solution was then subjected to γ-ray irradiation from a ^{60}Co source. At the same time, the solutions also contained 2-propanol to scavenge the hydroxyl radicals that were generated during the radiolysis of the aqueous solvent. However, the Ag NCs prepared in this way were short-lived and aggregated to form large nanoparticles after several hours.[14]

Polymers could also be used as stabilizers rather than templates for Ag NCs. Pradeep and coworkers found that the stability of bovine serum albumin

(BSA) templated Ag NCs was enhanced by the addition of polyvinylpyrrolidone (PVP) due to the coordination ability of PVP with metal clusters. By adding 1 mL of 2.5 mg mL^{-1} PVP, the photoluminescence properties of the Ag NCs at room temperature remained almost unchanged after one week.[15]

By using kinetic control, Lopez-Quintela and coworkers reported stable Ag NCs obtained in microemulsions for the first time (Figure 5.3). The Ag NCs displayed luminescence, molecular-like paramagnetic properties, and were very stable which had been confirmed by working with the procedure for over 2 years. The Ag NCs produced in the microemulsions showed a strong luminescence emission at 350 nm (3.54 eV). The excellent stability of the Ag NCs might be ascribed to their large HOMO–LUMO bandgap (*ca.* 2.3 eV). Comparing the bandgaps deduced by using optical and electrochemical techniques, a value of about 0.6 eV was obtained for the Coulomb charging energy, which corresponded to a cluster capacitance of about 0.3 aF. The MALDI-TOF-MS results showed that such prepared Ag NCs mainly consisted of Ag_3O_2, Ag_5, and Ag_9 which were consistent with the HOMO-LUMO bandgap result. By tuning the kinetic control, Ag NCs of other sizes and geometries with different properties could also be prepared. The results indicated that by using such a facile kinetic control approach in microemulsions scaffolds were not needed for the preparation of Ag NCs, as was reported in thermal-induced synthesis of Ag NCs.[16]

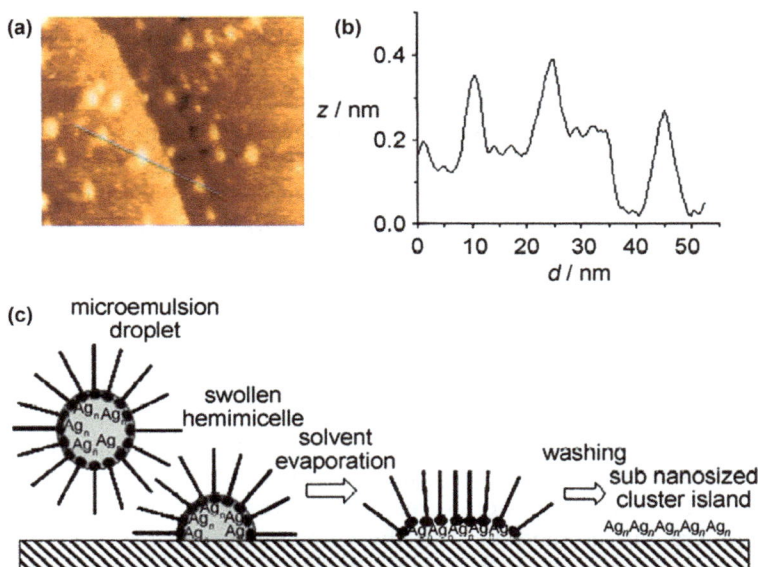

Figure 5.3   (a) STM image (86×65 nm) of silver nanoislands produced in microemulsions after deposition onto an Au(111) surface. (b) Height (z) profile along the line shown in (a). (c) Hypothetical scheme for the deposition of nanoislands composed of subnanosized clusters onto a substrate (reprinted with permission from ref. 16).

5.3 Protein- and Peptide-Protected Silver Nanoclusters and Their Related Applications

5.3.1 Protein-Protected Silver Nanoclusters and Their Related Applications

Mimicking biomineralization, Narayanan and Pal synthesized luminescent Ag NCs using protein as scaffolds. The Ag NCs were directly conjugated to an enzyme, bovine pancreatic α-chymotrypsin (CHT), using NaBH$_4$ as the reducing agent at ambient conditions. By using steady-state UV-vis absorption/photoluminescence spectroscopy and high-resolution transmission electron microscopy, the structural characterization of such synthesized Ag–CHT nanobioconjugates was carried out and the diameters of the Ag NCs were about 1 nm. The emission energy of CHT-conjugated Ag NCs was correlated with the numbers of atoms in the cluster and was found to be almost consistent with the experimental results using the spherical jullium model. By assuming a spherical shape and a uniform *fcc* structure, the average number of silver atoms per Ag NC was calculated to be 31. Circular dichroism studies demonstrated that the conformation of the enzyme changed little after the conjugation of Ag NCs. The enzymatic activity of the reconstituted Ag–CHT was found to reduce by about 2 times as compared to that of reconstituted CHT. This was consistent with previous studies on an analogous system (CdS-bound CHT).[17]

Deng *et al.* employed a unique and especially efficient synthetic platform eggshell membrane (ESM) to generate luminescent Ag NCs *via* various chemical routes. It was very difficult to obtain luminescent Au NCs by NaBH$_4$ reduction or UV irradiation, while Au NCs were successfully synthesized in a concentrated NaOH solution (4.5 M). Compared with Au NCs, Ag NCs could be synthesized much easier by NaBH$_4$ reduction, UV irradiation and in a concentrated NaOH solution (4.5 M). The as-obtained ESM protein templated Ag NCs had a brownish–yellow colour under room light and emitted red to near infrared luminescence centered at 702 nm with an excitation maximum at 480 nm. The ESM based platform was cost-effective, green and facile for the preparation of luminescent Ag NCs, which allowed for many applications in catalysis, surface enhanced Raman scattering (SERS), fluorescent patterning and chemical sensing.[18]

Guo and Irudayaraj reported a new water-soluble, stable, luminescent Ag NC using denatured bovine serum albumin (dBSA) as a stabilizing agent (Figure 5.4).[19] The dBSA had 35 free cysteine residues which could contribute to polyvalent interactions with the Ag NCs and serve as effective stabilizing agents for the Ag NCs. MALDI-TOF-MS and high-resolution ESI-MS spectroscopy were employed for the determination of the structure of such synthesized dBSA templated Ag NCs; yet, due to the natural existence of two silver isotopes, the MS spectra could not be assigned to any specific Ag NC species. The Ag NCs displayed a luminescence emission at 637 nm with a

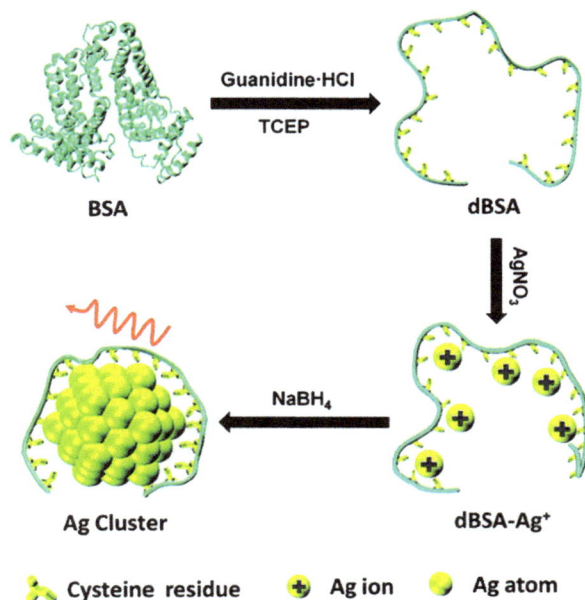

Figure 5.4 Schematic of the Denatured Protein Directed Synthesis of Fluorescent Ag clusters (reprinted with permission from ref. 19).

quantum yield of about 1.2%. The photoluminescence of the Ag NCs was quite stable and the intensity changed little even after 37 days dispersing in water. With the photoluminescence properties and stability of the dBSA-protected Ag NCs, they were utilized for the detection of Hg^{2+} with high sensitivity and selectivity. The detection limit was 10 nM in the linear range from 10 nM to 5 μM.[19] Another related work on BSA templated Ag NCs was reported by the Schneider group through a wet chemistry route by $NaBH_4$ reduction. The MALDI-TOF-MS results demonstrated that the BSA-protected Ag NCs consisted of Ag_8–BSA. The excitation and emission peaks of such synthesized Ag NCs were located at 490 nm and 690 nm with no shift or decreased intensity after 1 month in refrigerated storage. The quantum yield of the BSA templated Ag NCs was a little low with an estimated value of 1.46% which might be related to the low concentration of Ag NCs in BSA as was proved by XPS and infrared spectroscopy results. The lifetime of BSA-protected Ag NCs was 1.0 ns (84%) and 5.3 ns (16%).[20] Pradeep *et al.* utilized BSA as scaffolds to synthesize luminescent Ag NCs also through a wet chemical route. MALDI-MS results indicated that such synthesized Ag NCs mainly consisted of Ag_{15} clusters. The Ag NCs showed similar luminescence excitation and emission at 490 and 690 nm, respectively. The quantum yield of the Ag NCs was 10.71% in water and the luminescence was stable in a pH range of 1–12. In addition, it was found that after the addition of PVP, the luminescence excitation and emission of the Ag NCs changed to 510 nm and 740 nm, respectively and the stability of the clusters was enhanced.[15]

Cell matrices were considered to be rich libraries of various proteins for the production of luminescent Ag NCs. Yu *et al.* investigated an *in vitro* approach to generate both Au NCs and Ag NCs under the protection of proteins. Aqueous solutions of gold and silver ions were incubated with fixed cells for 2 to 24 h and then the unbound metal ions were washed off before irradiation of the cells with the light of a 100 watt mercury arc-discharge lamp passing through a bandpass filter (BP 420–440 nm). Interestingly, the Ag NCs emission in the cell nuclei was not stronger than that in cytoplasm, which was somewhat unexpected as DNA has a strong affinity for silver ions. However, it substantiated the understanding that the generation of luminescent Ag NCs depended on the sequence and conformation of DNA molecules. Through such photoactivation of Ag NCs or Au NCs in a cell matrix, proteins were proved to be suitable scaffolds for the generation of bright, stable noble metal nanoclusters.[21]

5.3.2   Peptide-Protected Silver Nanoclusters and Their Related Applications

Wright *et al.* had used the repeating consensus sequence AHHAHHAAD from the histidine-rich protein II of *Plasmodium falciparum* to mediate the aqueous self-assembly of several metal sulfide, metal oxide, and zerovalent metal clusters.[22] Dickson and coworkers reported a novel *in vitro* and intracellular approach for the production of peptide-encapsulated luminescent Ag NCs for the first time in 2007 (Figure 5.5). In contrast to the ordinary harsh silver staining conditions, the formation of luminescent Ag NCs could be initiated and used to stain cells at much lower AgNO$_3$ concentrations (20 mM) by photoactivation at ambient temperature. Such synthesized Ag NCs were mainly located in the regions of nucleoli rather than cytoplasm. Consisting of more than 700 amino acids, neceolin was identified as one of the major proteins to bind silver atoms in silver staining and nucleate the formation of luminescent Ag NCs. However, nucleolin was too large to be useful as a label and the silver binding site was unknown, a short peptide, KECDKKECDK-KECDK (P1), incorporating the specific amino acids most prevalent in nucleolin: glutamic acid (15.6%), lysine (12.7%), and aspartic acid (8.5%), was designed for the synthesized of Ag NCs. The Ag NCs protected with P1 in aqueous solution were only moderately stable at room temperature (chemical lifetime of 3 days) due to the short length of P1 peptide. By incorporating several hydrophobic amino acids, another two peptides HDCHLHLHKCHLHLHCDH (P2) and HDCNKDKHDCNKDKHDCN (P3) were designed to synthesize Ag NCs. P2 and P3 templated Ag NCs were much more stable (chemical lifetime of 2 weeks in deionized water and 5 weeks in PBS, respectively) than P1-protected Ag NCs. Nevertheless, all the peptide-protected Ag NCs were stable at −20 °C for at least one month. These

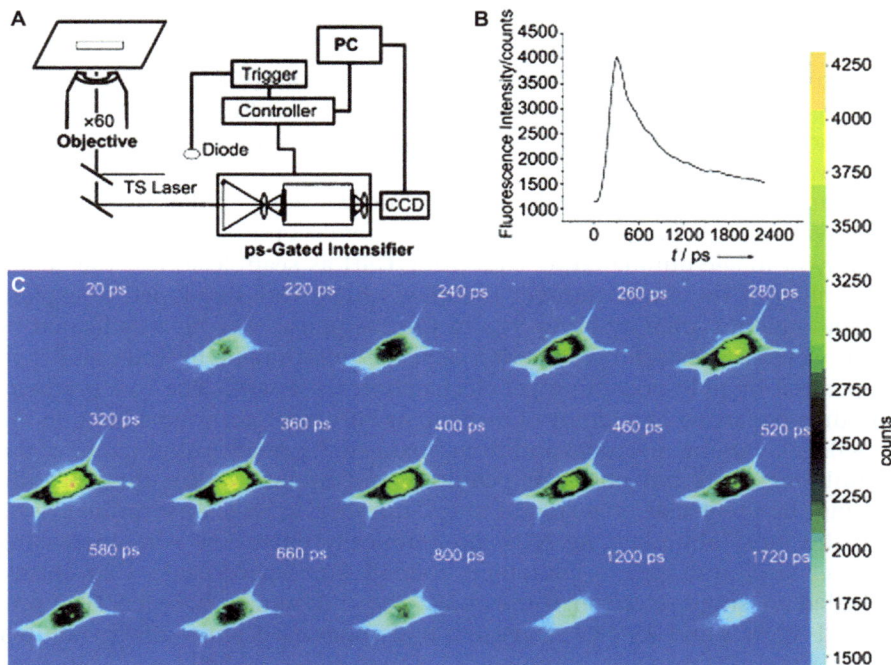

Figure 5.5   Time-gated images of NIH3T3 cells stained with silver nitrate. (A) Microscope setup integrated with ps-gated intensifier (LaVision Picostar HR) and CCD camera (Andor iXon). (B) Time slices of a cell stained with silver nitrate. (C) Time profile of the time series images showing the fast silver nanocluster emission at short times. Note that black indicates an intermediate intensity level in this colour scheme (reprinted with permission from ref. 23).

peptide-protected Ag NCs exhibited similar photophysics to those in the nucleolus, but with narrower emissions at 610 nm (P1-Ag NCs), 615 nm (P2-Ag NCs), and 630 nm (P3-Ag NCs). MALDI mass spectrometry confirmed the presence of Ag NCs, in which Ag_2, Ag_3, Ag_4, and Ag_5 were bound to a single peptide. The luminescence decay of peptide-protected Ag NCs were P1-Ag NCs (58% 260 ps, 42% 2300 ps), P2-Ag NCs (15% 73 ps, 85% 1900 ps), and P3-Ag NCs (36% 420 ps, 64% 2900 ps).[23]

The Bonacic-Koutecky group presented a theoretical study of the structure and optical properties of tripeptide-Ag NC hybrid systems by TDDFT using B3LYP functional and 11e-RECP with the corresponding AO basis set. The structural properties and stationary absorption spectra of Trp-$(Ala)_2$-Ag_n^+ ($n = 1$, 3, 5, 9) were determined by using the DFT theory with the hybrid B3LYP functional. The search for structures was performed by the simulated annealing procedure coupled with molecular dynamics simulations with use of the semiempirical AM1 method. The theoretical study of tripeptide-Ag NC hybrid systems showed that Ag NCs induced significant absorption enhancement in the spectral region between 225 and 350 nm with respect to the pure peptide, which allowed the use of clusters as chromophores for absorption enhancement of peptides and proteins and offered a potential for different applications in biosensing. The Ag NC binding could change the conformational preference for the secondary structure type possibly leading to new functional properties.[24]

Glutathione (GSH), as a kind of tripeptide, was also used as a scaffold for the synthesis of luminescent Ag NCs. Bigioni's group had synthesized a family of magic-numbered Ag NCs using GSH as scaffold to test the generalizability of the magic-number theories, which were developed to explain the anomalous stability of Ag NCs in the gas phase. The Ag NCs ligated by GSH were synthesized by reduction of silver glutathiolate in water and then separated by polyacrylamine gel electrophoresis (PAGE). The raw synthetic product consisted of a family of discrete Ag NCs, each forming a band in the PAGE gel. Varying reaction conditions changed the relative abundance of the family members but not their positions and colours within the gel, indicating the molecular precision of magic-number Ag NCs. Absorption onsets for the most abundant Ag NCs monotonically decreased with increasing cluster size, and spectra contained a small number of peaks that corresponded to single electron transitions. The differences between GSH-protected Ag NCs and Au NCs suggested that condensed phase magic-number cluster theories might be more complex than currently believed.[25] Lin *et al.* also used GSH as a scaffold to synthesize water-soluble, luminescent Ag NCs for the detection of Hg^{2+}. The pH of the solution containing $AgNO_3$ and GSH was adjusted to 6.45 by adding 1 M NaOH, after the addition of $NH_2NH_2 \cdot H_2O$, the GSH–$AgNO_3$ complex was reduced to GSH templated Ag NCs. The Ag NCs showed a strong fluorescence emission at 620 nm and were utilized for the detection of Hg^{2+} with a detection limit of 0.1 nM, which was the lowest detection limit among literature systems reported so far.[26]

Adhikari and Banerjee had used a short-peptide-based hydrogel as a template for the *in situ* synthesis of luminescent Ag NCs by using sunlight. *N*-terminally Fmoc-protected dipeptide, Fmoc-Val-Asp-OH, formed a transparent, stable hydrogel with a minimum gelation concentration of 0.2% w/v. The silver-ion-encapsulating hydrogel could efficiently and spontaneously produce luminescent Ag NCs under sunlight at physiological pH (7.46) by using a green chemistry approach. In the absence of any conventional reducing agent but in the presence of sunlight, silver ions were reduced by the carboxylate group of a gelator peptide that contained an aspartic acid residue. MALDI mass spectrometric analysis showed the presence of a few atoms in Ag NCs containing only Ag_2 and the Ag NCs showed a very narrow emission profile at 634 nm and large Stokes shift (>100 nm). The luminescent Ag NCs within the hydrogel were very stable even after 6 months storage in the dark at 4 °C and could be applied in antibacterial preparations, bioimaging and other purposes.[27]

Gao *et al.* had utilized an artificial peptide with the amino acid sequence of CCYRGRKKRRQRRR to biomineralize serial Ag NCs (Figure 5.6). Under

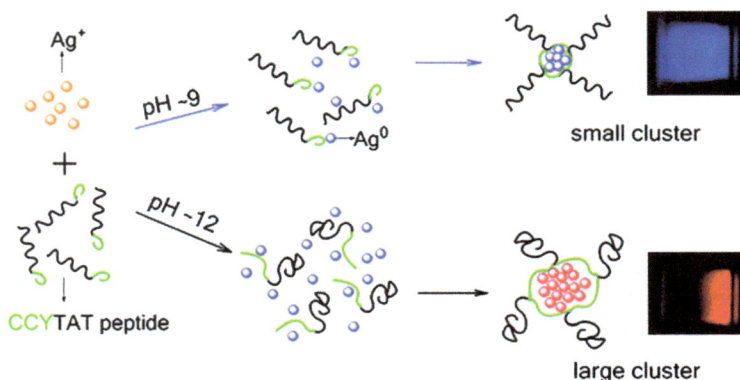

Figure 5.6 Illustration of peptide–Ag cluster formation. The peptide can biomineralize Ag^+ *in situ* and produce Ag clusters in aqueous solution. At pH values of ~ 9 and ~ 12, Ag clusters with blue and red emission are formed, respectively. Red- and blue-emitting clusters are denoted $C-1$ and $C-2$, respectively (reprinted with permission from ref. 28).

different alkaline conditions, Ag NCs with red and blue emission were biomineralized by the peptide, respectively. The MALDI-TOF MS spectra implied that the red-emitting Ag NC sample was composed of Ag_{28}, while the blue-emitting Ag NC sample was composed of Ag_5, Ag_6, and Ag_7. The UV-visible absorption and infrared spectra revealed that the peptide phenol moiety reduced Ag^+ ions and that Ag NCs were captured by the peptide thiol moieties. The phenol reduction potential was controlled by the alkalinity and played an important role in determining the Ag NC size. Circular dichroism observations suggested that the alkalinity tuned the peptide secondary structure, which might also affect the Ag NC size.[28]

Griffith and coworkers presented a general strategy to acquire mass spectra of fragile metal clusters with reliable mass assignments. By optimizing sample solution conditions, high-quality ESI mass spectra of a prototypical silver:glutathione (Ag:SG) cluster were obtained without significant fragmentation. By using gentle conditions and solution conditions designed to stabilize the clusters, fragmentation was dramatically reduced and mass spectra with isotopic resolution were measured. Using this strategy, the authors had made the first formula assignment for a ligand-protected Ag cluster of $Ag_{32}(SG)_{19}$.[29]

5.4 Short Molecule-Protected Silver Nanoclusters and Their Related Applications

In the above we have discussed the synthesis, properties and applications of polymer, protein- and peptide-protected Ag NCs. Whereas various scaffolds have been used for the synthesis of luminescent Ag NCs, it remains challenging to point out the regular pattern of luminescent Ag NCs due to the complexity of polymers, proteins and peptides. Note that polymers, proteins and peptides consist of different short molecules, thus it is very important to understand the fundamentals of short molecule-protected Ag NCs.

Natan *et al.* reported a self-assembly of monodisperse gold and silver colloid particles into monolayers on polymer-coated substrates yielding macroscopic surfaces that were highly active for SERS. Gold and silver nanoparticles were found to be bound to the substrate through multiple bonds between the colloidal metal and functional groups on the polymers such as the cyanide (CN), amine (NH$_2$), and thiol (SH) moieties. The rate of surface formation was SH $\approx$ NH$_2 \gg$ CN which indicated that the interactions between SH or NH$_2$ and silver atoms were stronger than that of CN and silver atoms. Various self-assembled electrodes were fabricated by the interactions between SH or NH$_2$ and gold or silver atoms.[30]

Jensen *et al.* developed a model system of a pyridine-Ag$_{20}$ cluster for studying SERS using the TDDFT method, based on a short-time approximation to the Raman scattering cross section. Photodepletion studies of small silver clusters (2–21 atoms) embedded in rare-gas matrices showed that the absorption spectrum of Ag$_{20}$ clusters in an argon matrix was dominated by a broad peak at 3.70 eV and a weaker one at 3.97 eV. Although this broad absorption feature could not be considered as a true collective excitation due to the small size of the cluster, it could be considered as a microscopic analogue to the plasmon excitation observed in nanoparticles. By adopting the model system, the authors showed that a consistent treatment of both the electromagnetic (EM) and the chemical enhancements could be achieved using the recently developed approach. In the Raman spectrum, both absolute and relative intensities depended not only on the local chemical environment of the molecule–metal binding site, but also on the incident excitation wavelength. Different contributions to the enhancement were analyzed and found to be of the order of static chemical enhancement (factor of 10) < charge-transfer enhancement (10^3) < EM enhancement (10^5). The degree of enhancement for the important normal modes could be rationalized by their vibrational motions and local chemical environments of the molecule which provided a fairly simple visual picture showing key information about the enhancement (Figure 5.7).[31] By using TDDFT, the size dependence of the absorption and Raman scattering properties of pyridine interacting with small Ag NCs Ag$_n$ (n = 2–8, 20) were studied by Jensen and coworkers. The authors simulated both the normal and the "surface"-enhanced Raman spectra by employing a recently developed short-time approximation for the Raman scattering cross-section. The absorption spectra of the small Ag NCs were studied both in the gas phase and embedded in rare gas matrices and both the absorption and Raman properties depended strongly on cluster size and adsorption site. The normal Raman spectra of Ag$_n$-pyridine complexes resembled that of isolated pyridine, where the enhancement increased as cluster size increased. The

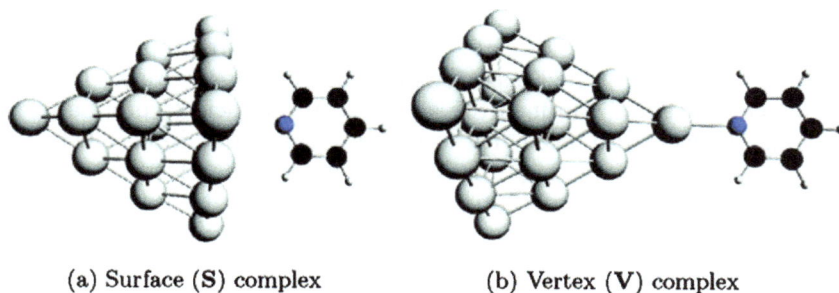

(a) Surface (**S**) complex   (b) Vertex (**V**) complex

Figure 5.7   Configurations of the two pyridine–Ag$_{20}$ complexes having C$_s$ symmetry (reprinted with permission from ref. 31).

total enhancement for the complexes was between 10^3 and 10^4 and quite surprisingly the strongest enhancement was found for the Ag$_2$-pyridine complex. The enhancement trends could be correlated with the distance of the molecule to the center of the metal cluster and with the resonance polarizability in a way that was suggestive of electromagnetic enhancement; the enhancement mechanism for these small clusters was similar to what was found for larger particles.[32] Wu and coworkers presented a quantum chemical study of pyridine interacting with copper, silver, gold and platinum metals based on chemical enhancement effects in SERS signals. The results showed that the relative Raman intensities of SERS spectra depended strongly on the binding interactions between pyridine and the SERS active centers, the electronic properties of metal materials, and the incident wavelengths. When the bonding between pyridine and a SERS site was very weak, analogous to physical adsorption, the Raman spectra of the adsorbed pyridine were similar to that of free pyridine. For pyridine interacting strongly with copper, gold, and platinum clusters, we found that the Raman intensities of the v_1, v_{6a}, v_{9a}, and v_{8a} modes of pyridine were enhanced, whereas the intensity of the v_{12} mode decreased. The preresonance Raman spectra were also calculated to check the enhancement effect of the charge-transfer mechanism and the results showed that the Raman spectral properties depended strongly on the properties of these excited states and the electronic structures of the metal materials.[33] Amino acids containing NH$_2$ were also used as scaffolds to protect Ag NCs due to the interactions between NH$_2$ and silver atoms. Pakiari and Jamshidi studied the binding of Au NCs and Ag NCs with amino acids (glycine and cysteine) using DFT. The authors optimized the geometries of neutral, anionic, and cationic amino acids with Au$_3$ and Ag$_3$ clusters using the DFT-B3LYP approach and found out that the interaction of amino acids with Au NCs and Ag NCs was governed by two major bonding factors: (a) the anchoring N–Au(Ag), O–Au(Ag), and S–Au(Ag) bonds and (b) the nonconventional N–H$\cdots$Au(Ag) and O–H$\cdots$Au(Ag) hydrogen bonds. Among the three forms of amino acids, anionic ones exhibited the strongest tendency to interact with the Au NCs and Ag NCs. The results showed that these bonds were partially electrostatic and partially covalent.[34] Bonacic-Koutecky and coworkers investigated the effect of the environment on doubly charged Ag NCs stabilized by tryptophan in the gas phase through a combined experimental and theoretical study. The nanohybrids composed of a unit of the aromatic amino acid tryptophan and a small Ag NC were produced in the gas phase by combining electrospray ionization and multiple stage mass spectrometry. The electrolyte solution was prepared by mixing a solution of silver nitrate salt (500 μM in H_2O : CH_3OH 1 : 1 (v/v)) and a solution of amino acid (500 μM in H_2O : CH_3OH 1 : 1 (v/v)) in a ratio of 1 : 1. The $[(Trp-H) + Ag_4]^+$ complex containing Ag_4^{2+} was formed in a trapping cell by collision-induced fragmentation of the precursor ion $[(Trp)_2 - 3H + 4Ag]^+$. The coupling between electronic excitations of the doubly charged metal core and charge-transfer-type excitations between the ligand and the metal moieties was responsible for a strong absorption below 250 nm. The authors demonstrated that singly and doubly charged metal species in hybrid systems could be discriminated by their different optical properties which would allow identification of seed precursors and provide mechanistic insights into the aggregation and growth of nanoparticles.[35] Scaiano's group also reported a facile photochemical method for the preparation of highly fluorescent Ag NCs using the interactions between NH_2 and silver atoms. The method made use of photogenerated ketyl radicals that reduced Ag^+ from silver trifluoroacetate in the presence of amines (hexadecylamine and cyclohexylamine). Such synthesized Ag NCs exhibited luminescence excitation and emission bands at about 450 nm and 530 nm, respectively. There was a weak but characteristic absorption peak at 390 nm which was considered to be the plasmon transitions. The luminescence quantum yields were measured to be about 0.11 and 0.15 for Ag NCs in toluene using hexadecylamine and cyclohexylamine as capping ligands, respectively. And the luminescence quantum yield for Ag NCs in THF with cyclohexylamine was 0.11. In the case of toluene, the decay was fit to a monoexponential function with a lifetime of 2.6 ± 0.1 ns and a goodness of fit $\chi^2 = 1.46$. The resulting nanoparticles were around 3 nm in diameter but their luminescence properties were derived from small Ag NCs, predominantly Ag_2, that were believed to be located at the nanoparticle surface or stabilization layer. Thus, the synthesis and characterization of Ag NCs showed the generality of these concepts and the fact that ketyl radicals were excellent tools for nanoparticle synthesis.[36]

Recently, aqueous synthesis of Ag NCs through the use of DNA oligomers as the stabilizer has also been developed. Due to the complexity of DNA molecule structures, the bonding and optical properties of even the simplest Ag NCs templated by DNA were still unknown. The large number of possible structures made a theoretical understanding of the stability of different isomers and their spectral properties particularly desirable. To initiate development of a predictive understanding of the properties of DNA-protected Ag NC species, Metiu and coworkers used quantum chemical calculations to examine the binding of few-atom, neutral Ag NCs to the DNA bases. The choice to investigate Ag NCs bound to the bases alone, rather than to the entire DNA molecule, was motivated by computational constraints and justified by experimental indications that Ag NCs bound to the DNA bases, and not to the sugar-phosphate backbone. Using DFT, the authors found that Ag NCs preferred to bind to the doubly bonded ring nitrogens, and that binding to nucleobase T is generally much weaker than to nucleobases of C, G, and A. Ag_3 and Ag_4 made stronger bonds. Bader charge analysis indicated a mild electron transfer from all bases except T to the Ag NCs. The donor bases (C, G, and A) bound to the sites on the Ag NCs where the lowest unoccupied molecular orbital had a pronounced protrusion and the site where Ag NCs bound to the base was controlled by the shape of the higher occupied states of the base. The results demonstrated that geometric constraints on binding, imposed by designed DNA structures, might be a feasible route to engineering the selection of specific cluster-base assemblies.[37] Wang *et al.* recently developed a high-yield synthesis of Ag NCs using DNA monomers as the scaffold for the first time (Figure 5.8). Most of the Ag NCs were formed with nine silver atoms, and only the Ag NCs protected by deoxycytidine monomers (dC) showed luminescence emission. The results provided basic evidence of the benefit of using cytosine-rich DNA strands for the synthesis of luminescent Ag NCs. The mechanism of the formation of DNA-protected Ag NCs, the roles of the four bases in the synthesis of Ag NCs with DNA scaffolds and the reason why cytosine-rich DNA strands were good scaffolds for luminescent Ag NCs were explained by the authors. DFT computations were carried out to predict the luminescence of dC-protected Ag NCs and the results were in very good agreement with the experiments. The results provided basic guidelines for further experimental and theoretical studies on DNA-protected luminescent Ag NCs and other scaffold-protected Ag NCs. The study might contribute to achieving programmed synthesis of DNA-stabilized Ag NCs with photoluminescence properties.[38] Wang's group is well-known in the synthesis and applications of DNA templated Ag NCs and they reported a new Ag NC-based luminescent assay capable of specifically identifying single nucleotide modifications in DNA. Their study showed an interesting dependence of Ag NC formation on a single nucleotide in a DNA sequence and even a single nucleotide mismatch located two bases away from the Ag NC formation site would prohibit the generation of luminescent Ag NCs. As an example, they demonstrated the capability of this strategy to identify the sickle cell anemia mutation in the haemoglobin beta chain (HBB) gene. This design suggested that a general protocol based on duplex DNA was possible for single-nucleotide mismatch detection.[39,40] They also reported luminescent oligonucleotide-stabilized Ag NCs as novel and environmentally-friendly fluorescence probes for the determination of Hg^{2+} ion and biothiols with a low detection limit and high selectivity. Due to the strong affinity of Hg^{2+} ions for DNA strands, the presence of Hg^{2+} would disturb the stability of DNA-encapsulated Ag NCs. The luminescence of the DNA-stabilized Ag NCs was quenched by Hg^{2+} with a detection limit down to 5 nM. However, the lifetime of the Ag NCs did not obviously change with the addition of Hg^{2+} ions, indicating that the quenching was due to the

A

B

C

D

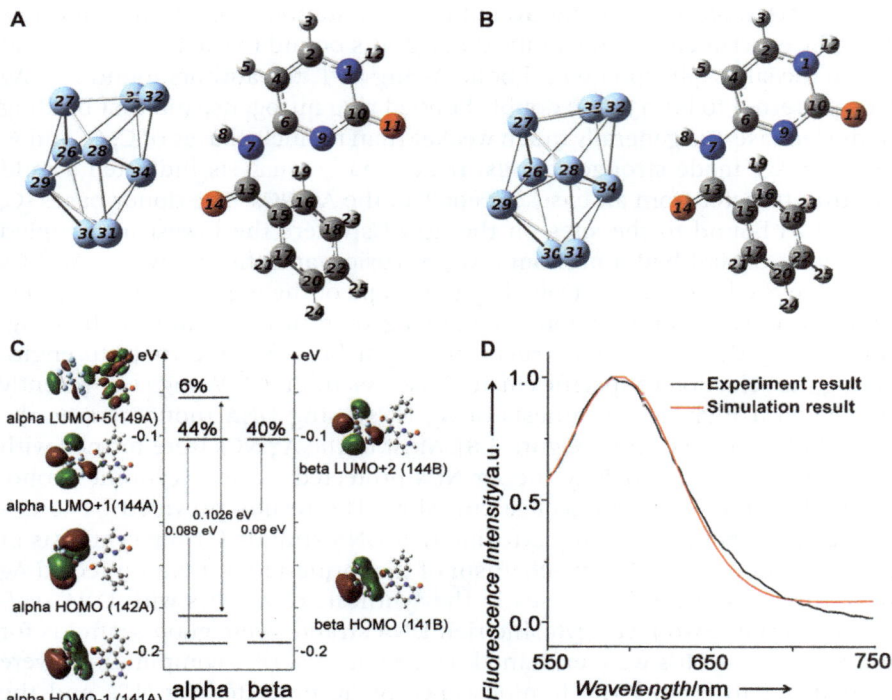

Figure 5.8 Optimal ground-state (A) and excited-state (B) geometries of the dC–Ag$_9$ complex. For clarity, only a benzoyl-protected cytosine base in proximity to the Ag$_9$ cluster is shown. (C) Molecular orbitals and electronic contributions of the relevant excited states, and (D) fluorescence spectra (black: experiment; gray: simulation) of the dC–Ag$_9$ complex (reprinted with permission from ref. 38).

formation of a non-emissive complex between DNA-encapsulated Ag NCs and Hg^{2+}, without energy transfer.[41–43] Recently, Wang's group reported photoinduced electron transfer (PET) for the first time between DNA–Ag NCs and G-quadruplex–hemin complexes, accompanied by a decrease in the fluorescence of DNA–Ag NCs. In the PET process, a parallel G-quadruplex and the sensing sequences were blocked by a duplex. The specific combination of targets with the sensing sequence triggered the release of the G-quadruplex and allowed it to fold properly and bind hemin to form a stable G-quadruplex–hemin complex. The complex proved favourable for PET because it made the G-quadruplex bind hemin tightly, which promoted the electron transfer from the DNA–Ag NCs to the hemin Fe^{III} center, thus resulting in a decrease in the fluorescence intensity of the DNA–Ag NCs. The novel PET system enabled the specific and versatile detection of target biomolecules such as DNA and ATP with high sensitivity based on the choices of different target sequences.[44]

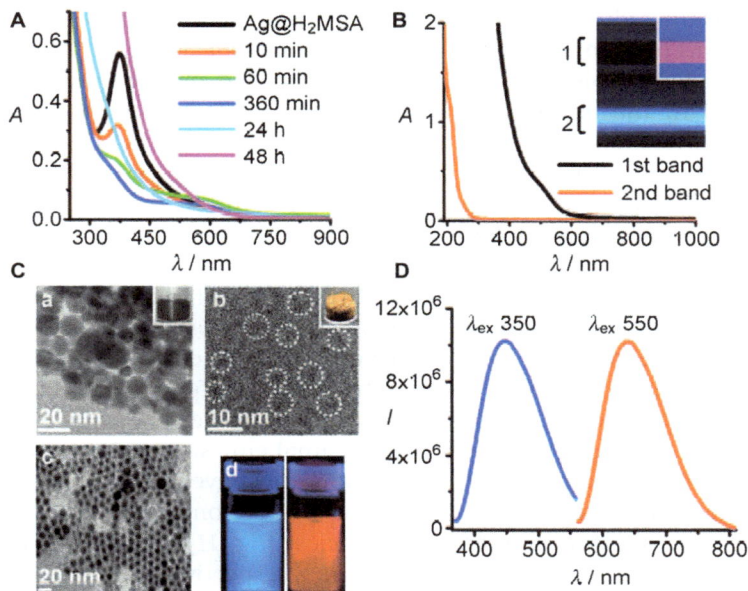

Figure 5.9 (A) Time-dependent UV-vis spectra of the clusters synthesized during interfacial etching at room temperature. (B) UV-vis absorption spectra of the clusters obtained from the two bands in PAGE. The inset shows a photograph of the wet gel after electrophoresis in UV light at room temperature, and the inset to the inset an image of the first band at 273 K. (C) HRTEM images of (a) as-synthesized Ag@(H$_2$MSA), (b) the product obtained after interfacial etching, and (c) particles in the blue layer at the interface. Individual clusters are not observable by TEM, but aggregates are seen faintly (b, shown in circles). Insets of (a) and (b) are photographs of Ag@MSA and crude cluster samples. (d) Photographs of aqueous cluster solutions of first (cluster 1) and second (cluster 2) PAGE bands at 273 K and room temperature, respectively. (D) Luminescence emission of cluster 1 and cluster 2 in water, excited at 550 and 350 nm, respectively (reprinted with permission from ref. 46).

As the interactions between SH and silver atoms were much stronger than that between NH$_2$ and silver, various mercapto derivatives have been utilized as scaffolds for the synthesis of Ag NCs. The Pradeep group utilized 2-mercaptobenzothiazole monolayers to prepare Au and Ag NCs and characterized them with various spectroscopic methods. The optical spectra showed features that might be assigned to charge-transfer excitation between the monolayer and the cluster, in addition to a red shift and reduction of the plasmon absorption. These clusters were stable in air for several months and the results showed that on different crystallographic planes the orientation of 2-mercaptobenzothiazole was different.[45] They also reported an interfacial approach for the preparation of luminescent Ag NCs using mercaptosuccinic acid (H$_2$MSA) as scaffolds (Figure 5.9). H$_2$MSA-protected Ag nanoparticles (Ag@H$_2$MSA) were synthesized through a wet chemical approach using

NaBH$_4$ as the reduction agent in water under ice-cold conditions. Interfacial etching was performed in an aqueous–organic biphasic system. H$_2$MSA thiol was partially dissolved in an organic solvent and parent Ag@H$_2$MSA was dispersed in the aqueous phase. An aqueous solution of as-synthesized Ag@H$_2$MSA nanoparticles was added to an excess of H$_2$MSA in toluene (1 : 2 water : toluene ratio). A weight ratio of 1 : 3 was used (Ag@(H$_2$MSA) : H$_2$MSA). The resulting mixture was stirred for 48 h at room temperature. As the reaction proceeded, the colour of the aqueous phase changed from reddish brown to yellow and finally to orange. The reaction product was precipitated by the addition of methanol and washed with methanol to remove excess H$_2$MSA. The crude mixture was separated into differently sized clusters by polyacrylamide gel electrophoresis (PAGE) and the electrophoresis showed two bands, which indicated the presence of two different clusters. The first band (cluster 1) was red and the second band (cluster 2) was light yellow in visible light. Cluster 1 and 2 showed maximal emission at 650 and 440 nm, with excitation maxima at 550 and 350 nm, respectively. The confirmation of the molecular formulas came from mass spectrometric studies and the MALDI MS spectra showed the species $[Ag_8(H_2MSA)_4(HMSA)_4]^-$ and $[Ag_7(H_2MSA)_5(HMSA)_2]^-$ for clusters 1 and 2, respectively. The luminescence decay of Ag$_8$ in the solid state was 35 ps (97%), 37.2 ns (0.6%), 37.2 ns (1.72%), and 5.68 ns (0.6%) and the lifetimes of Ag$_7$ in the solution were 12 ps (88.9%), 0.396 ns (4.8%), 2.10 ns (4.8%), and 8.31 ns (1.3%).[46] The Pradeep group also reported a solid-state route to prepare Ag$_9$ quantum clusters using H$_2$MSA as the scaffolds. The detailed process for synthesizing the Ag NCs has been described in Chapter 3. The ESI-MS spectrum of the PAGE-purified Ag NCs demonstrated that the Ag NCs were mainly composed of $[Ag_9(H_2MSA)_6(MSA)]^{2-}$. The ^{1}H and ^{13}C NMR spectra of H$_2$MSA, Ag(I)MSA and Ag$_9$(MSA)$_7$ were measured in D$_2$O. Two strong multiplets at 2.8 and 3.7 ppm that were assigned to CH$_2$ and CH were broadened in the Ag NCs and the CH protons were shifted downfield due to the proximity to the silver core.[47] Pradeep and coworkers also obtained Ag NCs in high yield from the reduction of a silver thiolate precursor, Ag-SCH$_2$CH$_2$Ph and the Ag NCs exhibited a single sharp peak near 25 kDa in the MALDI-MS spectrum and a well-defined metal core of ∼2 nm by TEM measurements. The Ag NCs yielded a single fraction in high-performance liquid chromatography (HPLC) study. Increased laser fluence fragmented the cluster until a new peak near 19 kDa predominated, suggesting that the parent cluster-Ag$_{152}$(SCH$_2$CH$_2$Ph)$_{60}$-evolved into a stable inorganic core-Ag$_{152}$S$_{60}$. Exploiting combined insights from investigations of clusters and surface science, a core-shell structure model was developed, with a 92-atom silver core having icosahedral–dodecahedral symmetry and an encapsulating protective shell containing 60 Ag atoms and 60 thiolates arranged in a network of six-membered rings resembling the geometry found in self-assembled monolayers on Ag(111). The structure was in agreement with small-angle X-ray scattering (SAXS) data. First-principles electronic structure calculations showed, for the geometry-optimized structure, the development of a ∼0.4 eV energy gap between the highest-occupied and the lowest-unoccupied states, originating from a superatom 90-electron shell-closure and conferring stability to the Ag NCs. The optical absorption spectrum of the Ag NCs resembled that of plasmonic Ag nanoparticles with a broad single feature peaking at 460 nm, but the luminescence spectrum showed two maxima with one attributed to the ligated shell and the other to the core.[48]

The Jin group developed a high yield, large scale synthesis of thiolate-protected Ag NCs using *meso*-2,3-dimercaptosuccinic acid (DMSA) as the scaffolds. And it was the first report that achieved a precise MS determination of the composition of silver thiolate clusters. In a typical experiment, $AgNO_3$ was dissolved in ethanol and the solution was cooled to ~ 0 °C, DMSA was then added to the solution. After the formation of $Ag_x(DMSA)_y$ intermediates, $NaBH_4$ (powder) was slowly added to the solution under vigorous stirring. The reaction mixture turned from yellowish green to deep brown, indicating the reduction of $Ag_x(DMSA)_y$ and the formation of Ag NCs (Figure 5.10). The suspension was centrifuged at 13 500 rpm for 10 min, and the resultant black precipitates were collected, washed thoroughly with methanol, then dissolved in water. Recrystallization for 2–3 times led to highly pure Ag NCs. The purification process was evaluated by PAGE analysis and the recrystallized clusters showed a well-defined band, indicating a high purity. The as-prepared Ag NCs showed a strong absorption at ~ 500 nm while they had no fluorescence at all. To determine the exact size of the Ag NCs (*i.e.*, the number of Ag atoms and ligands in the cluster), ESI-MS analysis was carried out. The ESI-MS results demonstrated that such prepared Ag NCs mainly consisted of $Ag_7(DMSA)_4$.[49] They also showed in detail that a unique series of silver sulfide cluster anions ($Ag_nS_4^-$) were observed sequentially from $n = 7$ to 1 when subjecting the thiolate-protected Ag_7 cluster to an MS/MS experiment. Random Ag NC anion distributions were not observed in a wide range of collision energies which indicated the special structure and stability of these gas phase $Ag_nS_4^-$ clusters. Global minimum search based on DFT-enabled basin hopping had yielded the most stable structures for $Ag_nS_4^-$ ($1 \leq n \leq 7$). The global minima showed a transition from three-dimensional to two-dimensional and then to one-dimensional

$$AgNO_3 \text{ (in } C_2H_5OH) \xrightarrow[\text{+DMSA}]{0 \text{ °C}} Ag_x(DMSA)_y \text{ (intermediate)} \xrightarrow[\substack{\text{(powders} \\ \text{added)}}]{\text{+NaBH}_4} Ag_7(DMSA)_4$$

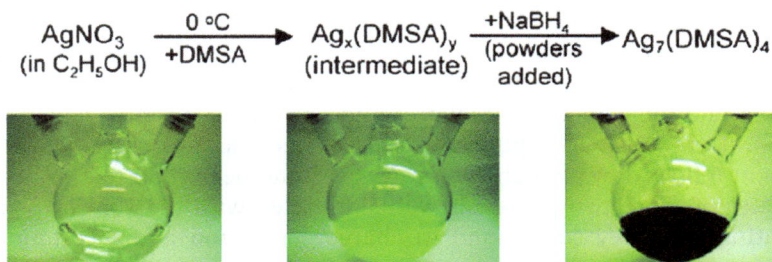

Figure 5.10 Colour change during the synthesis of silver clusters. (Left) ethanol solution of $AgNO_3$. (Middle) 4 h after addition of DMSA. (Right) 12 h after addition of $NaBH_4$ (reprinted with permission from ref. 49).

geometry with decreasing n for the $Ag_nS_4^-$ cluster. The joint experimental and computational effort provided a pathway to discover and elucidate metal sulfide clusters of unique stoichiometry, which were not accessible through conventional methods such as laser ablation of mixed metal and sulphur powders.[50]

Stellacci and coworkers reported a wet chemical synthesis of silver nanoparticles with broad multiband linear optical absorption. These intensely and broadly absorbing nanoparticles (IBANs) were prepared through the reduction of a silver salt solution in the presence of the capping ligand 4-fluorothiophenol (4FTP) in a procedure that deviated from the usual one phase nanoparticle synthesis. In a typical reaction, 4FTP was stirred with a silver salt (4FTP : Ag = 2 : 1) in N,N-dimethylformamide (DMF) for 15 min. A solution of $NaBH_4$ in DMF ($NaBH_4$: Ag = 4 : 1) was then added to the reaction vessel, after which the solution turned brown and darkened further with time. The reaction mixture was left to stir for 4 h until the colour of the solution faded to transparent yellow. The decay of the plasmon-like resonance indicated that particles had either been digested back into seeds or disassociated completely into silver ions or into the layered compound that was known to be an intermediate precursor for the synthesis of silver nanoparticles. A small amount of water was added, increasing the reducing power of $NaBH_4$. Within minutes, the reaction colour began to darken again and the reaction was left in the freezer (-4 °C) over several days. Such synthesized silver nanoparticles showed broad multiband linear optical absorption and an emission at 1375 nm. All the bands were considered to belong to one absorbing species. Various techniques were utilized for the characterization of the size of the IBANs indicating that the IBANs were likely to be smaller than 2 nm. TDDFT calculations of the electronic structure and an optical spectrum of a cluster of $[Ag_{25}(SH)_{18}]^-$ were performed to correlate qualitatively the structure of the nanoparticles with its optical properties and to illustrate further the similarity between the origins of the optical behaviour of previously reported Au NCs. The theoretical results indicated a significant sulfur contribution to all the molecular orbitals (MOs) responsible for the optical transitions, which suggested that changes to the charge density of the sulfur atom might result in perturbation to the MOs. The absorption spectra of IBANs synthesized from several other ligands showed that the absorption features were mostly governed by the structure of the metal core and the thiol–silver interaction, further confirming the results of the computational model.[51] Stellacci and coworkers recently reported the results of transient-absorption measurements and they observed a fast transient followed by a much slower decay as had been reported for Au NCs. The rates of these two processes were correlated with the solvent polarity, indicating that they correspond to charge separation and recombination within the ligand-stabilized clusters. Although the current experiments demonstrated the existence of a charge-separated state, they did not identify its nature and a full understanding would require detailed knowledge of the atomic structure of the clusters.[52]

Kitaev and coworkers developed a facile etching procedure in aqueous solutions to prepare chiral thiol-stabilized Ag NCs. Such chiral thiol-stabilized Ag NCs prepared through a novel approach of cyclic reduction in oxidative conditions (CROC) were a single dominant species. Three different thiols such as captopril, glutathione, and cysteine were used as scaffolds. Captopril and glutathione yielded the highest quality Ag NCs. The Ag NCs prepared by the CROC procedure exhibited quite prominent optical properties: well-defined absorption peaks in UV-vis spectra and strong characteristic CD bands in the visible range. The single most important characteristic of CROC synthesis of Ag NCs was that only one type of optical signature had been observed for all the ligands used. The CD spectra clearly demonstrated that the prepared Ag NCs were chiral based on their characteristic signals, which closely corresponded to UV-vis absorption peaks at *ca.* 650, 490, and 430 nm. All Ag NCs showed only weak fluorescence due to the presence of the thiol ligands with quantum yield estimated at 5×10^{-4}. All spectra featured the same excitation maxima at 440–445 nm (2.8 eV) and emission maxima at 625–630 nm (2.0 eV) for all clusters independent of ligands and Ag NC concentrations as well as corresponding emission and excitation wavelength. The size-selected Ag NCs produced were proposed to be comprised of 22–28 silver atoms (most likely 25 silver atoms, based on strong parallels in size selection and stability with $Au_{25}(SR)_{18}$).[53] Yao and coworkers also reported the synthesis and chiroptical properties of Ag NC enantiomers. The surface of the Ag NCs was covered with L/D-penicillamine (PA) or their racemate. The Ag NCs were separated by PAGE according to their charge and size into well-defined compounds. The mean core diameter was around 1 nm. The separated compounds of Ag NCs covered with L- or D-form thiol showed strong optical activity or circular dichroism with opposite sign (mirror image relationship) in the metal-based electronic transitions (the anisotropy factor obtained was in the order of 1×10^{-3} to 1×10^{-5}), suggesting that the Ag NCs had well-defined stereostructures as common chiral molecules. With a decrease in the mean cluster diameter, the anisotropy factor gradually increased at first, but a steep rise was observed when the diameter became smaller than ~ 1.5 nm. The chiroptical response of the Ag NCs was consequently several-fold larger than that of the Au NCs having the same ligand. The possibilities of a dissymmetric field model as well as an inherently chiral core model were discussed as the origin of the intense optical activity.[54]

Xiang and coworkers developed a new genetic algorithm approach to search for the global lowest-energy structures of ligand-protected metal clusters. In combination with the DFT theory, the genetic algorithm simulations showed that the ground state of $[Ag_7(DMSA)_4]^-$ had eight instead of four Ag–S bonds and had a much lower energy than the structure based on the $[Ag_7(SR)_4]^-$ cluster with a quasi-two-dimensional Ag_7 core. The simulated X-ray diffraction pattern of the $[Ag_7(DMSA)_4]^-$ was in good agreement with the experimental results. The lowest-energy structures of $[Ag_7S_4]^-$, $[Ag_6S_4]^-$, and $[Ag_5S_4]^-$ were predicted and the calculations for the $[Ag_7(SR)_4]^-$ and

$[Ag_7(DMSA)_4]^-$ clusters revealed for the first time that –RS–Ag–RS– could be a stable motif in thiolate-protected Ag NCs.[55]

The Watts group carried out DFT/TDDFT calculations for a series of Ag NCs and Au NCs (Ag_n, Au_n, $n = 12$–120) whose structures were of cigar-type. Pentagonal Ag_n clusters with $n = 49$–121 and hexagonal Au_n clusters with $n = 14$–74 were also calculated for comparison. The significant features of the experimental spectra of actual Ag NCs and Au NCs were well reproduced by the calculations on the clusters and the calculated spectra patterns were also in agreement with previous theoretical results on different-type Ag_n clusters. Many differences in the calculated properties were found between the Ag_n and Au_n clusters, which could be explained by relativistic effects.[56]

Ag NCs were also found to be very important in the formation of larger nanoparticles. Takesue *et al.* utilized the time dependence of small-angle X-ray scattering (SAXS) to observe the formation of silver nanoparticles *in situ* at a time resolution of 0.18 ms, which was 3 orders of magnitude higher than that used in previous reports (*ca.* 100 ms). The SAXS analyses showed that silver nanoparticles were formed in three distinct periods from a peak diameter of *ca.* 0.7 nm (corresponding to the size of a Ag_{13} cluster) during nucleation and the early growth period. The Ag_{13} clusters were most likely elementary clusters that agglomerate to form silver nanoparticles.[57]

There were also some reports on bimetallic ligand-protected clusters. Kumara and Dass synthesized $Au_{144-x}Ag_x(SR)_{60}$ alloy nanomolecules and used ESI mass spectrometry to characterize the structure of the clusters. Ag incorporation into the nanoalloy did not depend linearly on the Ag ratio of the starting material. Based on UV-visible spectroscopy, the Ag were hypothesized to be selectively incorporated into the symmetry equivalent 60-atom shell having Au_{12}, Au_{42}, Ag_{60} concentric shells with 30 –SR–Au–SR– protecting units.[58] Malola and Hakkinen utilized the DFT theory for electronic structures and bonding in $Au_{144-x}Ag_x(SR)_{60}$ based on a structure model of the icosahedral $Au_{144}(SR)_{60}$ that featured a 114-atom metal core with 60 symmetry-equivalent surface sites and a protecting layer of 30 RSAuSR units. In the optimal configuration the 60 surface sites of the core were occupied by silver in $Au_{84}Ag_{60}(SR)_{60}$. Silver enhanced the electron shell structure around the Fermi level in the metal core, which predicted a structured absorption spectrum around the onset (~ 0.8 eV) of electronic metal-to-metal transitions. The calculations also implied element-dependent absorption edges for Au(5d) to Au(6sp) and Ag(4d) to Ag(5sp) interband transitions in the "plasmonic" region with their relative intensities controlled by the Ag/Au mixing ratio.[59]

The Khanna group recently synthesized a bimetallic ligand-protected cluster and the structure of the cluster was determined to be $Ag_4Ni_2(DMSA)_4$ after characterization by ESI mass spectroscopy (Figure 5.11). Such bimetallic clusters involving a noble metal and a first-row transition metal had not been previously reported. Theoretical calculations revealed an octahedral structure with silver atoms occupying the corners of the square plane

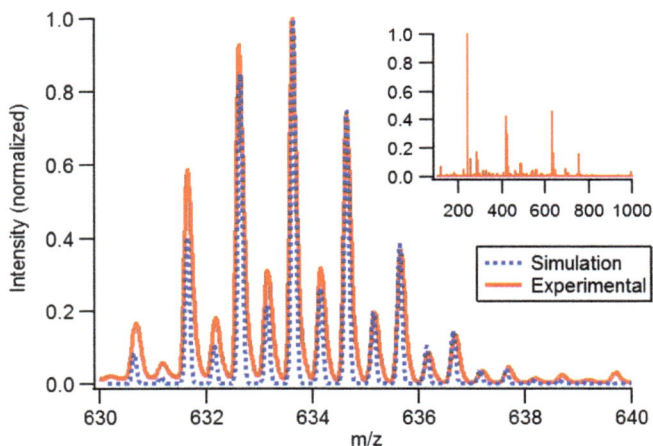

Figure 5.11 Experimental mass spectrum of $Ag_4Ni_2(DMSA)_4$ cluster overlaid with a simulation of the peak positions and isotope distribution. The inset shows a zoomed-out spectrum (reprinted with permission from ref. 60).

and the nickel atoms at the apexes. The predicted spectroscopic features showed good agreement with the observed spectroscopic features.[60]

The Pradeep group also utilized H_2MSA as the scaffolds to synthesize bimetallic 13-atom clusters Ag_7Au_6 which exhibited well-defined transitions in the absorption profile. The cluster was pure as confirmed by PAGE and the ESI-MS results showed that the cluster consisted of $Ag_7Au_6(H_2MSA)_{10}$.[61]

The Fujita group developed three-dimensional m×n arrays of metal ion clusters assembled as aromatic stacks of planar polynuclear metal complexes within columnar coordination cages. The polynuclear complexes and cage height program the final array structures of the metal ion clusters. The authors also synthesized a silver-sandwiched-hetero Au_3–Ag–Au_3 cluster by treating a hexanuclear Au_3–Au_3 cluster with Ag(I) ion.[62] They also synthesized Au_3–Ag–Au_3 and Au_3–Ag–Au_3–Ag–Au_3 multi-decker, discrete ion clusters within a self-assembled coordination cage (Figure 5.12). The multi-decker complexes were stable in aqueous solution because they were tightly encapsulated by the box-shaped cage and the triple-decker complex Au_3–Ag–Au_3–Ag–Au_3 could not be synthesized without the help of the cage. As the cage height could be systematically elongated, the family of box-shaped cages was a potential platform for the preparation of large multi-decker clusters.[63]

5.5 Conclusion and Perspective

In this chapter, we have summarized the recent research progress on Ag NCs protected by different scaffolds such as polymers, proteins, peptides and short molecules. We mainly focus on the synthesis methods, unique

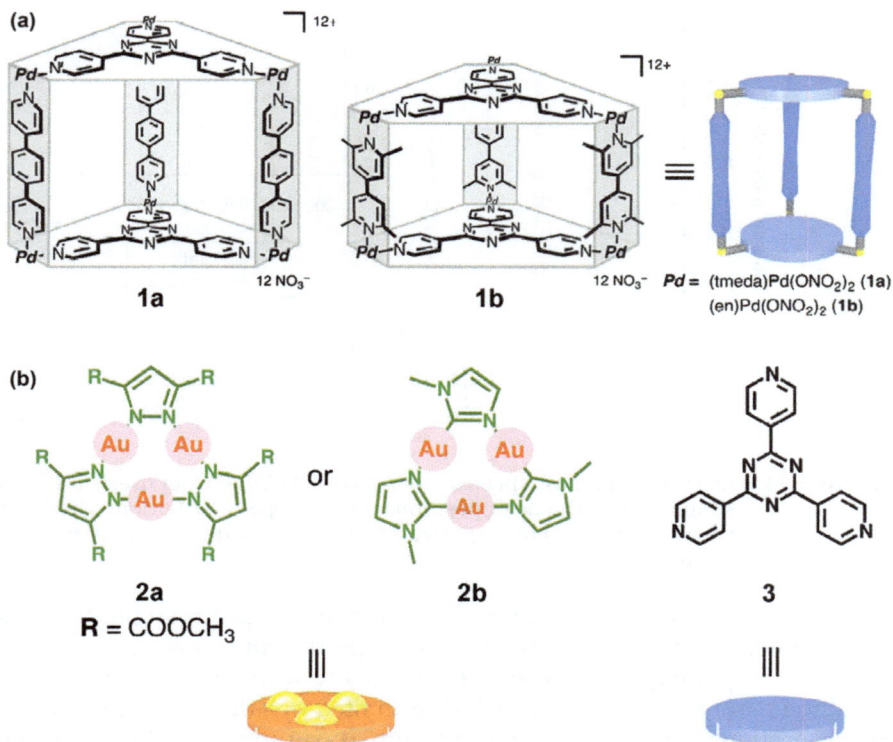

Figure 5.12   (a) Self-assembled box-shaped cage 1. (b) Accumulated trinuclear
                   Au^I complexes 2 and panel ligand 3 (reprinted with permission
                   from ref. 63).

properties and promising applications. As Ag NCs are in general less stable
than Au NCs, a suitable scaffold is essential to stabilize Ag NCs and the
interactions between the silver atoms and different functional groups are
utilized for the synthesis of Ag NCs. Moreover, various synthetic techniques
have also been developed such as "bottom-up" and "top-down" routes. It
should be noted that due to the size-dependent properties of Ag NCs, the
structures of Ag NCs are important for understanding the unique properties
of Ag NCs. Thus, different experimental techniques such as MS and SAXS
and computational simulations are developed for the determination of the
structures of Ag NCs and the prediction of the properties. To study the size-
dependent properties, various separation techniques are utilized for the
purification of Ag NCs, such as repeated fraction crystallization, chromatography and PAGE. Despite substantial progress in Ag NCs, there are still
great challenges in this fascinating field.

First, more scaffolds, especially molecules with good biocompatibility and
different synthetic approaches, should be developed for the synthesis of Ag
NCs. Of these, good biocompatibility will make the Ag NCs more suitable for

the applications of bioimaging and cancer therapy. Although various synthetic techniques have been developed recently, there is still a lack of efficient routes for preparing size-precise Ag NCs of high quality and high yield. The high monodispersity and purity are the critical points of the synthesis of Ag NCs due to their size-dependent properties. Thus, an efficient route for the preparation of size-controlled Ag NCs and easy separation methods will be challenging tasks in future synthetic work.

Second, as the unique properties of Ag NCs are strongly dependent on their small sizes, accurate determination of the structures of Ag NCs is very important to understand their unique properties. Although many unique optical and electronic properties have been discovered, the structures of most Ag NCs are still unknown. Limited by the difficulty of growing high quality single crystals of Ag NCs, there are very few Ag NCs which have been determined by SAXS. In recent years, more and more theoretical methods have been developed to reveal the relationships between the structures and the properties of Ag NCs.

Finally, Ag NCs are mainly applied for the detection of metal ions and bioimaging. Whereas a number of Ag NCs have been synthesized, the preparation of Ag NCs with good stability and biocompatibility is still a great challenge due to the complex environment of cells. The applications of Ag NCs based on the magnetic properties and catalytic activities are few and more work based on Ag NCs is needed to catalyze different reactions. As the scaffolds are necessary to synthesize stable and monodisperse Ag NCs and the active sites on the surface of Ag NCs are blocked by the capping ligands, it is a great challenge to synthesize surface-naked Ag NCs with high-stability and catalytic activity.

Overall, significant progress has been achieved in past decades in the preparation of Ag NCs, future work will mainly focus on the challenges in the synthesis, understanding the unique properties and applications of Ag NCs. With further systematic experimental and theoretical efforts, we believe that Ag NCs will find wide applications in environment, new energy, medicine, biology, *etc.*

Acknowledgements

This work was supported by the National Natural Science Foundation of China with the Grant Numbers of 21190040, 11174105 and 91227114 and 973 Projects 2010CB933600.

References

1. Z. X. Luo, G. U. Gamboa, J. C. Smith, A. C. Reber, J. U. Reveles, S. N. Khanna and A. W. Castleman Jr, *J. Am. Chem. Soc.*, 2012, **134**, 18973.
2. J. Zheng and R. D. Dickson, *J. Am. Chem. Soc.*, 2002, **124**, 13982.
3. W. Lesniak, A. U. Bielinska, K. Sun, K. W. Janczak, X. Y. Shi, J. R. Baker Jr. and L. P. Balogh, *Nano. Lett.*, 2005, **5**, 2123.

4. J. G. Zhang, S. Q. Xu and E. Kumacheva, *Adv. Mater.*, 2005, **17**, 2336.
5. L. Shang and S. J. Dong, *Chem. Commun.*, 2008, 1088.
6. L. Shang and S. J. Dong, *J. Mater. Chem.*, 2008, **18**, 4636.
7. I. Diez, M. Pusa, S. Kulmala, H. Jiang, A. Walther, A. S. Goldmann, A. H. E. Muller, O. Ikkala and R. H. A. Ras, *Angew. Chem., Int. Ed.*, 2009, **48**, 2122.
8. L. Koponen, L. O. Tunturivuori, M. J. Puska and Y. Hancock, *J. Chem. Phys.*, 2010, **132**, 214301.
9. H. X. Xu and K. S. Suslick, *ACS Nano*, 2010, **4**, 3209.
10. X. M. Wang, S. P. Xu and W. Q. Xu, *Nanoscale*, 2011, **3**, 4670.
11. S. H. Liu, F. Lu and J. J. Zhu, *Chem. Commun.*, 2011, **47**, 2661.
12. M. A. H. Muhammed, F. Aldeek, G. Palui, L. Trapiella-Alfonso and H. Mattoussi, *ACS Nano*, 2012, **6**, 8950.
13. T. Wen, F. Qu, N. B. Li and H. Q. Luo, *Anal. Chim. Acta*, 2012, **749**, 56.
14. B. G. Ershov and A. Henglein, *J. Phys. Chem. B*, 1998, **102**, 10663.
15. A. Mathew, P. R. Sajanlal and T. Pradeep, *J. Mater. Chem.*, 2011, **21**, 11205.
16. A. Ledo-Suarez, J. Rivas, C. F. Rodriguez-Abreu, M. J. Rodriguez, E. Pastor, A. Hernandez-Greus, S. B. Oseroff and M. A. Lopez-Quintela, *Angew. Chem., Int. Ed.*, 2007, **46**, 8823.
17. S. S. Narayanan and S. K. Pal, *J. Phys. Chem. C*, 2008, **112**, 4874.
18. C. Y. Shao, B. Yuan, H. Q. Wang, Q. Zhou, Y. L. Li, Y. F. Guan and Z. X. Deng, *J. Mater. Chem.*, 2011, **21**, 2963.
19. C. L. Guo and J. Irudayaraj, *Anal. Chem.*, 2011, **83**, 2883.
20. X. L. Guevel, B. Hotzer, G. Jung, K. Hollemeyer, V. Trouillet and M. Schneider, *J. Phys. Chem. C*, 2011, **115**, 10955.
21. S. Choi, R. M. Dickson, J. Lee and J. H. Yu, *Photochem. Photobiol. Sci.*, 2012, **11**, 274.
22. J. M. Slocik, J. T. Moore and D. W. Wright, *Nano Lett.*, 2002, **2**, 3.
23. J. H. Yu, S. A. Patel and R. M. Dickson, *Angew. Chem., Int. Ed.*, 2007, **46**, 2028.
24. A. Kulesza, R. Mitric and V. Bonacic-Koutecky, *J. Phys. Chem. A*, 2009, **113**, 3783.
25. S. Kumar, M. D. Bolan and T. P. Bigioni, *J. Am. Chem. Soc.*, 2010, **132**, 13141.
26. C. X. Wang, L. Xu, Y. Wang, D. Zhang, X. D. Shi, F. X. Dong, K. Yu, Q. Lin and B. Yang, *Chem. – Asian J.*, 2012, 7, 1652.
27. B. Adhikari and A. Baneriee, *Chem. – Eur. J.*, 2010, **16**, 13698.
28. Y. Y. Cui, Y. L. Wang, R. Liu, Z. P. Sun, Y. T. Wei, Y. L. Zhao and X. Y. Gao, *ACS Nano*, 2011, **5**, 8684.
29. J. S. Guo, S. Kumar, M. Bolan, A. Desireddy, T. P. Bigioni and W. P. Griffith, *Anal. Chem.*, 2012, **84**, 5304.
30. R. G. Freeman, K. C. Grabar, K. J. Allison, R. M. Bright, J. A. Davis, A. P. Guthrie, M. B. Hommer, M. A. Jackson, P. C. Smith, D. G. Walter and M. J. Natan, *Science*, 1995, **267**, 1629.

31. L. L. Zhao, L. Jensen and G. C. Schatz, *J. Am. Chem. Soc.*, 2006, **128**, 2911.
32. L. Jensen, L. L. Zhao and G. C. Schatz, *J. Phys. Chem. C*, 2007, **111**, 4756.
33. D. Y. Wu, X. M. Liu, S. Duan, X. Xu, B. Ren, S. H. Lin and Z. Q. Tian, *J. Phys. Chem. C*, 2008, **112**, 4195.
34. A. H. Pakiari and Z. Jamshidi, *J. Phys. Chem. A*, 2007, **111**, 4391.
35. A. Kulesza, R. Mitric, V. Bonacic-Koutecky, B. Bellina, I. Compagnon, M. Broyer, R. Antoine and P. Dugourd, *Angew. Chem., Int. Ed.*, 2011, **50**, 878.
36. L. Maretti, P. S. Billone, Y. Liu and J. C. Scaiano, *J. Am. Chem. Soc.*, 2009, **131**, 13972.
37. V. Soto-Verdugo, H. Metiu and E. Gwinn, *J. Chem. Phys.*, 2010, **132**, 195102.
38. X. Yang, L. F. Gan, L. Han, E. K. Wang and J. Wang, *Angew. Chem., Int. Ed.*, 2013, **52**, 2022.
39. L. Shang, S. J. Dong and G. U. Nienhaus, *Nano Today*, 2011, **6**, 401.
40. W. W. Guo, J. P. Yuan, Q. Z. Dong and E. K. Wang, *J. Am. Chem. Soc.*, 2010, **132**, 932.
41. W. W. Guo, J. P. Yuan and E. K. Wang, *Chem. Commun.*, 2009, 3395.
42. B. Y. Han and E. K. Wang, *Biosens. Bioelectron.*, 2011, **26**, 2585.
43. S. M. Choi, R. M. Dickson and J. H. Yu, *Chem. Soc. Rev.*, 2012, **41**, 1867.
44. L. B. Zhang, J. B. Zhu, S. J. Guo, T. Li, J. Li and E. K. Wang, *J. Am. Chem. Soc.*, 2013, **135**, 2403.
45. N. Sandhyarani and T. Pradeep, *J. Mater. Chem.*, 2000, **10**, 981.
46. T. U. B. Rao and T. Pradeep, *Angew. Chem., Int. Ed.*, 2010, **49**, 3925.
47. T. U. B. Rao, B. Nataraju and T. Pradeep, *J. Am. Chem. Soc.*, 2010, **132**, 16304.
48. I. Chakraborty, A. Govindarajan, J. Erusappan, A. Ghosh, T. Pradeep, B. Yoon, R. L. Whetten and U. Landman, *Nano Lett.*, 2012, **12**, 5861.
49. Z. K. Wu, E. Lanni, W. Q. Chen, M. E. Bier, D. Ly and R. C. Jin, *J. Am. Chem. Soc.*, 2009, **131**, 16672.
50. Z. K. Wu, D. E. Jiang, E. Lanni, M. E. Bier and R. C. Jin, *J. Phys. Chem. Lett.*, 2010, **1**, 1423.
51. O. M. Bakr, V. Amendola, C. M. Aikens, W. Wenseleers, R. Li, L. D. Negro, G. C. Schatz and F. Stellacci, *Angew. Chem., Int. Ed.*, 2009, **48**, 5921.
52. M. Pelton, Y. Tang, O. M. Bakr and F. Stellacci, *J. Am. Chem. Soc.*, 2012, **134**, 11856.
53. N. Cathcart, P. Mistry, C. Makra, B. Pietrobon, N. Coombs, M. Jelokhani-Niaraki and V. Kitaev, *Langmuir*, 2009, **25**, 5840.
54. N. Nishida, H. Yao, T. Ueda, A. Sasaki and K. Kimura, *Chem. Mater.*, 2007, **19**, 2931.
55. H. J. Xiang, S. H. Wei and X. G. Gong, *J. Am. Chem. Soc.*, 2010, **132**, 7355.
56. M. S. Liao, P. Bonifassi, J. Leszczynski, P. C. Ray, M. J. Huang and J. D. Watts, *J. Phys. Chem. A*, 2010, **114**, 12701.

57. M. Takesue, T. Tomura, M. Yamada, K. Hata, S. Kuwamoto and T. Yonezawa, *J. Am. Chem. Soc.*, 2011, **133**, 14164.
58. C. Kumara and A. Dass, *Nanoscale*, 2011, **3**, 3064.
59. S. Malola and H. Hakkinen, *J. Phys. Chem. Lett.*, 2011, **2**, 2316.
60. S. R. Biltek, S. Mandal, A. Sen, A. C. Reber, A. F. Pedicini and S. N. Khanna, *J. Am. Chem. Soc.*, 2013, **135**, 26.
61. T. Udayabhaskararao, Y. Sun, N. Goswami, S. K. Pal, K. Balasubramanian and T. Pradeep, *Angew. Chem., Int. Ed.*, 2012, **51**, 2155.
62. T. Osuga, T. Murase, K. Ono, Y. Yamauchi and M. Fujita, *J. Am. Chem. Soc.*, 2010, **132**, 15553.
63. T. Osuga, T. Murase and M. Fujita, *Angew. Chem., Int. Ed.*, 2012, **51**, 12199.

CHAPTER 6

Novel Synthetic Strategies for Thiolate-Protected Au and Ag Nanoclusters: Towards Atomic Precision and Strong Luminescence

XUN YUAN,[†] QIAOFENG YAO,[†] YONG YU,[†] ZHENTAO LUO[†] AND JIANPING XIE*

Department of Chemical and Biomolecular Engineering, National University of Singapore, 10 Kent Ridge Crescent, Singapore, 119260
*Email: chexiej@nus.edu.sg

6.1 Introduction

Thiolate-protected metal (*e.g.*, Au and Ag) nanoclusters or Au/Ag NCs for short, typically containing a few to hundreds of metal atoms, are a subclass of metal nanoparticles (NPs).[1–3] Thiolate-protected Au/Ag NCs are smaller than 2 nm and can be denoted as $M_n(SR)_m$, where M and SR represent Au/Ag and thiolate ligand, respectively. Particles in this sub-2 nm size range have strong quantum confinement of free electrons,[1] resulting in discrete electronic transitions[4–6] and unique atomic packing structures.[7,8] These two features are distinctively

---

[†]These four authors contributed equally to this chapter.

RSC Smart Materials No. 7
Functional Nanometer-Sized Clusters of Transition Metals: Synthesis, Properties and Applications
Edited by Wei Chen and Shaowei Chen
© The Royal Society of Chemistry 2014
Published by the Royal Society of Chemistry, www.rsc.org

different from that of large Au/Ag NPs (>2 nm). In particular, Au/Ag NPs show quasi-continuous electronic states and have a face-centered cubic (fcc) atomic packing, leading to a characteristic surface plasmon resonance (SPR) property.[9-11] In contrast, the discrete and size-dependent electronic transitions in NCs provide them with unique molecular-like properties such as HOMO–LUMO transitions,[5,12,13] optical chirality,[14-16] quantized charging,[12,17] magnetism,[18] and luminescence.[3,19,20] These interesting properties make Au/Ag NCs attractive for both fundamental studies and application explorations, which have also motivated a rapid development of efficient synthesis methods for high-quality Au/Ag NCs.[3,19-23]

In general, high-quality Au/Ag NCs must feature two particle attributes: good size monodispersity (with atomic precision) and/or strong luminescence. The physical and chemical properties of the NCs are highly sensitive to the cluster size (refers to the number of metal atoms and thiolate ligands per cluster) owing to their size-dependent electronic structures. The size effect is more significant in the sub-2 nm size region compared to that of large NPs. Minimal variations in the number of metal atoms and thiolate ligands in a NC (*e.g.*, one metal atom or ligand number difference) can lead to totally different physical and chemical properties of the NCs, reflecting that the properties of the NCs are highly size-dependent. One example is the optical absorption feature of the Au/Ag NCs. Thiolate-protected Au/Ag NCs show molecular-like optical absorptions, and such optical properties are highly size-dependent.[24,25] Another example is the catalytic property of the NCs. Recent studies have shown that the catalytic activities of the NCs are dictated by the cluster size.[26-28] Besides cluster size, luminescence properties are another important attribute of Au/Ag NCs, which have been extensively investigated in recent years. For example, luminescent Au/Ag NCs have shown great potential in a variety of applications including sensors for environment and healthcare, and biomedical (*e.g.*, diagnostic and therapeutic) applications.[19,22,29,30]

The development of efficient synthesis methods for high-quality Au/Ag NCs is particularly important for using them in both fundamental and applied research. In this chapter, we will summarize recent advances in the synthesis of monodisperse or luminescent Au/Ag NCs, with an emphasis on the underlying principles that can affect the size and luminescence properties of the NCs. In general, there are three types of precursors that can be used to produce monodisperse or luminescent Au/Ag NCs. They are Au(i)/Ag(i)-thiolate or Au(i)/Ag(i)–SR complexes, polydisperse Au/Ag NCs, and large Au/Ag NPs (>2 nm). In line with this, there are three major processes that can convert the precursors to monodisperse or luminescent Au/Ag NCs. They are (1) reductive decomposition of the Au(i)/Ag(i)–SR complexes by a particular chemical reductant, (2) conversion of polydisperse Au/Ag NCs with help from thiolate ligands, and (3) decomposition of large Au/Ag NPs by thiolate ligands. Since different synthetic chemistry may be involved in different conversion processes, we will organize our discussion according to the conversion process for a particular precursor, with an emphasis on critical factors in precursor tailoring (*e.g.*, size, structure, and surface properties of the precursors) and conversion process control (*e.g.*, synthetic environment and reduction/conversion/decomposition kinetics). We will present our understanding on the underlying chemistry in achieving monodisperse or luminescent Au/Ag NCs. Since there are more studies on Au, Au will be used as an example to illustrate the synthesis of monodisperse or luminescent NCs for most methods. There is a large body of excellent work on this subject. However, this chapter is not intended to be comprehensive, and therefore for each method we will only provide some exemplary papers to illustrate the method rather than list all the references where the method was applied. Hopefully this chapter can help pave the way for the advance of the newly-developed Au/Ag NCs for practical applications.

6.2 General Synthetic Routes for Thiolate-Protected Au/Ag NCs: Brust and Brust-Like Methods

In the last two decades, increasing interest in solution-phase synthesis of thiolate-protected Au NCs has been evoked by the seminal work of Brust *et al.* in 1994.[31] In a typical Brust synthesis, a Au(III) salt (*e.g.*, HAuCl$_4$) in aqueous solution is first transferred to the organic phase by a phase-transfer agent such as tetraoctylammonium bromide (TOAB), followed by the reduction of Au(III) to Au(I) by a hydrophobic thiolate ligand (*e.g.*, dodecanethiol or DDT). The thiolate ligands will then interact strongly with Au(I) to form strong Au(I)–S bonds, resulting in polymeric Au(I)–SR complexes in the organic phase. The as-formed polymeric Au(I)–SR complexes are then reduced by a strong reducing agent (*e.g.*, NaBH$_4$) *via* a reductive decomposition process, leading to the formation of small thiolate-protected Au NPs with core sizes of 2–2.5 nm. In brief, two primary reactions are involved in the Brust method for Au NP synthesis:

Reduction of Au(III) to form Au(I)–SR complexes:

$$Au(III) + H–SR \rightarrow Au(I)–SR \text{ complexes} \tag{6.1}$$

Reductive decomposition of Au(I)–SR complexes to form Au NCs:

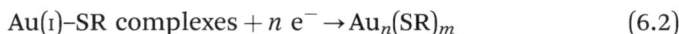

$$Au(I)–SR \text{ complexes} + n \text{ } e^- \rightarrow Au_n(SR)_m \tag{6.2}$$

The particle size in the Brust method can be further reduced to the sub-2 nm size regime. One effective way is to increase the ratio of protecting ligands (thiolate) to metals (*e.g.*, Au) or the thiol-to-Au ratio ($R_{[SR]/[Au]}$). For example, in a pioneering work of Whetten *et al.*, the size of Au NPs was reduced to ~1.5 nm by increasing $R_{[SR]/[Au]}$ to 2 (this value was 1 in the original Brust method).[32] Laser desorption ionization mass spectrometry (LDI-MS) was then used to characterize the size of as-synthesized Au NPs, indicating that the Au NPs had a distinct core mass of 28 kDa (in the form of Au$_n$S$_m$ due to the cleavage of the S–C bonds during the ionization process).[33,34] This Au NP species has been further confirmed to be Au$_{144}$ NCs.[34–37] Further increasing the $R_{[SR]/[Au]}$ to 3 reduced the particle size to 1.1–1.9 nm.[38] Similarly, the Murray group has synthesized Au NCs of ~1.5 nm *via* a modified Brust method by varying the $R_{[SR]/[Au]}$.[39]

Some early studies also revealed that the stability of thiolate-protected Au NCs in solution is largely dependent on the cluster size, where Au NCs with specific particle sizes show better chemical stability than Au NCs with other sizes. For example, the Whetten group obtained stable Au NCs in the organic phase in the size range 1.1–1.9 nm by using a solubility-based fractionation and liquid-gravity column chromatography.[38] Interestingly, they observed that Au NCs with discrete core masses (or sizes, in the form of Au_nS_m) of 8, 14, 22, 29, and 38 kDa were remarkably abundant in the NC solution [Figure 6.1(a)]. This observation suggests that Au NCs with the above core sizes show superior stability over Au NCs with other core sizes. In 2005, the Tsukuda group obtained another important series of stable thiolate-protected Au NCs in water.[24] The authors used polyacrylamide gel electrophoresis (PAGE) to separate Au NCs with different sizes in water; these Au NCs were protected by a tripeptide glutathione or GSH, and can be denoted as $Au_n(SG)_m$, where SG represents the GSH ligand. Nine distinct bands were obtained in the PAGE gel, and their molecular formulae were determined to be $Au_{10}(SG)_{10}$, $Au_{15}(SG)_{13}$, $Au_{18}(SG)_{14}$, $Au_{22}(SG)_{16}$, $Au_{22}(SG)_{17}$, $Au_{25}(SG)_{18}$, $Au_{29}(SG)_{20}$, $Au_{33}(SG)_{22}$, and $Au_{39}(SG)_{24}$ by electrospray ionization mass spectrometry (ESI-MS) [Figure 6.1(b)]. These Au NCs constitute a series of cluster sizes that show relatively high stability in solution. Further studies have shown that the stability of these Au NCs is highly size-dependent. For example, some Au NCs, such as Au_{15} and Au_{18} NCs, showed relatively poor stability in solution, and they were considered kinetic trapping products. Others, such as Au_{25},[40–44] Au_{38},[33,45,46] and Au_{144}[47,48] NCs, were ultrastable in solution even under harsh conditions such as in the presence of excess thiolate ligands. Such Au NCs were treated as thermodynamically stable products. The stability of thiolate-protected Au NCs could be attributed to the electronic factor (or "superatom electronic theory") and/or the geometric factor.[42,44,49,50] We will discuss this in more detail in section 6.4.1.

Besides the ultrasmall size feature (<2 nm), a good size monodispersity is another requirement for high-quality thiolate-protected Au NCs. The discussion above indicates that Au NCs of different sizes may show similar chemical stability in solution, and therefore the Brust-like methods often produced polydisperse Au NCs. High-resolution separation techniques are required to obtain Au NCs with a specific size (with atomic precision). For example, solubility-based fractionation,[36,38,51,52] liquid chromatography,[53–55] and PAGE,[24,56–59] have been successfully applied to isolate monodisperse $Au_n(SR)_m$ with $n = 10$, 15, 18, 22, 25, 29, 38,[24] 40,[55,60] 55,[53,61] 68,[62] 102,[52] 130,[49] 144,[36,38] and 187.[49] However, the practical applications of these analytical separation techniques for the production of monodisperse Au NCs are limited by the tedious separation processes involved and small production scale (often in several to tens of micrograms, and they are also difficult to scale up).

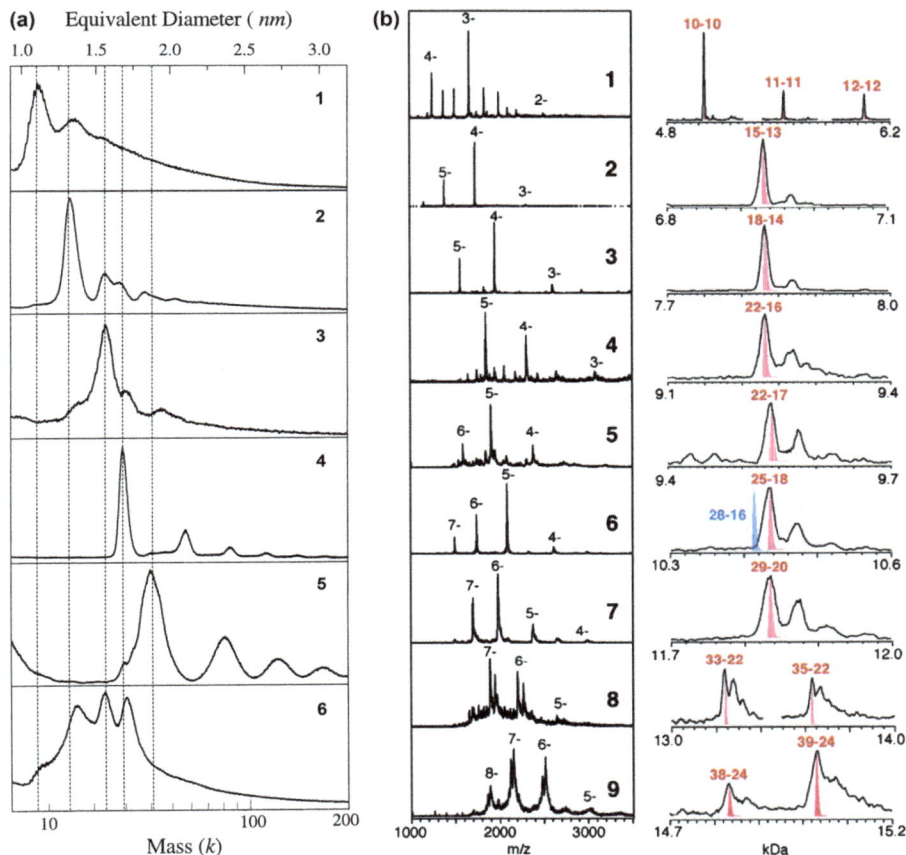

Figure 6.1 (a) Low-resolution LDI-MS spectra of thiolate-protected Au NCs (organic phase) with discrete core masses of 8 kDa (1), 14 kDa (2), 22 kDa (3), 29 kDa (4), 38 kDa (5), and crude product (6). Reproduced with permission from ref. 38. Copyright (1997) American Chemical Society. (b) Low-resolution (left) and high-resolution (right) ESI-MS spectra of $Au_n(SG)_m$ NCs (water phase) with discrete sizes of $n-m = 10-10$ (1), 15–13 (2), 18–14 (3), 22–16 (4), 22–17 (5), 29–20 (7), 33–22 (8), and 39–24 (9). Reproduced with permission from ref. 24. Copyright (2005) American Chemical Society.

From the large-scale production perspective, the direct synthesis of monodisperse Au NCs is attractive, which could also show higher efficiency or yield for Au NCs with a specific size. This has led to the development of a number of direct synthesis methods for monodisperse Au NCs in both organic and aqueous solutions. In contrast to the separation techniques, where the size-selection of the NCs is achieved by a physical separation, in the direct synthesis methods, the atomic precision control of the NCs is realized by a chemical selection. In particular, Au/Ag NCs with extraordinary stability in solution could be synthesized thermodynamically in a relatively harsh reaction environment, whereas those metastable NCs (less stable than the thermodynamically stable NCs) could be kinetically trapped during the synthesis in a mild and well-controlled reaction environment.

Besides ultrasmall size and good monodispersity (atomic precision), luminescence is another interesting feature of thiolate-protected Au/Ag NCs. Similar to the direct synthesis of monodisperse Au/Ag NCs, luminescent Au/Ag NCs could also be synthesized by manipulating either the thermodynamic or kinetic factor during the NC synthesis. The following are highlights of some good examples of applying thermodynamic and kinetic principles to tailor the size, monodispersity, and luminescence properties of Au/Ag NCs. The discussion is divided into three sections according to the types of precursors used to produce monodisperse or luminescent Au/Ag NCs. These sections are: reductive decomposition of Au(I)/Ag(I)–SR complexes, thiol etching of polydisperse Au/Ag NCs, and thiol etching of large Au/Ag NPs.

6.3 Reductive Decomposition of Au(I)/Ag(I)–SR Complexes

Au(I)/Ag(I)–SR complexes are ubiquitous precursors for the synthesis of thiolate-protected Au/Ag NCs.[31] In this section, we will use Au for the illustration. Au(I)–SR complexes are usually in polymeric forms as RS–[Au(I)–SR]$_n$, which result from the bridging effect of the thiolate ligands with the Au(I) centers.[63] A reductive decomposition process is generally used to convert Au(I)–SR complexes to thiolate-protected Au NCs, as illustrated in Equation (6.2). The reductive decomposition is dependent on the size and structure of Au(I)–SR complexes, which may be varied to tailor the size of Au NCs. In general, a uniform size and structure of Au(I)–SR complexes, and mild and controlled reduction kinetics are crucial for the direct synthesis of monodisperse Au NCs.[21] This principle is particularly important for the synthesis of Au NCs with metastable sizes, which are generally produced *via* a kinetic trapping method. In this section, we will highlight some recent studies on the synthesis of monodisperse Au/Ag NCs by tailoring the size and structure of Au(I)/Ag(I)–SR complexes and controlling the reductive decomposition kinetics during the synthesis.

6.3.1 Tailoring the Size of Au(I)/Ag(I)–SR Complexes

Au(I)–SR complexes are typically prepared by mixing a Au(III) salt (*e.g.*, HAuCl$_4$) with thiolate ligands, as illustrated in Equation 6.1. There are two common approaches to control the size of as-formed Au(I)–SR complexes. The first one is the kinetic method. In a typical kinetic method, a well-controlled reaction environment is created to kinetically trap Au(I)–SR complexes during their formation, where the as-formed complexes may not be the most stable species (or thermodynamically stable species) in the reaction solution. A slow reduction to convert Au(III) to Au(I) is generally preferred for the formation of

Au(I)–SR complexes with a narrow size distribution. The thermodynamic method is the second approach to control the size of Au(I)–SR complexes. Different from the kinetic trapping of Au(I)–SR complexes, in a typical thermodynamic method, only the most stable Au(I)–SR complexes are preserved in the solution, and other less stable (or metastable) complex species would be digested in the relatively harsh reaction conditions.

For example, Zhu *et al.* developed an efficient way to control the size of Au(I)–SR complexes by manipulating the kinetics of their formation [Figure 6.2(a)], where a particular aggregation state (or size) of the polymeric Au(I)–SR complexes was achieved by carefully controlling the reaction

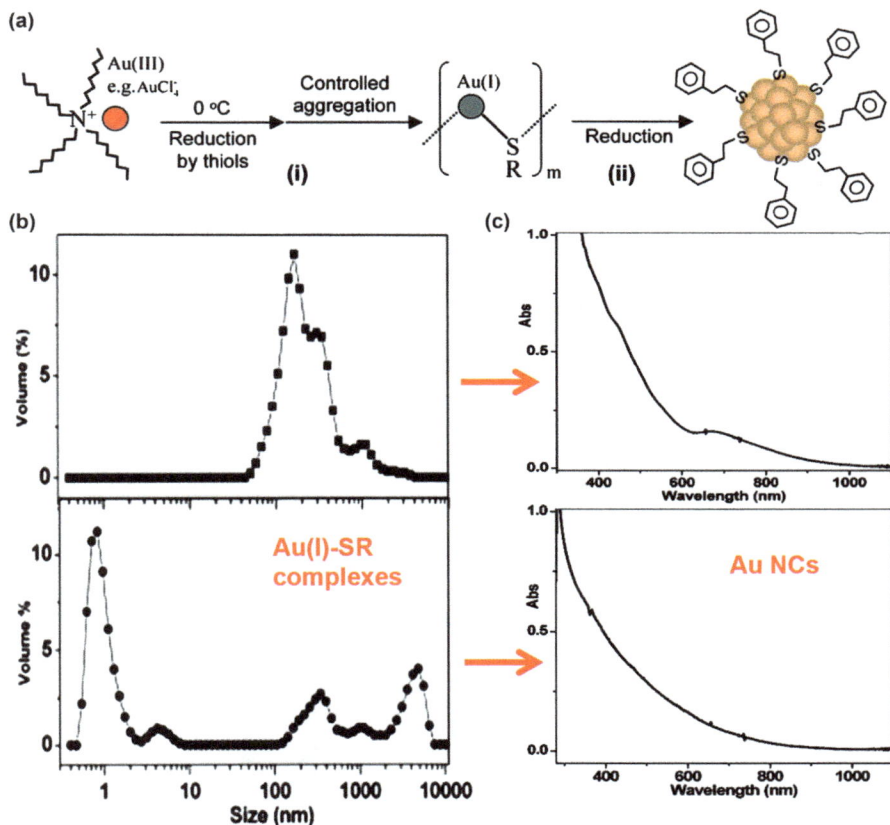

Figure 6.2 (a) Schematic illustration of the synthesis of Au$_{25}$(SR)$_{18}$ NCs *via* kinetically controlling the size of Au(I)–SR complexes. (b) Dynamic light scattering (DLS) spectra of Au(I)–SR complexes formed at 273 K (top) and room temperature (bottom); and (c) the corresponding UV-vis absorption spectra of the Au NCs synthesized by the NaBH$_4$ reduction of the Au(I)–SR complexes.
Reproduced with permission from ref. 64. Copyright (2008) American Chemical Society.

temperature (0 °C) and the extent of stirring (slow, ~ 30 rpm).[64] The as-formed polymeric Au(I)–SR complexes had a narrow size distribution of 100–400 nm (Figure 6.2(b), top panel), which, upon reductive decomposition, led to the formation of Au_{25} NCs in a good yield. The as-synthesized Au_{25} NCs showed a well-defined UV-vis absorption spectrum characteristic of that of thiolate-protected Au_{25} NCs (Figure 6.2(c), top panel). By comparison, the Au(I)–SR complexes prepared at room temperature showed a broad size distribution (*e.g.*, <2 nm, 100–400 nm, and >1 μm, Figure 6.2(b), bottom panel), and upon the reductive decomposition, they were converted to polydisperse Au NCs. These Au NCs showed a featureless UV-vis absorption spectrum (Figure 6.2(c), bottom panel), reflecting their polydisperse features.

The size tailoring of Au(I)/Ag(I)–SR complexes to synthesize monodisperse Au/Ag NCs can also be extended to the Ag NC system. For example, the size of Ag(I)–SR complexes can be controlled thermodynamically *via* a cyclic reduction–decomposition process, and upon reductive decomposition, the monodisperse complexes led to the formation of atomically precise $Ag_{16}(SR)_9$ NCs (Figure 6.3).[65] In a recent study, Yuan *et al.* first prepared Ag(I)–SR complexes (type-I complexes) by directly mixing $AgNO_3$ with the thiolate ligand GSH. The as-formed type-I complexes were a mixture of different-sized complex species, which were determined to be $Ag_{1–11}(SG)_{1–6}$ by matrix-assisted laser desorption/ionization time-of-flight (MALDI-TOF). A cyclic reduction–decomposition process was then used to thermodynamically select the ultrastable Ag(I)–SR complexes in solution, where other metastable

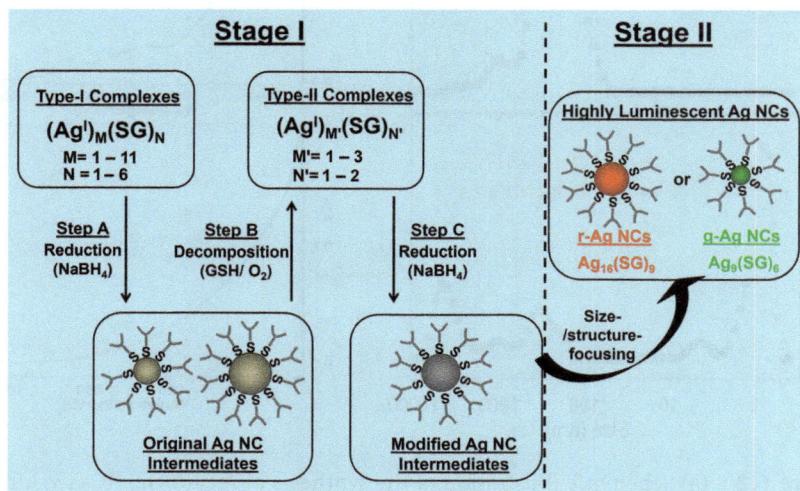

Figure 6.3   Schematic illustration of (Stage I) the cyclic reduction–decomposition process for narrowing the size distribution of Ag(I)–SR complexes and (Stage II) the subsequent synthesis of strongly emissive and atomically precise Ag NCs.
Reproduced with permission from ref. 65. Copyright (2013) Nature Publishing Group.

complex species were gradually decomposed. In a typical cyclic reduction–decomposition process, the type-I complexes were first reduced by a strong reducing agent $NaBH_4$, resulting in the formation of Ag NCs with a broad size distribution (6–19 Ag atoms, denoted as original Ag NCs). The original Ag NCs were not stable and gradually decomposed in the presence of excess GSH in solution, leading to the formation of a new family of $Ag(\textsc{i})$–SR complexes (type-II complexes). The type-II complexes had a narrow size distribution with a molecular formula of $Ag_{1-3}(SG)_{1-2}$. Further reduction of the type-II complexes led to the formation of Ag NCs with a narrow size distribution (18–20 Ag atoms, denoted as modified Ag NCs). A subsequent size-focusing process was used to finally convert the modified Ag NCs into atomically precise $Ag_{16}(SG)_9$ NCs. More interestingly, the $Ag_{16}(SG)_9$ NCs showed strong red emission with an emission wavelength of 647 nm. Moreover, this cyclic decomposition–reduction method can be used to tailor the size of Ag NCs. For example, smaller sized $Ag_9(SG)_6$ NCs were synthesized by fine-tuning the amount of the reductant $NaBH_4$. The as-synthesized $Ag_9(SG)_6$ NCs showed strong green emission with an emission wavelength of 595 nm.

6.3.2 Tailoring the Structure of Au(ı)/Ag(ı)–SR Complexes

The structure of $Au(\textsc{i})/Ag(\textsc{i})$–SR complexes is also pivotal to synthesizing monodisperse Au/Ag NCs. Thiolate ligands could affect the properties of $Au(\textsc{i})/Ag(\textsc{i})$–SR complexes. Steric hindrance and charge state of the thiolate ligands are two major factors that can affect the structure of $Au(\textsc{i})/Ag(\textsc{i})$–SR complexes. For example, the structure of $Au(\textsc{i})$–SR complexes could be modified by the steric hindrance provided by an extraneous surfactant such as cetyltrimethylammonium bromide (CTAB). Recently, Yuan *et al.* developed a new protection–deprotection method to synthesize $Au_{25}SR_{18}$ NCs [H–SR = cysteine (Cys) or 3-mercaptopropionic acid (MPA)]. In this study, a CTAB protecting shell was first grafted on the preformed $Au(\textsc{i})$–SR complexes *via* the formation of $(CTA)^+(COO)^-$ ion pairs between the COO^- anions in the complexes (each protecting ligand, Cys, has one carboxylic group) and CTA^+ cations in CTAB (Figure 6.4).[66] The CTAB-protected $Au(\textsc{i})$–SR complexes formed well-defined inverse micelles in toluene, which created a unique microenvironment for further reductive decomposition of the complexes. The additional steric effect of the CTAB protecting layer can significantly increase the liability of the encapsulated $Au(\textsc{i})$–SR complexes against reductive decomposition, leading to a fast synthesis (<10 min) of atomically precise $Au_{25}(SR)_{18}$ NCs. In addition, the CTAB protecting layer could be easily removed after the synthesis by using a counter ion to disrupt the $(CTA)^+(COO)^-$ ion pairs. The resultant surfactant free $Au_{25}(SR)_{18}$ could then be transferred back to the aqueous phase.

A pH-dependent approach for tailoring the aggregation extent of $Au(\textsc{i})$–SR complexes for the synthesis of $Au_{15}(SG)_{13}$ and $Au_{18}(SG)_{14}$ NCs was recently reported by Yao *et al.* (Figure 6.5).[67] The aggregation extent of $Au(\textsc{i})$–SR

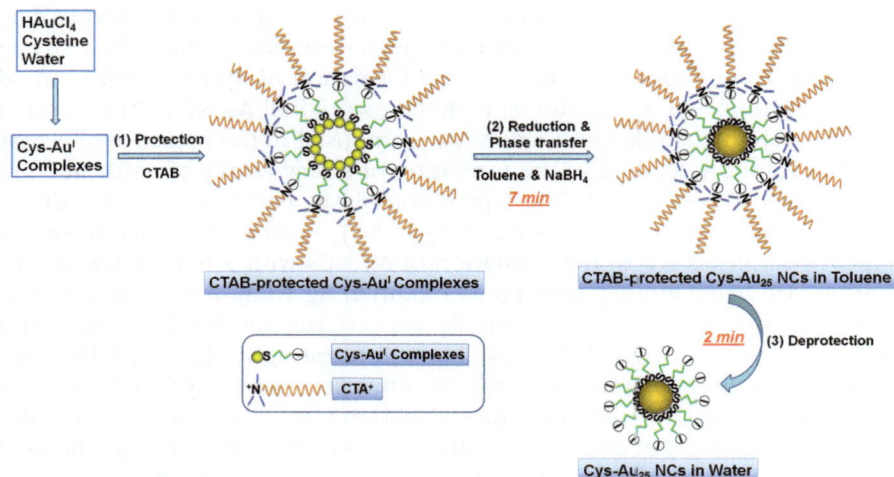

Figure 6.4   Schematic illustration of the synthesis of monodisperse Au$_{25}$ NCs by modifying the structure of the Au(I)–SR complexes with a CTAB protecting layer.
Reproduced with permission from ref. 66. Copyright (2012) American Chemical Society.

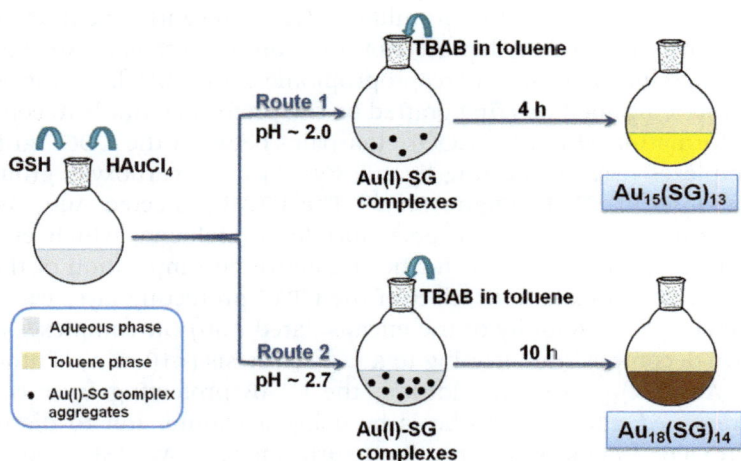

Figure 6.5   Schematic illustration of the synthesis of monodisperse Au$_{15}$ and Au$_{18}$ NCs by using the pH-sensitive aggregation–dissociation equilibrium to control the size of the Au(I)–SR complexes.
Reproduced with permission from ref. 67. Copyright (2013) Wiley-VCH.

complexes was controlled by delicate balancing of the attractive force [*e.g.*, Au(ɪ)···Au(ɪ) interactions] and the repulsive force (*e.g.*, electrostatic interactions). The thiolate ligand GSH possesses two carboxyl, one amine, and one thiol groups, corresponding to four pK_a: $pK_{a1} = 2.12$, $pK_{a2} = 3.53$, $pK_{a3} = 8.66$, and $pK_{a4} = 9.62$. Therefore, the charges of GSH (or H–SG) in the Au(ɪ)–SG complexes can be easily controlled by varying the solution pH, leading to the formation of Au(ɪ)–SG complexes with different aggregation extents (or size). At pH 2.7, a value above pK_{a1} and below pK_{a2}, –SG ligands were neutral in charge, therefore providing insufficient electrostatic repulsions for Au(ɪ)–SG complexes, which led to the formation of large Au(ɪ)–SG complexes. The large complexes resulted in a relatively low concentration of free Au(ɪ) in the reaction solution according to the aggregation-dissociation equilibrium. As a result, a mild reduction kinetics was created, which facilitated the formation of small $Au_{18}(SG)_{14}$ NCs with a high purity. Another intriguing finding in this study is that the aggregation extent of Au(ɪ)–SG complexes could affect the size of as-synthesized Au NCs. For example, by reducing the solution pH to ~2.0 (below all pK_a of GSH), much smaller Au(ɪ)–SG complexes were formed due to the increased electrostatic repulsion between the Au(ɪ)–SG complexes as most –SG ligands carry a positive charge at this pH. A mild reduction of such small Au(ɪ)–SG complexes led to the formation of $Au_{15}(SG)_{13}$ NCs.

6.3.3 Tailoring the Reductive Decomposition Kinetics

The reductive decomposition process converts Au(ɪ)/Ag(ɪ)–SR complexes to Au/Ag NCs, which could be engineered as such to control the size, monodispersity, and luminescence of thiolate-protected Au/Ag NCs. Delivering a mild reducing power and creating a uniform reduction environment are two major approaches to tailor the reductive decomposition kinetics. A mild reducing power is usually delivered by a proper selection and modification of the reducing agents, and a uniform reduction environment is generally made possible by applying external sources (*e.g.*, microwave) to facilitate the control.

6.3.3.1 *Delivering a Mild Reducing Power*

In the Brust-like methods, a strong reducing agent $NaBH_4$ was commonly used in the reductive decomposition process for the synthesis of thiolate-protected Au NCs. The strong reducing power of $NaBH_4$ and the fast reduction kinetics often lead to the formation of Au NCs with a broad size distribution because the newly-formed Au NCs have insufficient time for ripening or size-focusing. The monodispersity of as-synthesized Au NCs can be improved through decreasing the reducing power of the reducing agents, which could not only benefit the subsequent size-focusing of the newly-formed NCs but also facilitate the kinetic trapping of NCs with metastable sizes.

6.3.3.1.1   Decreasing the Reducing Power of the Reducing Agents.   A mild reducing power could be achieved with strong reducing agents (*e.g.*, $NaBH_4$) by decreasing their concentration (or reductant-to-Au ratio), limiting their diffusion, or inhibiting their reactivity. For example, Zhu *et al.* achieved a slow reduction of Au(I)–SR complexes by reducing the amount of $NaBH_4$ to 1 equivalent (relative to Au).[68] This value was 10 in the common Brust-like methods. The mild reduction kinetics in a low concentration of $NaBH_4$ led to the formation of atomically precise $Au_{24}(SR)_{20}$ NCs. This is in stark contrast to the use of a larger amount of $NaBH_4$ (10 equivalent relative to Au), where a thermodynamically stable $Au_{25}(SR)_{18}$ was formed.

The reducing power of $NaBH_4$ can also be controlled by limiting its diffusion. For example, Rao *et al.* developed a solid-state synthetic strategy to produce $Ag_9(SR)_7$ NCs (Figure 6.6).[69] The authors first prepared Ag(I)–SR complexes [H–SR = mercaptosuccinic acid (MSA)] by grinding a solid mixture of $AgNO_3$ and MSA. The as-formed Ag(I)–SR complexes were then crushed with 5 equivalent of solid $NaBH_4$ (relative to Ag), leading to the formation of a brownish–black powder. Water-extraction of such brownish–black powder finally formed $Ag_9(SR)_7$ NCs (The detailed preparation process can be found in Chapter 3). The authors attributed the successful preparation of $Ag_9(SR)_7$ NCs to the slow reduction kinetics made possible by the unique solid-state strategy.

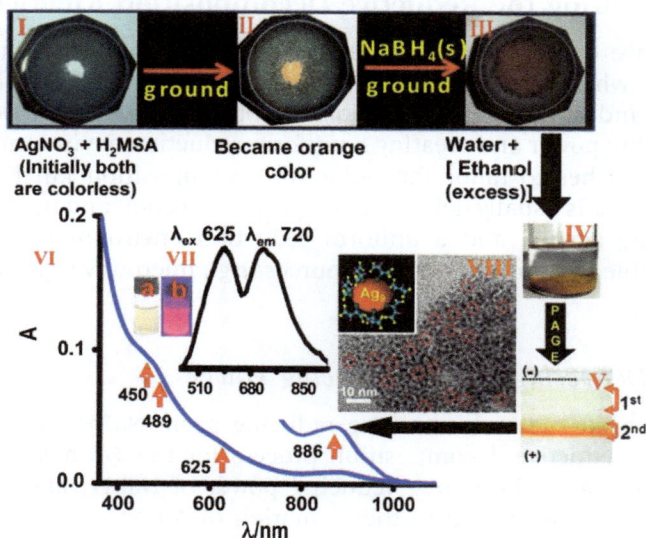

Figure 6.6   Schematic illustration (I–IV) of the synthesis of $Ag_9(SR)_7$ NCs *via* the solid-state approach, and the PAGE result (V), UV-vis absorption spectra (VI), luminescence spectra (VII) and TEM image (VIII) of as-synthesized $Ag_9(SR)_7$ NCs.
Reproduced with permission from ref. 69. Copyright (2010) American Chemical Society.

The reactivity of $NaBH_4$ is also dependent on the solvent conditions. For example, Bakr *et al.* used a mixed solvent of water and *N,N*-dimethylformamide (DMF) to control the reactivity of $NaBH_4$ for the synthesis of Ag NCs protected by 4-fluorothiophenolate (4-FTP). They found that when the ratio of DMF-to-water was high (7 : 1), the reactivity of $NaBH_4$ was low and a mild reduction environment was created, resulting in the formation of $Ag_{44}(SR)_{30}$.[70] The good monodispersity of as-synthesized Ag NCs was further confirmed by analytical ultracentrifugation (AUC) and ESI-MS.[71] Another good example is reported by Wu *et al.*, who used ethanol to suppress the reactivity of $NaBH_4$ and have successfully synthesized $Ag_7(SR)_4$ NCs [H–SR = 2,3-dimercaptosuccinic acid (DMSA)].[72]

6.3.3.1.2 Using Mild Reducing Agents.

A mild reducing environment can also be created by choosing a mild reducing agent. A number of mild reducing agents have been successfully applied to synthesize thiolate-protected Au/Ag NCs. Some good examples are borane-based reducing agents, reductive gases [*e.g.*, carbon monoxide (CO)], and reducing-cum-protecting thiolate ligands. The mild reducing agents can slow down the reductive decomposition kinetics and also facilitate the kinetic trapping of NCs with metastable sizes.

In organic synthesis, borane-based reducing agents with various reducing power and selectivities have been extensively exploited.[73–82] Recently, some of them [*e.g.*, borane-*tert*-butylamine complexes (TBAB) and sodium cyanoborohydride ($NaBH_3CN$)] have also been used to synthesize thiolate-protected Au/Ag NCs. Compared with $NaBH_4$, these borane-based reducing agents have weaker reducing power, and could provide better control for the NC synthesis. For example, Wu *et al.* used TBAB as the reducing agent and have successfully synthesized $Au_{19}(SR)_{13}$ NCs.[83] TBAB was also used by other researchers. Examples include the synthesis of biicosahedral Au_{25} NCs protected by mixed ligands (phosphine and thiolate ligands)[84] and two ultrasmall $Au_{15}(SG)_{13}$ and $Au_{18}(SG)_{14}$ NCs.[67] Besides TBAB, $NaBH_3CN$ has also been used to synthesize $Au_{18}(SG)_{14}$ NCs.[85]

Reductive gases, such as carbon monoxide CO, have recently emerged as a new class of promising reducing agents for the synthesis of monodisperse Au NCs. Besides the mild reducing capability, CO also shows unique coordination and catalysis features for Au. These features can help create a well-controlled growth process for Au NCs. In particular, CO can help stabilize the growing Au(0) core by its strong coordination with Au(0). In addition, the freshly-formed Au(0) surface can also catalyze the oxidation of CO, generating electrons on the Au(0) surface which will be taken by the Au(I) and resulting in the *in situ* deposition of Au atoms on the Au(0) surface. Taking all of these effects together, a unique reaction environment and a well-controlled growth process for Au NCs will be achieved in the CO-reduction system, which lead to the formation of monodisperse Au NCs. Yu *et al.* first developed this CO-reduction method and have successfully synthesized high purity $Au_{10-12}(SR)_{10-12}$, $Au_{15}(SR)_{13}$, $Au_{18}(SR)_{14}$, and $Au_{25}(SR)_{18}$

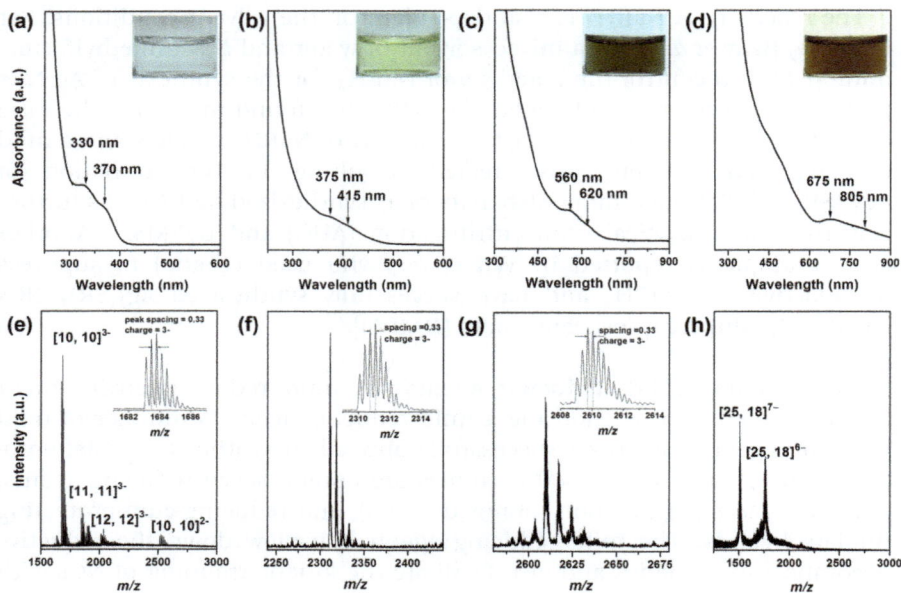

Figure 6.7   UV-vis absorption (top row) and ESI-MS (bottom row) spectra of (a) Au_{10-12}, (b) Au_{15}, (c) Au_{18}, and (d) Au_{25} NCs prepared by the CO-reduction method. The insets show digital photos of the respective products (top row) and the corresponding isotope patterns (bottom row).
Reproduced with permission from ref. 86. Copyright (2013) American Chemical Society.

NCs (Figure 6.7).[86,87] The size-selection synthesis of Au NCs was made possible by the pH-dependent control of reduction kinetics.[86] Large-scale production (gram scale) can also be achieved by using the CO-reduction method. Another attractive feature of the CO-reduction method is that the slow reduction process in the presence of CO can be used to monitor the formation process of Au NCs because the slow and well-controlled reduction kinetics allow the isolation and characterization (*e.g.*, *via* time-course UV-vis spectroscopy and mass spectrometry) of key NC intermediates in the formation of monodisperse Au NCs, which are, however, impossible to achieve in a fast $NaBH_4$-reduction process.[87]

Combining the mild reducing and protecting capability into the thiolate ligands is another interesting strategy to synthesize monodisperse Au/Ag NCs. Replacing extraneous reducing agents like $NaBH_4$ by thiolate ligands, especially those ligands with good biocompatibility (*e.g.*, GSH and Cys) is also attractive for the use of Au/Ag NCs in biomedical applications. Custom-designed peptides provide a good platform to test this concept because the thiol-containing amino acid (Cys or C) and reductive amino acid (*e.g.*, tyrosine or Y)[88] can be readily integrated into a custom-designed peptide. For example, Cui *et al.* designed an oligopeptide (CCYRGRKKRRQRRR)

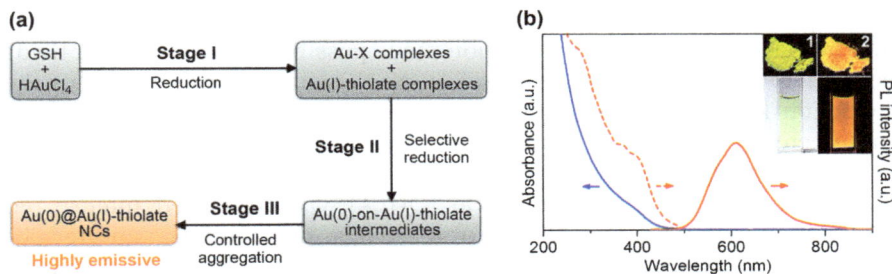

Figure 6.8 (a) Schematic illustration of the synthesis of luminescent Au(0)@Au(I)-thiolate NCs using a thiolate ligand as protecting-cum-reducing agent. (b) UV-vis absorption (blue line), photoexcitation (red dotted line) and photoemission (red solid line) spectra of the Au(0)Au(I)–thiolate NCs. Reproduced with permission from ref. 90. Copyright (2012) American Chemical Society.

containing C and Y, and the sequence of CCY can be used to reduce Ag ions to form Ag NCs.[89]

Besides protecting Au NCs, the thiol group (or the disulfide group formed during the synthesis) of thiolate ligands also possesses a mild reducing capability for the synthesis of Au NCs. Recently, Luo *et al.* demonstrated a facile and one-pot synthesis of highly luminescent Au NCs by using GSH as the reducing-cum-protecting agent [Figure 6.8(a)].[90] In a typical synthesis of luminescent Au NCs, HAuCl$_4$ was first reduced to Au(I) by using a relatively low thiol-to-Au ratio, which was 1.5 : 1 instead of the ratio of 3 : 1 commonly used in Brust-like methods. As a result, the GSH was insufficient to co-ordinate and stabilize all freshly-generated Au(I) to form Au(I)-SG complexes, leaving a small amount of Au(I) coordinated by non-thiolate groups (here-after referred to as X) to form Au(I)–X complexes. These Au(I)–X complexes were less stable than Au(I)–SG complexes, and could be reduced to form Au(0) nuclei by the disulfide groups of GS–SG (the oxidized GSH formed during the synthesis) at an elevated temperature (343 K). The *in situ* gener-ated Au(0) nuclei provided aggregation sites for Au(I)–SG complexes, leading to the formation of a new family of core-shell structured Au(0)@Au(I)-SG NCs. The as-synthesized Au NCs have a high content of Au(I)–SG complexes and showed a strong orange emission at ∼610 nm. They also have high quantum yield (QY) of ∼15%. The authors also found that the luminescence of the NCs originated from the aggregation-induced emission of the Au(I)–SG complexes on the NCs [Figure 6.8(b)].

6.3.3.2 *Creating a Uniform Reduction Environment*

Creating a uniform reduction environment is another efficient way to control the reduction kinetics. Besides stirring, applying an external physical field could also help create a homogeneous reduction environment. External fields like microwave can deliver high-intensity energy to the reaction system in a typical smooth and homogeneous manner. The high energy intensity and homogeneous energy distribution made possible by such external fields could help synchronize the growth of Au/Ag NCs and properly accelerate the reduction kinetics, leading to the formation of fairly monodisperse or luminescent Au/Ag NCs in a relatively short reaction time. For example, Shang *et al.* developed a microwave-assisted method to synthesize luminescent Au NCs protected by dihydrolipoic acid (DHLA), which showed intense red emission at 715 nm.[91] The uniform energy distribution delivered by microwave allowed the resultant Au NCs to have a narrow size distribution (1.6 ± 0.3 nm), which was corroborated by their sharp absorption features with three absorption peaks at 480, 580, and 690 nm.

6.4   Thiol Etching of Polydisperse Au/Ag NCs

Besides $Au(I)/Ag(I)$–SR complexes, polydisperse Au/Ag NCs can also be used as precursors to produce monodisperse Au/Ag NCs. Different from the $Au(I)/Ag(I)$–SR complexes whose metal centers have a higher oxidative state than those in the NCs, polydisperse NCs have a comparable metallic core (*e.g.*, in size and oxidation state) with the final NC product. Therefore, the conversion of polydisperse NCs to monodisperse counterparts is mainly achieved by size-focusing, which is typically assisted by thiolate ligands and can be referred to as thiol etching. In a typical thiol etching process, a mixture of different-sized NCs undergoes a size evolution and re-structuring, which is assisted by the strong metal–thiolate interaction in the presence of excess thiolate ligands. Since different-sized NCs have different stability in the reaction solution, after the thiol etching only the most stable NC species can be retained in the reaction solution. The thermodynamic factor dictates the thiol etching or size-focusing process, and has been widely used to synthesize monodisperse Au/Ag NCs.[92] Experimental conditions, such as the reaction temperature, solvents, and type and amount of reducing agents and thiolate ligands, have strong effects on the thiol etching process, and can be varied as such for the control of synthesis of Au/Ag NCs.

In this section, thiolate-protected Au NCs will be used as an example to illustrate the underlying chemistry for size-focusing. The well-documented $Au_{25}(SR)_{18}$ is chosen to show that size-focusing is a generic method for the synthesis of monodisperse Au NCs. We will also discuss some critical factors that can help improve the practicality of the size-focusing methods.

6.4.1   Size-Dependent Stability – the Driving Force for Size-Focusing

In a typical size-focusing process, polydisperse Au NCs with a broad size distribution are subjected to thiol etching, where only stable sizes are retained and less stable sizes are gradually decomposed or converted to more stable sizes (Figure 6.9). A number of experimental results revealed

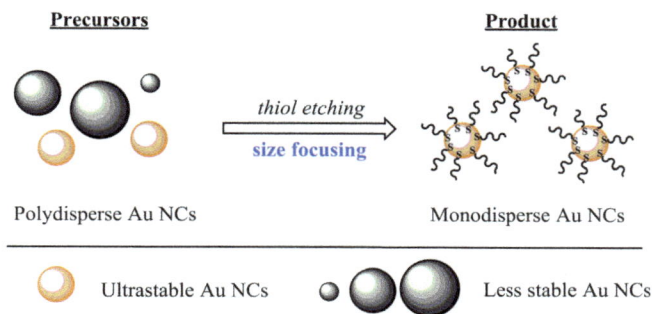

Figure 6.9 Schematic illustration of the size-focusing of Au NCs in the presence of excess thiolate ligands.

that Au NCs with some specific sizes, such as $Au_{25}(SR)_{18}$, $Au_{38}(SR)_{24}$, and $Au_{144}(SR)_{60}$, are more stable in solution against thiol etching than Au NCs of other sizes.

6.4.1.1 Stable $Au_{25}(SR)_{18}$ NCs

The $Au_{25}(SR)_{18}$ NC is one of the most stable NC species among a series of discrete-sized Au NCs. It was first identified as the predominant species among a series of $Au_n(SG)_m$ NCs isolated by PAGE.[56,57] Low-resolution mass spectrometric analysis based on MALDI-TOF and ESI-MS suggested that this NC species had a molecular weight of ~ 10.4 kDa.[56] Subsequent studies using PAGE and ESI-MS determined its formula as $Au_{25}(SG)_{18}$ along with several other Au NC species $[Au_n(SG)_m, n = 10–39]$ in a seminal work by the Tsukuda group (see section 6.2 for more details).[24] In a follow-up study, the stability of these $Au_n(SG)_m$ NCs was evaluated in the presence of excess GSH at an elevated temperature (328 K).[93] The UV-vis spectra of these Au NCs before and after the thiol etching were compared, and it was found that $Au_{25}(SG)_{18}$ was the most stable Au NC species in solution. In contrast, Au NCs with core sizes larger than Au_{25} were converted to Au_{25} NCs, while those Au NCs with core sizes smaller than Au_{25} were decomposed to Au(I)–SR complexes [Figure 6.10(a)].

The ultrastable feature of $Au_{25}(SR)_{18}$ was also suggested by monitoring the size evolution of $Au_n(SR)_m$ (H–SR = phenylethanethiol) in a prolonged aging process.[43] The time-course MALDI-TOF mass spectra clearly showed that $Au_{25}(SR)_{18}$ NCs were more stable in the presence of excess thiolate ligands than Au NCs of other sizes [Figure 6.10(b)]. The Au NCs formed at ~ 5 min after the reductive decomposition of Au(I)–SR complexes showed a broad size distribution, including Au_{25}, Au_{38}, Au_{68}, and Au_{102} species. During the aging process, the signals of Au_{38}, Au_{68}, and Au_{102} in the mass spectra gradually diminished. A prolonged aging to 3 days led to a distinct signal from Au_{25} NCs in the mass spectrum, and other Au NCs larger than Au_{25} were converted to Au_{25} NCs or decomposed to Au(I)–SR complexes.

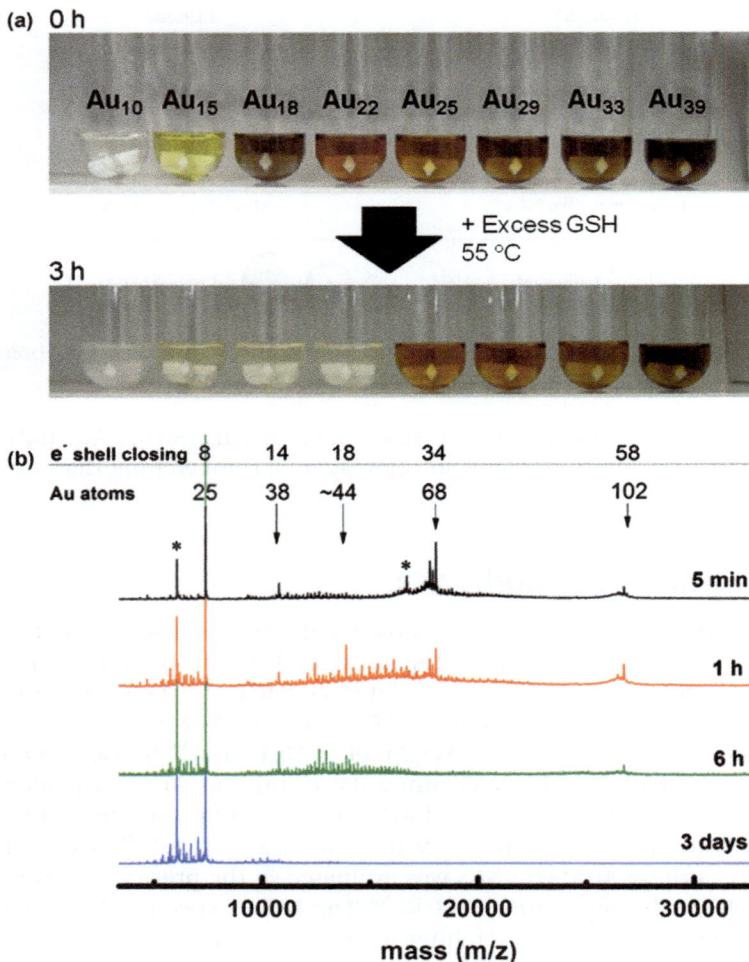

Figure 6.10   (a) Digital photos of $Au_n(SG)_m$ with $(n, m) = (10, 10)$, $(15, 13)$, $(18, 14)$, $(22, 16)$, $(22, 18)$, $(29, 20)$, $(33, 22)$, and $(39, 24)$ before (0 h) and after (3 h) the thiol etching process by using excess GSH at 328 K. Reproduced with permission from ref. 93. Copyright (2007) Wiley-VCH. (b) Time-course MALDI-TOF mass spectra of the reaction mixture in a one-pot synthesis of $Au_{25}(SR)_{18}$.
Reproduced with permission from ref. 43. Copyright (2009) American Chemical Society.

6.4.1.2   Stable Au_{38} $(SR)_{24}$ NCs

$Au_{38}(SR)_{24}$ is another stable Au NC species that can withstand the harsh thiol etching. Au_{38} NCs were first isolated by a repeated fractional crystallization. LDI-MS suggested its core size to be ~ 8 kDa.[38] ESI-MS further determined its formula as $Au_{38}(SR)_{24}$.[36,46,94] The thiol tolerant feature of Au_{38} NCs was first observed by Schaaff and Whetten. They performed a thiol etching experiment on Au NCs protected by hexanethiol (C_6S-H) with a core mass of ~ 14 kDa using neat dodecanethiol ($C_{12}S-H$) as the etchants. The size evolution of the Au NCs was monitored by LDI-MS and UV-vis spectroscopy. The authors observed that, after 40 h of thiol etching at 343 K, the 14 kDa NC species was finally converted to Au_{38}, which suggests that Au_{38} is more stable than the 14 kDa NC species.[33] Toikkanen *et al.* further confirmed the good stability of Au_{38} NCs against thiol etching by electrochemical and MALDI-TOF measurements of the reaction mixture before and after thiol etching. They found that larger Au NCs with core masses of 22 and 29 kDa were eliminated during the thiol etching process, whereas the Au_{38} NC species was retained in the harsh reaction conditions.[45]

6.4.1.3 Stable $Au_{144}(SR)_{60}$ NCs

Apart from $Au_{25}(SR)_{18}$ and $Au_{38}(SR)_{24}$, Au NCs with other sizes such as $Au_{144}(SR)_{60}$ are also stable in the presence of excess thiolate ligands. Similar to $Au_{38}(SR)_{24}$, $Au_{144}(SR)_{60}$ was also first isolated by fractional crystallization. Its core mass was determined to be ~ 29 kDa by LDI-MS.[38] Schaaff *et al.* further observed that the homogeneity of this 29 kDa NC species (prepared by fractional crystallization) can be improved by a thiol etching process at 333 K, where only the 29 kDa NC species was preserved and other NC species were eliminated.[35] Recently, Chaki *et al.* applied a thiol etching process (353 K, >22 h) to a mixture of Au NCs with a molecular weight of 8 and 29 kDa, and separated the product by solubility-based fractionation. The authors also assigned the 29 kDa NC species to $Au_{144}(SC_{12})_{59}$ by ESI-MS.[36] More recently, Qian and Jin developed a two-step method to synthesize the 29 kDa NC species, where a modified Brust method and a thiol etching process at 353 K were applied. The product was determined to be $Au_{144}(SC_2H_4Ph)_{60}$.[47]

Besides the aforementioned examples, the extraordinary stability of $Au_{25}(SR)_{18}$, $Au_{38}(SR)_{24}$, and $Au_{144}(SR)_{60}$ was also observed in a number of other studies.[8,44,46,48,94–97] The size-dependent stability forms the basis of size-focusing for the synthesis of monodisperse Au NCs.

6.4.1.4 Stability Origin of Thiolate-Protected Au NCs

It should be mentioned that although the stability differences among thiolate-protected Au NCs with different sizes are widely recognized, the origin of their stability is still not well understood. Superatom electronic theory and geometric factors are well-accepted in current developments. For example, Walter *et al.* proposed a superatom electronic theory to explain the stability of thiolate-protected Au NCs with some specific or magic sizes.[42] In this theory, the valence electrons of the NCs are counted, and those NCs with a total electron count representing a shell-closure of the superatom orbitals (*e.g.*, 1S, 1P, 1D..., corresponding to a total electron count of 2, 8, 18, 34,

58...) are considered to be stable.[6,42,98] The superatom electronic theory has been used to explain the good stability of $Au_{25}(SR)_{18}^{-}$ (total electron count of 8) and $Au_{102}(SR)_{44}$ (total electron count of 58). However, this theory may not be able to predict stable Au NCs with other non-shell-closure electron counts, such as $Au_{38}(SR)_{24}$ (total electron count of 14) and $Au_{144}(SR)_{60}$ (total electron count of 84).

Besides the electronic factor, a geometric factor may also affect the stability of thiolate-protected Au NCs, where the NC stability originates from their highly symmetric atomic packing structures. For example, Negishi *et al.* compared the stability of $Au_{25}(SR)_{18}^{x}$ (x is the overall charge of the cluster) carrying different charges of +1, 0 and −1. They found that all $Au_{25}(SR)_{18}^{x}$ species were stable in solution regardless of their cluster charge. This observation suggests that the geometric factor could be more significant for the good stability of $Au_{25}(SR)_{18}^{x}$ NCs than the electronic factor.[50]

6.4.2 Facile Size-Focusing Methods

The driving force in the size-focusing process is the stability difference among different-sized Au NCs. The experimental setups for size-focusing are facile if the reaction conditions meet two requirements: the stability difference of different-sized Au NCs and the presence of excess thiolate ligands. Polydisperse Au NCs are precursors for size-focusing, and these NC precursors can be protected by thiolate or non-thiolate ligands. In addition, the thiol etching process could involve one or two phases. In this section, we will use $Au_{25}(SR)_{18}$ NCs as an example to illustrate the facile feature of the size-focusing methods.

6.4.2.1 Thiol Etching of Thiolate-Protected Au NC Precursors

Thiolate-protected Au NCs with different sizes can be used as precursors for thiol etching. For example, on the basis of the size-dependent stability of $Au_n(SG)_m$ ($n = 10$–39) NCs against thiol etching,[93] a two-step synthetic method was proposed by Muhammed *et al.* to synthesize monodisperse $Au_{25}(SG)_{18}$ NCs.[40] The authors first prepared a mixture of $Au_n(SG)_m$ ($n = 10$–39) NCs, and treated the mixture with excess GSH at 328 K for ~12 h. $Au_{25}(SG)_{18}$ NCs with a high purity were then obtained.

6.4.2.2 Thiolation of Non-Thiolate-Protected Au NC Precursors

Non-thiolate-protected Au NCs, typically with a specific cluster size, can also be used as precursors for thiol etching to produce monodisperse thiolate-protected Au NCs. For example, phosphine-protected Au_{11} NCs were used to synthesize $Au_{25}(SG)_{18}$ *via* a thiol etching process by using GSH.[99] The as-synthesized $Au_{25}(SG)_{18}$ NCs have a good purity as confirmed by PAGE, UV-vis spectroscopy, and ESI-MS measurements. This method can produce $Au_{25}(SG)_{18}$ NCs in a large quantity of ~70 mg. Recent studies also demonstrated that PVP-protected Au NCs can be used as precursors to synthesize $Au_{25}(SR)_{18}$ NCs *via* a similar thiol etching process.[93]

6.4.3 Tailoring the Thiol Etching Process to Synthesize Metastable Au NCs

Au NCs with a full size spectrum are pivotal to establishing a precise size–property correlation of the NCs. In contrast to those ultrastable Au NCs discussed above, other Au NCs with metastable sizes cannot sustain a harsh thiol etching process. Therefore an elaborate control of the thiol etching process is required to use the size-focusing method to synthesize Au NCs with metastable sizes. For example, proper size control of the NC precursors may produce mixed-sized Au NCs consisting of only metastable sizes, which, upon size-focusing, will form the most stable Au NCs which are those NCs with a specific metastable size. The protection chemistry of different ligands can also be used to tailor the stability of different-sized Au NCs.

6.4.3.1 Tailoring the Size Distribution of Au NC Precursors

The size distribution of the Au NC precursors is an important factor for the synthesis of Au NCs with a predesigned size. If the size distribution of the Au NC precursors is confined to a narrow size range, within which a NC species with a specific metastable size is the most stable species, a subsequent well-controlled thiol etching will lead to the production of Au NCs with this metastable size. In this strategy, the experimental conditions in synthesizing Au NC precursors, such as the reaction temperature, solvents, type and amount of reducing agents, need to be controlled carefully, which may help produce Au NC precursors with a predesigned size distribution.

As demonstrated in section 6.3.3, a mild reduction environment for the reductive decomposition of Au(I)–SR complexes is particularly useful in narrowing the size distribution of the resultant Au NCs. A good example was recently reported by Qian *et al.*,[46] who synthesized $Au_{38}(SR)_{24}$ by circumventing the formation of other thermodynamically stable Au NCs. The $Au_{38}(SR)_{24}$ NC species was generally obtained by a separation technique in previous studies.[36,38,97] Thiol etching was also used to improve the yield of $Au_{38}(SR)_{24}$ NCs, but only a low yield was achieved in the final product.[45] This is because the as-prepared Au NC precursors had a broad size distribution, which, upon thiol etching, led to the formation of Au NCs with several stable sizes including $Au_{25}(SR)_{18}$ and $Au_{38}(SR)_{24}$. As discussed in 6.4.1, $Au_{25}(SR)_{18}$ is more stable than $Au_{38}(SR)_{24}$ in solution,[43] therefore it is very difficult to remove $Au_{25}(SR)_{18}$ from $Au_{38}(SR)_{24}$ by the thiol etching process. In a recent study, Qian *et al.* hypothesized that the solvent conditions can affect the size distribution of as-prepared Au NC precursors.[46] They used acetone as the solvent to replace conventional methanol for the preparation of Au NC precursors [Figure 6.11(a)]. The as-prepared $Au_n(SG)_m$ NCs in methanol were

(a)

(b)

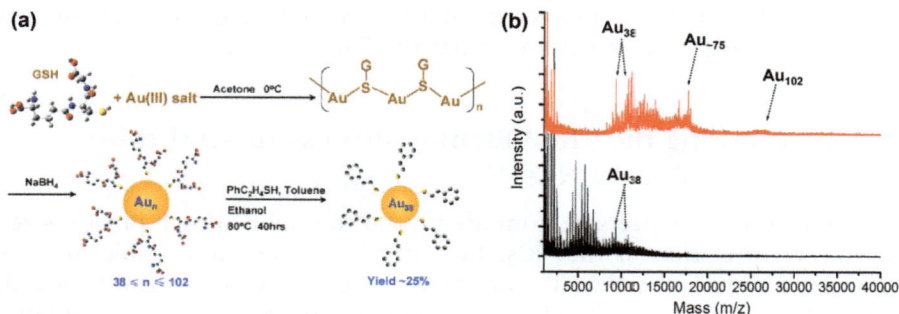

Figure 6.11 (a) Schematic illustration of a two-step approach to synthesize Au$_{38}$(SC$_2$H$_4$Ph)$_{24}$ NCs. (b) MALDI-TOF mass spectra of the polydisperse (raw) Au NCs prior to the thiol etching: (black) Au NC precursors prepared in methanol and (red) Au NC precursors prepared in acetone.
Reproduced with permission from ref. 46. Copyright (2009) American Chemical Society.

in the size range of $n = 10$–39.[24] In comparison, the as-prepared Au$_n$(SG)$_m$ NCs in acetone were much larger and were in the size range of $n = 38$–102 [determined by MALDI-TOF, Figure 6.11(b)]. A subsequent thiol etching of as-prepared Au$_n$(SG)$_m$ NC precursors in acetone by phenylethanethiol formed Au$_{38}$(SR)$_{24}$ NCs. Since Au$_{25}$(SR)$_{18}$ was not in the as-prepared Au$_n$(SG)$_m$ NC precursors, Au$_{38}$(SR)$_{24}$ was the most stable NC species in the size range of $n = 38$–102, and a high purity of Au$_{38}$(SR)$_{24}$ was obtained after the size-focusing.

6.4.3.2 Delicate Selection of Protecting and Etching Ligands

Besides manipulating the size distribution of the Au NC precursors, delicate selection of the protecting ligands of the Au NC precursors and etching ligands for thiol etching is also crucial for the synthesis of monodisperse Au NCs. Different ligands have different protection chemistry with the Au core. For example, the Au–P bond is weaker than the Au–S bond. In addition, the phosphine ligands usually protect Au NCs *via* a simple terminal mode, whereas the thiolate ligands can bridge neighbouring Au atoms to form RS–[Au–SR]$_x$ staples to protect the Au NCs.

6.4.3.2.1 Protecting Ligands of the Au NC Precursors. Thiol etching of phosphine-protected Au NCs (this process is typically termed thiolation) often form thiolate-protected Au NCs with different sizes from the parent phosphine-protected Au NCs. For example, the thiolation of Au$_{11}$(PPh$_3$)$_8$Cl$_3$[99] and Au$_{55}$(PPh$_3$)$_{12}$Cl$_6$[100] NCs produced thiolate-protected Au$_{25}$ and Au$_{75}$ NCs, respectively. The increase in size after thiolation was attributed to the agglomeration of Au cores during the ligand exchange process. Interestingly, the size of phosphine-protected Au NC precursors also dictates the final size of thiolate-protected Au NCs. For example, during the thiolation process, the $Au_{11}(PPh_3)_8Cl_3$ NCs would first aggregate to form thiolate-protected Au NCs in the size range of $n = 10-39$,[93] which, upon further thiol etching, led to the formation of $Au_{25}(SR)_{18}$ NCs. In contrast, the thiolation of $Au_{55}(PPh_3)_{12}Cl_6$ first formed mixed-sized Au NCs larger than Au_{55}, which, upon thiol etching, formed Au_{75} NCs, as this NC species was the most stable NC in the above size range.

Polymer-protected Au NCs can also be thiolated to produce thiolate-protected Au NCs. For example, Tsunoyama *et al.* obtained a new series of magic-sized Au NCs with a molecular weight of 8, 11, 21, and 26 kDa after the thiolation of PVP-protected Au NCs by neat octadecanethiol $(C_{18}S-H)$.[97] This size series is different from the previously reported 8, 14, 22, and 29 kDa Au NC species.[38] The 11 kDa NC species has been chromatographically isolated,[53] and was determined to be $Au_{55}(SC_{18}H_{37})_{31}$.[101] This finding is of fundamental importance as $Au_{55}(SC_{18}H_{37})_{31}$ is the thiolated analogue of the well-known phosphine-protected $Au_{55}(PPh_3)_{12}Cl_6$.[102,103]

6.4.3.2.2 Thiolate Ligands for Thiol Etching.

Thiolate ligands are generally used as the etchants in the size-focusing of Au NC precursors, and their physicochemical properties may also affect the size of the resultant Au NCs. In particular, the steric hindrance of thiolate ligands has been used effectively to control the size of Au NCs. For example, Nishigaki *et al.* used a bulky thiolate ligand, Eind–SH [1,1,3,3,5,5,7,7-octaethyl-s-hydrindacane-4-thiol, Figure 6.12(b)] to etch PVP-protected Au NCs, which led to the formation of $Au_{41}(S-Eind)_{12}$ NCs [Figure 6.12(a)].[104] Detailed analyses of as-synthesized $Au_{41}(S-Eind)_{12}$ NCs by X-ray absorption fine structure (EXAFS), X-ray photoelectron spectroscopy (XPS), and mass spectrometry suggested the presence of short Au(I)–(S–Eind) binding motifs on the NC surface, which could result from strong steric hindrance of the bulky arylthiol ligands. This motif is distinctively different from the SR–$[Au(I)–SR]_x$ ($x = 1, 2$) binding motifs in Au NCs protected by small thiolate ligands. In addition, the Au_{41} core was only protected by 12 bulky thiolate ligands, and the thiol-to-Au ratio in $Au_{41}(S-Eind)_{12}$ NCs was much lower than that of common thiolate-protected Au NCs [Figure 6.12(c)]. Similar findings were reported by other groups. For example, Krommenhoek *et al.* reported that two bulky thiolate ligands [hexanethiol (Cy–SH) and 1-adamantanethiol (Ad–SH)] can produce $Au_{65}(SCy)_{30}$, $Au_{67}(SCy)_{30}$, $Au_{30}(SAd)_{18}$, and $Au_{39}(SAd)_{23}$ NCs.[105] These Au NCs have different sizes from that of common thiolate-protected Au NCs. Another intriguing finding in this study is that the core sizes of Au NCs decreased as the size of the thiolate ligands increased (*e.g.*, Ad–SH is larger than Cy–SH). This finding provides an efficient way to synthesize different-sized Au NCs by using thiolate ligands with different sizes.

Another crucial factor that may affect the NC size is the electronic conjugation effect of the thiolate ligands, which may also affect the stability of

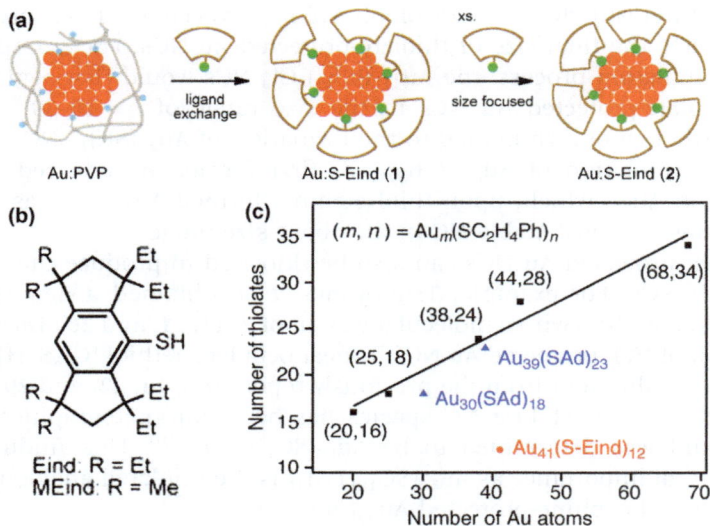

Figure 6.12   (a) Schematic illustration of the thiol etching of PVP-protected Au NPs by the bulky thiolate ligand (Eind–SH). (b) Molecular structure of Eind–SH. (c) The SR-to-Au ratios in Au NCs protected by different thiolate ligands.
Reproduced with permission from ref. 104. Copyright (2012) American Chemical Society.

Au NCs with different sizes. For example, Zeng *et al.* synthesized a new Au NC species with a molecular formula of $Au_{36}(SPh–tBu)_{24}$ *via* thiol etching of $Au_{38}(SC_2H_4Ph)_{24}$ by 4-*tert*-butylbenzenethiol (HSPh–tBu, where the −SH group is directly linked to the phenyl ring).[106] The total structure of $Au_{36}(SR)_{24}$ revealed that the NC contains a fcc Au_{28} core capped by four RS–[Au–SR]$_2$ staples and twelve terminating –SR ligands (Figure 6.13). This is the first experimental observation of the existence of a fcc core in thiolate-protected Au NCs. In addition, the terminating Au–SR motif is distinctively different from that of other reported thiolate-protected Au NCs, where the thiolate ligands are generally in the RS–[Au–SR]$_x$ ($x=1$ or 2) motifs.[13,14,16,107] The unique structure and excellent stability of $Au_{36}(SR)_{24}$ NCs were attributed to the electron conjugation effect of the phenyl rings with the Au core. This hypothesis was supported by a density functional theory (DFT) study. A similar Au_{36} NC species was also synthesized by Nimmala and Dass *via* the thiol etching of a mixture of PhC_2H_4S-protected Au_{68} and Au_{102} NCs by phenylthiol (HS–Ph).[108]

6.4.3.3   Tailoring the Thiol Etching Kinetics

Thiol etching kinetics are another key factor that could affect the size-focusing process. To synthesize Au NCs with ultrastable (or

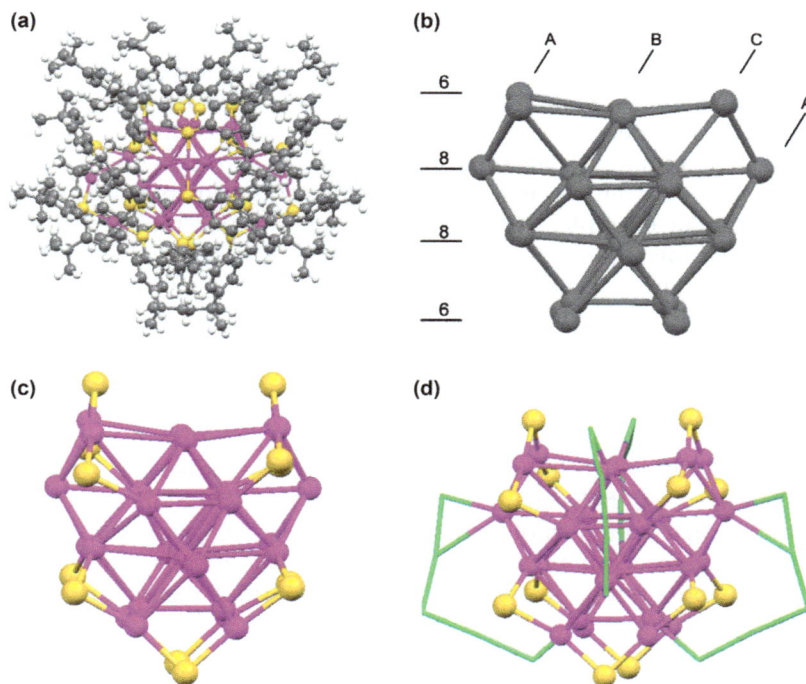

Figure 6.13 Cluster structure of $Au_{36}(SPh-tBu)_{24}$. (a) The entire particle (magenta, Au; yellow, S; grey, C; and white, H); (b) The fcc Au_{28} core; (c) The Au_{28} core plus 12 simple terminating ligands; and (d) The overall inorganic core ($Au_{36}S_{24}$, the four dimeric motifs are in green).
Reproduced with permission from ref. 106. Copyright (2012) Wiley-VCH.

thermodynamically stable) sizes, increasing the thiol etching kinetics can achieve a higher reaction conversion and also shorten the reaction time. Faster etching kinetics could be realized by increasing the thiol-to-Au ratio and/or the reaction temperature. For example, a slightly higher thiol-to-Au ratio (5 : 1) in conjunction with an elevated reaction temperature (318 K) can improve the yield of $Au_{25}(SR)_{18}$ NCs and also shorten the synthesis time to ~2 h.[41] The presence of oxygen may also increase the etching kinetics. For example, Parker *et al.* observed that oxygen can greatly improve the etching power of the thiolate ligands, and help produce $Au_{25}(SR)_{18}$ NCs in a higher yield (~50%).[109] Qian and Jin also found that the presence of oxygen can facilitate the elimination of unwanted side products (*e.g.*, Au_{102}) in the synthesis of $Au_{144}(SR)_{60}$ NCs.[48] The etching kinetics could also be tailored by varying the accessibility of the Au NC precursors toward the thiolate ligands. For example, by forming an interface between the Au NC precursors and thiolate ligands, Muhammed *et al.* have successfully synthesized Au_{23} and Au_{33} from $Au_{25}(SG)_{18}$ NCs.[110]

6.4.4 Versatile Size-Focusing Methods

The thiol etching (size-focusing) method can also be used to synthesize monodisperse Ag NCs. As Ag NCs are less stable than Au NCs in solution, slower etching kinetics are required in the Ag system because a fast etching process may completely decompose the Ag NCs to Ag(i)–SR complexes. There are several good strategies to generate a mild etching environment for Ag NC precursors. For example, Yuan *et al.* designed a phase-transfer strategy to synthesize thiolate-protected Ag NCs.[111] The authors used a cationic surfactant (CTAB) to coat the Ag NC precursors (protected by GSH and carrying negative charges in water) *via* the formation of ionic pairs of $(CTA)^+(COO)^-$. Coating of hydrophobic CTA^+ on the Ag NC surface can also transfer the Ag NCs from the aqueous to the toluene phase. A mild etching environment was then created in toluene due to the presence of a lower concentration of free thiolate ligands as well as a strong steric effect from the CTAB protecting layer, which led to the formation of monodisperse Ag NCs. The as-synthesized Ag NCs showed very strong red emission. It should be mentioned that a mild thiol etching of the Au/Ag NC precursors often produced luminescent Au/Ag NCs.[112–114] However, the detailed mechanism of the formation of luminescent Au/Ag NCs *via* thiol etching is presently unknown.

6.5 Thiol Etching of Large Au/Ag NPs

Besides Au(i)/Ag(i)–SR complexes and polydisperse Au/Ag NCs, large Au/Ag NPs can also be used as precursors to synthesize monodisperse Au/Ag NCs. This process typically uses thiolate ligands to digest large NPs, which is considered as a "top-down" approach.[115] The introduction of excess thiolate ligands in conjunction with an input of extraneous energy (*e.g.*, electron beam) can facilitate the etching process of large NPs. In this method, the selection and manipulation of the etchants and the NP precursors could directly affect the formation of monodisperse NCs. Therefore, these factors can be varied to control the NC synthesis. In addition, the etching environment (*e.g.*, temperature, pH, and solvent) is also crucial for the synthesis of monodisperse or luminescent Au/Ag NCs. In this section, we will focus our discussion on the effects of etchants, etching environment, and the NP precursors on the synthesis of monodisperse or luminescent Au/Ag NCs.

6.5.1 Delicate Selection of the Etchants

All molecules that have strong interactions with the large NP precursors (the metal core or protecting ligand) could be used as etchants. Thiolate ligands are the most popular etchants in the synthesis of Au/Ag NCs owing to their strong interaction with Au/Ag. Besides the chemical properties (*e.g.*, bonding, charge, and polarity) of the thiolate ligands which have shown strong effects on the synthesis of Au/Ag NCs, recent studies revealed that the physical properties (*e.g.*, steric hindrance) of the thiolate ligands could also affect the size of Au/Ag NCs.[105] Recently, Yuan *et al.* also found that the etching capability of thiolate ligands is also dictated by their size, where larger (or bulkier) thiolate ligands showed weaker etching capability due to the steric hindrance that affects the diffusion of the thiolate ligands to access the metal core.[116] Therefore, the steric hindrance of thiolate ligands, including their chain length and size, may affect the etching process and therefore determine the quality of as-synthesized Au/Ag NCs.

6.5.1.1 Chain Length of the Thiolate Ligands

Thiolate ligands with a longer chain can provide better protection for metal cores, which could minimize the nonradiative relaxation of the NCs and inhibit the luminescence quenching of the NCs by quenchers in solution. For example, Huang *et al.* prepared thiolate-protected Au NCs with emissions in the range of 500–615 nm (Figure 6.14) by using different alkanethiols, including 2-mercaptoethanol (2-ME), 6-mercaptohexanol (6-MH), and 11-mercaptoundecanoic acid (11-MUA), as etchants to etch ∼ 2.9 nm Au NPs protected by tetrakis(hydroxymethyl)phosphonium chloride (THPC).[117] As shown in Figure 6.14(a), the UV-vis absorption spectra of the resultant 2-ME-Au NCs, 6-MH-Au NCs, and 11-MUA-Au NCs were different from the THPC-Au NPs. In comparison with non-luminescent THPC-Au NPs, 2-ME–, 6-MH–, and 11-MUA–Au NCs showed red, yellow, and green emission,

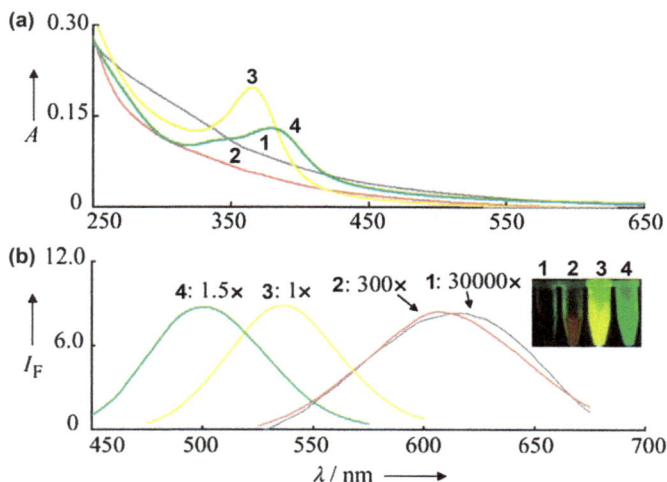

Figure 6.14 UV-vis absorption (a) and normalized luminescence (b) spectra of (1) THPC–Au NPs, (2) 2-ME–Au NCs, (3) 6-MH–AuNCs, and (4) 11-MU–AuNPs.
Reproduced with permission from ref. 117. Copyright (2007) Wiley-VCH.

respectively [Figure 6.14(b)]. In addition, as the length of the carbon chain in the thiolate ligands increased from 2-ME to 11-MUA, the sizes of the corresponding Au NCs decreased from 2.3 to 2.0 nm, and their QYs increased from 0.0062% to 3.1% accordingly. More recently, Zhang *et al.* reported that luminescent Au NCs with tunable emissions in the range of 530–630 nm can be synthesized by varying the ratio of two etchants [11-MUA and D-penicillamine (DPA)].[118]

6.5.1.2   Size of the Thiolate Ligands

Recent studies suggest that the thiolate ligands may affect the structure of SR–[M(I)–SR]$_x$ staples, which could subsequently determine the size and structure of the NCs.[104] In addition, the size and structure of the metal core are also strongly dependent on the size and structure of the thiolate ligands. For example, Qian *et al.* have synthesized two types of Au$_{25}$ NCs: rod-like and spherical, by etching polydisperse TOAB-protected Au NPs (1–3.5 nm) with two different thiolate ligands (PhC$_2$H$_4$S–H and GSH).[119] The size effect of thiolate ligands on the final NC size could be understood from the thermodynamic perspective of the etching process discussed in Section 6.4.3.2. The etchants are not limited to thiolate ligands. Other molecules such as polymers, proteins, peptides, and DNA can also be used as etchants to etch large NPs.[120,121]

6.5.2   Tailoring the Etching Environment

Reaction conditions such as temperature, pH, and solvent could affect the etching kinetics and therefore determine the attributes (*e.g.*, size and luminescence) of the final NC product. As discussed in Section 6.4, the size-focusing process is driven by the stability difference of different-sized NCs. In a relatively harsh etching environment, metastable NC species are eliminated, and only the most stable NC species are retained as the final product. In contrast, in a mild etching environment, some metastable NC species could be kinetically trapped and retained as the final NC product. There are two general strategies to control the etching environment of the NP precursors to form NCs: (1) tailoring the etching environment to thermodynamically select the most robust NCs; and (2) creating a mild etching environment to kinetically trap metastable NC species.

6.5.2.1   Thermodynamic Selection of the Ultrastable NC Species

A thermodynamic method typically produces the most thermodynamically stable NC species in the reaction solution, and variations in the etching kinetics have negligible effects on the final NC product. This principle is generally used to synthesize ultrastable NCs. Since the precursors (large NPs) and product (small NCs) have large size differences, the increase of the etching kinetics is an efficient way to produce NCs. Elevated temperature and excess thiolate ligands are typically used to increase the etching kinetics.

6.5.2.1.1 Temperature-Assisted Etching of the NP Precursors.

In 2004, Jin *et al.* developed a facile method to synthesize Au NCs by etching large Au NPs with dodecylthiol (DDT).[122] In this study, ~6 nm DDT-protected Au NPs were mixed with excess DDT and refluxed at 573 K, leading to the formation of Au_3 NCs after 50 min. The as-synthesized Au_3 NCs showed strong blue emission at 340 nm. The same strategy can be used to synthesize Ag and Pt NCs.[112,123] In general, a lower temperature was used in the etching of Ag NPs. For example, Dhanalakshmi *et al.* used 343 K and MSA to etch citrate-protected Ag NPs (30–70 nm) to synthesize red-emitting Ag NCs.[112]

6.5.2.1.2 pH-Assisted Etching of the NP Precursors.

The solution pH can affect the etching capability of thiolate ligands and the size/structure of $Au(I)$–SR complexes, which as such could be varied to control the synthesis of Au NCs from the Au NP precursors. For example, by varying the solution pH, Muhammed *et al.* obtained two different-sized Au NCs: Au_{25} and Au_8, from the same Au NP precursor, where 4–5 nm MSA-protected Au NPs were etched by excess GSH in aqueous solutions.[124] Au_{25} NCs were formed when the thiol etching was performed at pH 2.7–3, whereas smaller Au_8 NCs were formed when the solution pH was 7–8. The authors proposed two possible processes for the formation of Au NCs from Au NPs (Figure 6.15).

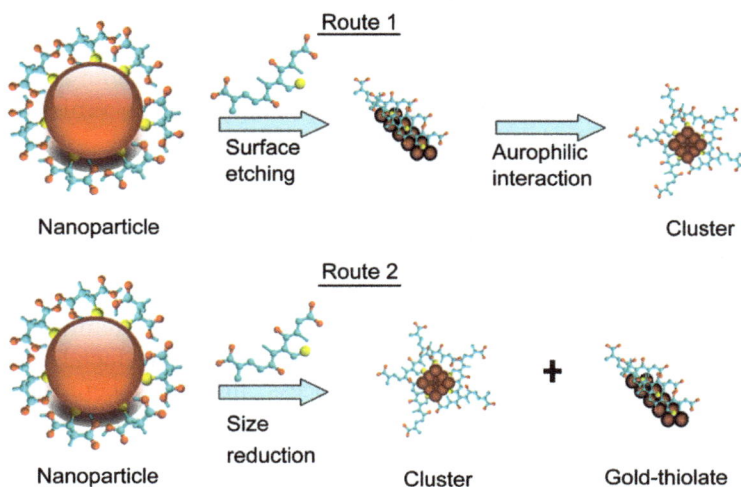

Figure 6.15 Schematic illustration of two possible processes for the formation $Au_{25}(SG)_{18}$ NCs from large Au NP precursors.
Reproduced with permission from ref. 124. Copyright (2008) Tsinghua University Press and Springer-Verlag Berlin Heidelberg.

One possible process is that during the thiol etching, Au atoms were first removed from the Au NP surface by free GSH to form Au(I)–SR complexes. These complexes were then aggregated to form $Au_{25}(SG)_{18}$ NCs. Another possible formation process is the direct etching of Au NPs to form $Au_{25}(SG)_{18}$ NCs. In this study, the authors also suggested that large and dense polymeric Au(I)–SR complexes were formed at low pH, leading to the formation of large Au_{25} NCs, whereas small and loose complexes were formed at high pH, leading to the formation of small Au_8 NCs.

6.5.2.2 *Kinetic Trapping of the Metastable NC Species*

Temperature- and pH-assisted thiol etching methods are effective in the synthesis of thermodynamically stable Au/Ag NCs from large Au/Ag NPs. However, they are less efficient in the synthesis of Au/Ag NCs with metastable sizes. Therefore a delicate control of the etching environment is required for a successful synthesis of metastable NCs. Several strategies have been developed to generate a mild etching environment for the NP precursors. Successful examples include decreasing the amount of etchants and constraining the etching reaction at the interface.

6.5.2.2.1 Decreasing the Amount of Etchants. The etching kinetics could be effectively decreased by using a smaller amount of etchants. The mild etching kinetics could facilitate the collection of metastable NC species during the etching of the NP precursors. For example, Guével *et al.* used a moderate amount of GSH to etch MSA-protected Ag NPs at 338 K, and successfully obtained Ag NCs with different emissions.[125] The authors observed that the solution color of Ag NPs changed from yellow to orange–brownish at day 1, to pale yellow at day 3, and finally to colorless at day 5. They collected three samples at 1, 3, and 8 days of etching, and these samples showed red, yellow, and blue emission with an emission wavelength of 720, 570, and 450 nm, respectively (Figure 6.16).

6.5.2.2.2 Constraining the Etching Reaction at the Interface. Owing to the unique reaction environment at the interface of two phases, interfacial etching can be readily used to synthesize Au/Ag NCs.[126] An additional advantage of the interfacial etching is that the resultant NCs will be spontaneously transferred to a single phase upon interfacial etching, therefore inhibiting further etching of as-formed NCs. The interfacial etching is therefore a mild and well-controlled process, and has been successfully used to produce Au/Ag NCs from the small NP precursors (*e.g.*, Au NPs with sizes below 8 nm and Ag NPs with sizes below 10 nm).[127,128] For example, Lin *et al.* used the interfacial etching method to synthesize luminescent Au NCs.[127] They used lipoic acid in water to etch DDAB-protected Au NPs in toluene with sizes of ~3.17 nm, and produced red-emitting Au NCs. The synthesis is illustrated in Figure 6.17. This method can also be used to synthesize luminescent Ag NCs.[129] There are a number of other

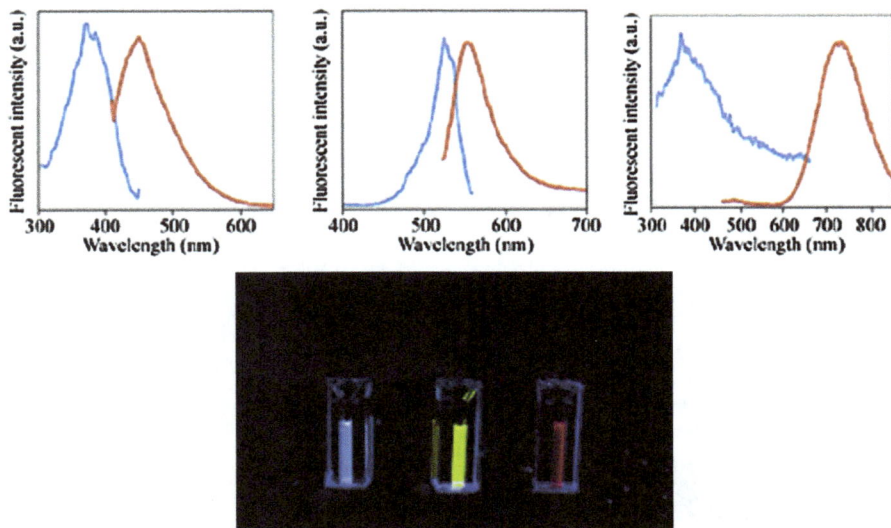

Figure 6.16 Luminescence spectra and photographs (taken under UV light) of blue-, yellow-, red-emitting Ag NCs.
Reproduced with permission from ref. 125. Copyright (2012) Tsinghua University Press and Springer-Verlag Berlin Heidelberg.

successful efforts at synthesizing luminescent Ag NCs *via* the interfacial etching.[130,131]

6.5.3 Tailoring the NP Precursors

Size, size distribution, and surface properties of the NP precursors can also affect the thiol etching of the NPs to form NCs, which will be discussed in this section.

6.5.3.1 *Size and Size Distribution of the NP Precursors*

Similar to the polydisperse NC precursors discussed in 6.4.3.1, the size and size distribution of the NP precursors will also affect the quality of the final NC product, which could be tailored by varying the synthesis conditions, such as the ligand-to-metal ratio, reaction temperature, and solvent conditions. For example, Huang *et al.* prepared a series of THPC-protected Au NPs with sizes from 2.2 to 3.6 nm by varying the THPC-to-Au ratio, and etched these Au NP precursors with MUA, leading to the formation of three Au NCs with a particle size of 2.1, 1.7, and 1.2 nm.[132] The size and size distribution of the NP precursors can also be tailored *via* post-synthesis methods. For example, Lin *et al.* used Au(III) to treat DDAB-protected Au NPs with sizes of 5.55 ± 0.68, and formed Au NPs with a narrow size distribution (3.17 ± 0.35 nm, Figure 6.17).[127]

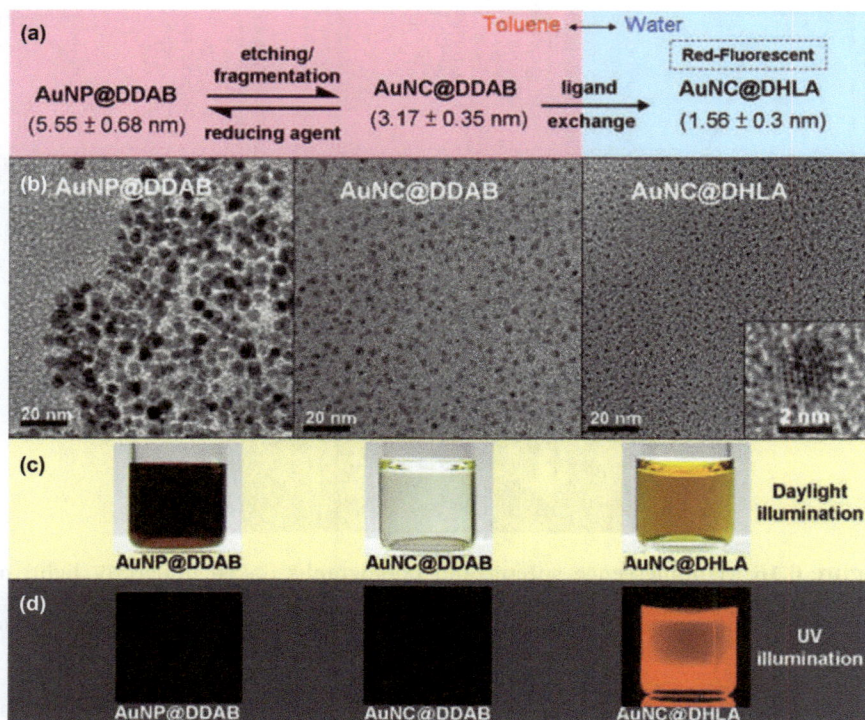

Figure 6.17   (a) Schematic illustration of the interfacial etching method for synthesizing luminescent Au NCs from Au NPs. (b) TEM images and digital photos under (c) normal light and (d) UV light of the corresponding Au NPs and NCs.
Reproduced with permission from ref. 127. Copyright (2009) American Chemical Society.

6.5.3.2 Surface Properties of the NP Precursors

The ligands on the NP precursors could also affect the quality of the final NC product. For example, a recent study showed that THPC-protected Au NPs with an average size of 3.4 nm can be etched by MUA, leading to the formation of Au NCs with an average size of 1.4 nm, whereas, if citrate-protected Au NPs with an average size of 3.3 nm were used as the precursors, Au NPs with an average size of 2.8 nm were formed by a similar MUA-etching process.[132]

6.6 Conclusions

In this chapter, we have summarized novel synthesis methods to produce thiolate-protected Au/Ag NCs with good monodispersity and/or luminescence. Three types of precursor can be used to produce monodisperse or luminescent Au/Ag NCs. They are Au(I)/Ag(I)-thiolate complexes, polydisperse Au/Ag NCs, and large Au/Ag NPs. We have discussed the underlying principles of converting these three precursors to monodisperse or luminescent Au/Ag NCs. If Au/Ag(I)–SR complexes are used as the precursors, a good uniformity of the precursors in terms of size and structure, and mild and controlled reduction kinetics are required for the production of high-quality Au/Ag NCs. If polydisperse Au/Ag NCs are used as the precursors, delicate control and use of the size-dependent stability of the NCs in the presence of excess thiolate ligands could help synthesize high-quality Au/Ag NCs. In addition, the size and size distribution of the NC precursors, the choice of protecting and etching ligands, and tailoring of the etching kinetics are crucial for the synthesis of monodisperse or luminescent Au/Ag NCs. Metastable NC species can also be synthesized *via* the size-focusing method if the size distribution of the NC precursors could be properly tailored. If large Au/Ag NPs are used as the precursors, several important factors including the size and size distribution and the protecting ligands of the NP precursors, the etchants, and the etching environment, are required to be well-controlled to produce high-quality Au/Ag NCs. We hope the underlying principles presented in this chapter could motivate the researchers to develop more efficient synthesis methods for monodisperse or luminescent Au/Ag NCs, and further pave the way for the practical applications of these newly-developed functional nanomaterials.

However, the underlying principles for producing monodisperse or luminescent Au/Ag NCs discussed in this chapter are mainly from phenomenological observations, and detailed understandings of the formation mechanisms, especially those at the molecular level, are presently lacking, which could be the major research interest in this research direction in the future. For example, some fundamental issues related to the size and structure of the precursors (*e.g.*, Au(I)/Ag(I)–thiolate complexes, polydisperse Au/Ag NCs, and large Au/Ag NPs), the underlying conversion chemistry or detailed formation mechanism from the precursors to the final NC product, and the precise correlation of physicochemical properties between the precursors and the final NC product, need to be addressed for further developing more efficient synthesis methods for high-quality Au/Ag NCs. This is a multidisciplinary topic and joint efforts from chemists, physicists, materials scientists, and engineers with both theoretical and experimental background are required. Several other issues may also need to improve to further advance this research direction: (1) standardizing and developing novel characterizing techniques to determine the size, shape, and composition of the NCs, especially to determine the chemical purity of the desired NC product; and (2) developing more efficient strategies to resolve the structure of Au/Ag NCs. The cluster structure is pivotal to understanding the fundamental issues related to the physicochemical properties of the Au/Ag NCs, and also of utmost importance in developing efficient synthesis methods for high-quality Au/Ag NCs.

References

1. R. C. Jin, *Nanoscale*, 2010, **2**, 343.
2. Q. B. Zhang, J. P. Xie, Y. Yu and J. Y. Lee, *Nanoscale*, 2010, **2**, 1962.
3. Y. Lu and W. Chen, *Chem. Soc. Rev.*, 2012, **41**, 3594.
4. Y. Gao, N. Shao and X. C. Zeng, *ACS Nano*, 2008, **2**, 1497.
5. Y. Li, G. Galli and F. Gygi, *ACS Nano*, 2008, 2, 1396.
6. C. M. Aikens, *J. Phys. Chem. Lett.*, 2010, **2**, 99.
7. Y. Pei and X. C. Zeng, *Nanoscale*, 2012, **4**, 4054.
8. J. Akola, M. Walter, R. L. Whetten, H. Häkkinen and H. Grönbeck, *J. Am. Chem. Soc.*, 2008, **130**, 3756.
9. J. Yang, E. Sargent, S. Kelley and J. Y. Ying, *Nat. Mater.*, 2009, **8**, 683.
10. Y. Yu, Q. Zhang, X. Lu and J. Y. Lee, *J. Phys. Chem. C*, 2010, **114**, 11119.
11. Q. Zhang, Y. Tan, J. Xie and J. Lee, *Plasmonics*, 2009, **4**, 9.
12. R. W. Murray, *Chem. Rev.*, 2008, **108**, 2688.
13. M. Zhu, C. M. Aikens, F. J. Hollander, G. C. Schatz and R. Jin, *J. Am. Chem. Soc.*, 2008, **130**, 5883.
14. P. D. Jadzinsky, G. Calero, C. J. Ackerson, D. A. Bushnell and R. D. Kornberg, *Science*, 2007, **318**, 430.
15. M. Zhu, H. Qian, X. Meng, S. Jin, Z. Wu and R. Jin, *Nano Lett.*, 2011, **11**, 3963.
16. H. Qian, W. T. Eckenhoff, Y. Zhu, T. Pintauer and R. Jin, *J. Am. Chem. Soc.*, 2010, **132**, 8280.
17. S. Chen, R. S. Ingram, M. J. Hostetler, J. J. Pietron, R. W. Murray, T. G. Schaaff, J. T. Khoury, M. M. Alvarez and R. L. Whetten, *Science*, 1998, **280**, 2098.
18. M. Zhu, C. M. Aikens, M. P. Hendrich, R. Gupta, H. Qian, G. C. Schatz and R. Jin, *J. Am. Chem. Soc.*, 2009, **131**, 2490.
19. L. Shang, S. J. Dong and G. U. Nienhaus, *Nano Today*, 2011, **6**, 401.
20. J. Zheng, C. Zhou, M. Yu and J. Liu, *Nanoscale*, 2012, **4**, 4073.
21. Y. Yu, Q. Yao, Z. Luo, X. Yuan, J. Y. Lee and J. Xie, *Nanoscale*, 2013, **5**, 4606.
22. X. Yuan, Z. Luo, Y. Yu, Q. Yao and J. Xie, *Chem. – Asian J.*, 2013, **8**, 858.
23. Y. Zhu, H. Qian and R. Jin, *J. Mater. Chem.*, 2011, **21**, 6793.
24. Y. Negishi, K. Nobusada and T. Tsukuda, *J. Am. Chem. Soc.*, 2005, **127**, 5261.
25. S. Kumar, M. D. Bolan and T. P. Bigioni, *J. Am. Chem. Soc.*, 2010, **132**, 13141.
26. W. Chen and S. Chen, *Angew. Chem., Int. Ed.*, 2009, **48**, 4386.
27. Y. Liu, H. Tsunoyama, T. Akita, S. Xie and T. Tsukuda, *ACS Catal.*, 2010, **1**, 2.
28. Y. Zhu, H. Qian, M. Zhu and R. Jin, *Adv. Mater.*, 2010, **22**, 1915.
29. I. Diez and R. H. A. Ras, *Nanoscale*, 2011, **3**, 1963.
30. Y.-C. Shiang, C.-C. Huang, W.-Y. Chen, P.-C. Chen and H.-T. Chang, *J. Mater. Chem.*, 2012, **22**, 12972.

31. M. Brust, M. Walker, D. Bethell, D. J. Schiffrin and R. Whyman, *J. Chem. Soc., Chem. Commun.*, 1994, 801.
32. R. L. Whetten, J. T. Khoury, M. M. Alvarez, S. Murthy, I. Vezmar, Z. L. Wang, P. W. Stephens, C. L. Cleveland, W. D. Luedtke and U. Landman, *Adv. Mater.*, 1996, **8**, 428.
33. T. G. Schaaff and R. L. Whetten, *J. Phys. Chem. B*, 1999, **103**, 9394.
34. T. G. Schaaff, *Anal. Chem.*, 2004, **76**, 6187.
35. T. G. Schaaff, M. N. Shafigullin, J. T. Khoury, I. Vezmar and R. L. Whetten, *J. Phys. Chem. B*, 2001, **105**, 8785.
36. N. K. Chaki, Y. Negishi, H. Tsunoyama, Y. Shichibu and T. Tsukuda, *J. Am. Chem. Soc.*, 2008, **130**, 8608.
37. O. Lopez-Acevedo, J. Akola, R. L. Whetten, H. Grönbeck and H. Häkkinen, *J. Phys. Chem. C*, 2009, **113**, 5035.
38. T. G. Schaaff, M. N. Shafigullin, J. T. Khoury, I. Vezmar, R. L. Whetten, W. G. Cullen, P. N. First, C. Gutiérrez-Wing, J. Ascensio and M. J. Jose-Yacamán, *J. Phys. Chem. B*, 1997, **101**, 7885.
39. M. J. Hostetler, J. E. Wingate, C.-J. Zhong, J. E. Harris, R. W. Vachet, M. R. Clark, J. D. Londono, S. J. Green, J. J. Stokes, G. D. Wignall, G. L. Glish, M. D. Porter, N. D. Evans and R. W. Murray, *Langmuir*, 1998, **14**, 17.
40. M. A. H. Muhammed, A. K. Shaw, S. K. Pal and T. Pradeep, *J. Phys. Chem. C*, 2008, **112**, 14324.
41. Z. Wu, J. Suhan and R. Jin, *J. Mater. Chem.*, 2009, **19**, 622.
42. M. Walter, J. Akola, O. Lopez-Acevedo, P. D. Jadzinsky, G. Calero, C. J. Ackerson, R. L. Whetten, H. Grönbeck and H. Häkkinen, *Proc. Natl. Acad. Sci. U. S. A.*, 2008, **105**, 9157.
43. A. C. Dharmaratne, T. Krick and A. Dass, *J. Am. Chem. Soc.*, 2009, **131**, 13604.
44. M. A. Tofanelli and C. J. Ackerson, *J. Am. Chem. Soc.*, 2012, **134**, 16937.
45. O. Toikkanen, V. Ruiz, G. Rönnholm, N. Kalkkinen, P. Liljeroth and B. M. Quinn, *J. Am. Chem. Soc.*, 2008, **130**, 11049.
46. H. Qian, Y. Zhu and R. Jin, *ACS Nano*, 2009, **3**, 3795.
47. H. Qian and R. Jin, *Nano Lett.*, 2009, **9**, 4083.
48. H. Qian and R. Jin, *Chem. Mater.*, 2011, **23**, 2209.
49. Y. Negishi, C. Sakamoto, T. Ohyama and T. Tsukuda, *J. Phys. Chem. Lett.*, 2012, **3**, 1624.
50. Y. Negishi, N. K. Chaki, Y. Shichibu, R. L. Whetten and T. Tsukuda, *J. Am. Chem. Soc.*, 2007, **129**, 11322.
51. M. M. Alvarez, J. T. Khoury, T. G. Schaaff, M. Shafigullin, I. Vezmar and R. L. Whetten, *Chem. Phys. Lett.*, 1997, **266**, 91.
52. Y. Levi-Kalisman, P. D. Jadzinsky, N. Kalisman, H. Tsunoyama, T. Tsukuda, D. A. Bushnell and R. D. Kornberg, *J. Am. Chem. Soc.*, 2011, **133**, 2976.
53. H. Tsunoyama, Y. Negishi and T. Tsukuda, *J. Am. Chem. Soc.*, 2006, **128**, 6036.
54. H. Qian and R. Jin, *Chem. Commun.*, 2011, **47**, 11462.

55. S. Knoppe, J. Boudon, I. Dolamic, A. Dass and T. Bürgi, *Anal. Chem.*, 2011, **83**, 5056.
56. T. G. Schaaff, G. Knight, M. N. Shafigullin, R. F. Borkman and R. L. Whetten, *J. Phys. Chem. B*, 1998, **102**, 10643.
57. T. G. Schaaff and R. L. Whetten, *J. Phys. Chem. B*, 2000, **104**, 2630.
58. Y. Negishi, Y. Takasugi, S. Sato, H. Yao, K. Kimura and T. Tsukuda, *J. Am. Chem. Soc.*, 2004, **126**, 6518.
59. Y. Negishi, Y. Takasugi, S. Sato, H. Yao, K. Kimura and T. Tsukuda, *J. Phys. Chem. B*, 2006, **110**, 12218.
60. H. Qian, Y. Zhu and R. Jin, *J. Am. Chem. Soc.*, 2010, **132**, 4583.
61. H. Kuang, W. Chen, W. Yan, L. Xu, Y. Zhu, L. Liu, H. Chu, C. Peng, L. Wang, N. A. Kotov and C. Xu, *Biosens. Bioelectron.*, 2011, **26**, 2032.
62. A. Dass, *J. Am. Chem. Soc.*, 2009, **131**, 11666.
63. C. F. Shaw, N. A. Schaeffer, R. C. Elder, M. K. Eidsness, J. M. Trooster and G. H. M. Calis, *J. Am. Chem. Soc.*, 1984, **106**, 3511.
64. M. Zhu, E. Lanni, N. Garg, M. E. Bier and R. Jin, *J. Am. Chem. Soc.*, 2008, **130**, 1138.
65. X. Yuan, M. I. Setyawati, A. S. Tan, C. N. Ong, D. T. Leong and J. Xie, *NPG Asia Mater.*, 2013, **5**, e39.
66. X. Yuan, Y. Yu, Q. Yao, Q. Zhang and J. Xie, *J. Phys. Chem. Lett.*, 2012, 2310.
67. Q. Yao, Y. Yu, X. Yuan, Y. Yu, J. Xie and J. Y. Lee, *Small*, 2013, **9**, 2696.
68. M. Zhu, H. Qian and R. Jin, *J. Phys. Chem. Lett.*, 2010, **1**, 1003.
69. T. U. B. Rao, B. Nataraju and T. Pradeep, *J. Am. Chem. Soc.*, 2010, **132**, 16304.
70. O. M. Bakr, V. Amendola, C. M. Aikens, W. Wenseleers, R. Li, L. Dal Negro, G. C. Schatz and F. Stellacci, *Angew. Chem., Int. Ed.*, 2009, **48**, 5921.
71. K. M. Harkness, Y. Tang, A. Dass, J. Pan, N. Kothalawala, V. J. Reddy, D. E. Cliffel, B. Demeler, F. Stellacci, O. M. Bakr and J. A. McLean, *Nanoscale*, 2012, **4**, 4269.
72. Z. Wu, E. Lanni, W. Chen, M. E. Bier, D. Ly and R. Jin, *J. Am. Chem. Soc.*, 2009, **131**, 16672.
73. C. A. Brown and V. K. Ahuja, *J. Org. Chem.*, 1973, **38**, 2226.
74. J. V. B. Kanth and M. Periasamy, *J. Org. Chem.*, 1991, **56**, 5964.
75. A. Aramini, L. Brinchi, R. Germani and G. Savelli, *Eur. J. Org. Chem.*, 2000, **2000**, 1793.
76. C. F. Lane, *Synthesis*, 1975, **1975**, 135.
77. R. F. Borch, M. D. Bernstein and H. D. Durst, *J. Am. Chem. Soc.*, 1971, **93**, 2897.
78. R. O. Hutchins, C. A. Milewski and B. E. Maryanoff, *J. Am. Chem. Soc.*, 1973, **95**, 3662.
79. R. J. Mattson, K. M. Pham, D. J. Leuck and K. A. Cowen, *J. Org. Chem.*, 1990, **55**, 2552.
80. G. C. Andrews and T. C. Crawford, *Tetrahedron Lett.*, 1980, **21**, 693.
81. B. Carboni and L. Monnier, *Tetrahedron*, 1999, **55**, 1197.
82. R. O. Hutchins, K. Learn, B. Nazer, D. Pytlewski and A. Pelter, *Org. Prep. Proced. Int.*, 1984, **16**, 335.

83. Z. Wu, M. A. MacDonald, J. Chen, P. Zhang and R. Jin, *J. Am. Chem. Soc.*, 2011, **133**, 9670.

84. S. Park and D. Lee, *Langmuir*, 2012, **28**, 7049.

85. A. Ghosh, T. Udayabhaskararao and T. Pradeep, *J. Phys. Chem. Lett.*, 2012, 1997.

86. Y. Yu, X. Chen, Q. Yao, Y. Yu, N. Yan and J. Xie, *Chem. Mater.*, 2013, **25**, 946.

87. Y. Yu, Z. Luo, Y. Yu, J. Y. Lee and J. Xie, *ACS Nano*, 2012, **6**, 7920.

88. J. Xie, Y. Zheng and J. Y. Ying, *J. Am. Chem. Soc.*, 2009, **131**, 888.

89. Y. Cui, Y. Wang, R. Liu, Z. Sun, Y. Wei, Y. Zhao and X. Gao, *ACS Nano*, 2011, **5**, 8684.

90. Z. Luo, X. Yuan, Y. Yu, Q. Zhang, D. T. Leong, J. Y. Lee and J. Xie, *J. Am. Chem. Soc.*, 2012, **134**, 16662.

91. L. Shang, L. Yang, F. Stockmar, R. Popescu, V. Trouillet, M. Bruns, D. Gerthsen and G. U. Nienhaus, *Nanoscale*, 2012, **4**, 4155.

92. R. Jin, H. Qian, Z. Wu, Y. Zhu, M. Zhu, A. Mohanty and N. Garg, *J. Phys. Chem. Lett.*, 2010, **1**, 2903.

93. Y. Shichibu, Y. Negishi, H. Tsunoyama, M. Kanehara, T. Teranishi and T. Tsukuda, *Small*, 2007, **3**, 835.

94. H. Qian, M. Zhu, U. N. Andersen and R. Jin, *J. Phys. Chem. A*, 2009, **113**, 4281.

95. S. Kumar and R. Jin, *Nanoscale*, 2012.

96. D.-e. Jiang, M. L. Tiago, W. Luo and S. Dai, *J. Am. Chem. Soc.*, 2008, **130**, 2777.

97. H. Tsunoyama, P. Nickut, Y. Negishi, K. Al-Shamery, Y. Matsumoto and T. Tsukuda, *J. Phys. Chem. C*, 2007, **111**, 4153.

98. A. Dass, *Nanoscale*, 2012, **4**, 2260.

99. Y. Shichibu, Y. Negishi, T. Tsukuda and T. Teranishi, *J. Am. Chem. Soc.*, 2005, **127**, 13464.

100. R. Balasubramanian, R. Guo, A. J. Mills and R. W. Murray, *J. Am. Chem. Soc.*, 2005, **127**, 8126.

101. R. Tsunoyama, H. Tsunoyama, P. Pannopard, J. Limtrakul and T. Tsukuda, *J. Phys. Chem. C*, 2010, **114**, 16004.

102. G. Schmid, *Chem. Soc. Rev.*, 2008, **37**, 1909.

103. Y. Liu, W. Meyer-Zaika, S. Franzka, G. Schmid, M. Tsoli and H. Kuhn, *Angew. Chem., Int. Ed.*, 2003, **42**, 2853.

104. J.-i. Nishigaki, R. Tsunoyama, H. Tsunoyama, N. Ichikuni, S. Yamazoe, Y. Negishi, M. Ito, T. Matsuo, K. Tamao and T. Tsukuda, *J. Am. Chem. Soc.*, 2012, **134**, 14295.

105. P. J. Krommenhoek, J. Wang, N. Hentz, A. C. Johnston-Peck, K. A. Kozek, G. Kalyuzhny and J. B. Tracy, *ACS Nano*, 2012, **6**, 4903.

106. C. Zeng, H. Qian, T. Li, G. Li, N. L. Rosi, B. Yoon, R. N. Barnett, R. L. Whetten, U. Landman and R. Jin, *Angew. Chem., Int. Ed.*, 2012, **51**, 13114.

107. M. W. Heaven, A. Dass, P. S. White, K. M. Holt and R. W. Murray, *J. Am. Chem. Soc.*, 2008, **130**, 3754.

108. P. R. Nimmala and A. Dass, *J. Am. Chem. Soc.*, 2011, **133**, 9175.
109. J. F. Parker, J. E. F. Weaver, F. McCallum, C. A. Fields-Zinna and R. W. Murray, *Langmuir*, 2010, **26**, 13650.
110. M. A. H. Muhammed, P. K. Verma, S. K. Pal, R. C. A. Kumar, S. Paul, R. V. Omkumar and T. Pradeep, *Chem. – Eur. J.*, 2009, **15**, 10110.
111. X. Yuan, Z. Luo, Q. Zhang, X. Zhang, Y. Zheng, J. Y. Lee and J. Xie, *ACS Nano*, 2011, **5**, 8800.
112. L. Dhanalakshmi, T. Udayabhaskararao and T. Pradeep, *Chem. Commun.*, 2012, **48**, 859.
113. M. A. H. Muhammed, F. Aldeek, G. Palui, L. Trapiella-Alfonso and H. Mattoussi, *ACS Nano*, 2012, **6**, 8950.
114. F. Aldeek, M. A. H. Muhammed, G. Palui, N. Zhan and H. Mattoussi, *ACS Nano*, 2013, 7, 2509.
115. S. Eustis and M. A. El-Sayed, *Chem. Soc. Rev.*, 2006, **35**, 209.
116. X. Yuan, Y. Tay, X. Dou, Z. Luo, D. T. Leong and J. Xie, *Anal. Chem.*, 2012, **85**, 1913.
117. C.-C. Huang, Z. Yang, K.-H. Lee and H.-T. Chang, *Angew. Chem., Int. Ed.*, 2007, **46**, 6824.
118. Z. Zhang, L. Xu, H. Li and J. Kong, *RSC Adv.*, 2012.
119. H. Qian, M. Zhu, E. Lanni, Y. Zhu, M. E. Bier and R. Jin, *J. Phys. Chem. C*, 2009, **113**, 17599.
120. H. Duan and S. Nie, *J. Am. Chem. Soc.*, 2007, **129**, 2412.
121. R. Zhou, M. Shi, X. Chen, M. Wang and H. Chen, *Chem. – Eur. J.*, 2009, **15**, 4944.
122. R. Jin, S. Egusa and N. F. Scherer, *J. Am. Chem. Soc.*, 2004, **126**, 9900.
123. X. Le Guével, V. Trouillet, C. Spies, G. Jung and M. Schneider, *J. Phys. Chem. C*, 2012, **116**, 6047.
124. M. H. Muhammed, S. Ramesh, S. Sinha, S. Pal and T. Pradeep, *Nano Res.*, 2008, **1**, 333.
125. X. Guével, C. Spies, N. Daum, G. Jung and M. Schneider, *Nano Res.*, 2012, **5**, 379.
126. J. Yang, J. Y. Lee and J. Y. Ying, *Chem. Soc. Rev.*, 2011, **40**, 1672.
127. C.-A. J. Lin, T.-Y. Yang, C.-H. Lee, S. H. Huang, R. A. Sperling, M. Zanella, J. K. Li, J.-L. Shen, H.-H. Wang, H.-I. Yeh, W. J. Parak and W. H. Chang, *ACS Nano*, 2009, **3**, 395.
128. W. Guo, J. Yuan and E. Wang, *Chem. Commun.*, 2012, **48**, 3076.
129. S. Huang, C. Pfeiffer, J. Hollmann, S. Friede, J. J.-C. Chen, A. Beyer, B. Haas, K. Volz, W. Heimbrodt, J. M. Montenegro Martos, W. Chang and W. J. Parak, *Langmuir*, 2012, **28**, 8915.
130. T. Udaya Bhaskara Rao and T. Pradeep, *Angew. Chem., Int. Ed.*, 2010, **49**, 3925.
131. K. V. Mrudula, T. U. Bhaskara Rao and T. Pradeep, *J. Mater. Chem.*, 2009, **19**, 4335.
132. C.-C. Huang, H.-Y. Liao, Y.-C. Shiang, Z.-H. Lin, Z. Yang and H.-T. Chang, *J. Mater. Chem.*, 2009, **19**, 755.

CHAPTER 7

Noble Metal Clusters in Protein Templates

THALAPPIL PRADEEP,* ANANYA BAKSI AND
PAULRAJPILLAI LOURDU XAVIER

DST Unit of Nanoscience and Thematic Unit of Excellence, Department of
Chemistry, Indian Institute of Technology Madras, Chennai 600036, India
*Email: pradeep@iitm.ac.in

7.1 Introduction

7.1.1 General Properties of Clusters

Clusters are composed of a group of atoms (also known as nanoclusters, atomic clusters, quantum clusters, superatoms, *etc.*) and are considered to be the bridge between atoms/molecules and nanoparticles.[1,2] Cluster science deals with the study of atomically precise materials and their application areas where each atom counts. In this chapter, we are concerned with the clusters of noble metals, which have been reviewed before.[3,4] A number-dependence is reflected in all the properties of the materials derived from such forms of matter, made either in the gas phase or in the condensed phase. Clusters can be of several types like metal clusters, semiconductor clusters, ionic clusters, rare gas clusters, molecular clusters or cluster molecules. Clusters are present in the outer space and even in biological entities. Clusters have specific molecule-like absorption features and have certain fascinating properties which strongly depend on the core size. Hence, knowing the atomic composition of clusters becomes crucial.

RSC Smart Materials No. 7
Functional Nanometer-Sized Clusters of Transition Metals: Synthesis, Properties and Applications
Edited by Wei Chen and Shaowei Chen
© The Royal Society of Chemistry 2014
Published by the Royal Society of Chemistry, www.rsc.org

Rationally, there are two ways to find their composition: (1) Imaging and counting (using advanced imaging techniques such as electron microscopy (EM) and scanning probe microscopy (SPM)) and (2) Using mass spectrometry to decipher the mass and thereby the composition. This method possesses several advantages over imaging such as studying isotope distribution, relative abundance of electronically closed-shell systems, *etc.* Imaging techniques to characterize clusters were popularized by Palmer *et al.*[5] Palmer and co-workers have extensively worked on characterizing deposited clusters using high angle annular dark field scanning transmission electron microscopy (HAADF STEM). Nevertheless, there are limited reports on ligand-stabilized clusters characterized by TEM.[6,7] The core size cannot be measured properly by high resolution transmission electron microscopy alone for the smallest analogues due to the inherent limitation of electron beam-induced aggregation. However, it could be overcome using a suitable support matrix.[6,7] When imaging facilities could not be used to characterize the core in the early days, laser desorption ionization mass spectrometry (LDI MS) played an important role to get the overall mass of the entity, although precise assignments were not possible due to the broad nature of the mass spectrum. So in the early days, clusters were given names like 28 kDa clusters and so on.[8] The nomenclature of these clusters is also very important and several terminologies have been adopted by the cluster community like nanoclusters, quantum clusters, quantum sized/confined clusters, nanomolecules, monolayer protected clusters, ultra small nanoparticles, faradaurrates, and several others.[9] It may also be stated that some of the above terminologies have inherent problems and are against the accepted principles of naming. A cluster core can be homoatomic, having a formula like M_a ($a > 2$), or it can have different atoms, with a formula like $M_a N_b$ ($a > 2$, $b > 1$), known as mixed or alloy clusters. These names are valid when we talk about gas phase clusters, which do not have any protecting agent. But in solution, due to high reactivity and a high degree of aggregation, protecting agents are necessary. This results in monolayer protected clusters which contain a cluster core and a ligand shell around it. Various types of ligands can be used for this purpose. For noble metals like Au and Ag, phosphine and sulfur containing molecules have normally been used. For this specific class, the molecular formula can be assigned as $M_x L_y$, where x corresponds to the number of atoms constituting the core and y corresponds to the number of ligand molecules surrounding the core. At the size regime of clusters, surface plasmon is absent; as the size reaches the de Broglie wavelength of electrons at the Fermi energy of metals, it can no longer support the plasmon absorption.[4] Some of the cluster cores having number of core atoms n^*, where $n^* = 13, 25, 38, 55, 102, 147$, *etc.* are extremely stable and known as "magic number" clusters (MN).[10] Their extreme stability is attributed to their shell closing electronic structure, where the HOMO state of the cluster closes a shell. Such spherical clusters are comprised of $n^* = 2, 8, 18 (20), 34 (40), 58, 92$ delocalized electrons. In certain cases, a stable structure is achieved at the magic numbers in parenthesis depending on the degree of deviation of the cluster geometry from the spherical nature. Several theoretical groups have studied the structure and other properties like absorption behavior, luminescence, stability, reactivity and catalytic properties. Several models have also been developed in the past few decades among which the Jellium model is being used in most of the cases.[11] Each of the clusters has a specific stable geometry; for instance, Au_{13} has an icosahedral structure,[12] whereas thiolated Au_{25} has an icosahedral Au_{13} core and 12 Au-atom shell on it.[13] Along with this, each core has specific absorption features due to d–sp, sp–sp, inter and intra band transitions.[14–32] Luminescence from these clusters cores (although in the NIR region) has drawn great attention, whereas water soluble monolayer protected clusters can also emit in the visible region.[17] These clusters can also show unusual magnetism as seen in the case of $Au_{25}SR_{18}$.[33] Since the focus of this review is not about such monolayer protected clusters (MPCs), with this briefing of properties of noble metal clusters in general, we would like to discuss the different ligands used for cluster synthesis and how the trend has been progressing. There are several excellent review articles and readers are requested to refer to them to learn about the properties of clusters.[9,34–37]

7.1.2   Trends in the Choice of Ligands for Cluster Synthesis

Controlled synthesis of such MPCs is a crucial step. Choice of ligands having specific affinity towards a specific metal is important. Based on the ligand affinities, in the early period of the literature, phosphine protected gold clusters were prepared like $Au_{24}(PPh_3)_{10}(SC_2H_4Ph)_5(X_2)^+$ (where X = Cl, Br),[38] $[Au_{20}(PPhpy_2)_{10}(Cl_4)]Cl_2$,[39] *etc.* After that, taking advantage of the high affinity of sulfur towards gold, organic thiols were used and even now new thiol containing ligands of varying complexity and functions are getting introduced for better understanding of the role of ligands. Whetten and Murray developed thiol based cluster synthesis by introducing a water soluble peptide, glutathione (GSH) as the ligand.[8] Tsukuda and his group extended the study and purified the clusters to get a series of clusters with different nuclearity.[17] Oragnic thiols like phenyl ethane thiol (PET), hexanethiol, octanethiol, dodecanethiol, tertiary butyl benzyl mercaptan, *etc.* were used as ligands in organic synthesis. The advantage of these ligand protected clusters is their good mass spectral signature. Mercaptosuccinic acid (MSA), another water soluble ligand had been introduced by Chen and Kimura for both nanoparticle and cluster synthesis.[40,41] Other water soluble ligands like D/L peniciallamine,[42] captopril,[43] *etc.* are also used as ligands. Several synthetic methods like direct synthesis by reducing metal precursors using strong reducing agents like sodium borohydride ($NaBH_4$) in the presence of ligands, core etching from nanoparticles and ligand exchange are followed by cluster chemists.[14–32] Recently, a new method called 'solid state synthesis'[44,45] has been introduced by Pradeep's group, where ligand and metal ion precursor are ground together with a reducing agent and then extracted in a preferred solvent. Several monolayer protected clusters of gold and silver[46–49] have been synthesized to date along with very few efforts on clusters of copper[50–53] and platinum.[54] Although gold cluster systems are plentiful now, very few of the thiolated systems have been crystallized namely, $Au_{102}(SR)_{40}$,[31] $Au_{25}(SR)_{18}$,[13,18] $Au_{38}(SR)_{24}$[30] and most recently, $Au_{36}(SR)_{24}$.[32] Three alloy clusters $Au_{13}Cu_x$ (where x is 2, 4 and 8)[55] and Au_{28} protected with mixed monolayers have been crystallized recently.[56] Silver clusters are relatively less stable when compared to gold clusters due to their high reactivity.[46–49] This is reflected in the number of known crystal structures of silver clusters. Some examples of Cu clusters with crystal structures are also available. They are produced using $LiBH_4$ as the reducing agent in inert atmosphere where hydride is the reducing agent, whereas for solution phase, normally nascent hydrogen does this job. In this Cu cluster, the smallest tetrahedral Cu moiety contains H^- encapsulated in it, which has been proven by crystal structure as well as neutron diffraction.[57] Several monolayer protected clusters of Au and Ag are known from several groups and some of them are thoroughly studied both experimentally and theoretically. Several of these clusters are known to show extraordinary catalytic properties towards many unusual reactions. These monolayer protected Au and Ag clusters can be made in the alloy form as reported for the Ag_7Au_6 cluster.[58] It is also possible to dope a single foreign atom like Ag to a Au core to make an alloy. Several other metals are also known to form alloys with Au in bulk. Negishi and his group could successfully incorporate Pd into Au_{25} and Au_{38} systems.[59,60] A maximum of five Cu atoms can also be exchanged with Au in the Au_{25} system as shown by the same group.[61] With the passage of time, there is an increasing need to find new protecting agents with added functionalities making the material more robust and useful. A new window of opportunity has opened up by using macromolecular templates like DNA,[62] dendrimer, *etc.*[63] Proteins as ligands are rather a recent entry to this fascinating family.[64] These clusters exhibit stable luminescence, enhanced biocompatibility due to the presence of a biological scaffold and proteins add functionality to these clusters which makes them highly attractive. Xavier *et al.* have written a short review on 'protein protected metal clusters' in 2012 terming them as an emerging trend in atomic cluster science, due to their fascinating properties and a rapid increase is seen in the number of research articles published on this topic.[9] At the time of this book chapter, there are significant improvements in the overall understanding of this new class of system. New questions exist, however. Several new protein containing clusters have been introduced and attractive applications have been developed. The birth of a new system always comes up with several intriguing questions and the combination of two complex systems (complex functional macromolecules and quantum confined metal clusters) with umpteen existing choices and rational designs would continue to evolve and attract the attention of curious minds. With this abridged account of the choice of ligands for cluster synthesis, we would like to give a brief overview of how protein protected quantum clusters (QCs@proteins) connect between different disciplines.

7.1.3 Protein Protected Metal Clusters—Conglomeration of Disciplines

Proteins are molecules, end product in the central dogma of molecular biology and they can be considered as soft nanomachines. They are composed of 20 different amino acids connected through peptide bonds having a primary linear sequence, a secondary structure through hydrogen bonding and tertiary folding and quaternary structure through a combination of different protein sub units. Earlier, there were attempts to conjugate MPCs through cysteines of macromolecules.[65,66] Palmer and co-workers have worked on the interaction of size selected clusters with proteins such as GroEL, a chaperone and horse radish peroxidase, *via* cysteine residues.[67]

In 2009, Xie *et al.* reported the protein directed synthesis of gold clusters by a protein, bovine serum albumin (BSA).[64] Apart from BSA, lactotransferrin (Lf),[68] human transferrin (Tf),[69] horseradish peroxidase (HRP),[70] lysozyme (Lyz)[71] and several proteins are used as ligands for cluster synthesis which reveals the versatility of this system in synthesizing clusters. The availability of different proteins has been continuously explored by researchers which throw in more and more new systems to be studied thoroughly, as each protein may behave differently. The main idea of choosing proteins for making clusters is the availability of sulfur containing amino acid, cysteine (Cys) in the protein, although it may not be the only reason. QCs@proteins may likely fall in the category of thiolated metal clusters as the metallic cores may be protected with highly folded complex macromolecules with multiple thiols. Proteins having a high molecular weight with high Cys content and other proteins having a low molecular weight with low Cys content are also used to make Au and Ag clusters (see Table 7.1). Unlike thiol ligands, where the number of core atoms depends on the proportions of metal and thiol precursor used, the size of the protein may play a vital role in this particular case. It has been believed that the growth of clusters in proteins is associated with the biomineralization process. Living organisms biomineralize for functional purposes or to escape toxicity from toxic metal ions. Biomineralization occurs through the binding of metal ions by specialized vesicles, peptides and proteins.[72,73] One may ask whether such biomineralization of gold happens in nature. Indeed, bacterial biomineralization of gold occurs in nature; the bacteria, *Cupriavidus mettalidurans* biomineralize toxic gold ions into metallic gold and are responsible for the formation of gold nuggets.[73] Several efforts have been carried out to understand these processes. Wei *et al.* have grown Au nanoparticles in single crystals of lysozyme and studied nanoparticle growth inside the crystals with time.[74] A few groups have already started doing first principle studies on how growth of gold particle occurs in peptide templates.[75]

Mass spectrometry (MS) has been an indispensable tool for the investigation of clusters especially when precise molecular composition can be deciphered though soft ionization methods like matrix assisted laser desorption ionization mass spectrometry (MALDI MS). Efforts have been

Table 7.1 Noble metal clusters in different protein templates and their applications.

Protein/peptide	Mol. wt. kDa	Number of cysteine residues	Metal	Assigned core size M_xM = Metal; x = core size	Study/Applications	Ref.
Bovine serum albumin	67	35	Au, Ag, Cu	$Au_{4,13,25,20-25,16}$, Ag_{15}, $Cu_{5,13}$	Hg^{2+}, Cu^{2+}, Pb^{2+}, S^{2-} pyrophosphate, querectin, uric acid sensing, logic gate, bio-imaging	81,92,134,142,144,155, 172,174,198,218,223
Lysozyme	17	08	Au, Ag	$Au_{25,8,10,12}$	Hg^{2+} sensing, cluster growth, deriving gas phase clusters	74,81,86,140,220
Lactotransferrin	83	35	Au	$Au_{13,25,34,40}$	Cluster evolution, FRET, conjugation with RGO, Cu^{2+} sensing, bio-imaging	68,76,219
Serum transferrin	78	40	Au	Au_{20-25}	Bio-imaging, conjugation with RGO	69,212a
Bovine pancreatic ribonuclease	18	08	Au		Bio-imaging	201
Human cells (in nucleolin)			Ag		*In vitro* synthesis of clusters in cells	80
Egg shell membrane and bird feathers[a]			Au, Ag		Solid platform synthesis, metal ion sensing	84
Human hair[a]			Au		Solid platform synthesis	85
Human serum albumin	67	35	Au, Ag	$Ag_{9,14}$	NOx sensing	92,125

Protein			Metal	Application	Ref.
Lactalbumin	18	08	Au	Anticancer activity	102
Chymotrypsin	26	10	Ag	Enzyme activity	79
Trypsin	26	12	Au, Ag	Hg^{2+} sensing, bio-imaging	142,180
Ferritin complex			Au	Bio-imaging	221
Papain	39	08	Au	Cu^{2+} sensing	222
Pepsin	35	06	Au	Hg^{2+} sensing, blue, green and red emitting Au_{QC}	141
Insulin/insulin fibrils	51	06	Au	Bioactivity, bio-imaging, cluster growth in crystals, solid state platform, tracking metabolism	77,106,213
Horse radish peroxidase	39	10	Au	H_2O_2 sensing	200
Ovalbumin	43	06	Au		94
Cellular retinoic acid binding protein II	16	03	Au		71
Egg white[a]			Au,Pt	H_2O_2 sensing, metal ion sensing	159
Glucose oxidase	66	03	Au	Enzyme activity, glucose sensing	154b
Bovine hyaluronidase-3	14	04	Au		112

[a]Mixture of proteins.

made to understand the mechanism of cluster growth by Pradeep's group.[76] They have shown how the Au^+ intermediate is converted to metallic clusters in proteins and suggested a novel inter-protein metal ion transfer mechanism which has raised several intriguing questions (see Figure 7.1).[76] Also, the difference between the core size present in small and large proteins suggests that core sizes largely depend upon the protein used (see section 7.2.4). How conformational changes in proteins facilitate the formation of clusters and affect the luminescence have also been investigated (see section 7.2.4). As mentioned earlier, bright luminescence in these clusters has been the most attractive property. However, the photophysics of the origin of luminescence is yet to be clearly known (see section 7.3). The retention of the bioactivity of proteins[77] after cluster formation adds functional attributes to the clusters made. This has potential applications as optical probes in, but not limited to, biology and medicine and also opens up many fundamental studies in this area (see section 7.4). When metal precursor is added to a protein solution, a metal ion–protein adduct forms; without proceeding further, if we stop the reaction at the adduct formation step and introduce the complex to laser desorption and ionization, a new class of gas phase bare clusters form, as reported by Baksi *et al.* recently (see section 7.5),[78] revealing that proteins can also be used to prepare gas phase clusters. As one may observe from the above, protein protected clusters bring several molecular disciplines/concepts together such as clusters from solution phase to gas

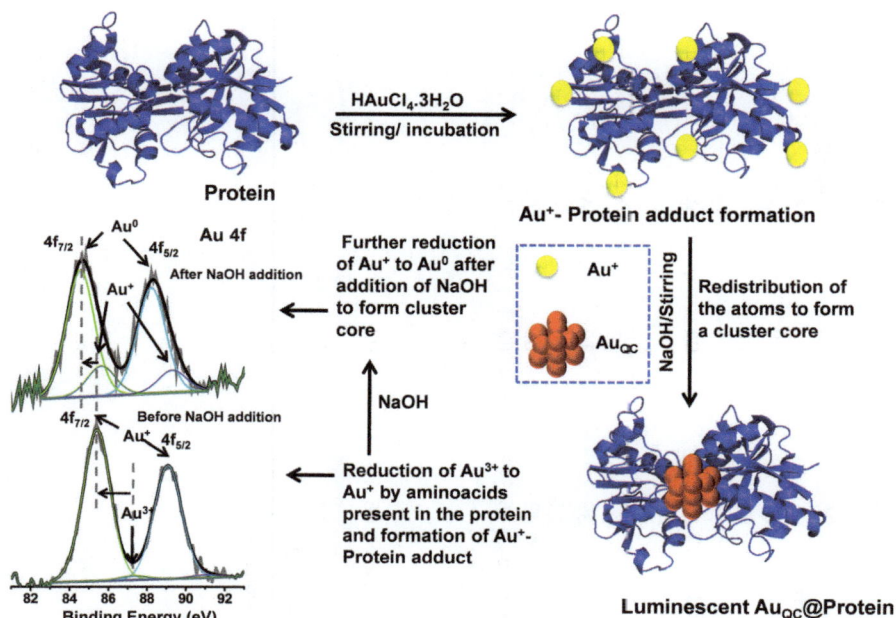

Figure 7.1   General synthetic scheme for protein protected luminescent gold clusters.
Adapted from different papers.

phase, biomineralization, photophysics, structural and molecular biology, quantum mechanics and medicine at the nano-biointerface, making it a conglomeration of disciplines.

Let us discuss how these molecular hybrids are prepared, characterized (including their evolution) and applied in the coming sections.

7.2 Synthesis and Characterization

7.2.1 General Synthetic Route and Separation of Protein Protected Clusters

Following biomineralization, Narayanan and Pal have established a method to prepare Ag clusters in an enzyme, α-chymotrypsin using $AgNO_3$ as the Ag precursor and $NaBH_4$ as the reducing agent.[79] Yu *et al.* have transferred the as-synthesized Ag nanoclusters to another biological scaffold by a shuttle based method (switching between scaffolds) and also prepared Ag cluster intracellularly in NIH 3T3 cells.[80] All these syntheses required an external reducing agent. In 2009, Xie *et al.* first reported one step synthesis of a luminescent Au_{25} cluster in a BSA template without using any external reducing agent.[64] They have reported that when Au^{3+} is mixed with the protein and the pH is maintained at 12, aromatic amino acids donate electrons to reduce Au^{3+} and the broken disulfides at elevated pH stabilize the nucleus through Au–sulfur linkages. Several groups still follow the same procedure (see Figure 2.16 in Chapter 2 for details). There are some reports where ascorbic acid is used for reduction at pH 3[81] and several modified paths including core etching of Au nanoparticles with BSA, as reported by Muhammed *et al.*,[82] exist. A general synthetic procedure being used in most of the cases (See Figure 7.1) is given below:

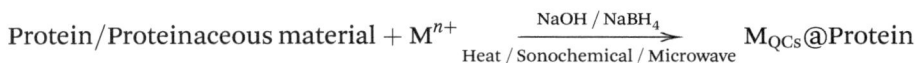

$$\text{Protein/Proteinaceous material} + M^{n+} \xrightarrow[\text{Heat / Sonochemical / Microwave}]{\text{NaOH / NaBH}_4} M_{QCs}@\text{Protein}$$

Dansyl chloride is a well-known fluorescent tag for biological samples which does not change chemical and biological activity of biomolecules. Blue luminescent dansylated BSA (dBSA) was used for making clusters which are red emitting when excited with 365 nm radiation. G-75 sephadex gel was used for column separation purpose. Separation of $Au_{QCs}@$dBSA from dBSA can be seen visually under 365 nm UV lamp.[83] In the beginning, a mixture of both cluster and free protein is observable and with time red luminescent clusters start separating. This separated cluster solution was collected and freeze dried to get dry nanoclusters and was further purified to get nearly 85% pure $Au_{QCs}@$dBSA from a mixture of it and dBSA. After purification, the properties of the cluster did not change as proved by several characterization techniques. Another type of cluster synthesis involves the use of solid platforms as reported by Deng *et al.* for luminescent Au and Ag cluster synthesis using egg shell membrane (ESM) as template.[84] ESM is a double layer water

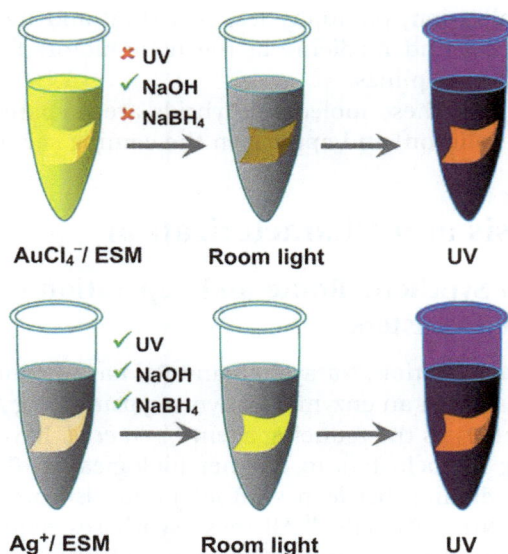

Figure 7.2   Schematic representation of an ESM based multimodal platform for the synthesis of fluorescent gold and silver nanoclusters.
© Royal Society of Chemistry. Reproduced from ref. 84 with permission.

insoluble membrane with fibrous structure which is attached inside a natural hen eggshell. ESM is composed of highly cross linked and cysteine-abundant proteins like keratin, collagen or elastin. In a typical ESM based heterogeneous synthesis, the freshly stripped membrane is soaked in double distilled water for 48 hours and then it is incubated in $HAuCl_4$ or $AgNO_3$ solution followed by incubation with suitable reducing agent solution to get fluorescent Au or Ag nanoclusters. In ESM, Ag clusters can be made with both UV and $NaBH_4$. They have also synthesized clusters even in bird feathers (see Figure 7.2). Haveli *et al.* have synthesized luminescent gold clusters using hair fibers as the solid state platform.[85]

7.2.2   General Characterization

Unlike monolayer protected gold clusters which have distinct absorption features due to the core, the absorption spectrum of a gold cluster core protected with protein is rather ill defined in most cases. Hence, deciphering the cluster composition from the spectra becomes highly limited. Mostly featureless absorption spectra have been reported with a peak nearly at 290 nm due to the absorption of protein. Another hump is seen at 350 nm and is attributed to oxidized aromatic amino acid residues.[9,68,76,86] In the case of silver clusters protected with BSA, absorption features at 415 and 485 nm have been reported by Mathew *et al.*[87] TEM shows a core size of nearly 1 nm as reported in maximum cases although true size cannot be obtained from such studies due to electron beam induced aggregation. However, such core sizes below 1.5 nm suggest the presence of quantum confined clusters. Normally these clusters have two excitation (nearly 365 nm and 500 nm) and emission maxima (450 nm and 650–690 nm) in the photoluminescence spectra. When excited at 365 nm, QCs show two emissions, one near 450 nm which is attributed to emission from protein (also reported to be due to Au_8 by different groups) and another NIR emission centered around 630–690 nm (depending upon the core and the protein), due to emission from the cluster core.[9] Au_{QCs}@proteins are having good quantum efficiency. The values vary from protein to protein; Au_{QCs}@Lf is 6%,[68,76] Au_{QCs}@BSA is 7%[88] and Au_{QCs}@Lyz is 15%[86] (note that core sizes are also different for these cases). This luminescence can be used to sense specific ions or molecules and also in imaging purposes which will be discussed in detail (sections 7.3.1 and 7.4.1). Au clusters have longer fluorescence decay compared to Ag_{QCs} as reported for different proteins. For Au_{QCs}@Lf, lifetime components are 0.18 ns (58%), 0.82 ns (22%), 3.58 ns (11%) and 110.70 ns (9%),[68] for Ag_{QCs}@BSA these values are 0.28 ns (60%), 1.17 ns (30%) and 4.10 ns (10%), for Au_{QCs}@Lyz these are 0.1 ns (40%), 1.1 ns (55%) and 19.0 ns (5%) and for $(AuAg)_{QCs}$@BSA the values are 0.12 ns (71%), 1.20 ns (22%) and 11.8 ns (7%).[88] The origin of longer lifetime and other properties of luminescence will be discussed in section 7.3.

Presence of a nearly metallic core is often proven by X-ray photoemission spectroscopy. Almost all protein protected Au_{QCs} have Au $4f_{7/2}$ binding energy nearly at 84.0 eV confirming a Au^0 oxidation state. A small amount of Au^{1+} is also present; as we know that reduction of Au^{3+} to Au^0 proceeds *via* a Au^{1+} intermediate.[76] For Ag_{QCs} we also see the Ag $3d_{5/2}$ binding energy to be 368.0 eV which corresponds to the Ag^0 binding energy value.[87] Metal–sulfur bonding can also be proved by XPS. The S $2p_{3/2}$ binding energy ranges from 162.0–16.0 eV which is attributed to thiolate formation. Unlike monolayer protected clusters which are sensitive to X-ray induced damage and show sulfate or sulfonate formation, protein protected clusters, due to complete encapsulation, do not show this kind of effect in most of the cases. Atomic ratios calculated from XPS prove the presence of excess protein in the system, as reported for Au and Ag clusters. Besides that, the structural and electronic properties of Au_{QCs}@BSA have been studied using X-ray absorption spectroscopy (XAS).[89] Studies using extended X-ray absorption fine structure (EXAFS) suggest that a Au_{25} core is present with a Au–thiolate staple motif ($-SR(Au–SR)_x-$). They have used X-ray absorption near-edge structure (XANES) and Au 4f XPS to probe the electronic behavior of Au–BSA. The Au d-electron density of Au–BSA was found to decrease by 0.047 e^- relative to that of the bulk.

7.2.3 Mass Spectrometry and Clusters

As we have already discussed, general spectroscopic methods commonly practiced in nanoscience are not enough to understand the core size for protein protected clusters since they do not exhibit any specific absorption feature. Mass spectrometry has become an indispensable tool for such kinds of systems. In the early days, mass spectrometry was limited to small thermostable molecules as there were no soft ionization techniques which could transfer ionized molecules from condensed phase to gas phase without fragmentation. In the late 1980s two path changing techniques were discovered, namely ESI MS and MALDI MS. These two methods changed the whole situation and made protein analysis even quicker and easier. Extension of these techniques led to new mass analyzers like multistage hybrid quadruple time of flight and tandem time of flight mass spectrometers. Although ESI is more accurate and gives good resolution, due to large charge distribution and fragmentation, it becomes difficult to analyze protein protected clusters by ESI MS. Moreover, there are issues about poor ionization in solution phase mass spectrometric analysis. Till now there is no report on protein protected clusters analyzed by ESI MS. In this scenario, MALDI MS is the commonly used technique which is a soft ionization method and capable of showing the parent singly charged ion known as the molecular ion. However, laser desorption ionization often leads to S–C bond breakage and aggregate formation. In MALDI MS, the matrix is used to enhance the ionization. MALDI matrices are such organic molecules which form crystals easily. Typically sinapic acid (SA), α-cyano-α-hydroxy cinnamic acid (CHCA), *etc.* are used for proteins. A minimum quantity of cluster solution is mixed well with a larger volume of matrix solution and spotted on the MALDI plate to yield a dried droplet. In this technique, it is assumed that ions are generated instantly upon laser irradiation. The time width of a typical 337 nm N_2 laser used in most of the MALDI TOF instruments is the order of few nanoseconds. When laser intensity is high enough to exceed the ion generation threshold, ions may continue to be generated even after the completion of laser irradiation. So in this process ions are lost. To avoid such circumstances and to improve resolution, a delay time is given between ion generation and extraction to the detector. It is well-known from the time of fullerenes that by controlling delay time, stable ions can be generated.[90]

Xie *et al.* showed that a 25-atom core of Au is stabilized by BSA.[64] MALDI MS data show a shift of about 5 kDa from the parent BSA peak in the case of Au_{QCs}@BSA confirming the presence of Au_{25}@BSA in the system. These data were further supported by thermogravimetric analysis data. Muhammed *et al.* prepared a Au cluster in BSA from nanoparticles by a core etching process.[82] Mass spectra showed a clear shift of 38 Au atoms from the native BSA peak and no other peaks were observed in higher mass range. Therefore, they have concluded that mostly Au_{38} core is present. Xavier *et al.* have introduced lactotransferrin (Lf) as a ligand to this luminescent cluster family. From positive ion MALDI MS using SA as matrix, they have shown the presence of Au_{13} and Au_{25} cores in Lf.[68] Chaudhari *et al.* have further extended the Au_{QCs}@Lf system to understand the growth mechanism. They suggested the presence of Au_{25} and Au_{13} clusters apart from the free protein.[76] They also reported the presence of 89% Au_{25}, 4% Au_{13} and 7% Lf by assuming all the parameters affecting the mass spectral intensity to be the same. Schneider *et al.* proposed the formation of a blue luminescent Au_8 core under mild basic conditions (pH 8) and red luminescent Au_{25} core after reducing with ascorbic acid at pH 3.[91] MALDI TOF using a CHCA matrix showed a Gaussian type distribution having a series of peaks. Two types of peaks were present, major peaks were separated by *m/z* 197 due to Au and minor peaks were separated by *m/z* 32 (from the major peaks) due to sulfur attachment assuming $Au_nS_m^+$ aggregate formation. The peaks with maximum intensity were related to the high population of Au_{22}–Au_{25} cores. As discussed earlier in section 7.2.1, Wu *et al.* separated free BSA from Au_{QCs}@BSA by using the fluorescent tag, dansyl chloride.[83] They also showed that before separation, a small hump appears due to dBSA, but after separation, the hump was absent confirming separation of excess protein. The cluster peak shifted by the mass of 25 Au atoms from the parent protein peak. And the cluster peak position remained constant before and after separation suggesting no change in the nuclearity of the core during the separation process (see Figure 7.3). Human transferrin has also been used for Au cluster synthesis. Guevel *et al.* through MALDI TOF MS showed a broad distribution of Au nanoclusters ranging from Au_{10}–Au_{55}.[69] Peak maxima for the Gaussian distribution were obtained for 22–33 gold atoms covalently bound to the thiol-bearing cysteine residues of human transferrin. For this case also, major peaks were separated by *m/z* 197 and minor peaks by 32, due to Au and S, respectively.

So far, we have discussed larger proteins having a large number of cysteine residues. A few proteins with a limited number of cysteines are also used for cluster synthesis, namely insulin, lysozyme, trypsin and α chymotrypsin.

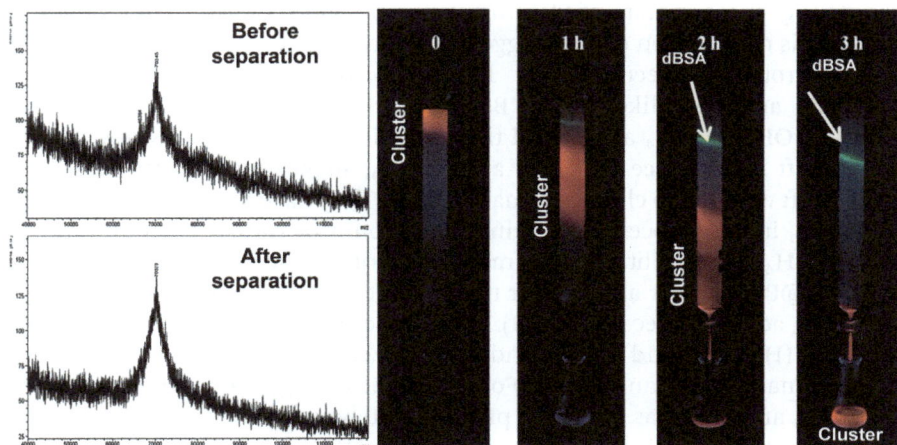

Figure 7.3 MALDI MS of the Au_{NC}@dBSA before (top) and after (bottom) gel column separation. Fluorescence photographs of the G-75 Sephadex gel column separation procedure of Au_{NC}@dBSA from dBSA under 365 nm irradiation. Photos represent the elution at the indicated time periods. © Royal Society of Chemistry. Reproduced from ref. 83 with permission.

In the case of insulin, Liu *et al.* described the formation of red luminescent Au clusters; however, they did not see any mass shift from the parent protein peak.[77] Also, disulfide bonds were intact as confirmed by the Raman spectrum proposing a different type of cluster growth in the Au_{QC}@insulin system. Chen and Tseng have made blue luminescent Au clusters in lysozyme in acidic conditions and showed the presence of Au_8 by MALDI MS.[81] Baksi *et al.* have proposed the presence of small Au clusters inside a single Lyz and its aggregates.[86] Detailed mass spectrometric studies revealed that only small cluster cores (Au_{10}–Au_{12}) can be stabilized in a single Lyz molecule. Unlike large proteins like BSA or Lf, the mass spectrum of Lyz consists of several conspicuously observable aggregates along with the monomer. After cluster formation, a shift in the parent peak by a total mass of 10 Au atoms was observed for a $1:4$ ratio of Lyz : Au^{3+} and the cluster was assigned to Au_{10}@Lyz. Similar to Lyz^+, aggregates like Lyz_2^+, Lyz_3^+,... clusters also show the presence of $(Au_{10}@Lyz)_2^+$, $(Au_{10}@Lyz)_3^+$, ... confirming the presence of cluster cores inside a single protein molecule. Au_{11} and Au_{12} cores could also be achieved by using different Lyz : Au^{3+} ratios. For the lowest concentration of Au^{3+} to Lyz, cluster growth was also observable among multiple protein molecules.

While protein protected Au clusters are plentiful, only a few reports are available on Ag clusters in proteins. Among the reported Ag_{QCs}@protein systems, only a few have been assigned with core size using MS. Mathew *et al.* have reported luminescent Ag clusters in BSA by $NaBH_4$ reduction in basic medium.[87] They have done a detailed mass spectrometric study using MALDI MS to find out the nuclearity of the cluster core. BSA shows a peak around m/z 66.7 kDa and the as-prepared cluster showed a mass shift of 1.6 kDa suggesting a Ag_{15} core inside. The difference is half (m/z 800 Da) for the doubly charged state of BSA and is around 500 Da for the triply charged state. This observation clearly suggests the presence of cluster cores inside a single protein molecule. They have also done several control mass spectral analyses like BSA, BSA + $AgNO_3$, BSA + $AgNO_3$ + NaOH and BSA + NaOH + $NaBH_4$ and found that BSA alone does not show any drastic mass shift in presence of NaOH and $NaBH_4$, suggesting that the observed mass shift was due to cluster formation only. They have also checked the role of $NaBH_4$ in this process by varying the amount of $NaBH_4$ used and found that $NaBH_4$ only facilitates the transformation of the Ag–BSA conjugate to the Ag_{15}@BSA cluster as no other new product was formed with an increase in $NaBH_4$ addition (See Figure 7.4). Mukherjee *et al.* have used human serum albumin (HSA) instead of BSA and synthesized two different types of silver clusters, namely Ag_9 and Ag_{14}.[92] For blue emitting Ag_9 the synthesis mixture of $AgNO_3$ and HSA was stirred at pH 11 for 10 hours while red emitting Ag_{14} clusters were synthesized by rapid reduction with $NaBH_4$. Mass spectra show a visible shift to 67.4 kDa from 66.4 kDa for Ag_9@HSA and to 67.9 kDa for Ag_{14}@HSA. Similar shifts were also observed for the doubly charged state. Recently, Mohanty *et al.* have reported Ag_{31}@BSA and used the same for

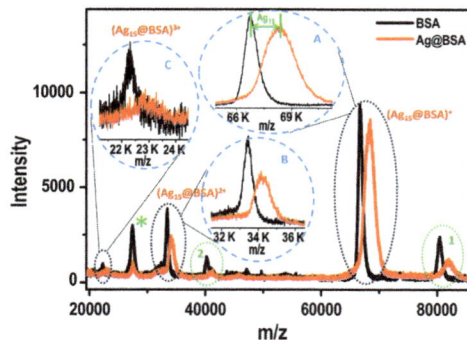

Figure 7.4 MALDI MS of pure BSA solution collected in linear positive ion mode using sinapic acid as matrix and that of the as-prepared red emitting Ag_{15}@BSA. The peaks due to singly, doubly and triply charged ions of Ag_{15}@BSA are expanded in the inset marked A, B and C, respectively. Peaks marked 1 and 2 are singly and doubly charged species of conalbumin which is an internal standard used in BSA protein. The impurity peak at m/z 27.5 kDa is marked with an '*'.
© Royal Society of Chemistry. Reproduced from the ref. 87 with permission.

alloy cluster formation with Au.[88] They have prepared Ag_{31} and Au_{38} protected with BSA and mixed both the clusters in different ratios namely, Ag_{31} : $Au_{38} = 90:10$, $50:50$ and $10:90$ and found systematic shift from both the parent clusters and proposed tunable alloy formation by this process. For the $90:10$ ratio, only a few Ag atoms were replaced by Au whereas for $10:90$, some Au atoms have been replaced by Ag. These tunable alloy clusters showed well-defined compositions suggesting the formation of $Au_{1-x}Ag_x$ clusters across the entire compositional window and there was no signature of parent clusters in the mass spectra. After the alloy formation, the presence of free protein was also observable suggesting inter-protein metal ion transfer. In another approach, they have tried to make alloy clusters through a galvanic exchange method assuming that if both preformed Au and Ag clusters react with each other to form the alloy, there must be reactivity of individual Au/Ag ions towards the other cluster, like Au ion should interact with preformed Ag clusters. To prove it, they have added various amounts of Au^{3+} to Ag cluster solution and found a systematic mass shift revealing alloy formation. XPS data showed the formation of Au^0 after 4 hours of stirring. Inspired by these findings, they have studied time dependent changes upon the addition of Au^{3+} to Ag clusters and found well-defined and well characterized AgCl formation confirming galvanic exchange. Since chloride is added to the solution in the form of $HAuCl_4$, the replaced Ag ions combine with Cl^- to form AgCl and get precipitated. They have shown the presence of AgCl by SEM-EDAX proving the galvanic exchange process.

7.2.4 Mass Spectrometric Studies on the Growth of Clusters in Protein Templates

7.2.4.1 $Au_{QCs}@Lf$

Chaudhari *et al.* have reported a detailed study on the growth of Au_{QCs} in Lf templates.[76] They have added Au^{3+} and Lf together to get a final concentration of 2.5 mM and 150 μM, respectively. Just after the addition of Au^{3+} to Lf, it uptakes Au^{3+} and at a pH of 4–5, Au^{3+} is reduced to Au^{1+} as supported by XPS study. Native Lactoferrin shows a broad peak at m/z 83 kDa and its dimer at around 166 kDa through salt bridge interaction. When Au^{3+} is added to the solution, Au^{1+}–Lf conjugates are immediately formed and a peak around 86.2 kDa is observed suggesting about 16 Au attachments to the parent protein through interaction with several amino acids and no further change is observed in the absence of NaOH (See Figure 7.5). After NaOH is added, the pH of the solution is raised to 12.4 at once and at such elevated pH, aromatic amino acids are able to reduce Au^{1+} to Au^0 and cystine bonds are broken. Sulfurs of cysteine stabilize the cluster core. After 4 hours, the cluster core forms, as it starts showing luminescence in the NIR region and a bright red color is shown under UV light. Mass spectral data show a peak at 87.6 kDa after 4 hours corresponding to $Au_{22-23}@NLf$ which emerges with time and after 12 hours Au_{13} appears along with reemergence of free protein (parent NLf peak). After 25 hours, $Au_{25}@NLf$ is observable and after 48 hours there is no change in the spectrum and it also remains the same after 3 months. There are possibilities of formation of the cluster core surrounded by mulitiple proteins which are ruled out as the dimension of the protein is far higher than the ∼1 nm cluster core. To revalidate this, monomer and dimer regions of the protein and cluster were checked carefully. Just after addition of Au^{3+} to Lf, the dimer region showed around double the uptake of the monomer region. After 12 hours of cluster formation, free protein was regenerated along with Au_{13} and Au_{25} as seen in the monomer region. The dimer exhibits several possibilities like one protein with Au_{13} and another with Au_{25}, or both the units with Au_{13} or Au_{25} altogether. Due to all these, we see a broad distribution suggesting cluster core formation inside a single protein molecule. Various concentrations of Au^{3+} were used for keeping protein concentration the same and at low concentration, small cores nucleate slowly whereas for higher Au^{3+} concentration, bigger clusters form suggesting a distinctly different growth mechanism in these cases. This whole process is a single step process. The authors tried to re-feed the free protein with the addition of Au^{3+} ions and named this path a two step approach. In this process, first clusters were prepared using 2.5 mM Au^{3+} and after 24 hours the final concentration was adjusted to 3–5 mM. Under these conditions, free proteins again uptake Au and form clusters: which is supported by two-fold increase in the luminescence intensity suggesting an increase in cluster concentration. This is also proven by MS where Au_{25} intensity increases and free protein intensity decreases. Also, the Au_{13} peak is

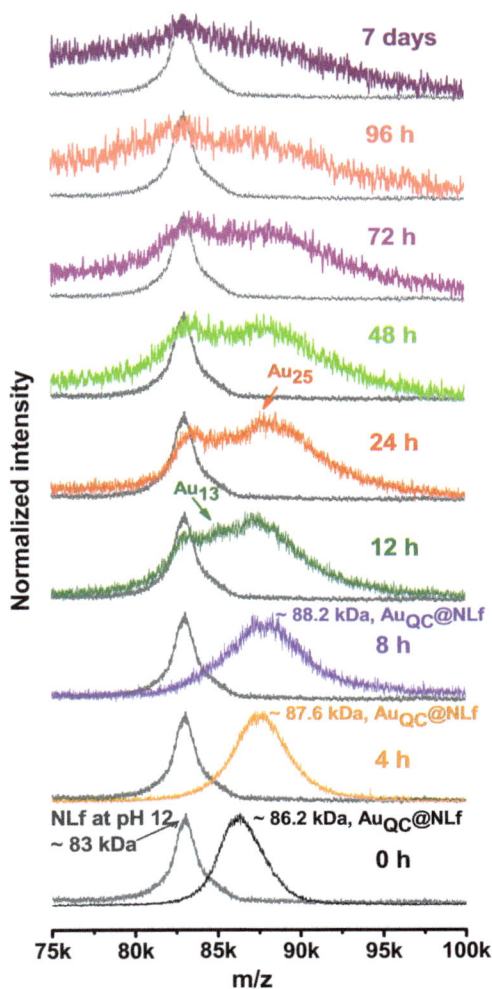

Figure 7.5 MALDI MS data of Au$_{QC}$@NLf monitored with time. Reaction was carried out with 150 μM NLf and 2.5 mM Au^{3+} at pH ~12.4 (5% NaOH v/v). Shift in mass was compared against the mass of NLf at pH ~12.4.
© ACS Publishing. Reproduced from the ref. 76 with permission.

completely absent suggesting complete transformation to Au$_{25}$. A two step approach is seemingly preferential as we can only get one type of mono-dispersed cluster with enhanced luminescence. Regeneration of free protein can be explained in terms of redistribution of Au ions through inter-protein metal ion transfer. When a cluster core starts growing inside one protein, it attracts gold, attached to other proteins due to aurophilic interaction. This allows preferential attraction of inter or intra protein gold to aggregate around the nucleus; this autophilic interaction is explained as due to the relativistic expansion of the gold d orbitals. Growth of Au_{QCs}@Lf has also been monitored by CD and photoluminescence.

7.2.4.2 Au_{QCs}@BSA

To check if the growth mechanism can be further extended to other larger proteins or not, BSA has been chosen by the same group to probe the cluster growth mechanism.[76] In this case also, just after the addition of Au^{3+} to the system, the BSA peak shifts to 68.4 kDa and no free BSA is observed. This entity evolves and shifts to higher mass suggesting more and more Au attachment to the protein. After eight hours, a small Au_8 cluster core was observed along with the emergence of free protein. This suggests a common mechanism of intra protein metal transfer as described before for Lf. Au_8 is known to be blue emitting as reported by some groups although it is very hard to differentiate the emission from Au_8 as protein alone has 450 nm emission. Other spectroscopic studies like UV-Vis absorption and photoluminescence also support the mass spectral observation. After 12 hours, Au_{25} can be seen which further grows and forms Au_{38} and other cores with higher nuclearity. Recently, by MALDI MS studies, Yu *et al.* have reported the formation of Au 4, 8, 10, 13 and 25 in BSA through CO mediated synthesis.[93]

7.2.4.3 Au_{QCs}@Lyz

Although cluster growth inside larger proteins has been proven, there was no direct proof whether a cluster core can be stabilized inside small proteins or not.[71,77,94] It is known and already discussed that the cysteine residues of proteins play a vital role in the stabilization of the cluster core by Au–S bonding. Small proteins contain a limited number of cysteines, so stabilization of bigger cores is unlikely in a single protein. Chou *et al.* have proposed a different type of cluster growth in insulin as luminescent clusters can be grown inside protein crystals.[77]

7.2.4.3.1 Single Cluster Core in a Protein. Computational investigations of thiolated dendrimers interacting with Au_{55} have shown that single dendrimers could wrap entire clusters by anchoring multiple thiol groups on them.[95] MS and DLS studies, in the case of larger proteins have also shown that clusters are likely to be present in a single protein.[64,69,76] To investigate cluster growth mechanism in smaller proteins, Baksi *et al.* have conducted a detailed mass spectrometric study for Lyz protected Au clusters.[86] Unlike other larger proteins like Lf or BSA, Lyz also shows a well-defined mass spectral signature due to its aggregates. Lyz shows a sharp molecular ion peak at 14.3 kDa and along with that, peaks at 28.8, 42.9, 57.2, 71.5 kDa also appeared which can be attributed to dimer, trimer, tetramer and pentamer of Lyz and assigned as Lyz_2^+, Lyz_3^+, Lyz_4^+ and Lyz_5^+. Different ratios of Lyz : Au^{3+} were used; namely, 1 : 2.5, 1 : 4, 1 : 5

and 1 : 8 for making clusters. All the clusters were made by mixing Lyz and Au^{3+} followed by the addition of NaOH and incubation. All these clusters were studied using MALDI MS and fluorescence spectroscopy for 15 days. For the 1 : 4 ratio, a well-defined cluster feature appeared along with the free protein separated by nearly 2 kDa due to 10 Au atoms and the cluster was assigned as Au_{10}@Lyz (see Figure 7.6). This cluster feature was almost stable up to 15 days. However, a small change in the nuclearity was observed after 2 days which is supported by the luminescence spectrum where emission from the cluster shows a blue shift of 25 nm. A similar type of growth was also observed in the case of 1 : 5 and 1 : 8 ratios of $Lyz : Au^{3+}$ where Au_{11} and Au_{12} cores are seen, respectively. Although a shift in mass is observable in the monomer region, it is not fully clear whether clusters are forming inside a single protein or not. Here, the changes observed in the aggregate region helped them understand the mechanism. In aggregates, there are many possibilities, like one protein with a core and another free protein, two proteins with two different cores or two proteins with the same core. Interestingly it is the cluster feature which is prominent in the whole mass range studied. All the aggregates show similar patterns of Au binding as seen for the monomer where the total number of Au atoms

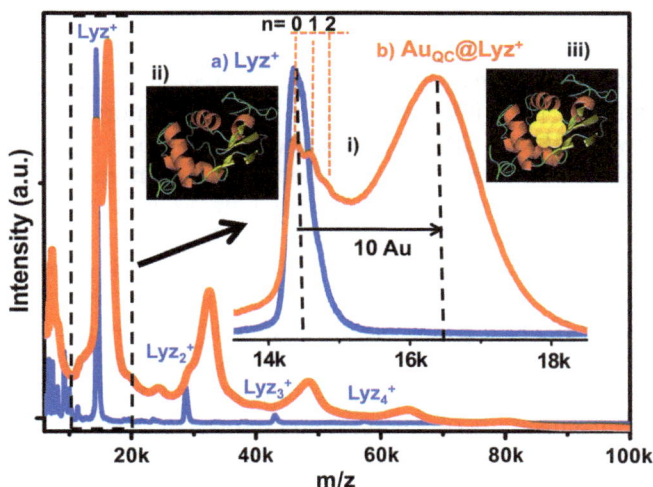

Figure 7.6 Positive ion MALDI MS of Lyz at pH 12 in linear mode (a) and Au_{QC}@Lyz after 24 hours of incubation (b). All the spectra were measured in the linear positive mode over the *m/z* range of 2000–100 000. Both Lyz and Au_{QC}@Lyz showed aggregate formation. The expanded monomer region in inset (i) clearly shows a separation of 10 Au atoms from the parent protein. In the dimer, trimer, tetramer and pentamer regions, the separations are of 20, 30, 40 and 50 Au atoms, respectively. In insets (ii) and (iii), schematic representations of Lyz and Au_{QC}@Lyz, respectively, are shown.
© Royal Society of Chemistry. Reproduced from ref. 86 with permission.

divided by the number of proteins in the aggregates confirms a 10 Au atom distribution in each case. This suggests that each protein entity consists of 10 strongly bound Au atom core in it and the aggregates are likely to be aggregates of proteins containing the same core rather than aggregates of protein molecules bound to a single multiatom gold species. In the latter case (multiprotein–single gold cluster) the number of attached Au atoms cannot increase systematically in multiples of 10 with the number of proteins in the aggregate. For the other ratios like 1 : 5 and 1 : 8, the same kind of aggregation behavior is observed confirming complete protection of the cluster core by a single protein. In the whole mass range, peaks are assigned as $Au_{10-12}@Lyz^+$, $(Au_{10-12}@Lyz)_2^+$, $(Au_{10-12}@Lyz)_3^+$, $(Au_{10-12}@Lyz)_4^+$, *etc.* showing a linear dependence in number of Au atoms *vs.* number of proteins in aggregates. This observation implies that aggregate formation is not a gas phase effect as there are several association possibilities in the gas phase which are not observed. Therefore, these entities are present in solution although there is no signature of their presence in the luminescence spectrum which can be explained in terms of longer inter-protein distance for efficient energy transfer.

7.2.4.3.2 Single Cluster @ Multiple Small Proteins. Time dependent change is more dramatic in the case of the lowest exposure of Au ions to protein (Lyz : $Au^{3+} = 1 : 2.5$). Upon incubation for a day, a small hump corresponding to Au_{10} appears in the monomer region of the mass spectrum while the shift in the dimer region was neither 10 nor 20 Au, it is somewhat in-between 10 and 20 Au. The same sample upon 7 days of exposure shows a shift of 12 Au atoms in the monomer region and in the dimer region, two distinct peaks appear, separated by 12 and 20 Au atoms from Lyz_2^+. The same observation is found for the trimer and the tetramer where, along with 12 Au separation, peaks corresponding to 22 and 26 Au atoms, respectively appear. Because of broadness and low intensity, it is hard to identify if more peaks are there or not in higher mass aggregates. As explained earlier, for other proteins there is also a gradual emergence of free protein with time in protein protected cluster systems. And this is also the lowest concentration of Au used; therefore, there is always some free protein in the system. Together, these increased concentrations of free proteins facilitate the formation of $Au_{10-12}@Lyz–Lyz^+$ and $(Au_{10-12}@Lyz)_2^+$ with characteristic mass shift. As free Lyz keeps on increasing in the system upon exposure time, only the $Au_{10-12}@Lyz–Lyz^+$ feature becomes prominent in the case of all the aggregates, along with protein alone aggregates. This observation explained that for smaller proteins like Lyz under specific conditions it is possible to get the cluster core stabilized by multiple proteins.

7.2.5 Conformational Changes in Proteins Upon Cluster Synthesis

Protein structure can be understood by its primary, secondary, tertiary and quaternary structure as mentioned earlier. Primary structure deals with only the linear sequence of amino acids and the secondary structure deals with hydrogen bonds between the backbone amino acids and the carboxylic groups. Rough secondary structure is that the % of α-helix, β-sheets or random coils can be measured spectroscopically by far ultraviolet circular dichroism where a double minimum at 208 and 222 nm indicates α-helical structure, whereas a single minimum at 217 nm reflects β-sheets and 204 nm reflects random coil structures. The tertiary structure of proteins (stabilized by S–S bonds) can be characterized by near UV-CD. Infrared spectroscopy is also used to probe the secondary structure of proteins but less commonly. Most of the proteins are highly sensitive to their environments and fragile, apart from the proteins present in extremophiles which can withstand harsh conditions. Conformational changes in proteins due to interactions with heavy metal ions and nanomaterials and upon synthesis of nanomaterials are well-known in the literature.[96–100] In cluster synthesis, interaction with heavy metal ions and elevated pH causes conformational changes which in turn facilitate nanomaterial formation. It is reported by most of the groups that gold and silver clusters would be bound to the cysteine residues of proteins which would likely cost the native structure of the proteins. Because, in proteins, cysteine residues form disulfide bonds with another cysteine which is in proximity and stay as a cysteine dimer, known as cystine and addition of Au^{3+} ions would cleave the S–S bond of cystines which is well documented in the literature.[101] Cleavage of disulfides which preserve the tertiary structure of proteins would likely affect the secondary structure in the vicinity. Here, we will discuss briefly the change in the secondary structure of BSA, Lf and Lyz due to cluster formation; however, there are also studies available on the conformational changes upon cluster formation in a few other proteins.[69,102] It is known that at elevated pH, disulfides which preserve the tertiary structure of proteins break. In BSA, the majority of the 35 cysteine residues are situated in the α-helix regions. Though only pH change can induce change in the percentage of α-helix, the breakage of disulfide bonds in the α-helical vicinity would likely aggravate the change in the total α-helix content. Similarly Lf consists of 691 amino acids and 35 of them are cysteine. If we consider small proteins like Lyz, it contains 129 amino acids[103] among which 8 are cysteines forming 4 disulfides. Alpha-helix content can be calculated from circular dichroism data by the formula proposed by Chen *et al.*[104]

$$\alpha\text{-helix } (\%) = -(\theta + 3000)/39\,000$$

where $\theta = MRW \times \theta_{222}/10lc$, where MRW is mean residual weight (114 g/residue for BSA and 110.9 g/residue for Lyz), θ_{222} is the elipticity at 222 nm, l is the path length of the cell and c is the concentration of the sample used for the measurement. In the case of Au_{QCs}@Lf, there are two valleys at 208 and 222 nm indicating the presence of an α-helical structure which changes with time as reported by Chaudhari *et al.*[76] With time, the 208 nm valley shifts to 202 nm suggesting emergence of random coils and the valley at 222 nm becomes shallower suggesting unfolding of protein with time, facilitating cluster formation. For BSA, CD data show remarkable loss in α-helical content after cluster formation. Au_{QC}@BSA contains 13.8%, Ag_{QCs}@BSA contains 20.7% and $(AuAg)_{QCs}$@BSA contains 16.5% helical structure as reported by Mohanty *et al.*[88] For Lyz, 28% loss in helical structure was observed as calculated from the CD spectrum.[86] Although CD is a very useful tool to understand protein secondary structure and is often practiced by biologists, infrared spectrum can also be affected by a change in protein secondary structure. There are significant changes observable in the amide I, II, III and amide A regions. A strong band near 1650 cm^{-1} arises mainly from C=O stretching and a small contribution arises from out of plane C–N stretching. This part is attributed to amide I. Another band near 1550 cm^{-1} is due to the out of phase combination of NH in plane bending and a small contribution is from C=O in plane as well as C–C and C–N stretching. The 1400–1200 cm^{-1} region is assigned as the amide III region. A significantly broad band arises at 3500–2900 cm^{-1} which is due to O–H, N–H and C–H vibrations. Another band near 700 cm^{-1} can be attributed to $-NH_2$ and –NH wagging. After cluster formation, changes in the gross spectrum are observable for all the proteins. Second derivative IR is more sensitive to changes in the secondary structure. The 1600–1700 cm^{-1} region is attributed to amide I region which shows maximum changes due to cluster formation. In this range, α-helix ranges from 1651–1658 cm^{-1}, β-sheets from 1618–1642 cm^{-1}, random coils from 1640–1650 cm^{-1} and turns from 1666–1688 cm^{-1}; the α-helix region shows huge changes for all kinds of proteins and cluster combinations. A clear change can be seen in the α-helix feature at 1654 cm^{-1} which is completely absent in the case of all the clusters studied so far. In the majority of cases, cysteines are located in the α-helix region and, due to disulfide bond breakage, there is a perturbation in the α-helical region at the time of cluster formation and secondary structure changes as protein loses its helicity.

Recently Zhnag *et al.*[105] have reported pressure-induced conformational changes of ligand BSA in Au_{QCs}@BSA, which in turn are responsible for luminescence enhancement. They have used IR to probe changes in the total helical structure. As pressure increases, fluorescence intensity also increases almost linearly and they have suggested that there may be some role of BSA and its secondary structure may be changing during this process. As mentioned above, BSA has no β-sheets; it contains only α-helix and random coils. Bands centered at 1629, 1654, 1640 and 1673 cm^{-1} are assigned to presence of exposed α-helix, buried α-helix, random coils and turn structure, respectively. They have used *in situ* IR spectroscopy of both BSA and Au_{NCs}@BSA in D_2O to monitor pressure-induced conformational changes in the secondary structure of the protein. The band at 1653 cm^{-1}, which is characteristic of the high content of buried α-helix in native BSA, shifts to a lower wavenumber and gets broadened suggesting significant loss of the buried α-helix structure. For Au_{NCs}@BSA, at normal atmospheric pressure, a major broad band near 1646 cm^{-1}, which is different from native BSA appears, which is attributed to variation of the protein structure due to the extreme basic conditions and cluster synthesis process. With increase in pressure, the amide I band increases and slightly shifts to lower wavenumbers. A drastic sigmoidal trend beginning from 100 MPa is observable to the maximum intensity with respect to pressure indicating occurrence of big changes in the conformation of BSA in Au_{NCs}@BSA but for native BSA, changes are observable starting from 400 MPa indicating that metastable BSA in Au_{NCs}@BSA is more sensitive towards pressure-induced changes. By calculating the area under the curves through Gaussian curve fitting in the range of 1600–1700 cm^{-1}, they found that native BSA has 19.9% exposed α-helix, 46.6 buried α-helix, 20.5% random coil and 12.9% turn structure at 0.1 MPa pressure. Upon exposure to 1000 MPa, there is an increase in exposed α-helix (31.7% from 19.9%) and random coil (27.6% from 20.5%), whereas a huge decrease occurs in the case of buried α-helix (27.4% from 46.6%). Similar studies on Au_{NCs}@BSA suggest that at ambient pressure, ligand BSA has 28.2% exposed α-helix, 20.7% buried α-helix, 28.2% random coil and 22.4% turn structure. But when the pressure is raised to 1000 MPa, the structural elements of BSA in Au_{NCs}@BSA rearrange themselves which is reflected in the exposed (31.9%) and buried (32.1%) α-helix content whereas random coil (17.3%) and turn structure (18.7%) content decreases. Therefore, high pressure facilitates acquisition of ordered structure for BSA in Au_{NCs}@BSA and gets closer to the structure of native BSA at the identical pressure. Net α-helix content in native BSA is 66.5% which decreased to 48.9% after cluster formation at ambient pressure. After inducing high pressure in the system, BSA regained its α-helix structure (69% for native BSA at 1000 MPa and 65% for BSA in Au_{NCs}@BSA at the same pressure). These structural modifications influenced the luminescence of clusters (see section 7.3.1).

7.2.6 Peptide Protected Metal Clusters

Though peptides are not equivalent to proteins (in terms of size and structure), metal clusters protected by oligo and polypeptides cannot be missed in this chapter, hence we give a brief account of them in this paragraph. The definition for transition from polypeptide to protein is rather inexact. It is known that several polypeptides alone have specific functions, for instance, insulin having 51 aa is a peptide hormone. Use of functional peptides adds functionality to the synthesized clusters and in a way makes them smart materials. Most people have used peptides with cysteine residues for synthesizing clusters. Glutathione, a tripeptide (ECG) is the most used for the synthesis of MPCs. Then several other oligopeptides and polypeptides have been used. So far, the longest polypeptide used for cluster synthesis is insulin and insulin fiber, which is an example of peptide assembled into a fibrous structure, has also been used for gold cluster synthesis.[77,106] Fabris *et al.* had shown that peptides, when bound to Au_{38}, remained conformationally constrained.[107] Gao and co-workers synthesized Au_{25} by using a peptide containing nuclear targeting sequence, CCYRGRKKRRQRRR, and they also synthesized a series of Ag clusters using the same peptide.[108,109] Recently, Cu_{14} has been synthesized by using a nuclear targeting peptide.[110a] Yuan *et al.* synthesized Ag_{QCs} using GSH and designed peptides DCD, ECE, and SCS for red emitting Ag_{QCs} and another tripeptide (KCL) for blue emitting Ag_{QCs}. Recently, Feng *et al.* have synthesized wavelength tunable Au_{QCs} in different peptide templates. They used CALNN, DDCAGGEYDTF-PYWDD and CDDDDD polypeptides to synthesize gold clusters emitting at 611, 668, and 550 nm, respectively.[111] Söptei *et al.* have attempted to elucidate the design rule of Au_{QC} synthesis by different peptides. They used a few different peptides containing Cys at different positions, one peptide without Cys and revealed the importance of the presence of Cys residues in cluster formation.[112] Clusters synthesized by only amino acids such as histidine and proline have also been reported.[113,114] L-Cystenyl-L-cysteine, a dipeptide, has also been used to synthesize gold clusters.[115] Morozov *et al.* used three different coiled-coil metal binding peptides which have Cys–X–X–Cys; incorporated in their hydrophobic core (AQLIC16C19, a trimer peptide; TETC17C20, a tetramer peptide; and HEXC17C20, a hexamer peptide) to synthesize blue emitting silver clusters.[116] These clusters protected with peptides may find applications where lesser hydrodynamic diameter is needed when compared to protein protected clusters. Right now, there is no clear transition from monolayer (very short peptides or ligands) to macro-molecule (or increasingly complex ligands) protected clusters in terms of ligand size. We speculate that rationally designed peptides with different compositions and increasing length would fill the gap and throw more light on the role of ligands (on influencing luminescence, chirality *etc.*).

7.3 Origin and Properties of Luminescence in Protein Protected Noble Metal Clusters

In the expanding nanocosmos of metallic clusters, luminescence is one of the most attractive optical properties and the intriguing aspect about QCs@proteins. Their 'bright luminescence' has various applications as optical probes. In this section, we discuss the optical properties of clusters in general; absorption, excited state ultrafast spectroscopic studies, fluorescence anisotropy and two photon studies and other excited state interactions.

Complete understanding of the origin and properties of luminescence of QCs@proteins is not yet available, as their total structures are yet to be solved. Fluorescence from metallic clusters has been calculated with the free electron model, $E_{Fermi}/N^{1/3}$ where E_{Fermi} is the Fermi energy of the metal and N is the number of atoms in the cluster. However, the general applicability of the Jellium model in the case of liganded clusters in solution is yet to be verified. Talking about the luminescence from liganded noble metal particles, for instance, luminescence from gold nanoparticles can be due to (1) particle size (free electron model), (2) the surface ligand effect, (3) valence states.[117] Dickson and co-workers have demonstrated the validity of the Jellium model in the case of dendrimer protected Au clusters.[63] However, in other thiolated gold clusters, the free electron model could not totally account for the emission mechanism and the influence of ligands was suggested. In the case of ligand protected clusters, the groups of Murray, Pradeep and Jin have shown the influence of ligands on the luminescence.[16,118,119]

Here, we have attempted to document the efforts to date to understand the optical absorption and nature of luminescence in QCs@proteins. As mentioned above, though different reasons for the luminescence from gold nanoparticles are known to exist,[117] MS studies,[64,76] the presence of Au–thiol staple motifs[89] and the observation of sub-nanometer sized particles in TEM strongly suggest the presence of quantum sized nanoclusters in proteins to be responsible for the luminescence (however, which aforementioned model they follow is yet to be understood). Regarding the optical properties of Au_{QCs}@proteins, for instance, though cluster core has been assigned as Au_{25}@BSA/Lf by MS studies, no characteristic feature of $Au_{25}SR_{18}$ at 670 nm is seen. Several groups have not observed any distinct features in the UV-Vis analysis, while a few have reported absorption peaks around 510–540 and 350–370 nm and the origin of these observed features is still not solved, while a few assume that they are from the core and shell of the clusters, which are the same features seen in the excitation spectra of protein protected clusters. Zhou *et al.* have reported the formation of luminescent gold particles with mixed valence states generated from polymeric Au(I) thiolates with a featureless absorption spectrum with a slight hump around 520 nm.[120] It has also been reported that when the prepared clusters are heterogeneous, one may obtain a featureless absorption spectrum. These reports on MPCs may indicate that proteins might contain a series of clusters (largely heterogeneous) leading to a featureless absorption spectrum in certain reported cases, however it may not be the only reason for the observed behavior. Overall, optical absorption studies suggest that these clusters are different from MPCs.

Ultrafast excited state spectroscopic studies are indispensable in studying the nature of luminescence in the metallic cluster system.[35,121] Readers may be directed to a review by Goodson and co-workers for a more comprehensive view on ultrafast processes in Au nanoclusters.[121] In the case of QCs@proteins, Muhammed *et al.* reported a long lifetime in the case of Au_{38}@BSA.[82] Xavier *et al.* reported a long lifetime of more than 100 ns for protein protected clusters and suggested the possibility of energy transfer between the protein and the cluster. In the same study, excited state quenching of tryptophan upon cluster formation was observed. However, due to the larger number of tryptophans in the lactotransferrin system, the cluster's position could not be located.[68] Le Guével *et al.* studied the lifetime of clusters synthesized at various pH in BSA and suggested that till pH 9 it is Au_8 which is forming predominantly and above pH 9, Au_{25} species are forming. They showed a linear increase in the long lifetime component, as a function of pH from 8 to 12. However, they reported that they could not get MS for Au_8@BSA.[91] Tang and co-workers conducted temperature dependent time resolved luminescence experiments on drop-casted Au_{25}@BSA films and correlated their results with the known $Au_{25}SR_{18}$ model, since the structure of Au_{25}@BSA is not yet available.[122] They resolved the emission spectra of Au_{25}@BSA into two bands *i.e.* band I and band II. They also assigned that band I arises from a 13-atom icosahedral core comprised of Au(0) alone and that band II arises predominantly from the [–S–Au(I)–S–Au(I)–S–] staple motif. They also suggested a larger energy barrier between band I and band II (*i.e.* the core and the semi ring), from the temperature independent intensity ratio.[122] In another study, they investigated the nature of luminescence in Au_{25}@BSA clusters using time resolved photoluminescence and transient absorption techniques. They proposed that the red luminescence from the Au_{25}@BSA consists of both prompt fluorescence in the nanosecond timescale and a dominant thermally activated delayed fluorescence in the microsecond timescale (where reverse intersystem crossing is seen). They also proposed the HOMO–LUMO gap of Au_{25}@BSA to be 2.34 eV, while it is known that it is 1.37 eV in the case of $Au_{25}SR_{18}$. However, they have qualified their claims by citing the unavailability of a detailed electronic structure of Au_{25}@BSA.[123] Mali *et al.* have carried out fluorescence lifetime imaging (FLIM) based studies on Au_{QCs}@BSA and HSA. They have also reported that the delayed emission spectra matching with the fluorescence spectra suggest classical α-type delayed fluorescence and phosphorescence.[124] In a recent study by Xie and co-workers,[125] it has been shown that in the case of aggregation-induced formation of a Au(0)–Au(I) core–shell emitting cluster system using small peptides, which is likely to be compared with protein encapsulated cluster systems, there were a higher number of thiols per cluster molecule, as found from the MS and TGA studies, against the $Au_{25}SR_{18}$ system. And their system was also not consistent with the Jellium model, suggesting that protein protected clusters may also behave in a similar way and their electronic structure can be totally different from the well-established $Au_{25}SR_{18}$ system.[125] Zang *et al.* observed pressure-induced fluorescence enhancement in the case of Au_{QCs}@BSA. They have reported that above 250 MPa applied pressure, the fluorescence intensity increases linearly. They have correlated this increasing luminescence intensity with the corresponding conformational changes in the protein (see section 7.2.4).[105] Recently, Raut *et al.* have carried out fluorescence anisotropic studies on Au_{QCs}@BSA. Fluorescence anisotropy measurements reveal whether a fluorophore is polarizable or not and give a rotational correlation time from which the tumbling motion timescale of the fluorophores can be understood. It depends upon the viscosity of the media, size and polarizability of the fluorophores. They have reported a correlation time below 0.2 μs in water and 1.84 μs in glycerol.[126] In another study, Raut *et al.* have carried out two photon luminescence studies on Au_{25}@BSA and observed that optical properties did not suffer any change since the emission maximum is around 650 nm, which showed a quadratic relationship between excitation power and emission intensity, revealing that it could be a potential probe for two photon induced luminescence studies, especially useful in deep tissue imaging.[127] Das *et al.* studied the effect of molecular oxygen (MO) on the luminescence of blue and red emitting gold clusters in BSA and found that MO enhanced the blue luminescence while red luminescence decreased. They hypothesized that such an effect was a likely outcome of superoxo and peroxo binding of MO on the surface of blue and red emitting clusters, respectively.[128] In another study, Lystvet *et al.* have carried out time resolved characterization of blue and red emitting clusters in various proteins.[129] Xavier *et al.* reported a pH dependent shift in the emission wavelength of Au_{25}@NLf; at neutral pH the emission maxima of the cluster blue shifted when compared to the same cluster in basic pH.[68] Wen *et al.* reported a pH-dependent shift in the fluorescence spectra of Au_{25} and Au_8 and attributed them to the quantum confined Stark effect. In the case of Au_8@BSA, the shift was around 63 meV from pH 2.14 to 12. The dual fluorescence band, as mentioned above, band I and band II which they observed in Au_{25}@BSA has two different shifts around 79 and 52 meV. They have suggested that these shifts may indicate the presence of a linear polar component in both clusters due to their asymmetric structure.[124b] However, they have not provided any CD, MS and DLS data correlating with the pH dependent studies to know whether any other mechanism may also be present, since protein conformational change and aggregation may happen in such varying pH conditions. Unlike gold clusters, silver clusters showed relatively much shorter lifetime in protein templates.[87,88] Baksi *et al.* have shown that Au_{10} clusters formed in smaller protein molecules like Lysozyme had a smaller lifetime compared to the clusters in BSA and Lf.[86] In lysozyme, Au_8 has also been reported.[81] Dragan *et al.* have carried out synchronous spectral analysis on the fluorescence of Au_{QCs}@BSA.[130] Nienhaus and co-workers have reported that when MPCs are conjugated with proteins, their luminescence intensity and lifetime are increased. However, a detailed mechanism for such behavior is yet to be known.[131] Belina *et al.* have carried out time dependent density functional theory calculations of S–Au–S and S–Ag–S staple motifs[132] and have concluded that transitions in Ag–S motifs are more pronounced than in Au–S motifs. Li *et al.* observed the fine tuning of emission maxima upon addition of silver ions.[133] Wang *et al.* reported static quenching of Au_{QCs}@BSA by CdTe quantum dots (QDs) due to formation of a non fluorescent complex of QCs@BSA and the QDs.[134]

Unlike Au_{QCs}@protein, ultrafast studies are limited in the case of Ag_{QCs}@proteins at present. Mathew *et al.* has reported a slow component of 4 ns in the case of Ag_{15}@BSA suggesting a shorter lifetime when compared to Au_{QCs}@proteins.[87] Anand *et al.* have synthesized red emitting Ag_{14}@HSA and oxidized them to form Ag_9@HSA blue emitting silver clusters by H_2O_2. They reported that rotational correlation time obtained from anisotropy measurements was different between blue and red emitting silver clusters. Further, they have carried out two photon studies with NIR excitation.[92] Mohanty *et al.* have shown that the alloy clusters had different lifetimes compared to that of gold and silver clusters prepared in BSA templates.[88]

Shibu *et al.* have reported that the long lived excited state of QCs would sensitize dissolved oxygen to form $^1O_2$. Photosensitized oxygen was tracked in the NIR region around 1270 nm which is due to the relaxation of the singlet oxygen produced. The thermally activated delayed emission from the clusters which has slow components in the range of µs effectively facilitates singlet oxygen production. Upon purging with N_2, the lifetime of the clusters

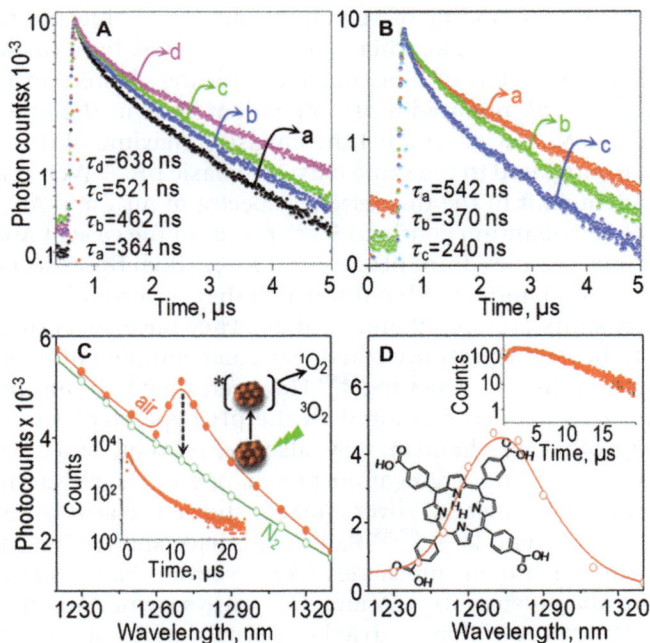

Figure 7.7 (A) PL decay profiles of QCs under (a) saturation by air and (b)–(d) purging by N₂ for (b) 30, (c) 45, and (d) 120 minutes. (B) Power dependent PL decay profiles of QCs under (a) 0.6, (b) 6, and (c) 60 W cm⁻² at 532 nm excitation. (C, D) Luminescence spectra of ¹O₂ generated by (C) QCs and (D) TCPP. Insets: PL decay profiles of ¹O₂ produced by QCs and TCPP.
©Wiley-VCH Publishing. Reproduced with permission from ref. 135.

gradually increased, suggesting the interaction of the long lived state with molecular oxygen. Also the lifetime decreased as a function of the power intensity (from 0.6 to 60 W cm⁻²) from 542 ns to 240 ns suggesting increased production of singlet oxygen. They have compared the luminescence observed at 1270 nm with the standard, TCPP (*meso*-tetra(4-carboxyphenyl))porphyrin and found that upon N₂ purging, the emission at 1270 nm from the clusters dissolved in D₂O vanished, further supporting the presence of singlet oxygen production. They have also monitored the singlet oxygen production using the green fluorescence from singlet oxygen sensor dye (SOSG) upon uncaging to form endoperoxide during interaction with the singlet oxygen produced (see Figure 7.7).[135]

7.3.1 Mechanism of Metal Ion Induced Quenching of Luminescence

As in any luminescent system, the influencing agents (which cause increase/decrease in emission intensity/shift in the emission) often become target agents (if they are of potential interest) to be detected. In the case of Au_{QCs}@proteins, Hg^{2+} sensing has been a well-established one, while copper sensing has also been reported. There are several other quenching organic molecules which we describe in the application section (see section 7.4.1). Here we will emphasize the mechanism of luminescence quenching by metal ions, especially Hg^{2+}. Interaction of Hg^{2+} with gold has been well-established in the literature. The observation of mercury-induced luminescence quenching in Au_{25}@BSA was first reported by Ying and co-workers.[136] They proposed metallophilic interaction to be the reason for such quenching and their XPS data have shown that the oxidation state of mercury changed upon interacting with gold clusters.[136] Recently, Tang and co-workers have investigated the quenching of delayed fluorescence in clusters by mercury ions and copper ions by time resolved fluorescence and transient absorption measurements.[137] They have reported that metallophilic electron transfer from the triplet state to the metal causes fluorescence quenching in the case of Hg^{2+} with Au_{QCs} (d^{10} system). The quenching of the slow component which is assigned to the delayed fluorescence suggests that electron transfer occurs from the triplet state to the metal (see Figure 7.8a). Quenching of the slow component occurs as a function of the Hg^{2+} concentration used (see Figure 7.8b). They have claimed that Hg^{2+} ions interact with the outer semi ring of Au_{25} clusters. They have also reported that luminescence in Au_8@BSA and Au_{10}@histidine does not quench, since such semi rings are not present; however, the reason for not quenching could be completely verified only when the total crystal structures are available. In metallophilic interaction-induced quenching, Dexter energy transfer is likely to be the reason, since the distance between the quencher and the probe is within a few angstroms and there also lies the likelihood of having a double-electron exchange process, as mixing of orbitals would occur.

There are several fundamental questions yet to be answered regarding the nature of clusters in protein templates. What would be the correct model to explain the luminescence and enhanced quantum yield of QCs@proteins? There is a school of thought that Au clusters with a core size less than ten atoms may be complexes rather than metallic clusters. Also, to know the veracity of the presence of Au_8 in protein templates requires detailed studies, since protein has intrinsic luminescence and similar spectral and lifetime profiles to the claimed Au_8 in protein templates. Several groups have reported that the observed blue luminescence is from proteins while a few have reported it to be from Au_8 clusters. Though MS studies are helpful in deciphering the cluster core in proteins, due to the inherent complexities that arise at the higher mass range in MALDI MS studies of QCs@proteins, high resolution MS studies need to be conducted like Q-TOF or HRESI to prove the cluster cores unambiguously. Since proteins themselves contain chiral aminoacids, whether protein protected clusters would be chiral or not is an interesting question. Once the total structures of QCs@protein are solved, we believe that questions on chiroptical and luminescence properties would be clarified.

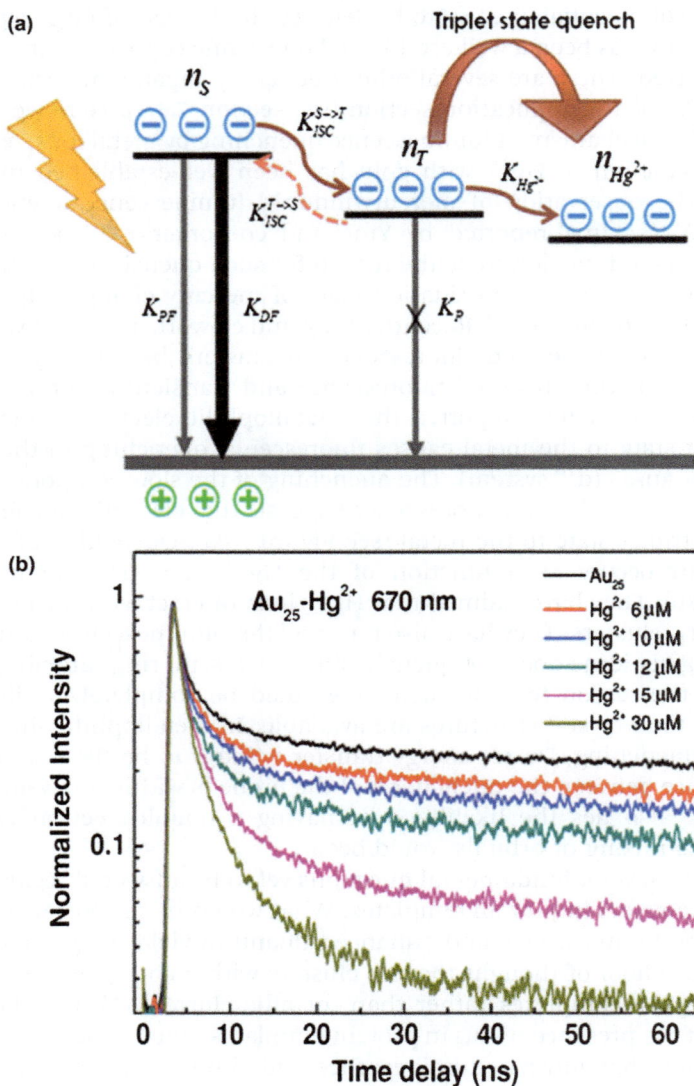

Figure 7.8 (a) The proposed mechanism of triplet electron transfer and Hg^{2+} quenching in Au_{25} nanoclusters. The n_S and n_T represent the singlet and triplet states of Au_{25}@BSA; ISC (S→T) and ISC (T→S) represent the intersystem crossing and reverse intersystem crossing, respectively; K Hg^{2+} is the electron transfer rate from the triplet state of NCs to Hg^{2+} ions. (b) Quenching of delayed fluorescence upon mercury ion interaction with Au_{25}@BSA while prompt fluorescence is not affected.
© Wiley-VCH Publishing. Reproduced with permission from ref. 137.

7.4 Applications of NMQCs@Proteins

Akin to the monolayer protected clusters and many other luminescent nanomaterials, 'the bright luminescence' in these clusters and reactivity of clusters have been exploited as a potential optical probe in various applications such as sensing, bio-imaging and molecular imaging guided delivery of therapeutics.[4,9] In many cases, QCs@protein have been made as composites with other materials, for added functionality.

7.4.1 Sensing

7.4.1.1 Sensing Metal Ions

Xie *et al.* demonstrated for the first time that Au_{25}@BSA could be used to detect ultra low concentrations of Hg^{2+} ions.[136] Subsequently, several others have used Au_{QCs}@BSA for mercury sensing (see above section 7.3.1 for the proposed mechanism of Hg ion sensing).[138,139] Muhammed *et al.* have used Au_{38}@BSA for the detection of Cu^{2+} ions and showed that addition of EDTA could reverse the quenching.[82] Wei *et al.* used Au_{QCs}@Lyz for Hg^{2+} sensing.[71] Au_{QCs}@Lyz type VI has also been used for Hg^{2+} and CH_3Hg^{2+} sensing.[140] Xavier *et al.* used Au_{QCs}@Lf for Cu^{2+} sensing.[68] Kawasaki *et al.* used Au_{QCs}@pepsin for Hg^{2+} sensing by fluorescence quenching and the detection of Pb^{2+} by fluorescence enhancement.[141] They also used Au_{QCs}@Trypsin for Hg^{2+} sensing.[142] Liu *et al.* reported Cu^{2+} sensing by Ag@Au$_{QC}$–BSA.[143] Pu *et al.* made composites of blue luminescent cationic-oligofluorene-substituted polyhedral oligomeric silsesquioxane (POSSFF) and red emitting Au_{QCs}@BSA, where the former acts as donor and the latter acts as acceptor due to emission–excitation spectral overlap between them; energy transfer between them led to enhancement in red emission from the clusters. Upon interaction with mercury, the red emission from Au_{QCs}@BSA was quenched and only blue luminescence from the POSSFF was visible leading to a dual emitting naked eye sensor for Hg^{2+} sensing.[144] Wu and co-workers used red emitting Au_{16}@BSA for sensing of silver ions which upon addition led to a blue shift from 604 nm to 567 nm and enhancement of luminescence at 567 nm. They have suggested the likely reason to be the formation of Ag^0 upon introducing silver ions due to reduction by Au clusters and formation of alloy species.[145,146] Shao *et al.* used Au_{QCs} made in eggshell membrane for Hg^{2+} sensing.[84] Goswami *et al.* used blue emitting $Cu_{5,13}$@BSA for the sensing of Pb^{2+} ions; they have proposed Pb^{2+} mediated aggregation to be the reason for the quenching of Cu cluster luminescence.[147] Su *et al.* made composites of Au_{QCs}@BSA complexed with polyelectrolytes, positively charged polydiallyldimethylammonium (PDDA) and negatively charged polystyrenesulfonate (PSS) and immobilized on glass slides as film. Upon introduction of Cu^{2+} ions, fluorescence was quenched and could be recovered by treatment with EDTA leading to a recyclable off-on sensor.[148] Zou *et al.* made composites of AuNPs/Au$_{QCs}$@Trypsin and used them for the detection of Pb^{2+} ions.[149] Paramanik *et al.* made composites of

Au_{QCs}@BSA and CdTe QDs and used for the sensing of Hg^{2+} and F^- ions. Upon addition of Hg^{2+} ions, they observed luminescence quenching at 533 nm to be nearly 74% which they assigned to aggregation induced by Hg^{2+}. Upon addition of F^- ions, they observed fluorescence enhancement of 128%.[150] Gao and Irudayaraj used Ag_{QCs}@BSA as a Hg^{2+} sensor.[151] Au_{16}@BSA was used for fluorescence enhanced sensing of silver ions.[146] Lu *et al.* used Ag_{QCs}@BSA for the selective detection of Hg^{2+} ions.[152] Electrospun fibers of Au_{QCs}@BSA have been used for the detection of Hg^{2+} sensing.[153]

7.4.1.2   *Sensing Anions, Organic, Biomolecules, Cancer Cells and as Optical Probe in Assays*

Several organic, biomolecules and cancer cells have been detected using QCs@proteins. In several cases, hybrid composites with QCs@proteins have been used.

7.4.1.2.1   Detection of Gaseous, Small and Biologically Important Molecules.   Wen *et al.* used red emitting Au_{QCs}@HRP (horseradish peroxidase) for the detection of hydrogen peroxide with a limit of detection (LOD) around 30 nM. Upon interaction with H_2O_2, the red luminescence (at 650 nm) decreased and blue luminescence (at 450 nm) increased.[70] Liu *et al.* used Au_{QCs}@BSA for cyanide sensing with a LOD around 200 nM. The mechanism for cyanide detection was proposed to be etching of the cluster core to Au ions by cyanide.[154a] Xia *et al.* synthesized Au_{QCs}@glucose oxidase and employed it for glucose sensing. Thioctic acid modified glucose oxidase was used to synthesize red emitting Au_{QCs}. After the synthesis of clusters, glucose oxidase still remained active and could catalyze the reaction of glucose and dissolved O_2 leading to the production of H_2O_2. The evolution of H_2O_2 quenched the fluorescence of Au_{QCs} which gives a potential platform for the sensing of glucose. The LOD of glucose was around 0.7 μM.[154b] BSA protected gold nanoclusters act as a fluorescent sensor for the selective and sensitive detection of pyrophosphate.[155] Jin *et al.* detected glucose by the fluorescence quenching of Au_{QCs}@BSA by H_2O_2 generated during the glucose oxidase–glucose interaction with a LOD of 5 μM.[156] Similarly, Wang *et al.* detected xanthine by the quenching of the cluster by H_2O_2 evolved during xanthine and xanthine oxidase interaction with a detection limit of 5×10^{-7} M.[157] Yan *et al.* used Au_{QCs}@HSA for NOx sensing.[158] Li *et al.* synthesized clusters in egg white and used for H_2O_2 sensing.[159] Chen and Tseng used Au_{QCs}@Lyz for GSH sensing.[81] Shiang *et al.* have conjugated protein-A to Au_{QCs}@BSA and used it for the detection of immunoglobulin G (IgG) in plasma.[160] Wang *et al.* used Au_{QCs}@BSA for the identification of aminoacids based on metal ion modulated fluorescence of gold clusters.[161] Wang *et al.* used Au_{QCs}@BSA for sensing glutaraldehyde.[162] Chen *et al.* used Au_{QCs}@BSA for the detection of ciproflaxacin.[163] Wang *et al.* used Au_{QCs}@BSA for the detection of proteases.[164] Park *et al.* used Au_{QCs}@BSA for the detection of biological thiols.[165] Yang *et al.* and Wang *et al.* used Au_{QCs} and Ag_{QCs}@BSA for the detection of ascorbic acid.[166,167] Chen *et al.* used Au_{QCs}@BSA for the detection of methotrexate.[168] Wang *et al.* used Au_{QCs}@histidine for iodide sensing.[169] Like gold clusters, Hu *et al.* have used Cu_{QCs}@BSA as peroxidise mimics to sense H_2O_2.[170] Tao *et al.* made composites of Au_{QCs}@protein and graphene oxide and used it as a H_2O_2 sensor.[171] Zao *et al.* used Au_{QCs}@BSA for the detection of uric acid.[173] Chen *et al.* used Au_{QCs}@BSA for querectin sensing.[174] Cui *et al.* used Au_{QCs}@BSA for detecting cystine.[175] Samari *et al.* used Au_{QCs}@BSA for detecting Vitamin B12.[176] Wang *et al.* developed barcodes using Au_{QCs}@BSA.[177] Au_{QCs}@BSA have been used for detecting nitrite.[178] Au_{QCs}@BSA were used to detect bisphenol in plastic samples[179] Liu *et al.* have made composites of cysteamine protected AuNPs decorated with Au_{QCs}@Trypsin which led to the quenching of fluorescence due to surface plasmon enhanced energy transfer (SPEET) since the excitation spectra of clusters match with that of the SPR of AuNPs. Upon addition of heparin, AuNPs aggregate leading to the dissociation of Au_{QCs}@trypsin from AuNPs which can be monitored by the recovery of luminescence due to decreased SPEET.[180] Hu *et al.* used Au_{QCs}@BSA for the detection of trypsin based on the proteolytic activity.[181] Similarly, Wang *et al.* used Au_{QCs}@BSA to detect trypsin activity in biological samples.[182] Kong *et al.* observed that Au_{QCs} made in different proteins get quenched differently based on the interacting protein species and quenching could be differentiated in a principal component analysis leading to effective protein discrimination based on Au_{QCs}@proteins.[183] Zhang *et al.* used Au_{QCs}@BSA as a fluorescent staining agent of serum proteins in polyacrylamide gel electrophoresis (PAGE[184] Chan *et al.* used Au_{QCs}@BSA for the detection of *Staphylococus aureus*.[185] Chan and Chen have used Au_{QCs}@Lyz for the detection of bacteria by mass spectrometry combined with principal component analysis.[186] Denatured Au_{QCs}@BSA have been used to detect acetylcholinesterase activity.[187]

Hu *et al.* have used Au_{QCs}@BSA microspheres deposited on Au electrodes in an electrochemical impedence set up and attached with monoclonal CEA antibody to detect carcinoembryogenic tumor cells (BXPC-3) which are CEA^{+ve} with a detection limit of 18 cells mL^{-1}.[188] Peng *et al.* made composites of calcium carbonate and Au_{QCs}@BSA and used the composite for the assembly of HRP/antibody. They could detect the cancer biomarker neuron specific enolase (NSE) effectively. The LOD were 2.0 and 0.1 pg mL^{-1} for fluorescence and electrochemical detection, respectively.[189] In another immunoassay Au_{QCs}@BSA has been used to detect cystatin c.[190]

Ag_{QCs}@BSA has been used in the detection of explosive trinitrotoluene by Mathew *et al.* They have coated silica coated mesoflowers with Ag clusters and FITC (fluorescein isothio cyanate) dye molecules. When TNT binds to Ag_{QCs}@BSA, it forms a Meisenheimer complex leading to quenching of the red luminescence of clusters, FITC alone would fluoresce leading to green emission alone. They have observed these changes in a dark field fluorescence microscope coupled with spectrophotometer, enabling them to detect zepto molar concentrations of TNT. They have also used the same method to detect Hg^{2+} ions.[191]

7.4.1.2.2 Electrochemiluminescence (ECL) Based Detections. ECL based sensing by QCs@Proteins has also been demonstrated.[192] Li *et al.* demonstrated that indium tin oxide (ITO) coated Au_{QC}@BSA exhibited ECL and reported that ITO played a significant role in enhancing ECL. They reported that in the presence of anionic co-reactant $S_2O_8^{2-}$, ECL was enhanced and demonstrated its application to detect bio molecule dopamine. In another study, Fang *et al.* showed the generation of ECL from Au_{QC}@BSA in the presence of tetraethyl amine (TEA) and showed that ECL is differently influenced by the metal ions; here they showed it to be affected by Pb^{2+}.[192] Wu *et al.* have conjugated Au_{QCs}@BSA with silica nanoparticles and coated on a glassy carbon electrode (GCE) and used it for a solid state electrochemiluminescence sensor for the determination of hydrogen peroxide.[193] Au_{QCs}@BSA have been employed in composite nanostructures (made of magnetic beads, DNA and Au_{QCs}@BSA) and used as a chemiluminescent probe to detect lysozyme in cancer cells (K562 and B lymphoma cells) by nicking endonuclease signal amplification technology with a LOD of 2×10^{-13} M.[194] Chen *et al.* used graphene–Au_{QCs}@BSA composites as an electrochemiluminescence probe.[195]

7.4.2 Bio-imaging

Bio-labelling of proteins with atomically precise clusters and using them for imaging started in the 1980s.[65] Safer *et al.* imaged the biotin binding site on avidin.[196] Yonath and co-workers used undecagold clusters protected by phosphine monolayers to study t-RNA binding to ribosomal subunits.[66] However, all these studies were dependent on the X-ray/electron scattering properties of the heavy metal atoms. Now, the recent attraction is the high luminescence observed in the cluster system which has potential applications in bio-labeling in addition to their high electron scattering properties. The red luminescence observed in Au/Ag$_{QCs}$@proteins has been advantageous in bio-imaging since it is out of the auto fluorescence window of biological tissues. Unlike the bare/monolayer protected gold nanoparticles,[99,197] proteins provide biologically compatible scaffolds and reduce the cytotoxicity. Several reports are available on both *in vitro* and *in vivo* imaging. In most of the cases, these clusters are reported to be biologically safe.

7.4.2.1 In vitro *Imaging*

Archana *et al.* have shown for the first time that Au_{QC}@BSA can conjugate with folic acid (FA) and can be used in folate targeted imaging of cancer cells exploiting the biological phenomenon that certain cancer cells overexpress folate receptors (FR).[198] They conjugate FA with Au_{QCs} by the well-established EDC coupling method. They have demonstrated receptor-targeted cancer detection using Au_{QCs} conjugated with FA on FR^{+ve} oral squamous cell carcinoma (KB) and breast adenocarcinoma cell MCF-7, where the FA-conjugated Au_{25} clusters were found to be internalized in significantly higher concentrations compared to the lung carcinoma cell lines (A549) which are FR^{-ve} (see Figure 7.9).[198] Muhammed *et al.* also conjugated FA to Au_{38}@BSA synthesized by the nanoparticle etching method and used it for folate-targeted imaging.[82] Wang *et al.* have conjugated Au_{QC}@BSA to herceptin, a humanized monoclonal antibody in the treatment of breast cancer to nuclear target Erb2 over expressing HER2 + breast cancer cells. They have also studied the nuclear targeting dynamics of herceptin conjugated Au_{QCs}@BSA. The endocytosed Au_{QCs}@BSA escaped the endosomal pathway and entered the nucleus which resulted in enhanced efficacy of herceptin against breast cancer cells by inducing DNA damage.[199] Xavier *et al.* proposed that QCs made in the transferrin family proteins would be useful in imaging and guided therapeutic delivery.[68] Later, Le Guével *et al.* synthesized red emitting Au_{QCs} in serum transferrin and used them for imaging

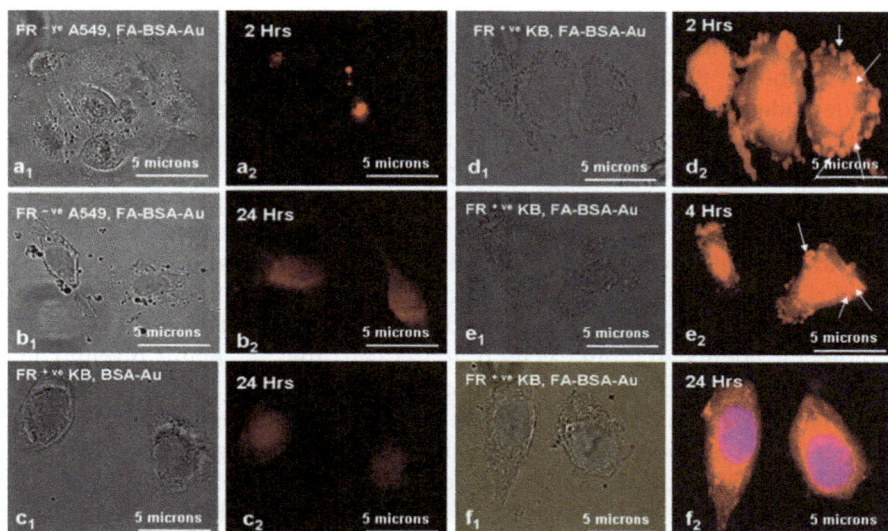

Figure 7.9 Molecular receptor specific uptake of Au_{QC}@BSA conjugated to folic acid (FA). Fluorescent microscopic images showing interaction of Au_{QC}@BSA–FA with different types of cell lines: (a1)–(a2) FR^{-ve} lung carcinoma A549, (b1)–(b2) FR-depressed oral cell carcinoma, KB, (c1)–(c2) FR^{+ve}KB cells with unconjugated Au clusters, (d1)–(d2) FR^{+ve} KB cells with FA-conjugated Au clusters at 2 h, (e1)–(e2) 4 h and (f1)–(f2) 24 h of incubation.
©IOP Publishing. Reproduced with permission from ref. 198.

A549 cells.[69] Although there were earlier suggestions [such as clusters in iron loaded lactotransferrin reported by Pradeep and co-workers,[68] binding of iron by serum transferrin and recognition by antibodies after cluster synthesis as reported by Le Guével and co-workers[69] (binding of Fe^{3+} ions is one of the signs of retention of bio-functionality in the case of transferrin proteins) and retention of the activity of horseradish peroxidase after Au_{QCs} synthesis, as reported by Zhang and co-workers[200]], retention of the bioactivity of functional proteins was not proven rigorously with various biological experiments until the synthesis of insulin protected clusters.[77] Insulin protected clusters became one of the landmark reports in this newly developing field. In order to prove the retention of bioactivity, Liu *et al.* have injected intraperitoneally (i.p.) 1.0 unit kg^{-1} of Au_{QCs}@insulin and commercially available human insulin (Humulin R) in different C57BL/6J mice and compared the activity of both. Both Au_{QCs}@insulin and humulin injected mice showed similar levels of blood glucose reduction, implying that insulin's bioactivity was retained. They have used Au_{QCs}@insulin to image C2C12, mouse myoblast cells. Uptake of insulin protected clusters was clearly seen in differentiated myoblast cells since differentiated myoblast cells have more insulin receptors than undifferentiated cells which may also explain the preservation of the selective receptor binding functionality in Au_{QCs}@insulin. Having proven the preservation of the biological metabolism in Au_{QCs}@insulin, they have treated Au_{QCs}@insulin with insulin degrading enzyme (IDE) and found that the luminescence is quenched. Then they introduced IDE inhibiting factors such as racecadotril and thiorphan, preventing insulin degradation which resulted in the retention of luminescence. In the same work, it was reported for the first time that Au_{QCs}@proteins can be used in computer tomography (CT) imaging as a contrast agent. Thus the Au_{QCs}@insulin system resulted in intrinsically bio-functional gold clusters which could be used in imaging insulin metabolic pathways and interactions providing an alternate for radio labelled tracking systems.[77] Kong *et al.* have used ribosomal endonuclease for gold cluster synthesis and conjugated with Vitamin B12 (cobalamine) and used it for imaging of human epithelial colorectal adenocarcinoma cells (Caco-2) cell lines.[201] Acute myeloid leukaemia (AML) cells overexpress CD33 myeloid antigen. Exploiting this aspect, Archana *et al.* have conjugated CD33 monoclonal antibodies with Au_{QCs}@BSA resulting in a ~ 12 nm sized hybrid tracking agent and used it for the detection of acute myeloid leukaemia cells by flow cytometry and confocal imaging. Target specific staining of the cells showed detection of 95.4% AML cells which overexpress CD33 compared to relatively low CD33 expressing non AML human peripheral blood cells (8.2% detection) which get stained non specifically.[202] Durgadas *et al.* have used Au_{QCs}@BSA to sense the copper ions in live HeLa cells and they have reported that glycine could be used to reverse the quenching.[203] Durgadas *et al.* in another report have conjugated Au_{QCs}@BSA with super paramagnetic iron oxide particles. They have used the hybrid nanosystem to separate and image cancer cells in the blood.[204,205] Amino acid and peptide protected clusters have also been used in imaging. Histidine protected green emitting gold clusters have been synthesized by Haiyan *et al.* and used in *in vitro* and *in vivo* imaging.[206] Biocompatible Gd[III]-functionalized fluorescent gold nanoclusters have been used for optical, magnetic resonance and X-ray imaging.[207,208] In another direction of research, researchers have started analyzing the interactions between MPCs with proteins and subsequent effects. These pre-formed clusters are adsorbed on to the proteins. Shang *et al.* have studied the consequences of the adsorption of silver clusters with human serum albumin and biological responses.[209]

Shibu *et al.* have conjugated biotinylated Au_{QCs}@BSA with avidin functionalized Fe_2O_3 nanoparticles and followed it with the uncaging of fluorescein by singlet-oxygen ($^1O_2$) in the methylanthracene-fluorescein-caged conjugate. They have used it for bio-imaging of lung adenocarcinoma cells, coupled with monitoring intracellular production of singlet oxygen. For receptor specific delivery, they have conjugated it with epidermal growth factor (EGF), which would enhance the uptake through G-protein coupled EGF receptors. They have also demonstrated the composite to be an efficient MRI contrast agent.[135]

7.4.2.2 In vivo *Imaging*

Wu *et al.* used Au_{QCs}@BSA for *in vivo* imaging in tumor bearing mice for the first time (see Figure 7.10).[210] Hu *et al.* have synthesized Au_{QCs}@BSA and conjugated them with Gd^{3+}–DTPA and used the system for *in vivo* imaging of tumor in mice, as tumor tissue would uptake more clusters because of the enhanced permeability and retention (EPR) effect which is due to the absence of lymphatic drainage leading to reduced clearance. The EPR effect allowed them to do triple modal imaging (fluorescence, computer tomography (CT) and magnetic resonance imaging (MRI)) on tumor containing tissue. They have tested the *in vivo* toxicity by histological analysis after 15 days of injection and also through enzyme assays; tests on aspartate aminotransferase (AST) and lactate dehydrogenase (LDH) revealed that no considerable toxicity was induced by Au_{QCs}–Gd^{3+}–DTPA composites.[211,212] Sun *et al.* synthesized clusters in ferritin (Ft) protein (a cage like protein complex composed of 24 sub-units composed of heavy chains (H) and light chains (L), with a 12 nm diameter hollow cage and a 8 nm cavity) in a controlled site specific manner at the ferroxidase active centre (in H–Ft chain) which has histidine (His), aspartic (Asp), glutamine (Gln) and glutamic acids (Glu).[221] The horse spleen Ft they used contained 22 L-Ft subunits and 2 H-Ft subunits. They could synthesise a controlled pair of Au_{QCs} in a single protein molecule with blue, green and red emission in a well calculated manner at the ferroxidase active centre with enhanced fluorescence due to energy transfer between assembled pairs of clusters. They used them for *in vitro* imaging of Caco-2 cells which have Ft receptors which could uptake the Au_{QCs}@Ft (HepG2 cells, employed as negative control, did not uptake the Au_{QCs}@Ft, as they do not express Ft receptors, proving the

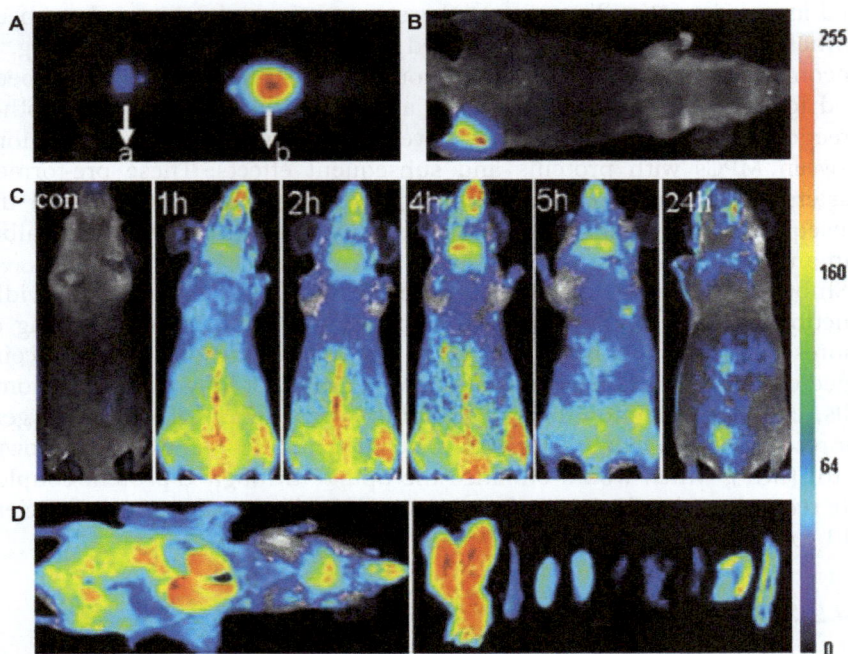

Figure 7.10   *In vivo* fluorescence image of 100 mL Au NCs injected subcutaneously [(a) 0.235 mg mL^{-1}, (b) 2.35 mg mL^{-1}] (A) and intramuscularly (2.35 mg mL^{-1}) into the mice (B). (C) Real-time *in vivo* abdomen imaging of intravenously injected mice with 200 mL of Au NCs (2.35 mg mL^{-1}) at different time points, post injection. (D) *Ex vivo* optical imaging of anatomized mice with injection of 200 mL of AuNCs (2.35 mg mL^{-1}) and some dissected organs during necropsy at 5 h pi. The organs are liver, spleen, left kidney, right kidney, heart, lung, muscle, skin and intestine from left to right.
© Royal Society of Chemistry. Reproduced with permission from ref. 210.

selectivity). Then they have used Au$_{QCs}$@Ft to image the kidneys of nude female mice. The presence of Scara-5 receptors of L-Ft subunits in the proximal tubule cells has been considered to facilitate the receptor specific uptake of Au$_{QCs}$-Ft in kidneys combined with the uptake processes in the retico-endothelial system (RES) of clearance leading to enhanced accumulation in the kidneys. Accumulation in the liver was also observed since liver cells express Ft receptors. Uptake in both organs strongly suggests receptor specific uptake.[221] Liu *et al.* conjugated folic acid to Au$_{QCs}$@Trypsin and used it for folate targeted *in vivo* imaging.[180] Wang *et al.* fabricated Au$_{QCs}$@Tf–graphene oxide (GO) nanocomposites where they observed energy transfer between Au$_{QCs}$@Tf to GO leading to reduction in the emission of Au$_{QCs}$@Tf. Upon interaction with Tf receptor (TfR), Au$_{QCs}$@Tf dissociate from GO, aggregate and bind with TfR leading to the recovery of NIR luminescence. They have used these turn-on NIR luminescent composites for TfR targeted imaging of HeLa tumors in nude mice. Control experiments *in vitro* and cell specific imaging (HeLa cells (TfR^{+ve}) were used as positive control while HepG2/3T3 cells expressing low levels of TfR were used as negative control) have confirmed the retention of activity of Tf.[212a] Sun *et al.* have fabricated a hybrid probe made of Au$_{QCs}$@BSA with Gd$_2$O$_3$ by a one step approach and used it for multimodal imaging *in vivo*. They have conjugated it with RGD peptides and used it for tumor imaging.[212b]

In a recent study by Liu *et al.* investigating the *in vivo* metabolic mechanism of Au$_{QCs}$@insulin, *in vivo* uptake of Au$_{QCs}$@insulins could be visualized through a least-invasive harmonic generation and two photon fluorescence (TPF) microscope. Mice ear and *ex vivo* assays on human fat tissues revealed that cells with rich insulin receptors have higher uptake of administrated insulin. They have also found a unique phenomenon that the Au$_{QCs}$@insulin can even permeate into lipid droplets (LDs) of adipocytes. Further, they found that enlarged adipocytes in type II diabetes mice have higher adjacent/LD concentration when compared to the small-sized ones in wild type mice. In the case of human clinical samples, the epicardial adipocytes of patients with diabetes and coronary artery disease (CAD) also show elevated adjacent/LD concentration contrast. Extrapolating these results they have suggested that subcellular insulin metabolism in model animals or patients with metabolic or cardiovascular diseases could be studied using multiphoton fluorescence based metabolic imaging of Au$_{QCs}$@insulin.[213] Zhang *et al.* reported that Au$_{QCs}$@BSA gets accumulated leading to retention in liver and is not cleared faster by the RES when compared to monolayer protected Au$_{25}$ and suggested that it could be toxic. Nevertheless they proposed it to be useful in liver targeted therapy.[214]

7.4.3 Molecular Imaging Guided Delivery of Therapeutics

Quantum clusters have been integrated into molecular medicine to act as tracking agents, and in addition the functional groups of proteins facilitate easy loading of therapeutics. Chen *et al.* have conjugated folic acid (FA) and doxorubicin to Au$_{QCs}$@BSA for molecular imaging guided delivery of folate receptor targeted anti-cancer therapy.[215a] Chen *et al.* have synthesized core–shell structured multifunctional nanocarriers consisting of gold nanoclusters. They used Au$_{QCs}$ as core and folate (FA)-conjugated amphiphilic hyperbranched block copolymer as shell. The shell consisted of an inner arm made of poly(L-lactide) (PLA) and outer arm made of FA-conjugated sulfated polysaccharide (GPPS–FA). The final hybrid product was Au NCs–PLA–GPPS–FA; which they used for targeted anticancer therapeutics. They have conjugated the anticancer drug camptothecin (a hydrophobic drug) to the hybrid carrier. These hybrid nanocarriers delivered the drug in a controlled manner starting with rapid release in the first 1 hour then sustained release (up to 15 h) and finally reached a plateau. The release studies were carried out under different pH conditions, at pH 5.3, 7.4 and 9.6. The release was better in the case of pH 9.6 than the other two former pH conditions. HeLa cells having folate receptors were effectively targeted and killed combined with tracking by imaging.[215b] Retnakumari *et al.* have developed hybrid protein nanoparticles for drug delivery with a combination of antibodies, polymeric linkers and Au_{QCs}@HSA and they utilized the drug binding properties of HSA to include an anticancer drug, sorefenib in the composite. Their focus was to inhibit the deregulated kinases, (often called the cancer kinome) in the cancer cells, thereby inhibiting the cancer proliferation. Deregulated protein kinases play a very critical role in tumorigenesis, metastasis, and drug resistance of cancer. Targeted inhibition by small molecule kinase inhibitors (SMI) is effective against many types of cancer; however, point mutations in the kinase domain induce drug resistance. An example is chronic myeloid leukemia (CML) caused by the BCR-ABL fusion protein, where a BCR-ABL kinase inhibitor, imatinib (IM) fails in the advanced stage of the treatment. To overcome the above-said condition, Retnakumari *et al.* have rationally designed hybrid protein anticancer drug conjugates to efficiently inhibit kinases and deliver drugs. They have developed a human serum albumin (HSA) based nanomedicine, loaded with STAT5 inhibitor (sorafenib), and surface conjugated the same with transferrin (Tf) ligands for TfR specific delivery. This dual-targeted transferrin conjugated albumin bound sorafenib nanomedicine (Tf-nAlb-Soraf) had uniform spherical morphology with an average size of ~ 150 nm and drug encapsulation efficiency of $\sim 74\%$. TfR specific uptake and enhanced antileukaemic activity of the nanomedicine were found to be maximum in samples having the highest level of STAT5 and TfR expression. The nanomedicine induced down-regulation of key survival pathways such as pSTAT5 and anti-apoptotic protein MCL-1.[216] Chandran *et al.* have simultaneously inhibited the deregulated cancer kinome using rationally designed protein based nanomedicine which has Au_{QCs}@BSA as its molecular tracking agent. They have developed a polymer–protein core–shell nanomedicine to inhibit critically aberrant pro-survival kinases such as mTOR, MAPK and STAT5 in primitive $CD34^+/CD38^-$ AML cells. The hybrid agent (~ 290 nm) was made of a poly-lactide-co-glycolide core (~ 250 nm) loaded with mTOR inhibitor, everolimus, and an albumin shell (~ 25 nm thick) loaded with MAPK/STAT5 inhibitor, sorafenib and the whole construct was surface conjugated with monoclonal antibody against CD33 receptor overexpressed in AML (see above). Flow cytometry and confocal studies showed enhanced cellular uptake of targeted nanomedicine. They demonstrated that kinases were inhibited along with the synergistic lethality against human leukaemic cells simultaneously with the aid of a single protein–drug composite. Healthy cells were not affected by the composites which showed specificity towards leukaemic cells. They have also reported that the hybrid agent showed a better inhibition effect on cancer cells than commercially available cytarbine and daunorubicin. The presence of luminescent Au_{QCs} assisted in tracking the uptake of the composite.[217] Conformationally modified lactalbumin (LA) is lethal to tumor cells. Lystvet *et al.* have made Au_{QCs}@LA and observed that the cluster mediated conformational change in LA was lethal to HeLa cells. They proposed that gold clusters not only induce conformational change but also help to retain the modified conformation.[102]

7.4.4 Other Applications

Zhang *et al.* have made a Boolean NAND logic gate using the detection capacity of nitrite (NO_2^-) ions by Au_{QCs}@BSA. It is known that H_2O_2 also quenches the emission of gold clusters, but neither of them completely quenches the cluster individually but when both are present cluster emission gets quenched completely. They have used H_2O_2 and HNO_3 as logical inputs and complete fluorescence quenching only in the presence of both of them makes the clusters suitable for designing a NAND gate. The inputs were a function of H_2O_2 and HNO_3 and the outputs were a function of color under UV radiation and fluorescence intensity. The logical inputs were only cluster (0,0), cluster + H_2O_2 (1,0), cluster + HNO_3 (0,1) and cluster + H_2O_2 + HNO_3 (1,1) which had the outputs (1,1), (1,1), (1,1) and (0,0), respectively.[218]

Au_{QCs}@Lyz were used as antibacterial materials which enabled killing of pan-drug-resistant *Acinetobacter baumannii* and vancomycin-resistant *Enterococcus faecalis*.[94] Sreeprasad *et al.* made composites of QCs@protein with graphene oxide for the first time. They made luminescent patternable composites of Au_{QCs}@NLf, chitosan and GO and also used them as antimicrobial material.[219]

7.5 Gas Phase Clusters Derived From Protein Templates

Till now, we have discussed clusters protected with proteins which exist in the solution phase, where the metal cluster core is in the zero valent state. There is another class of materials where clusters form in the gas phase.[78] As discussed before, cluster formation occurs *via* metal–protein adduct formation, wherein for Au, we have a Au^+–protein complex. When these adducts are subjected to laser desorption and ionization, bare clusters form in the gas phase. Cluster aggregation and growth from the precursor require release of aggregation and stabilization energy. Proteins having a large number of degrees of freedom, can act as an energy relaxation medium and heat bath reservoir during cluster growth. Recently Pradeep and co-workers reported the formation of stable gas phase clusters with specific nuclearities using protein templates experimentally through mass spectrometry and theoretically by first principles quantum mechanical calculations. Here the proteins act as selective cluster nucleation templates and allow their confined growth and electronic stabilization. In most of the studies they have used a small protein Lyz which allows acquisition of high quality mass spectra over the entire mass range covered.

Different concentrations of Au^{3+} were added to Lyz and the mixture was incubated to get a precursor where Au^{3+} is reduced to Au^+ by the amino acid residues or by the oxidation of disulfide bonds of cystine residues (dimeric cysteine). The Lyz mass spectrum shows a series of peaks including monomer (Lyz^+) and its aggregates (Lyz_2^+, Lyz_3^+, Lyz_4^+, *etc.*) in the high mass range, whereas in the lower mass region (<10 kDa), only Lyz^{2+} is visible, no other fragments are there at that condition. When the Au^+–Lyz adduct is subjected to laser desorption and ionization, along with the monomer, all the aggregates show multiple Au attachments (see Figure 7.11). In the monomer region, a maximum of 10 Au attachments are observable with highest concentration of Au used where the peaks are separated by m/z 197. Similar trends are found for aggregates. In the case of Lyz^{2+}, the separation is $\sim m/z$ 99, due to Au^{2+}. However, Au^+–Lyz adducts show a completely different type of distribution in the lower mass region (3–9 kDa), where

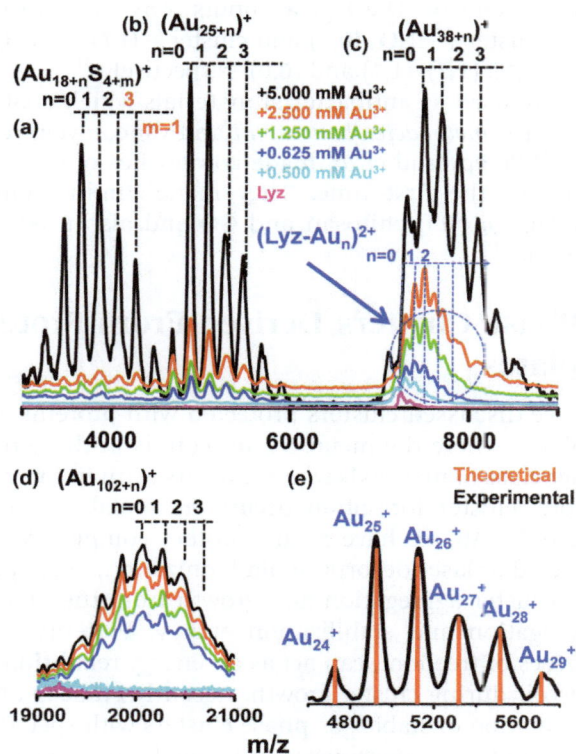

Figure 7.11 MALDI MS of Lyz–Au adduct in the linear positive mode showing distinct features of Au clusters. The spectrum of parent Lyz is also shown. The bare cluster series seen are separately shown: (a) $Au_{18}S_4$, (b) Au_{25} and (c) Au_{38} and (d) Au_{102}. The peaks show a separation of m/z 197. The circled region of (c) shows a Au uptake of Lyz^{2+} with a separation due to the Au^{2+} series. In (e) the experimental spectrum in the Au_{25} region is compared with the calculated peak positions.
© Wiley-VCH Publishing. Reproduced from ref. 78 with permission.

peaks are separated mostly by m/z 197 due to Au. These features keep increasing in intensity with increase in Au^{3+} concentration. Three different envelopes appear with maxima corresponding to $Au_{18}S_4^+$, Au_{25}^+, Au_{38}^+. The Au_{38}^+ region overlaps with the Lyz_2^+ region and its $(Au\text{-}Lyz)^{2+}$ peaks, but the latter are lower in intensity at the highest concentration of Au^{3+} used. At this concentration, the mass spectrum is spaced by $m/z = 197$, due to gold, while for $(Au_n\text{-}Lyz)^{2+}$, the peaks are separated by $m/z = 99$. Nearly 100 Da mass difference between these two species is easy to resolve at this mass range by the instrument. In all cases, the peaks also come along with many neighboring peaks, among which $Au_{18}S_4^+$, Au_{25}^+ and Au_{38}^+ have the highest intensity in the specific envelope and the envelope can be assigned as $Au_{18 \pm n}S_{4+m}^+$, $Au_{25 \pm n}^+$, $Au_{38 \pm n}^+$. At a higher mass region like 20 kDa, Au_{102}^+ is also observable although the intensity is less than the other bare clusters. In this region, Au_{102}^+ appears above a certain concentration of Au^{3+} used. Adjacent peaks form almost with equal intensity. Although most of the clusters appear without any protecting agent, Au_{18} shows four sulfur attachments to it. Starting from $Au_{16}S_4^+$ the envelope continues to $Au_{19}S_4^+$ and the very next peak appears with one gold and one sulfur separation from the previous one confirming the presence of sulfur in the cluster. In turn this observation suggests interaction of protein with the cluster cores through sulfur containing amino acid, cysteine.

One may ask the question whether it is possible to accommodate larger cores like 25, 38 or 102 Au inside such a small protein Lyz or are there some other factors. As mentioned above, Au uptake is observable throughout the monomer and aggregate regions (maximum 10 Au in the Lyz monomer at the highest concentration of Au used). But it is very unlikely that with increase in Au^{3+} concentration, protein will uptake more and more Au due to the presence of a much lower number of cysteines (8 cysteines per Lyz molecule). Therefore, it is almost impossible that one single molecule of Lyz will have 25, 38 or 102 Au atoms attached to it although aggregates are possible. These clusters appear well when analyzed in positive ion mode of MALDI MS, but in negative mode, intensity as well as peak shape are poor. Cluster formation chemistry does not depend on the photon flux as seen in laser intensity dependent study. Even at the lowest laser power, the same clusters form and with the highest laser power, the $Au–S$ bond does not break as seen for the $Au_{18}S_4^+$ region. This is a very important observation as normally for monolayer protected clusters, bond breakage is always there. As seen for monolayer protected as well as solution phase protein protected cluster synthesis, there is a specific change in the clusters with incubation time. An incubation time dependent study shows enhanced intensity for Au_{18} and Au_{25} regions without any change in peak position. But the Au_{38} region is somehow dependent on the incubation time. Enhanced intensity can be attributed to more and more Au consumption and reduction of Au^{3+} to Au^{1+} by the protein. In the case of Au_{38}, just after mixing of Au^{3+}, only $(Au_n\text{-}Lyz)^{2+}$ peaks are there. However, when incubated for longer time, Au_{38} peaks start appearing. After 6 hours, the intensity is comparable to Au_{25}. Upon passage of time, Au_{25} forms in the solution from a mixture of clusters. Time dependent study again proves that it is hard to accommodate and sustain growth of larger clusters by Lyz due to its size restriction. Sensitivity of Au_{38} to the incubation time of the solution can be attributed to specific structural characteristics of certain Au–Lyz adducts that require more time to form which subsequent to laser desorption–ionization, react favorably to form Au_{38} clusters. These observations further prove that proteins act as a gold source and facilitate the formation of specific clusters by allowing nucleation, growth and confinement.

The main question is still whether these clusters form in solution or in the gas phase by laser ablation. Gas phase growth is supported by solution phase mass spectrometric data where at all concentrations of Au^{3+} used, one can see a maximum of up to 3 Au attachment in any specific charge state, especially in the dimer region where the separation is clear unlike in MALDI MS where a maximum of 10 Au attachments is seen. This can be understood by charge-induced desorption of metal ions at high charge state ($+10$, $+11$, *etc.* in ESI MS) and merging of the peaks with the next charge state. There is no signature of any bare cluster formation in the solution phase data in the whole range of ESI MS (m/z 100–4000). Therefore, it can be said that these clusters are forming only in the gas phase.

If Lyz, being a small protein, can allow formation of such bare clusters, why not use other larger proteins known to make clusters in solution? To answer this question, Pradeep *et al.* have tried different proteins, namely BSA, Lf and Lyz for gas phase cluster synthesis. Although BSA and Lf show cluster formation in such conditions, Lyz is the most efficient among all three. It should be noted that none of the proteins have any fragments in the mass range studied. Unusual selectivity of Lyz may be due to its lower mass and structural flexibility which allows it to interact with the clusters in the gas phase.

From several observations it is clear that laser desorption is the main step of such cluster formation. However, how is it actually helping the formation? To explain the observations, a plasma reaction is proposed which allows generation of specific clusters. When laser is fired, a plasma containing ions, molecules, neutrals, aggregates, and electrons is formed. Several reactions can take place before the extraction of the ions to the detector. These may be categorized as:

(1) $Au_n + Au_m^+ \rightarrow Au_{n+m}^+ \quad n, m = 1,2,3,\ldots,$

(2) $(Lyz-Au_n)^+ + Au_m \rightarrow (Lyz-Au_{n+m})^+ \quad n = 0,1,2,3,\ldots; m = 1,2,3,\ldots,$

(3) $(Lyz-Au_n)^+ + (Lyz-Au_m) \rightarrow (Lyz-Au_{n+k})^+ + (Lyz-Au_{m-k}) \quad n, m = 1,2,3,\ldots;$
$k(<m) = 1,2,3,\ldots$

In the case of evaporation of metal salts, due to photoreduction, M_n species form as known from other studies. In the case of equation (3), inter-protein metal ion transfer occurs. All these reactions can happen if there is sufficient time lag between ionization and extraction to the detector. In the case of MALDI MS study, ions are extracted after a finite delay time and the process is known as delayed extraction which is of the order of several hundreds of nanoseconds and this allows the desorbed species in the gaseous plasma to interact with each other. This is a well-known method since the time of fullerenes, whereby by varying the delay time one can get the most stable ion. In this case, the above mentioned reactions are facilitated by the abundance of protein molecules and their gold adducts in the reaction zone, since their large mass slows down their movement and separation from the plasma cloud. The whole process is associated with heat change where due to a large number of degrees of freedom, host protein molecules can interact and help in the relaxation of cluster aggregation energy. Evolution of gold clusters with specific nuclearities and enhanced stability occurs following the above mentioned process and they eventually detach from the protein template and come to the gas phase and get detected as bare clusters.

All the clusters discussed in this section are known to be highly stable in solution and have electronic stability due to a shell closing electronic structure. This kind of closed shell cluster, where the HOMO of the cluster closes a shell, possess enhanced stability and are known as magic number (MN) clusters with $n^* = 2, 8, 18$ (20), 34 (40), 58, 92,... the number of delocalized electrons depending on the structure and degree of deviation from spherical symmetry. Free Au_{38} is not a magic number cluster on its own as there are four extra delocalized electrons (having 38 electrons whereas magic number stability comes with 34 electrons) and therefore the Au_{38} core has no stabilization gap. Still this cluster forms with high intensity in the gas phase. To find out the type of interaction the protein may have with the cluster, a theoretical model based on first principle DFT calculations was done using the partial Jellium model. Au_{38} exhibits a HOMO–LUMO gap of $\Delta_{HL} = 0.21$ eV due to the presence of four extra delocalized electrons. If by any means four electrons are removed from the system, it becomes a 34 electron system and hence gets a magic number and opens up a HOMO–LUMO gap of $\Delta_{HL} = 0.48$ eV. This 34 electron super atom gap is a result of a $1S^2 | 1P^6 | 1D^{10} | 2S^2 | 1F^{14}$ shell closure associated with the population of 17 delocalized orbitals containing 34 electrons. If the core interacts with the cystine groups of a protein it can lose an electron and achieve enhanced stability. To validate the speculation, dissociative binding of cystine on the cluster core was considered. For simplicity, only cystine $(SCH_2CHNHCOOH)_2$ binding was considered as it is difficult to simulate the whole protein. Au_{38} has a distorted truncated octahedral structure (see Figure 7.12). When a cystine binds to the d-TO Au_{38} with a binding energy of 0.58 eV there is no effect on the atomic arrangement, electronic structure and stability of the cluster. However, upon dissociation of two cystines to four cysteines, they adsorb strongly on the surface with a binding energy of 1.97 eV per cystine molecule. Thus for dissociative binding of two cystines we get: $E[Au_{38}(d\text{-}TO) + 2E[Cystine] - E[Au_{38}(Cysteine)_4] = 3.94$ eV. Where, $E[X]$ denotes the overall energy of the species. This cystine binding has a profound effect on the structure and electronic stability of the cluster as four delocalized 1G electrons are now engaged in Au–S bonding and most importantly it opens a shell closing gap of $\Delta_{HL} = 0.48$ eV. The main observation of this theoretical calculation is that none of the clusters are magic number

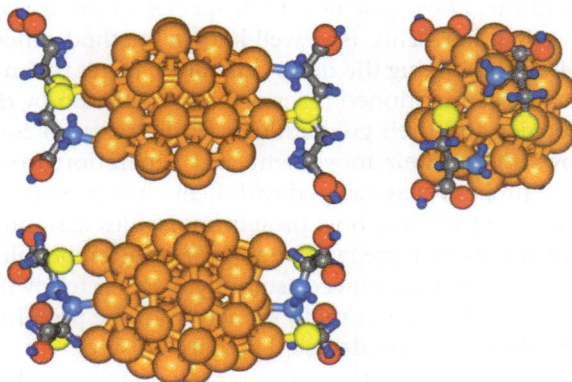

Figure 7.12   Three views of the structure of $Au_{38}(cysteine)_4$ with four adsorbed cysteine [$HO_2CCH(NH_2)CH_2S-$] residues, resulting from two dissociated cystine units. The S–S distances, between two neighboring adsorbed cysteine residues, are $d(S-S) = 4.50$ Å and 4.42 Å, and the average S–Au bond length is 2.36 Å. For an undissociatively adsorbed cystine molecule $(S-S) = 2.21$ Å and $d(Au-S) = 2.545$ Å and 2.495 Å.
© Wiley-VCH Publishing. Reproduced from ref. 78 with permission.

clusters as the number of delocalized 6s electrons in each of these clusters is rather close to a magic number, that is, $n^* = 92, 34$ and 20 for Au_{102}^+, Au_{38}^+ and Au_{25}^+ and they have $n-n^* = 10, 4$ and 5 excess delocalized electrons, respectively. All these clusters can achieve magic number stability through the interaction with cystine groups of Lyz and its aggregates. The mechanism presented here is completely different from the one known to apply for monolayer protected clusters. Here very few number of cystines are needed to achieve magic number stability unlike those present in solution phase namely, $Au_{25}(SR)_{18}$, $Au_{38}(SR)_{24}$ and $Au_{102}(SR)_{40}$ where a large number of thiol groups cover the surface of the cluster core. This is in good agreement with limited number of cystines (8 per Lyz) present in the protein. These size selected clusters may be used in various applications like in nanocatalysis, soft landing, *etc.*

From the above discussion it is clear that it is possible to make size selected magic number clusters in the gas phase from protein templates. Theoretical calculations and experimental findings suggest interaction of protein with the cluster occurs through the sulfur containing dimeric amino acid, cystine. These observations are also useful to elucidate the solution phase cluster growth mechanism in protein templates.

7.6   Outlook

In atomic cluster science, creating noble metal clusters in functional protein templates has opened up new avenues. It, like any new system, inherently has several unanswered questions which we hope will be answered as this field evolves. Although there is significant understanding about QCs@proteins, total crystal structures of protein protected clusters would solve most of the fundamental questions raised. However, it may require Herculean efforts (also unconventional methods) to reap success, for structural biologists to break this puzzle along with others (clusters grown in crystals of proteins are known; however, cluster containing proteins producing diffractable crystals are not yet reported). Advanced imaging methods may also be of help. So far, from the above discussion, one may understand that the role of MS has been crucial in understanding the system and, in future, high resolution MS studies need to be done. The quantum yield of NMQCs has to be improved on a par with semiconductor QDs for which detailed understanding of the underlying photophysics is necessary. Proteins are biological nanomachines; rationally designed functional proteins with luminescent metal clusters would mean trackable nanomachines with permanent indicators. Clusters stable at high temperatures would likely be useful in probing high temperature reactions and also are likely to be useful as probes in polymerase chain reaction (PCR). Clusters made in liquid proteins would have potential applications in future. The bright luminescence, biocompatibility and functionality have made this new system find a place quickly in several studies of biology and medicine. How the cell machinery perceives these hybrid molecules and what is the intracellular fate of the system are a subject of future research. Clusters made in peptide based aptamers (akin to nucleic acid based aptamers) would be of interest in biology and medicine. We speculate that several other interesting properties and some non-trivial quantum effects may also be present in this system. Gas phase quantum clusters from protein templates open a new area for understanding the structure of clusters in the gas phase and how protein is playing a role there to grow a few specific magic number clusters. From the ongoing research and future possibilities, we believe that QCs@proteins will leave a prominent mark in the annals of cluster science.

Acknowledgements

We thank the Department of Science and Technology, Govt. of India for constantly supporting our research program on nanomaterials. The research presented here has been due to the intense efforts of several colleagues whose names appear in the reference list.

References

1. J. A. Alonso and Editor, *Structure and Properties of Atomic Nanoclusters*, World Scientific, 2005.
2. D. P. Woodruff and Editor, *Atomic Clusters: from Gas Phase to Deposited. [In: The Chemical Physics of Solid Surfaces, 2007; 12]*, Elsevier B. V., 2007.

3. G. Schmid and Editor, *Clusters and Colloids: From Theory to Applications*, VCH, 1994.
4. M. A. H. Muhammed and T. Pradeep, in *Advanced Fluorescence Reporters in Chemistry and Biology II*, ed. A. P. Demchenko, Springer Berlin Heidelberg, 2010, pp. 333–353.
5. Y. Chen, J. A. Preece and R. E. Palmer, *Ann. N. Y. Acad. Sci.*, 2008, **1130**, 201–206.
6. A. Bruma, F. R. Negreiros, S. Xie, T. Tsukuda, R. L. Johnston, A. Fortunelli and Z. Y. Li, *Nanoscale*, 2013, **5**, 9620–9625.
7. Y. Han, D. S. He, Y. Liu, S. Xie, T. Tsukuda and Z. Y. Li, *Small*, 2012, **8**, 2361–2364.
8. R. S. Ingram, M. J. Hostetler, R. W. Murray, T. G. Schaaff, J. Khoury, R. L. Whetten, T. P. Bigioni, D. K. Guthrie and P. N. First, *J. Am. Chem. Soc.*, 1997, **119**, 9279–9280.
9. P. L. Xavier, K. Chaudhari, A. Baksi and T. Pradeep, *Nano Rev.*, 2012, **3**, 14767 and the references cited there in.
10. M. K. Harbola, *PNAS*, 1992, **89**, 1036–1039.
11. B. Yoon, P. Koskinen, B. Huber, O. Kostko, B. von Issendorff, H. Häkkinen, M. Moseler and U. Landman, *ChemPhysChem*, 2007, **8**, 157–161, and the references cited therein.
12. M. Gruber, G. Heimel, L. Romaner, J.-L. BrÃ©das and E. Zojer, *Phys. Rev. B*, 2008, 77, 165411.
13. M. Zhu, C. M. Aikens, F. J. Hollander, G. C. Schatz and R. Jin, *J. Am. Chem. Soc.*, 2008, **130**, 5883–5885.
14. M. Zhu, E. Lanni, N. Garg, M. E. Bier and R. Jin, *J. Am. Chem. Soc.*, 2008, **130**, 1138–1139.
15. Y. Shichibu, Y. Negishi, T. Tsukuda and T. Teranishi, *J. Am. Chem. Soc.*, 2005, **127**, 13464–13465.
16. E. S. Shibu, M. A. H. Muhammed, T. Tsukuda and T. Pradeep, *J. Phys. Chem. C*, 2008, **112**, 12168–12176.
17. Y. Negishi, K. Nobusada and T. Tsukuda, *J. Am. Chem. Soc.*, 2005, **127**, 5261–5270.
18. M. W. Heaven, A. Dass, P. S. White, K. M. Holt and R. W. Murray, *J. Am. Chem. Soc.*, 2008, **130**, 3754–3755.
19. A. Dass, *J. Am. Chem. Soc.*, 2009, **131**, 11666–11667.
20. Z. Wu, J. Suhan and R. Jin, *J. Mater. Chem.*, 2009, **19**, 622–626.
21. J. Akola, K. A. Kacprzak, O. Lopez-Acevedo, M. Walter, H. Gronbeck and H. Hakkinen, *J. Phys. Chem. C*, 2010, **114**, 15986–15994.
22. L. A. Angel, L. T. Majors, A. C. Dharmaratne and A. Dass, *ACS Nano*, 2010, **4**, 4691–4700.
23. O. Toikkanen, V. Ruiz, G. Ronnholm, N. Kalkkinen, P. Liljeroth and B. M. Quinn, *J. Am. Chem. Soc.*, 2008, **130**, 11049–11055.
24. Y. Pei, Y. Gao and X. C. Zeng, *J. Am. Chem. Soc.*, 2008, **130**, 7830–7832.
25. H. Qian, M. Zhu, U. N. Andersen and R. Jin, *J. Phys. Chem. A*, 2009, **113**, 4281–4284.
26. R. L. Donkers, D. Lee and R. W. Murray, *Langmuir*, 2008, **24**, 5976.

27. V. L. Jimenez, D. G. Georganopoulou, R. J. White, A. S. Harper, A. J. Mills, D. Lee and R. W. Murray, *Langmuir*, 2004, **20**, 6864–6870.
28. J. Kim, K. Lema, M. Ukaigwe and D. Lee, *Langmuir*, 2007, **23**, 7853–7858.
29. O. Lopez-Acevedo, H. Tsunoyama, T. Tsukuda, H. Hakkinen and C. M. Aikens, *J. Am. Chem. Soc.*, 2010, **132**, 8210–8218.
30. H. Qian, W. T. Eckenhoff, Y. Zhu, T. Pintauer and R. Jin, *J. Am. Chem. Soc.*, 2010, **132**, 8280–8281.
31. P. D. Jadzinsky, G. Calero, C. J. Ackerson, D. A. Bushnell and R. D. Kornberg, *Science*, 2007, **318**, 430–433.
32. C. Zeng, H. Qian, T. Li, G. Li, N. L. Rosi, B. Yoon, R. N. Barnett, R. L. Whetten, U. Landman and R. Jin, *Angew. Chem., Int. Ed.*, 2012, **51**, 13114–13118.
33. M. Zhu, C. M. Aikens, M. P. Hendrich, R. Gupta, H. Qian, G. C. Schatz and R. Jin, *J. Am. Chem. Soc.*, 2009, **131**, 2490–2492.
34. J. Zheng, C. Zhang and R. M. Dickson, *Phys. Rev. Lett.*, 2004, **93**, 077402/077401–077404.
35. J. Zheng, P. R. Nicovich and R. M. Dickson, *Annu. Rev. Phys. Chem.*, 2007, **58**, 409–431.
36. R. Jin, *Nanoscale*, 2010, **2**, 343–362.
37. G. C. Bond, C. Louis, D. T. Thompson and Editors, *Catalysis by Gold*, Imperial College Press, London, 2006.
38. A. Das, T. Li, K. Nobusada, Q. Zeng, N. L. Rosi and R. Jin, *J. Am. Chem. Soc.*, 2012, **134**, 20286–20289.
39. X.-K. Wan, Z.-W. Lin and Q.-M. Wang, *J. Am. Chem. Soc.*, 2012, **134**, 14750–14752.
40. S. Chen and K. Kimura, *Chem. Lett.*, 1999, **28**, 1169–1170.
41. S. Chen and K. Kimura, *Langmuir*, 1999, **15**, 1075–1082.
42. N. Nishida, H. Yao, T. Ueda, A. Sasaki and K. Kimura, *Chem. Mater.*, 2007, **19**, 2831–2841.
43. N. Cathcart and V. Kitaev, *J. Phys. Chem. C*, 2010, **114**, 16010–16017.
44. T. Udayabhaskararao and T. Pradeep, *J. Phys. Chem. Lett.*, 2013, **4**, 1553–1564.
45. T. U. B. Rao, B. Nataraju and T. Pradeep, *J. Am. Chem. Soc.*, 2010, **132**, 16304.
46. T. Udayabhaskararao, M. S. Bootharaju and T. Pradeep, *Nanoscale*, 2013, **5**, 9404–9411.
47. M. S. Bootharaju and T. Pradeep, *Langmuir*, 2013, **29**, 8125–8132.
48. I. Chakraborty, A. Govindarajan, J. Erusappan, A. Ghosh, T. Pradeep, B. Yoon, R. L. Whetten and U. Landman, *Nano Lett.*, 2012, **12**, 5861–5866, and the references cited therein.
49. T. Udaya Bhaskara Rao and T. Pradeep, *Angew. Chem. Int. Ed.*, 2010, **49**, 3925–3929.
50. M. Kabir, A. Mookerjee and A. K. Bhattacharya, *Phys. Rev. A*, 2004, **69**, 043203.
51. W. Wei, Y. Lu, W. Chen and S. Chen, *J. Am. Chem. Soc.*, 2011, **133**, 2060–2063.

52. N. Vilar-Vidal, M. C. Blanco, M. A. Loì pez-Quintela, J. Rivas and C. Serra, *J. Phys. Chem. C*, 2010, **114**, 15924–15930.
53. A. Ganguly, I. Chakraborty, T. Udayabhaskararao and T. Pradeep, *J. Nanopart. Res.*, 2013, **15**, 1522.
54. S.-I. Tanaka, J. Miyazaki, D. K. Tiwari, T. Jin and Y. Inouye, *Angew. Chem. Int. Ed.*, 2010, **50**, 431–435.
55. H. Yang, Y. Wang, J. Lei, L. Shi, X. Wu, V. Mãkinen, S. Lin, Z. Tang, J. He, H. Hãkkinen, L. Zheng and N. Zheng, *J. Am. Chem. Soc.*, 2013, **135**, 9568–9571.
56. C. Zeng, T. Li, A. Das, N. L. Rosi and R. Jin, *J. Am. Chem. Soc.*, 2013, **135**, 10011–10013.
57. R. S. Dhayal, J.-H. Liao, Y.-R. Lin, P.-K. Liao, S. Kahlal, J.-Y. Saillard and C. W. Liu, *J. Am. Chem. Soc.*, 2013, **135**, 4704–4707.
58. T. Udayabhaskararao, Y. Sun, N. Goswami, S. K. Pal, K. Balasubramanian and T. Pradeep, *Angew. Chem., Int. Ed.*, 2012, **51**, 2155–2159.
59. Y. Negishi, K. Igarashi, K. Munakata, W. Ohgake and K. Nobusada, *Chem. Commun.*, 2012, **48**, 660–662.
60. Y. Negishi, W. Kurashige, Y. Niihori, T. Iwasa and K. Nobusada, *Phys. Chem. Chem. Phys.*, 2010, **12**, 6219–6225.
61. Y. Negishi, K. Munakata, W. Ohgake and K. Nobusada, *J. Phys. Chem. Lett.*, 2012, **3**, 2209–2214.
62. J. T. Petty, J. Zheng, N. V. Hud and R. M. Dickson, *J. Am. Chem. Soc.*, 2004, **126**, 5207.
63. J. Zheng, J. T. Petty and R. M. Dickson, *J. Am. Chem. Soc.*, 2003, **125**, 7780–7781.
64. J. Xie, Y. Zheng and J. Y. Ying, *J. Am. Chem. Soc.*, 2009, **131**, 888–889.
65. D. Safer, L. Bolinger and J. S. Leigh Jr, *J. Inorg. Biochem.*, 1986, **26**, 77–91.
66. S. Weinstein, W. Jahn, M. Laschever, T. Arad, W. Tichelaar, M. Haider, C. Glotz, T. Boeckh, Z. Berkovitch-Yellin, F. Franceschi and A. Yonath, *J. Cryst. Growth*, 1992, **122**, 286–292.
67. L. Carl and E. P. Richard, *J. Phys.: Condens. Matter*, 2008, **20**, 353001/353001–353010.
68. P. L. Xavier, K. Chaudhari, P. K. Verma, S. K. Pal and T. Pradeep, *Nanoscale*, 2010, **2**, 2769–2776.
69. G. Xavier Le, D. Nicole and S. Marc, *Nanotechnology*, 2011, **22**, 275103.
70. F. Wen, Y. Dong, L. Feng, S. Wang, S. Zhang and X. Zhang, *Anal. Chem.*, 2011, **83**, 1193–1196.
71. H. Wei, Z. Wang, L. Yang, S. Tian, C. Hou and Y. Lu, *Analyst*, 2010, **135**, 1406–1410.
72. M. B. Dickerson, K. H. Sandhage and R. R. Naik, *Chem. Rev.*, 2008, **108**, 4935–4978.
73. F. Reith, S. L. Rogers, D. C. McPhail and D. Webb, *Science*, 2006, **313**, 233–236.

74. H. Wei, Z. Wang, J. Zhang, S. House, Y.-G. Gao, L. Yang, H. Robinson, L. H. Tan, H. Xing, C. Hou, I. M. Robertson, J.-M. Zuo and Y. Lu, *Nat Nano*, 2011, **6**, 93–97.

75. D. Toroz and S. Corni, *Nano Lett.*, 2011, **11**, 1313–1318.

76. K. Chaudhari, P. L. Xavier and T. Pradeep, *ACS Nano*, 2011, **5**, 8816–8827.

77. C.-L. Liu, H.-T. Wu, Y.-H. Hsiao, C.-W. Lai, C.-W. Shih, Y.-K. Peng, K.-C. Tang, H.-W. Chang, Y.-C. Chien, J.-K. Hsiao, J.-T. Cheng and P.-T. Chou, *Angew. Chem., Int. Ed.*, 2011, **50**, 7056–7060.

78. A. Baksi, T. Pradeep, B. Yoon, C. Yannouleas and U. Landman, *ChemPhysChem*, 2013, **14**, 1272–1282.

79. S. S. Narayanan and S. K. Pal, *J. Phys. Chem. C*, 2008, **112**, 4874.

80. J. Yu, S. A. Patel and R. M. Dickson, *Angew. Chem. Int. Ed.*, 2007, **46**, 2028–2030.

81. T.-H. Chen and W.-L. Tseng, *Small*, 2012, **8**, 1912–1919.

82. M. A. Habeeb Muhammed, P. K. Verma, S. K. Pal, A. Retnakumari, M. Koyakutty, S. Nair and T. Pradeep, *Chem. - Eur. J.*, 2010, **16**, 10103–10112.

83. H.-W. Li, K. Ai and Y. Wu, *Chem. Commun.*, 2011, **47**, 9852–9854.

84. C. Shao, B. Yuan, H. Wang, Q. Zhou, Y. Li, Y. Guan and Z. Deng, *J. Mater. Chem.*, 2011, **21**, 2863.

85. S. D. Haveli, P. Walter, G. Patriarche, J. Ayache, J. Castaing, E. E. Van, G. Tsoucaris, P.-A. Wang and H. B. Kagan, *Nano Lett.*, 2012, **12**, 6212–6217.

86. A. Baksi, P. L. Xavier, K. Chaudhari, N. Goswami, S. K. Pal and T. Pradeep, *Nanoscale*, 2013, **5**, 2009–2016.

87. A. Mathew, P. R. Sajanlal and T. Pradeep, *J. Mater. Chem.*, 2010, **21**, 11205–11212.

88. J. S. Mohanty, P. L. Xavier, K. Chaudhari, M. S. Bootharaju, N. Goswami, S. K. Pal and T. Pradeep, *Nanoscale*, 2012, **4**, 4255–4262.

89. G. A. Simms, J. D. Padmos and P. Zhang, *J. Chem. Phys.*, 2009, **131**, 214703–214709.

90. H. W. Kroto, J. R. Heath, S. C. O'Brien, R. F. Curl and R. E. Smalley, *Nature*, 1985, **318**, 162–163.

91. X. Le Guei vel, B. Hoìtzer, G. Jung, K. Hollemeyer, V. Trouillet and M. Schneider, *J. Phys. Chem. C*, 2011, **115**, 10955–10963.

92. U. Anand, S. Ghosh and S. Mukherjee, *J. Phys. Chem. Lett.*, 2012, **3**, 3605–3609.

93. Y. Yu, Z. Luo, C. S. Teo, Y. N. Tan and J. Xie, *Chemical Communications*, 2013, **49**, 9740–9742.

94. W.-Y. Chen, J.-Y. Lin, W.-J. Chen, L. Luo, D. E. Wei-Guang and Y.-C. Chen, *Nanomedicine*, 2010, **5**, 755–764.

95. D. Thompson, J. P. Hermes, A. J. Quinn and M. Mayor, *ACS Nano*, 2012, **6**, 3007–3017.

96. M. Murariu, E. Stela Dragan, A. Adochitei, G. Zbancioc and G. Drochioiu, *J. Pept. Sci.*, 2011, **17**, 512–519.

97. F. Mohr, ed., *Gold Chemistry: Applications and Future Directions in the Life Sciences*, Wiley-VCH, Weinheim, 2009.
98. N. R. S. Krishna, V. Pattabhi and S. S. Rajan, *Protein and Peptide Letters*, 2011, **18**, 457–466.
99. D. Choudhury, P. L. Xavier, K. Chaudhari, R. John, A. K. Dasgupta, T. Pradeep and G. Chakrabarti, *Nanoscale*, 2013, **5**, 4476–4489.
100. A. E. Nel, L. Mädler, D. Velegol, T. Xia, E. M. V. Hoek, P. Somasundaran, F. Klaessig, V. Castranova and M. Thompson, *Nat. Mater.*, 2009, **8**, 543–557.
101. P. L. Witkiewicz and C. F. Shaw, *J. Chem. Soc., Chem. Commun.*, 1981, **0**, 1111–1114.
102. S. M. Lystvet, S. Volden, G. Singh, I. M. Rundgren, H. Wen, Ø. Halskau and W. R. Glomm, *J. Phys. Chem. C*, 2013, **117**, 2230–2238.
103. R. Diamond, *J. Mol. Biol.*, 1974, **82**, 371–391.
104. Y.-H. Chen, J. T. Yang and K. H. Chau, *Biochemistry*, 1974, **13**, 3350–3359.
105. M. Zhang, Y.-Q. Dang, T.-Y. Liu, H.-W. Li, Y. Wu, Q. Li, K. Wang and B. Zou, *J. Phys. Chem. C*, 2013, **117**, 639.
106. A. R. Garcia, I. Rahn, S. Johnson, R. Patel, J. Guo, J. Orbulescu, M. Micic, J. D. Whyte, P. Blackwelder and R. M. Leblanc, *Colloids Surf., B: Biointerfaces*, 2013, **105**, 167–172.
107. L. Fabris, S. Antonello, L. Armelao, R. L. Donkers, F. Polo, C. Toniolo and F. Maran, *J. Am. Chem. Soc.*, 2005, **128**, 326–336.
108. Y. Cui, Y. Wang, R. Liu, Z. Sun, Y. Wei, Y. Zhao and X. Gao, *ACS Nano*, 2011, **5**, 8684–8689.
109. Y. Wang, Y. Cui, Y. Zhao, R. Liu, Z. Sun, W. Li and X. Gao, *Chem. Commun.*, 2012, **48**, 871–873.
110. (a) Y. Wang, Y. Cui, R. Liu, Y. Wei, X. Jiang, H. Zhu, L. Gao, Z. Chai, Y. Zhao and X. Gao, *Chemical Communications*, 2013, **49**, 10724–10726; (b) X. Yuan, Z. Luo, Q. Zhang, X. Zhang, Y. Zheng, J. Y. Lee and J. Xie, *ACS Nano*, 2011, **5**, 8800–8808.
111. J.-J. Feng, H. Huang, D.-L. Zhou, L.-Y. Cai, Q.-Q. Tu and A.-J. Wang, *J. Mat. Chem. C*,**2013**1, 4720–4725.
112. B. Söptei, L. Naszályi Nagy, P. Baranyai, I. Szabó, G. Mező, F. Hudecz and A. Bóta, *Gold Bull*, 2013, 1–9.
113. X. Mu, L. Qi, P. Dong, J. Qiao, J. Hou, Z. Nie and H. Ma, *Biosens. Bioelectron.*, 2013, **49**, 249–255.
114. X. Yang, M. Shi, R. Zhou, X. Chen and H. Chen, *Nanoscale*, 2011, **3**, 2596–2601.
115. S. Roy, G. Palui and A. Banerjee, *Nanoscale*, 2012, **4**, 2734–2740.
116. V. A. Morozov and M. Y. Ogawa, *Inorg. Chem.*, 2013, **52**, 9166–9168.
117. J. Zheng, C. Zhou, M. Yu and J. Liu, *Nanoscale*, 2012, **4**, 4073–4083.
118. Z. Wu and R. Jin, *Nano Lett.*, 2010, **10**, 2568–2573.
119. G. Wang, R. Guo, G. Kalyuzhny, J.-P. Choi and R. W. Murray, *J. Phys. Chem. B*, 2006, **110**, 20282.
120. C. Zhou, C. Sun, M. Yu, Y. Qin, J. Wang, M. Kim and J. Zheng, *J. Phys. Chem. C*, 2010, **114**, 7727–7732.

121. S. H. Yau, O. Varnavski and T. Goodson, *Acc. Chem. Res.*, 2013, **46**, 1506–1516.
122. X. Wen, P. Yu, Y.-R. Toh and J. Tang, *J. Phys. Chem. C*, 2012, **116**, 11830–11836.
123. X. Wen, P. Yu, Y.-R. Toh, A.-C. Hsu, Y.-C. Lee and J. Tang, *J. Phys. Chem. C*, 2012, **116**, 19032–19038.
124. (a) B. Mali, A. I. Dragan, J. Karolin and C. D. Geddes, *J. Phys. Chem. C*, 2013, **117**, 16650–16657; (b) X. Wen, P. Yu, Y.-R. Toh and J. Tang, *J. Phys. Chem. C*, 2013, **117**, 3621–3626.
125. Z. Luo, X. Yuan, Y. Yu, Q. Zhang, D. T. Leong, J. Y. Lee and J. Xie, *J. Am. Chem. Soc.*, 2012, **134**, 16662–16670.
126. S. Raut, R. Chib, R. Rich, D. Shumilov, Z. Gryczynski and I. Gryczynski, *Nanoscale*, 2013, **5**, 3441–3446.
127. S. L. Raut, D. Shumilov, R. Chib, R. Rich, Z. Gryczynski and I. Gryczynski, *Chem. Phys. Lett.*, 2013, **561–562**, 74–76.
128. T. Das, P. Ghosh, M. S. Shanavas, A. Maity, S. Mondal and P. Purkayastha, *Nanoscale*, 2012, **4**, 6018–6024.
129. S. M. Lystvet, S. Volden, G. Singh, M. Yasuda, O. Halskau and W. R. Glomm, *RSC Adv.*, 2013, **3**, 482–495.
130. A. I. Dragan, B. Mali and C. D. Geddes, *Chem. Phys. Lett.*, 2013, **556**, 168–172.
131. L. Shang, S. Brandholt, F. Stockmar, V. Trouillet, M. Bruns and G. U. Nienhaus, *Small*, 2012, **8**, 661–665.
132. B. Bellina, I. Compagnon, F. Bertorelle, M. Broyer, R. Antoine and P. Dugourd, *J. Phys. Chem. C*, 2011, **115**, 24549–24554.
133. B. Li, J. Li and J. Zhao, *J. Nanosci. Nanotechnol.*, 2012, **12**, 8879–8885.
134. H. Wang, C. Zheng, T. Dong, K. Liu, H. Han and J. Liang, *J. Phys. Chem. C*, 2013, **117**, 3011–3018.
135. E. S. Shibu, S. Sugino, K. Ono, H. Saito, A. Nishioka, S. Yamamura, M. Sawada, Y. Nosaka and V. Biju, *Angew. Chem. Int. Ed.*, 2013, 10559–10563.
136. J. Xie, Y. Zheng and J. Y. Ying, *Chem. Commun.*, 2010, **46**, 961–963.
137. P. Yu, X. Wen, Y.-R. Toh, J. Huang and J. Tang, *Part. Part. Syst. Charact.*, 2013, **30**, 467–472.
138. D. Hu, Z. Sheng, P. Gong, P. Zhang and L. Cai, *Analyst*, 2010, **135**, 1411–1416.
139. F. Chai, T. Wang, L. Li, H. Liu, L. Zhang, Z. Su and C. Wang, *Nanoscale Res. Lett.*, 2010, **5**, 1856–1860.
140. Y.-H. Lin and W.-L. Tseng, *Anal. Chem.*, 2010, **82**, 9194–9200.
141. H. Kawasaki, K. Hamaguchi, I. Osaka and R. Arakawa, *Adv. Funct. Mater.*, 2011, **21**, 3508–3515.
142. H. Kawasaki, K. Yoshimura, K. Hamaguchi and R. Arakawa, *Anal Sci.*, 2011, **27**, 591.
143. H. Liu, X. Zhang, X. Wu, L. Jiang, C. Burda and J.-J. Zhu, *Chem. Comm.*, 2011, **47**, 4237–4239.
144. K.-Y. Pu, Z. Luo, K. Li, J. Xie and B. Liu, *J. Phys. Chem. C*, 2011, **115**, 13069–13075.

145. Y. Yue, T.-Y. Liu, H.-W. Li, Z. Liu and Y. Wu, *Nanoscale*, 2012, **4**, 2251–2254.
146. H.-W. Li, Y. Yue, T.-Y. Liu, D. Li and Y. Wu, *J. Phys. Chem. C*, 2013, **117**, 1619–16165.
147. N. Goswami, A. Giri, M. S. Bootharaju, P. L. Xavier, T. Pradeep and S. K. Pal, *Anal. Chem.*, 2011, **83**, 9676–9680.
148. L. Su, T. Shu, Z. Wang, J. Cheng, F. Xue, C. Li and X. Zhang, *Biosens. Bioelectron.*, 2013, **44**, 16–20.
149. L. Zou, W. Qi, R. Huang, R. Su, M. Wang and Z. He, *ACS Sustainable Chem. Eng.*, 2013, **1**, 1398–1404.
150. B. Paramanik, S. Bhattacharyya and A. Patra, *Chem. -Eur. J.*, 2013, **19**, 5980–5987.
151. C. Guo and J. Irudayaraj, *Anal. Chem.*, 2011, **83**, 2883–2889.
152. D. Lu, C. zhang, L. Fan, H. Wu, S. Shuang and C. Dong, *Anal. Methods*, 2013, 5, 5522–5527.
153. Y. Cai, L. Yan, G. Liu, H. Yuan and D. Xiao, *Biosens. Bioelectron.*, 2013, **41**, 875–879.
154. (a) Y. Liu, K. Ai, X. Cheng, L. Huo and L. Lu, *Adv. Funct. Mater.*, 2010, **20**, 951–956; (b) X. Xia, Y. Long and J. Wang, *Anal. Chimi. Acta*, 2013, **772**, 81–86.
155. J.-M. Liu, M.-L. Cui, S.-L. Jiang, X.-X. Wang, L.-P. Lin, L. Jiao, L.-H. Zhang and Z.-Y. Zheng, *Anal. Methods*, 2013, **5**, 3942–3947.
156. L. Jin, L. Shang, S. Guo, Y. Fang, D. Wen, L. Wang, J. Yin and S. Dong, *Biosens. Bioelectron.*, 2011, **26**, 1965–1969.
157. X.-X. Wang, Q. Wu, Z. Shan and Q.-M. Huang, *Biosens. Bioelectron.*, 2011, **26**, 3614–3619.
158. L. Yan, Y. Cai, B. Zheng, H. Yuan, Y. Guo, D. Xiao and M. M. F. Choi, *J. Mater. Chem.*, 2012, **22**, 1000–1005.
159. M. Li, D.-P. Yang, X. Wang, J. Lu and D. Cui, *Nanoscale Res. Lett.*, 2013, **8**, 182.
160. Y.-C. Shiang, C.-A. Lin, C.-C. Huang and H.-T. Chang, *Analyst*, 2011, **136**, 1177–1182.
161. M. Wang, Q. Mei, K. Zhang and Z. Zhang, *Analyst*, 2012, **137**, 1618–1623.
162. X. Wang, P. Wu, Y. Lv and X. Hou, *Microchem. J.*, 2011, **99**, 327–331.
163. Z. Chen, S. Qian, J. Chen, J. Cai, S. Wu and Z. Cai, *Talanta*, 2012, **94**, 240–245.
164. Y. Wang, Y. Wang, F. Zhou, P. Kim and Y. Xia, *Small*, 2012, **8**, 3769–3773.
165. K. S. Park, M. I. Kim, M.-A. Woo and H. G. Park, *Biosens. Bioelectron.*, 2013, **45**, 65–69.
166. X.-H. Yang, J. Ling, J. Peng, Q.-E. Cao, L. Wang, Z.-T. Ding and J. Xiong, *Spectrochimica Acta Part A: Molecular and Biomolecular Spectroscopy*, 2013, **106**, 224–230.
167. X. Wang, P. Wu, X. Hou and Y. Lv, *Analyst*, 2013, **138**, 229–233.
168. Z. Chen, S. Qian, X. Chen, W. Gao and Y. Lin, *Analyst*, 2012, **137**, 4356–4361.

169. Y. Wang, H. Zhu, X. Yang, Y. Dou and Z. Liu, *Analyst*, 2013, **138**, 2085–2089.
170. L. Hu, Y. Yuan, L. Zhang, J. Zhao, S. Majeed and G. Xu, *Anal. Chim. Acta*, 2013, **762**, 83–86.
171. Y. Tao, Y. Lin, Z. Huang, J. Ren and X. Qu, *Adv. Mater.*, 2013, **25**, 2594–2599.
172. J.-M. Liu, M.-L. Cui, X.-X. Wang, L.-P. Lin, L. Jiao, Z.-Y. Zheng, L.-H. Zhang and S.-L. Jiang, *Sens. Actuators, B: Chemical*, 2013.
173. H. Zhao, Z. Wang, X. Jiao, L. Zhang and Y. Lv, *Spectrosc. Lett.*, 2012, **45**, 511–519.
174. Z. Chen, S. Qian, J. Chen and X. Chen, *J. Nanopart. Res.*, 2012, **14**, 1–8.
175. M.-L. Cui, J.-M. Liu, X.-X. Wang, L.-P. Lin, L. Jiao, L.-H. Zhang, Z.-Y. Zheng and S.-Q. Lin, *Analyst*, 2012, **137**, 5346–5351.
176. F. Samari, B. Hemmateenejad, Z. Rezaei and M. Shamsipur, *Anal. Methods*, 2012, **4**, 4155–4160.
177. Y.-Q. Wang, X.-W. He, W.-Y. Li and Y.-K. Zhang, *J. Mater. Chem. C*, 2013, **1**, 2202–2208.
178. Q. Yue, L. Sun, T. Shen, X. Gu, S. Zhang and J. Liu, *J Fluoresc*, 2013, **23**, 1313–1318.
179. N. Vasimalai and S. A. John, *Anal. Methods*, 2013.
180. J.-M. Liu, J.-T. Chen and X.-P. Yan, *Anal. Chem.*, 2013, **85**, 3238–3245.
181. L. Hu, S. Han, S. Parveen, Y. Yuan, L. Zhang and G. Xu, *Biosens. Bioelectron.*, 2012, **32**, 297–299.
182. X. Wang, Y. Wang, H. Rao and Z. Shan, *J. Braz. Chem. Soc.*, 2012, **23**, 2011–2015.
183. H. Kong, Y. Lu, H. Wang, F. Wen, S. Zhang and X. Zhang, *Anal. Chem.*, 2012, **84**, 4258–4261.
184. J. Zhang, M. Sajid, N. Na, L. Huang, D. He and J. Ouyang, *Biosens. Bioelectron.*, 2012, **35**, 313–318.
185. P.-H. Chan and Y.-C. Chen, *Anal. Chem.*, 2012, **84**, 8952–8956.
186. P. H. W. Chan, S. H. SY Lin and Y. C. Chen, *Rapid Commun. Mass. Spetrom.*, 2013, **27**, 2143–2148.
187. H. Li, Y. Guo, I. Xiao and B. Chen, *Analyst*, 2013, **139**, 285–289.
188. C. Hu, D.-P. Yang, Z. Wang, L. Yu, J. Zhang and N. Jia, *Anal. Chem.*, 2013, **85**, 5200–5206.
189. J. Peng, L.-N. Feng, K. Zhang, X.-H. Li, L.-P. Jiang and J.-J. Zhu, *Chem. Eur. J.*, 2012, **18**, 5261–5268.
190. H. Lin, L. Li, C. Lei, X. Xu, Z. Nie, M. Guo, Y. Huang and S. Yao, *Biosens. Bioelectron.*, 2013, **41**, 256–261.
191. A. Mathew, P. R. Sajanlal and T. Pradeep, *Angew. Chem., Int. Ed.*, 2012, **51**, 9596–9600.
192. (a) Y.-M. Fang, J. Song, J. Li, Y.-W. Wang, H.-H. Yang, J.-J. Sun and G.-N. Chen, *Chem. Commun.*, 2011, **47**, 2369–2371; (b) L. Li, H. Liu, Y. Shen, J. Zhang and J.-J. Zhu, *Anal. Chem.*, 2011, **83**, 661–665.
193. Y. Wu, H. Jinhua, T. Zhou, M. Rong, Y. Jiang and X. Chen, *Analyst*, 2013, **138**, 5563–5565.

194. X. Hun, H. Chen and W. Wang, *Biosens. Bioelectron.*, 2010, **26**, 248–254.
195. Y. Chen, Y. Shen, D. Sun, H. Zhang, D. Tian, J. Zhang and J.-J. Zhu, *Chem. Commun.*, 2011, **47**, 11733–11735.
196. H. J. Safer D, J. S. Wall and J. E. Reardon, *Science*, 1982, **218**, 290–291.
197. C. A. Simpson, B. J. Huffman, A. E. Gerdon and D. E. Cliffel, *Chem. Res. Toxicol.*, 2010, **23**, 1608–1616.
198. R. Archana, S. Sonali, M. Deepthy, R. Prasanth, M. Habeeb, P. Thalappil, N. Shantikumar and K. Manzoor, *Nanotechnology*, 2010, **21**, 055103.
199. Y. Wang, J. Chen and J. Irudayaraj, *ACS Nano*, 2011, **5**, 9718–9725.
200. F. Wen, Y. Dong, L. Feng, S. Wang, S. Zhang and X. Zhang, *Analytical Chemistry*, 2011, **83**, 1193–1196.
201. Y. Kong, J. Chen, F. Gao, R. Brydson, B. Johnson, G. Heath, Y. Zhang, L. Wu and D. Zhou, *Nanoscale*, 2013, **5**, 1009–1017.
202. R. Archana, J. Jasusri, C. Parwathy, M. Deepthy, N. Shantikumar, M. Ullas and K. Manzoor, *Nanotechnology*, 2011, **22**, 285102.
203. C. V. Durgadas, C. P. Sharma and K. Sreenivasan, *Analyst*, 2011, **136**, 933–940.
204. C. V. Durgadas, C. P. Sharma and K. Sreenivasan, *Nanoscale*, 2011, **3**, 4780–4787.
205. X. Guével, E.-M. Prinz, R. Müller, R. Hempelmann and M. Schneider, *J. Nanopart Res*, 2012, **14**, 1–10.
206. C. Haiyan, L. Bowen, W. Chuan, Z. Xin, C. Zhengqi, D. Xi, Z. Rui and G. Yueqing, *Nanotechnology*, 2013, **24**, 055704.
207. A. Zhang, Y. Tu, S. Qin, Y. Li, J. Zhou, N. Chen, Q. Lu and B. Zhang, *J. Colloid Interface Sci.*, 2012, **372**, 239–244.
208. G. Sun, L. Zhou, Y. Liu and Z. Zhao, *New J. Chem.*, 2013, **37**, 1028–1035.
209. L. Shang, R. Dörlich, V. Trouillet, M. Bruns and G. Ulrich Nienhaus, *Nano Res*, 2012, **5**, 531–542.
210. X. Wu, X. He, K. Wang, C. Xie, B. Zhou and Z. Qing, *Nanoscale*, 2010, **2**, 2244–2249.
211. D.-H. Hu, Z.-H. Sheng, P.-F. Zhang, D.-Z. Yang, S.-H. Liu, P. Gong, D.-Y. Gao, S.-T. Fang, Y.-F. Ma and L.-T. Cai, *Nanoscale*, 2013, **5**, 1624–1628.
212. (a) Y. Wang, J.-T. Chen and X.-P. Yan, *Anal. Chem.*, 2013, **85**, 2529–2535; (b) S.-K. Sun, L.-X. Dong, Y. Cao, H.-R. Sun and X.-P. Yan, *Anal. Chem.*, 2013, **85**, 8436–8441.
213. C.-L. Liu, T.-M. Liu, T.-Y. Hsieh, H.-W. Liu, Y.-S. Chen, C.-K. Tsai, H.-C. Chen, J.-W. Lin, R.-B. Hsu, T.-D. Wang, C.-C. Chen, C.-K. Sun and P.-T. Chou, *Small*, 2013, **9**, 2103–2110.
214. X.-D. Zhang, D. Wu, X. Shen, P.-X. Liu, F.-Y. Fan and S.-J. Fan, *Biomaterials*, 2012, **33**, 4628–4638.
215. (a) H. Chen, S. Li, B. Li, X. Ren, S. Li, D. M. Mahounga, S. Cui, Y. Gu and S. Achilefu, *Nanoscale*, 2012, **4**, 6050–6064; (b) T. Chen, S. Xu, T. Zhao, L. Zhu, D. Wei, Y. Li, H. Zhang and C. Zhao, *ACS Appl. Mater. Interfaces*, 2012, **4**, 5766–5774.

216. A. P. Retnakumari, P. L. Hanumanthu, G. L. Malarvizhi, R. Prabhu, N. Sidharthan, M. V. Thampi, D. Menon, U. Mony, K. Menon, P. Keechilat, S. Nair and M. Koyakutty, *Mol. Pharmaceutics*, 2012, **9**, 3062–3078.
217. P. Chandran, N. Gupta, A. P. Retnakumari, G. L. Malarvizhi, P. Keechilat, S. Nair and M. Koyakutty, *Nanomedicine : nanotechnology, biology, and medicine*, 2013, **9**, 1317–1327.
218. J. Zhang, C. Chen, X. Xu, X. Wang and X. Yang, *Chem. Commun.*, 2013, **49**, 2691–2693.
219. T. S. Sreeprasad, M. S. Maliyekkal, K. Deepti, K. Chaudhari, P. L. Xavier and T. Pradeep, *ACS Appl. Mater. Interfaces*, 2011, **3**, 2643–2654.
220. T. Zhou, Y. Huang, W. Li, Z. Cai, F. Luo, C. J. Yang and X. Chen, *Nanoscale*, 2012, **4**, 5312–5315.
221. C. Sun, H. Yang, Y. Yuan, X. Tian, L. Wang, Y. Guo, L. Xu, J. Lei, N. Gao, G. J. Anderson, X.-J. Liang, C. Chen, Y. Zhao and G. Nie, *J. Am. Chem. Soc.*, 2011, **133**, 8617–8624.
222. Y. Chen, Y. Wang, C. Wang, W. Li, H. Zhou, H. Jiao, Q. Lin and C. Yu, *J. Colloid Interface Sci.*, 2013, **396**, 63–68.
223. S. Su, H. Wang, X. Liu, Y. Wu and G. Nie, *Biomaterials*, 2013, **34**, 3523–3533.

Metal(0) Clusters in Catalysis

NOELIA VILAR-VIDAL,[a,b] JOSÉ RIVAS[b] AND
M. ARTURO LÓPEZ-QUINTELA*[a]

[a] Nanomag Laboratory, Research Technological Institute, University of
Santiago de Compostela, E-15782, Santiago de Compostela, Spain;
[b] INL – International Iberian Nanotechnology Laboratory, 4715-310, Braga,
Portugal
*Email: malopez.quintela@usc.es

8.1   Introduction

Over the last decade, much attention has been paid to the field of "nano-catalysis". Within this new field the unique characteristics of metal clusters (Au, Ag, Cu, Pd, *etc.*) make them strong candidates to replace the old catalysts. Metal clusters (CLs) can be defined as aggregates of metal atoms with sizes below approximately 1 to 2 nm.[1] They represent a novel state of matter, located between the classical bulk (or nanoparticle) behaviour and the different behaviour of the corresponding atoms. Different studies show that the new electronic/geometrical properties displayed by these atom clusters depend on their size due to quantum confinement. The quantized electronic structures of clusters and the extraordinary high stability of magic-number clusters, M_N ($N = 8, 18, 20, 34, 58...$) due to electronic shell closure,[2] could explain the chemical reactivity and novel properties of these clusters. One of the most important consequences of this regime is the presence of discrete energy levels and the appearance of a HOMO–LUMO gap (highly dependent on the cluster size, increasing with decreasing size) similar to that found in semiconductors (Figure 8.1). It has to be pointed out that this

RSC Smart Materials No. 7
Functional Nanometer-Sized Clusters of Transition Metals: Synthesis, Properties and
Applications
Edited by Wei Chen and Shaowei Chen
Published by the Royal Society of Chemistry, www.rsc.org

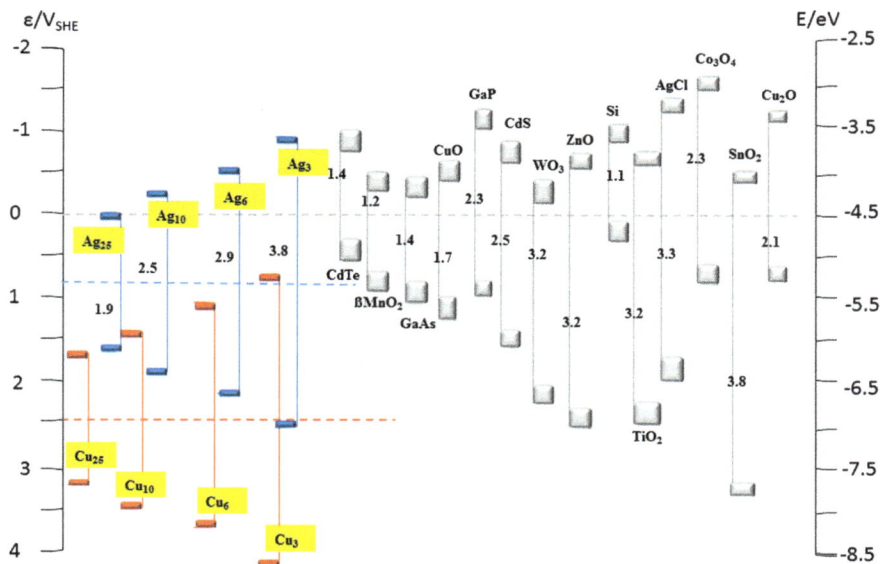

Figure 8.1 Energy diagram showing a comparison between the different calculated band gaps by the Jellium model, for Ag and Cu clusters, compared with other well-known semiconductor band gaps.

size-dependence gives the chance of obtaining different band gaps with only one metal element, and therefore, different specific clusters of the same metal could be used for different catalytic applications.[3] The discrete electronic states in clusters provide very different properties than those of their counterparts (atoms and bulk/nanoparticles), such as well-defined luminescence, magnetism and catalysis,[4] among others, providing a wide range of attractive applications,[5] like in biological labelling, optical sensing, *etc*. The correlation of the catalytic properties of metal clusters with their atomic structure is still under investigation and will provide important guidelines for the future design of new catalysts for specific chemical processes.

The topic of this chapter is focused on the use of metal clusters as "catalyst". A catalyst is a material which decreases the free energy of activation that controls the chemical reaction rate, therefore giving an increase in the rate without being consumed. The use of nano-sized materials for catalysis leads to an obvious improvement due to the small amount required, because the surface/volume ratio increases by decreasing the particle size. Additionally, it has been found that nanocatalysts not only increase the reaction rate, but also lead—in many cases—to the formation of specific products (increase of selectivity). Therefore, one of the main goals of this article will be to highlight the intrinsic catalytic activities exhibited by some metal clusters (supported and unsupported), for some specific chemical processes, such as the selective oxidation of alcohols and alkenes, homocoupling reactions, hydrogenation, degradation reactions, among others. Although there are many examples where metal clusters were used as catalysts, here we will show only those examples for which clusters have been well characterized. Furthermore, we have also made a compilation of different mechanisms proposed so far for the explanation of the novel catalytic properties displayed by clusters. It is important to note that, because this is a new emerging area, the catalytic mechanisms are still not well understood. Because of this, we have also tried in this revision to introduce a new *energy approach* to contribute to a better understanding of the cluster catalysis.

The chapter will be organized according to the catalysis category: homogeneous catalysis (using metal clusters in solution) and heterogeneous catalysis (a metal cluster supported on another material). Within each category different catalytic reactions will be presented (see Table 8.1).

Finally, in conclusion we will summarize some of the more important points extracted from the analysis carried out for the different catalytic studies mediated by metal clusters.

8.2 Metal Cluster Mediated Catalysis

8.2.1 Homogeneous Catalysis

A homogeneous catalyst can be defined as a soluble material in the reaction phase. It has the advantage of having all the catalytic sites accessible for the reactants. But, it requires proper removal from the reaction medium. In spite of this, homogeneous catalysts remain more effective than their heterogeneous counterparts.

8.2.1.1 Oxidation Reactions

8.2.1.1.1 Alcohol Oxidation. Aldehydes and ketones are particularly useful in the production of flavours, fragrances, and biologically active compounds. Conventionally, they were produced by selective alcohol oxidations using chromium salts, oxalyl chloride or hypervalent iodides.[6] Since discovery in 1989 by Haruta and co-workers,[7] the investigation about new nanocatalysts for aerobic oxidation of alcohols has become a great challenge. Comotti *et al.*[8] were the first researchers to demonstrate the intrinsic effect of cluster size on the catalytic activity for aerobic oxidation. Later, Tsukuda *et al.*[9–12] studied the catalytic properties of different Au : PVP cluster sizes. Au : PVP CLs with diameters in the range 1.3 to 9.5 nm were obtained for different concentration ratios of Au(III)/Au(0), using poly(N-vinyl-2-pyrrolidone) (PVP) as stabilizer. We have to point out that the smallest Au : PVP CLs (1.3 nm) correspond to Au_{55} clusters with a cuboctahedral motif. The same authors also reported the effect of the cluster size on the oxidation catalysis of 4-(hydroxymethyl)phenol[10] in basic media [Figure 8.2(A)], where p-hydroxybenzaldehyde was selectively obtained. In order to determine the origin of this size-specific catalysis

Table 8.1 List of metal clusters discussed in this chapter. Reactions appear in the text in the same order as in the table.

Cluster	Size/nm	Stabilizer	Support	Catalysis	Reference
Au:PVP	1.5–9	PVP	—	Alcohol oxidation	9,10,12,17
Pd:PVP	1.5–2.2	PVP	—	Alcohol oxidation	9
$Au_{25}(SR)_{18}$	≈1.1	SC_2H_4Ph	—	Styrene oxygenation	19a
Au:PVP	≈1.3	PVP	—	Homocoupling and Toluenesulfonamide addition	31
Au:PVP	≈1.3	PVP	—	Intramolecular heterocyclization of γ-hydroxyalkenes	32
$Au_{25}(SR)_{18}$	≈1.1	SC_2H_4Ph	—	Hydrogenation of α,β-unsaturated alcohols and ketones	19,22,39
Cu:PAMAM		PAMAM	—	Alkene/aldehyde hydrogenation	40
Cu_5/Cu_{13}	0.4/0.6	$TBANO_3$	—	MB reduction	3a
Cu_{9-13}	—	—	—	MB reduction	43
Au_{25}	—	—	$—/Al_2O_3$	4-nitrophenol reduction	44
Ag_9	<1	MSA		CCl_4 degradation	49
Au_9	<1			Hydration of alkynes	51
$Au_{25}(SR)_{18}$	≈1	SC_2H_4Ph	CeO_2	CO oxidation	55
Au_{11}^{3+}:TPP	0.8	TPP	SiO_2	Alcohol oxidation	59
Au_{55}	—	—	SiO_2	Styrene epoxidation	61
$Au_{25}(SG)_{18}$		Gluthatione	HAP	Styrene epoxidation	62
$Au_{25}(SR)_{18}$, $Au_{38}(SR)_{24}$, and $Au_{144}(SR)_{60}$	0.98, 1.3, 1.6	SC_2H_4Ph	HAP, SiO_2	Styrene oxidation	20
$Au_{25}(SR)_{18}$–$Pt_1Au_{24}(SR)_{18}$		SC_2H_4Ph	TiO_2	Styrene oxidation	52d
Cu_{13}		MPEG	SiO_2	Styrene oxidation	66
AuCLs	≤2		TS-1	Propylene epoxidation	70,71
Au_6/Au_{10}	<1		Al_2O_3	Propylene epoxidation	72
Ag_2/Ag_4	<1		Al_2O_3	Propylene epoxidation	73
Au_{10}–Au_{18}–Au_{25}–Au_{39} $(SG)_x$		Gluthatione	HAP	Cyclohexene oxidation	77
$Au_{25}(SR)_{18}$		SC_2H_4Ph	CeO_2–TiO_2–SiO_2–Al_2O_3	Aryl halides homocoupling	78
Au_{5-10}	<1		CNT's	Oxidation of Thiophenol	79
$Au_{25}(SR)_{18}$		SC_2H_4Ph	Fe_2O_3– TiO_2–SiO_2	Selective hydrogenation ketones	19b
Au	<1		La_2O_3–CeO_2–Al_2O_3	Aldehyde hydrogenation	81
Ag_{7-8}	<1	H_2SA	SiO_2–TiO_2–Fe_2O_3– Al_2O_3	Nitro group reduction	83

with Au:PVP clusters, they studied the electronic structures of the Au cores by X-ray absorption near-edge spectroscopy (XANES), Fourier-transform infrared (FTIR) spectroscopy and X-ray photoelectron spectroscopy (XPS). They observed a more negatively charged trend in smaller AuCLs than in larger ones studying the vibrational peak of CO adsorbed onto the

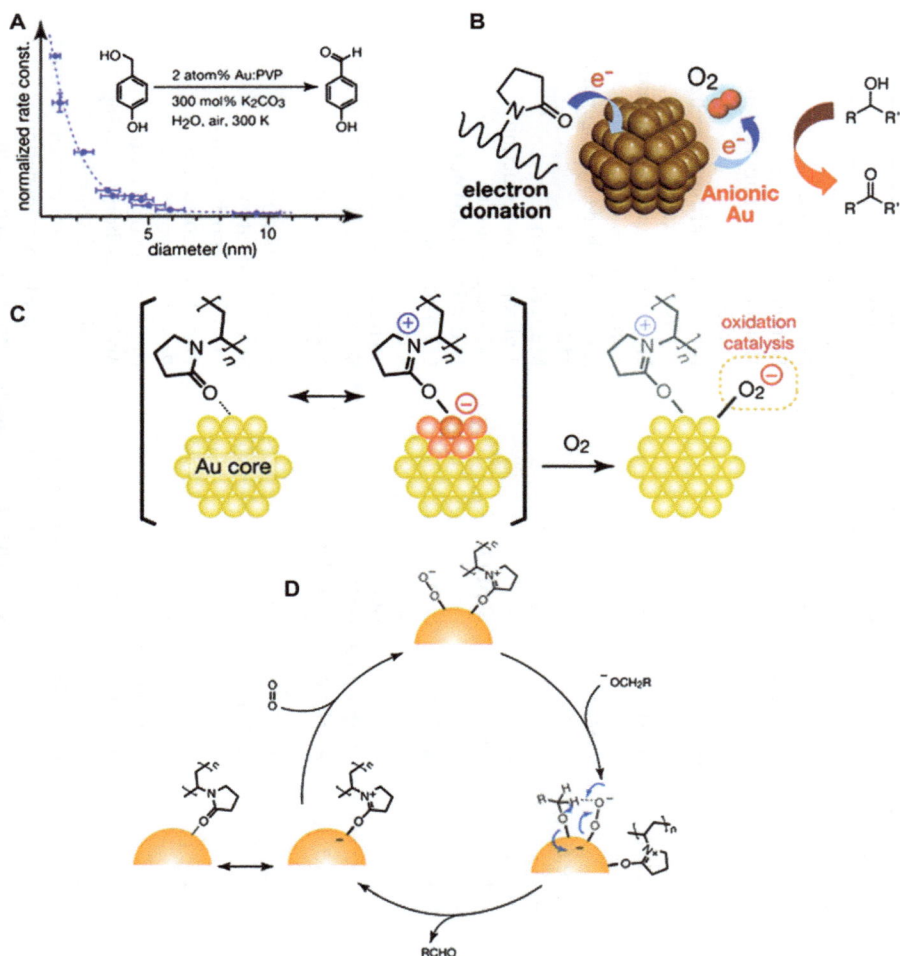

Figure 8.2   Dependence of catalytic activity on the size of Au:PVP clusters for the oxidation of *p*-HBA (*p*-hydroxybenzylalcohol). (A) Proposed mechanism for aerobic oxidation of alcohols by Au:PVP clusters based on the activation of molecular oxygen. (B) Mechanism for the activation of molecular oxygen by Au:PVP clusters. (C) Complete mechanism for the alcohol oxidation mediated by Au:PVP cluster catalysts (D).
(Reprinted with permission from ref. 12, Copyright © 2009 American Chemical Society and ref. 16, Copyright © 2011 WILEY-VCH Verlag GmbH & Co. KGaA, Weinheim.)

Au : PVP clusters. A red shift in the ν_{co} when decreasing the cluster size, as well as a PVP concentration dependence was noticed compared to that observed for the CO adsorbed on free Au_n^- clusters. An efficient e^- transfer from Au_n^- to the LUMO (π^*) of CO[13] could explain this behaviour. Moreover, XPS studies revealed a decrease of the position of the Au 4f band (82.7 eV) compared to that corresponding to bulk Au (84.0 eV). These results indicate that excess PVP donate e^-/electronic charge to the Au core, probably through interaction with –N–C=O groups [Figure 8.2(B)]. This hypothesis was also supported by theoretical calculations where an e^- is transferred from the pyrrolidone group to the Au_{13} core.[14] On the basis of these observations, the authors describe the oxidation catalytic mechanism by Au : PVP clusters as follows: the Au core is negatively charged by e^- donation from PVP (energetically and spatially favourable overlapping of the high-lying LUMO molecular orbitals only on small AuCLs[11]), then the e^- will be transferred to O_2 molecules (to the antibonding orbital–LUMO) to form superoxo or peroxo-like species in the gas phase [Figure 8.2(C)].[15,16]

Finally, the O_2^{n-} species could remove the α-hydrogen from the alcohol [Figure 8.2(D)] to form the corresponding aldehyde. Although this is a very plausible mechanism it is incapable of explaining some facts, such as the inactivity of these clusters under acidic pHs and a clear cluster size increase (around 1.5 nm after the fifth run) after the catalytic reaction.[17] Therefore, an alternative mechanism could be proposed (Figure 8.3).

Figure 8.3 shows an energy diagram for the LUMO energy levels of different cluster sizes (for the calculation of the energy levels we have assumed a dependence of the E_{Fermi} level on the cluster size, similar to that obtained before for CuCLs).[3a] As the reaction only takes place under basic conditions the O_2 redox potential $E^0(O_2/O_2^-) = -0.2$ V is also shown. The proposed

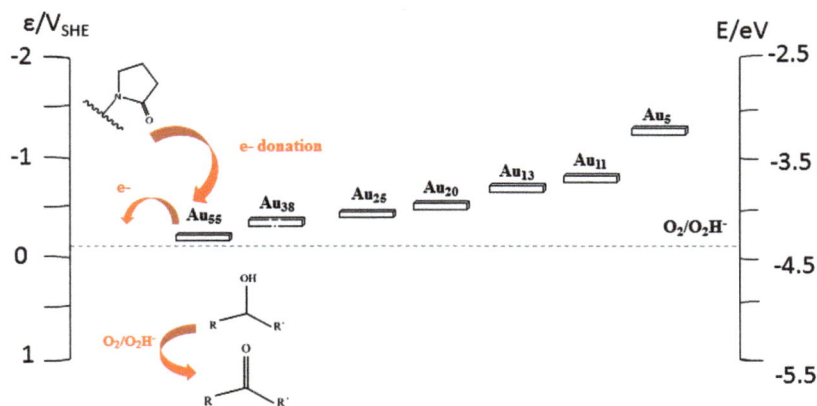

Figure 8.3 Alternative energy diagram mechanism for the e^- transfer from PVP to Au core and, then, to O_2 based on the redox position of the LUMO orbitals of Au : PVP CLs.

mechanism consists of an e^- donation from the PVP protected ligands to the Au clusters. Then, as the LUMO orbital for Au : PVP (1.3nm, Au_{55} cluster) is located above the redox potential for the oxygen superoxide formation, an e^- transfer from the cluster to the O_2 will take place with the formation of a superoxide, which will be responsible for the alcohol oxidation. This alternative mechanism can be supported by the theoretical studies performed by Haruta and co-workers[14] about the hetero-junction effect in PVP-stabilized Au_{13} clusters. They carried out hybrid density functional calculations for Au_{13}–PVP clusters and their interaction with O_2. The results suggested a dramatic change of the interactions of the Au_{13}–PVP–O_2 with the PVP amount, concluding that the negative Au atoms on the cluster surface play an important role in the activation of the oxygen resulting in negatively charged O_2 onto Au_{13}–PVP. These findings suggest that PVP acts, not only as a stabilizer to prevent the aggregation of Au clusters, but also as an e^- donor to Au clusters. The proposed mechanism could rationally explain the—until now—unresolved facts: the reaction should not proceed in the absence of a base because, due to the change of the O_2 redox potential with pH, the position of the cluster's LUMO orbital would be below the O_2 redox potential and the e^- transfer could not take place. The observed fact that there is a gradual decrease of the catalytic activity with increase of the cluster size can also be explained by PVP degradation (due to e^- transfer from PVP to the Au core) with subsequent unprotected gold cluster aggregation. Tsukuda *et al.*[9] also studied the catalytic activity of Pd:PVP clusters of a similar size to Au : PVP clusters, observing a higher activity for the gold clusters under the same reaction conditions. On the basis of the above proposed alternative mechanism, these results could be explained by the different positions of the E_{Fermi} levels of Pd and Au (-7.07 *vs.* -5.3 eV), which would locate the LUMO orbital of PdCLs below the O_2 redox potential, in this way preventing the reaction. The origin of the small catalytic activity observed could come from the polydispersity of the sample, which could include some smaller PdCLs, for which the position of the LUMO orbital should be conveniently located.

8.2.1.1.2 Alkene Oxidation. Supported Au catalysts have been found to be effective for the epoxidation of olefins,[18] however, one of their disadvantages is their polydispersity. The use of well-defined AuCLs would offer a unique opportunity for a deeper understanding of fundamental aspects, such as: how the size, the core-shell structure and the electronic properties of different AuCLs have influence on their catalytic performance. Following this direction, Zhu and co-workers[19] have reported the selective catalytic oxidation of styrene by different monodispersed, thiolate-protected AuCLs of different sizes: Au_{25} (1.0 nm in diameter), Au_{38} (1.3 nm) and Au_{144} (1.6 nm).[20] They demonstrated that the catalytic activity increases when the $Au_m(SR)_n$ cluster size decreases. The styrene oxidation gives three main products: benzaldehyde, styrene epoxide and acetophenone. Better conversion values were obtained using $Au_{25}(SR)_{18}$ (R = C_2H_4Ph) rather than larger AuCLs, such as Au_{38} or Au_{144}, although the selectivity was always the same: benzaldehyde. The ligand effect on the catalytic activity was found to be negligible, attributing the activity mainly to the metal core and the cluster structure.[20] For simplicity, we will describe here only the catalysis with Au_{25} clusters. The structure of $Au_{25}(SCH_2CH_2Ph)_{18}$ clusters is formed by a Au_{13} kernel [with eight ($q = -1$) or seven ($q = 0$) delocalized valence e^- originated from Au(6s)] and a partial positive charged ($Au^{\delta+}$ $0 < \delta < 1$) Au_{12} shell[21] (six $Au_2(SR)_3$), where only 12 of the 20 faces of the icosahedral core are capped, leaving eight catalytically active open faces. The proposed catalytic mechanism[22] consists of the formation of oxygen activated species (such as O_2^-) by the e^--rich Au_{13} kernel through an e^- transfer,[20] followed by the conversion of $[Au_{25}(SR)_{18}]^-$ into neutral $[Au_{25}(SR)_{18}]^0$ and the posterior activation of the nucleophilic $-CH=CH_2$ group by the electrophilic (positive charged) Au surface. Then, the activated $C=C$ bonds react with the O_2(ad) species through a side-by-side interaction on the Au_{25} surface sites. Finally, the oxidized $[Au_{25}(SR)_{18}]^0$ catalyst can be reduced to the anionic $[Au_{25}(SR)_{18}]^-$ by gaining one e^- when the $C=C$ bond leaves the Au_{25} cluster, and the catalytic cycle is completed [Figure 8.4(A)].

The authors also studied the possibility that the ligand shell falls off during the catalysis reaction by X-ray absorption spectroscopy. The results confirmed that the thiolate ligands remain on the clusters and the cluster size is not altered during the oxidation process, which should be expected to occur for temperatures below the thiolate desorption $\approx 200\ °C$.[20] Therefore, an alternative mechanism (considering again the HOMO/LUMO position of the AuCLs orbitals) could again be proposed for the alkene oxidation reaction (see Figure 8.5). For this purpose we have taken into account the following facts:

(1) The proposed band gap for the Au_{25} cluster is 1.3 eV, whereas that for the Au_{38} cluster is 0.9 eV. By using the Kohn–Sham orbital energy level diagram for a model compound $Au_{25}(SH)_{18}$ [Figure 8.4(B)] the higher LUMO band would be located around -1 V (*vs.* SHE) (LUMO $+ 2$), being more negative when the band gap increases (with an e^- donation increase by the ligand).[21]

(2) The O_2 activation in presence of $Au_{25}(SR)_{18}$ clusters could be a photomediated process. Recent results carried out by the same research group[23] suggest the possibility that the O_2 activation could be carried out by an e^- transfer from the Au cluster to the O_2 (accompanied by oxidative conversion of $[Au_{25}(SR)_{18}]^-$ into neutral $[Au_{25}(SR)_{18}]^0$, which has already been previously studied[24]) by a photomediated process.

(3) Zhu *et al.* located the position of the O_2/O_2^- redox couple around -0.6 V (*vs.* SHE), lying between the LUMO (-1.3 V) and the HOMO (0.3 to 0.6 V) of the $Au_{25}(SR)_{18}$ cluster [see Figure 8.4(C)].

(4) Electrochemical studies carried out with $Au_{38}(SCH_2CH_2Ph)_{24}$ clusters using DPV (differential pulse voltammetry) reveal the presence of an oxidation peak, located around -0.6 V (*vs.* Ag/AgCL) (-0.4 V *vs.* SHE), which was attributed to the $-1/0$ oxidation of the cluster.[25]

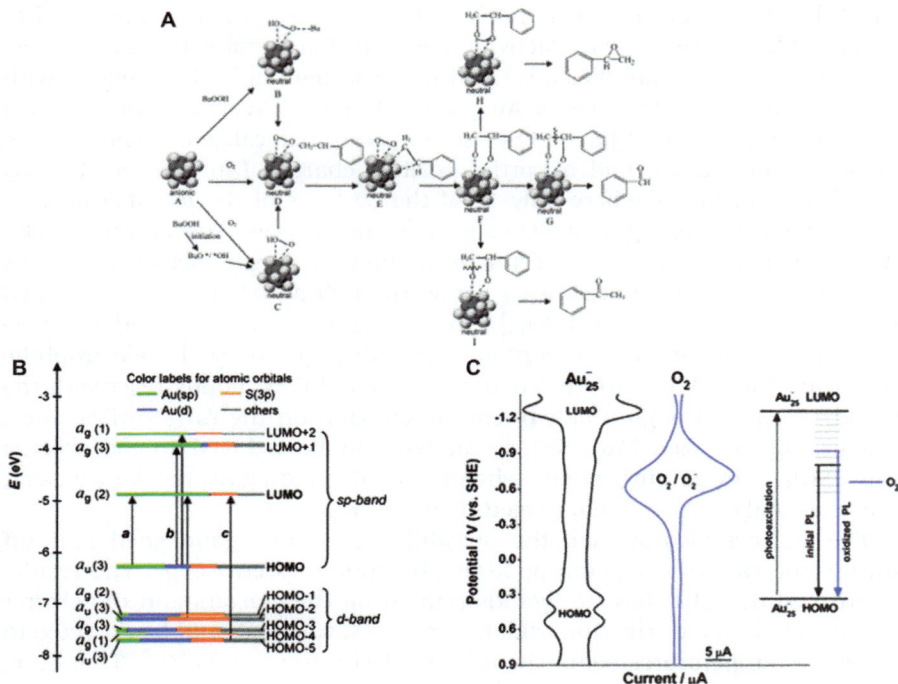

Figure 8.4 (A) Proposed mechanism for the selective oxidation of styrene catalysed by $[Au_{25}(SR)_{18}]^-$ clusters. (Magenta: core Au atoms, blue: shell Au atoms). [Reprinted from ref. 20.] (B) Kohn–Sham orbital energy level diagram for a model compound $Au_{25}(SH)_{18}{}^-$. [Reprinted with permission from ref. 21. Copyright © 2008 American Chemical Society.] (C) Position of the $Au_{25}{}^-$ HOMO and LUMO orbitals in comparison with the O_2 level (the $O_2/O_2{}^-$ redox couple). [Reprinted from ref. 23. Copyright © 2013 American Chemical Society.]

One can see that the results cannot be explained assuming e^- transfer through the LUMO orbitals because they are located below the O_2 redox potential. Therefore, we assume that the reaction proceeds through the LUMO + 1 orbitals.[26] If we further assume that the Au_N : SR clusters would be easily activated by visible light, then an e^- of the negative charged clusters could be promoted to the LUMO + 1 and from there to the O_2 (E^0 for $O_2/$ O_2H^- would be located around -0.6 V *vs.* SHE in organic solvents), which could easily accept this e^- (Figure 8.5). Then, this superoxide would be the oxidant species (strong oxidant power of the superoxide anion[27]) capable of oxidizing the alkene to the corresponding products. This alternative mechanism could explain the fact that larger clusters, such as Au_{34} and Au_{144}, display less catalytic activity than the smaller clusters because of their smaller driving force. At the same time, this mechanism can also explain the inalterability of the cluster size after the catalytic process (contrary to the

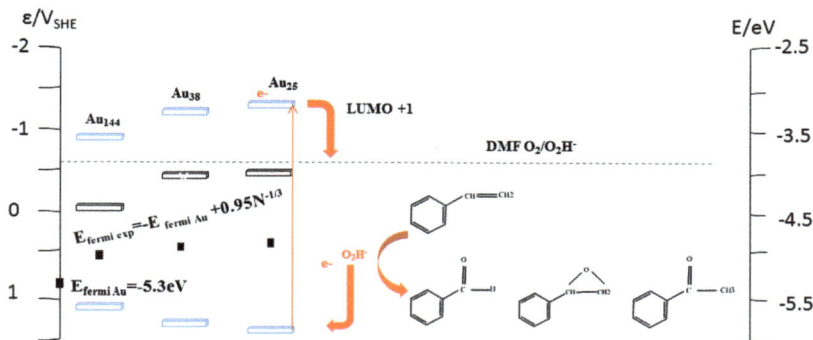

Figure 8.5 Alternative mechanism for the e$^-$ transfer from the HOMO orbital to the LUMO + 1 orbital by photoactivation. The e$^-$ can be further transferred to the O$_2$ forming the corresponding superoxide compound. Finally, this superoxide would be responsible for the alkene oxidation.

results obtained by Tsukuda *et al.* for the alcohol oxidation by Au : PVP clusters, see section 8.1.1.1) because no oxidation of the capping agents is involved in the process. It should be interesting to carry out experiments in the dark to confirm the photomediated hypothesis.

8.2.1.1.3 Cyclization Reactions

8.2.1.1.3.1 Oxidative Coupling Reactions. Symmetrical biaryls can be prepared by selective homocoupling of organoboron compounds.[28] Traditionally Pd(II) was used, although other oxidants such as Cu(II)[29] or Mn(III)[30] have also been utilized. Tsukuda and co-workers studied the catalytic properties of Au : PVP clusters through the homocoupling reaction of potassium aryltrifluoroborates in water.[31] For this purpose clusters of two sizes were employed: Au : PVP-1 (1.3 nm ≈ Au$_{55}$) and Au : PVP-7 (9.5 nm). The smaller Au : PVP clusters were found to have the highest activity whereas the bigger ones did not catalyse the reaction at all. The authors explained the results assuming a similar mechanism to that proposed for the alcohol oxidation (see section 8.1.1.1). A positively charged site on the Au cluster surfaces (generated by oxygen adsorption) generates a superoxo-complex-like intermediate with the consequent attack of nucleophiles, such as PhBF$_3$K, generating an Au–C intermediate. However, some questions, like the pH and size-dependent catalytic activity, as well as the low reusability of Au : PVP clusters remain to be explained. Trying to look for possible explanations to such questions, an alternative mechanism, similar to that introduced before (section 8.1.1.1), based on energy considerations about the positions of the cluster HOMO/LUMO orbitals, is depicted in Figure 8.6.

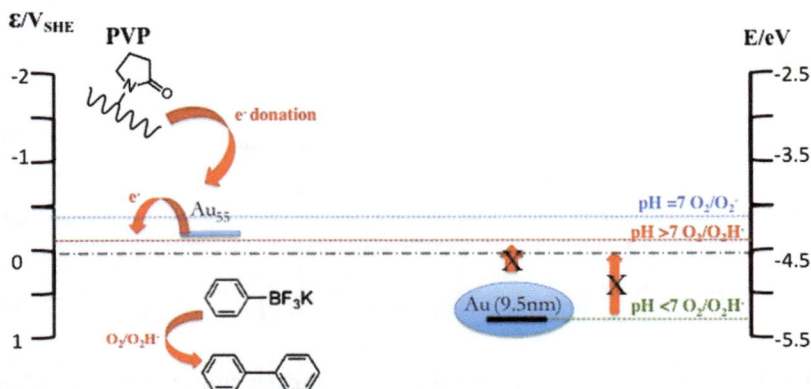

Figure 8.6 Alternative energy diagram mechanism for the e^- transfer from the PVP to the Au_{55} metal core and, then, to the O_2 forming superoxide species in basic media. Such superoxides can then attack the aryltrifluoroborate and give the homocoupling reaction.

This mechanism allows explanation of the previous unresolved questions, as follows:

(1) The redox potential for O_2 in acidic media is located around 0.6 V (*vs.* SHE). Therefore, although the Au : PVP clusters (Au_{55}) can transfer the e^- to the O_2 no oxidation of the organic compounds is possible because their potentials lie above the O_2 redox potential. On the other hand, no e^- transfer can be possible to the O_2 from the Fermi energy of Au (-5.3 eV) in the Au : PVP-7 nanoparticles (9.5 nm).
(2) The observed size growth after the first cycle (from 1.3 ± 0.3 nm to 2.3 ± 0.7 nm) corroborates the PVP e^- transfer to the Au core with corresponding oxidation of the capping agent and aggregation of clusters forming inactive big nanoparticles (>6 nm).

8.2.1.1.3.2 Lewis Acidic Reactions. Taking into account the hypothesis previously described of a possible e^--deficient site at the cluster surfaces by O_2 adsorption on metal clusters, some authors compared the clusters with Lewis acid species (species able to accept a lone pair of e^-). Therefore, such results could indicate that clusters could be active sites for transmetallation of aryl moieties. In line with this, Kamiya and co-workers[32] evaluated the intramolecular heterocyclization of γ-hydroxyalkenes and γ-aminoalkenes as suitable targets, to check this possible acid behaviour (conventionally the reaction is promoted by Lewis acid catalysts). Au : PVP CLs catalysts were synthesized similarly to the ones described in section 8.1.1.1. They used Au : PVP CLs with 1.3 nm and 9.5 nm sizes for the cyclization of 1,1-diphenyl-4-penten-1-ol. The cyclization proceeds with the smaller AuCLs in basic aqueous media and DMF, but not under

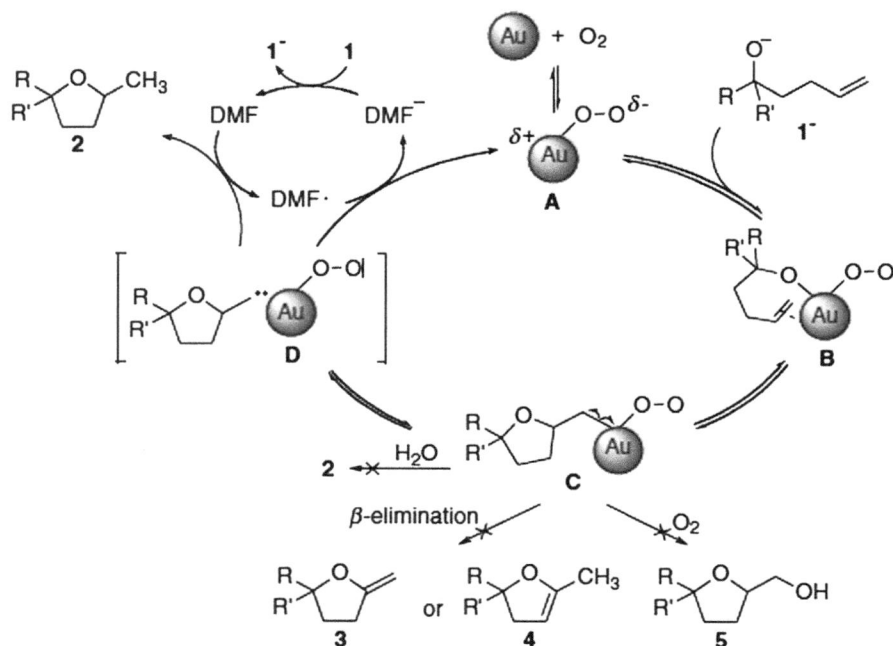

Figure 8.7 Au:PVP CL-catalysed hydroalkoxylation mechanism. (Reprinted from ref. 32.)

degassed conditions. The reaction mechanism suggested by the authors is analogous to the one described in section 8.1.1.1 (see Figure 8.7). Initially the formation of an Au cluster:O_2 intermediate **A** (with an e⁻-deficient site) would occur. Such an intermediate would then act as a Lewis acid, activating both the alkoxide and alkene **B** by adsorption onto the surface, and giving **C** by insertion of an alkene into the O–Au bond. Then, homolytic dissociation would take place, generating the radical intermediate **D**, which afforded the product *via* hydrogen removal from DMF (H source) accompanied by the regeneration of free Au clusters.

Kitahara and Sakurai[33] also reported similar studies for the intramolecular heterocyclization of γ-toluenesulfonylaminoalkenes. In this case the authors observed that no cyclization occurs under degassed conditions, in absence of basic media or increasing the cluster size (Au:PVP, 9.5 nm). The reason why the cyclization does not proceed under acidic conditions or by increasing the cluster size was not explained by the authors. Therefore, we could also apply the previously described alternative energy mechanism to this kind of reaction (see section 8.1.1.3.1). One can see that with such mechanism one could explain the results obtained by Kitahara and Sakurai just taking into account the pH redox potential dependence for the O_2 superoxide formation.

8.2.1.2   Hydrogenation Reactions

Hydrogenation is a common process employed in industry to reduce or saturate organic compounds. The reaction can be carried out in absence of a catalyst, but it requires high temperatures. In order to avoid this, catalysts are commonly used. Traditional homogeneous catalysts for hydrogenation include Ni, Pd, Ru and Pt.[34] Nowadays, however, selective hydrogenation remains a challenging objective, because a multifunctional molecule can potentially bind to the catalyst surface in a variety of ways and react through any functional groups. To attain selective hydrogenation many efforts have been devoted in the last few years to design and synthesis of new catalysts.[35]

8.2.1.2.1   Unsaturated Ketone and Aldehyde Hydrogenation.   The chemoselective hydrogenation of α,β-unsaturated ketones into unsaturated alcohols are of particular importance to the perfume and flavour industries. Although conventional AuNPs were used in the hydrogenation of α,β-unsaturated ketones, they cannot achieve 100% chemoselectivity for the unsaturated alcohol.[36] However, Zhu *et al.*[19] reported the capability of unsupported $Au_{25}(SR)_{18}$ ($R = CH_2CH_2Ph$ or $C_{12}H_{25}$)[37] for the preferential hydrogenation of the C=O bond in unsaturated ketones. They used benzalacetone at low temperatures (0 °C) under atmospheric pressure obtaining a conversion yield around 20%, and a high chemoselectivity towards the unsaturated alcohol [100%, Figure 8.8(A)], which was not possible with conventional AuNPs. No ligand effect was found on the catalytic activity. As was mentioned before, Au_{25} clusters seem to be formed by a structure $Au_{13(core)}/Au_{12(shell)}$. The Au_{13} core activates the C=O bond by an e^- transfer from Au_{13} to the O atom of the C=O group, and the Au_{12} shell provides sites for H_2 adsorption/dissociation (low coordinated Au atoms can dissociate H_2 to $2H$;[38] here the Au_{12} shell has a coordination number, $CN = 3$). Then, the Au_{13} core favours the selective hydrogenation of the C=O bond over the C=C bond. Subsequently, the weakly nucleophilic hydrogen attacks the activated C=O group, leading to the unsaturated alcohol product [Figure 8.8(A)].

The diastereoselective catalytic capability of the $Au_{25}(SR)_{18}$ clusters was also reported for the hydrogenation of a bicyclic ketone [Figure 8.8.(B)].[39] Complete selectivity for one isomer of the alcohol was obtained. Currently, industrial syntheses of cyclic alcohols are limited to metal hydride reducing agents, which only give rise to a moderate selectivity towards the more thermodynamically favourable conformation. The proposed catalytic mechanism agrees with that proposed for the selective hydrogenation of α,β-unsaturated ketones. The stereoselectivity is believed to be due to the second H addition to the C=O group. The reaction could proceed through attack to the axial position (leading to the *exo*-alcohol), through the equatorial position (*endo*-alcohol) or through both positions forming both stereoisomers. A preferred direction along the axial position is clearly seen,

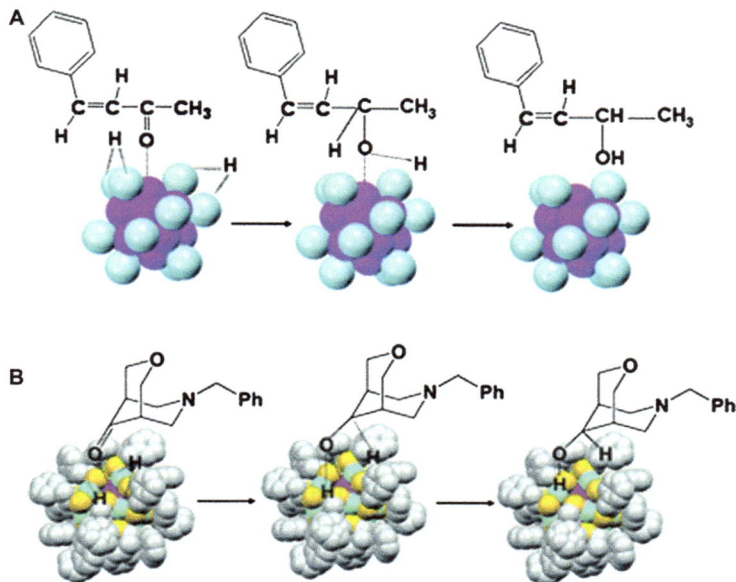

Figure 8.8 Proposed mechanism for the chemoselective hydrogenation of α,β-unsaturated ketone to unsaturated alcohol catalysed by $Au_{25}(SR)_{18}$ clusters. For clarity, the thiolate ligands are not shown. Pink: Au atoms of the core; cyan: Au atoms of the shell (A). Proposed mechanism of the stereoselective hydrogenation of bicyclic ketone using $Au_{25}(SR)_{18}$. First, activation of C=O bond and H_2. Second, H atom addition to the activated C–O group in a particular direction, and finally formation of *exo*-alcohol isomer (B).
(Reprinted from ref. 19b. Copyright © 2010 WILEY-VCH Verlag GmbH & Co. KGaA, Weinheim.)

which was associated with a spatial restriction imposed by the clusters, as well as with the activated geometry of the ketone.

8.2.1.2.2 Aldehyde and Alkene Hydrogenation. Recently, it was discovered by Maity *et al.*[40] that copper clusters (CuCLs) encapsulated into dendrimers (poly(amidoamine) dendrimers) can also catalyse the selective hydrogenation of different functional groups (C=C, C=O) in water under mild conditions (Figure 8.9). Cu_{30}@PAMAM–OH(G6) efficiently catalyses the hydrogenation of a range of carbonyl substrates (88–99% conversion with a selectivity of 100% for the alcohol) and olefins (15–22% conversion with 100% selectivity for the saturated alkanes). Due to the different catalytic activities of CuCLs they studied deeper the selective hydrogenation of cinnamaldehyde (CAL) to cinnamyl alcohol (COL). Thermodynamically, the C=C bond is easier hydrogenated than the C=O bond, but they obtained COL as a major product with 86% selectivity, and with only a 13% selectivity for 3-phenyl propanol (HCOL). They also observed that by decreasing

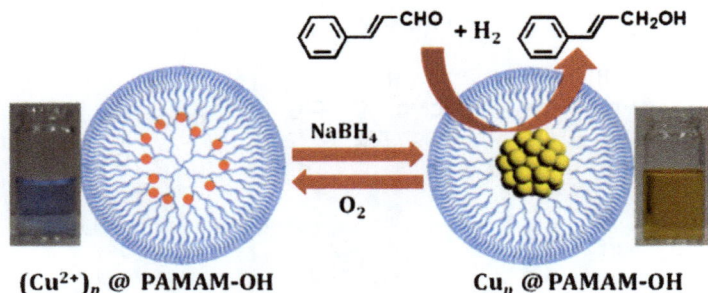

Figure 8.9 Dendrimer-encapsulated copper clusters as chemoselective and regenerable hydrogenation catalysts.
(Reprinted with permission from ref. 40. Copyright © 2013 American Chemical Society.)

the temperature the COL selectivity could be increased (from 86% to 93%). These results suggest that CuCLs can activate molecular hydrogen (*via* dissociative adsorption on the surface) displaying a similar behaviour to AuCLs. Theoretical studies corroborate this assumption predicting that H_2 can be dissociated at a corner of Cu atoms forming hydrides on the cluster surface.[41] Similar to previously discussed results for other reactions, the catalytic activity was dependent on the cluster size and, in this particular case, also on the generation of dendrimer.

8.2.1.3 Reduction Reactions

The first report about the size-dependent catalytic properties of unsupported Cu(0) CLs for a reduction reaction was carried out by Vilar-Vidal *et al.*[3a] They synthesized different CuCL sizes (from 5 to 20 Cu core atoms) by an electrochemical method and studied their catalytic properties using the reduction reaction of methylene blue (MB) by hydrazine (N_2H_4) as reaction model. A good catalytic performance was observed using smaller cluster sizes, such as Cu_5CLs (very efficient) and $Cu_{13}CLs$ (less efficient), whereas larger $Cu_{20}CLs$ were found to be catalytically inactive. The observed size-dependent catalytic behaviour was explained by an energy mechanism—similar to those proposed before (see Figure 8.3)—taking into account that the e^- transfer from the donor (N_2H_4) to the acceptor (MB) proceeds through the conduction band of the CuCLs, as can be schematically seen in Figure 8.10. They also showed that the position of the HOMO/LUMO frontier orbitals for the three clusters depend not only on the E_{gap} but also on the position of their Fermi levels. Assuming that the Fermi level of CuCLs can be described by the expression: E_F (eV) $= -E_F$ (Cu_{bulk}) $+ CN^{-1/3}$ ($C = 0.95$), which agrees with theoretical calculations,[42] the LUMO of Cu_5 and Cu_{13} would be located above the MB redox potential and below the N_2H_4, allowing the clusters to be catalytically active.

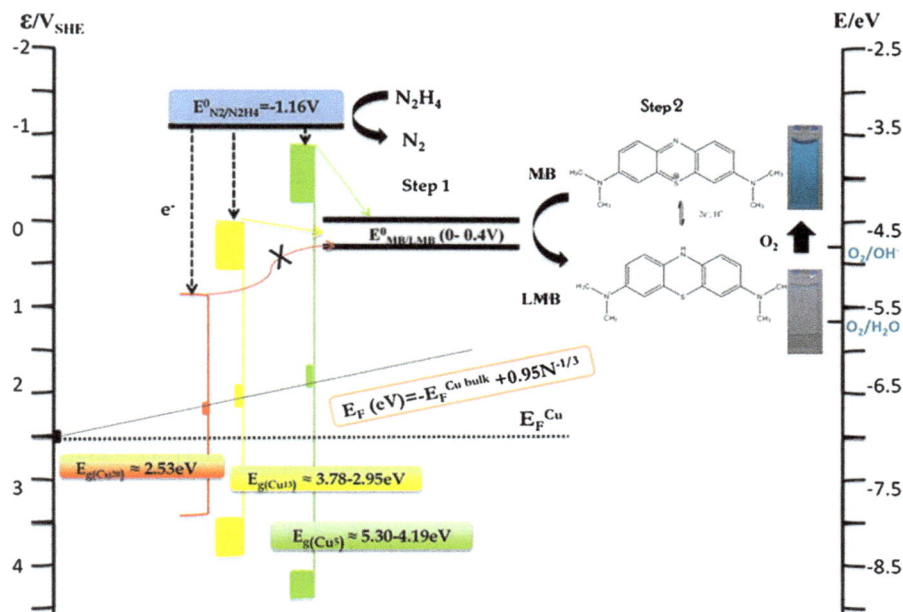

Figure 8.10 Schematic energy diagram showing the catalytic activity of different CuCLs (Cu_5, Cu_{13} and Cu_{20}) used for MB reduction by N_2H_4. (Reprinted with permission from ref. 3a. Copyright © 2012 American Chemical Society.)

However, the largest Cu_{20} clusters are inactive because their LUMO frontier orbitals are located below the MB redox potential. The same reaction was also studied by Shen *et al.*[43] in supramolecular hydrogels. They attribute the observed catalytic effects to the presence of small CuCLs ($\sim Cu_9$–Cu_{13} cores) in the samples identified by MALDI-TOF mass spectrometry. Unsupported gold clusters[44] ($Au_{25}L_{18}$ where L = different thiol size chains) were also studied for the reduction reaction of 4-nitrophenol using $NaBH_4$. Results showed higher rate constants with a decrease of the chain lengths, attributing this effect to an easy access for the substrate (4-nitrophenol) to the cluster surface. They also reported the quenching of the catalytic activity by O_2 due to the scavenging of e^-. This corroborates the results of Jin *et al.*[21] about the easy access of O_2 to the cluster surface, allowing e^- transfer from the clusters to the oxygen.

8.2.1.4 Degradation Reactions

Nowadays there are many pollutants in the environment, such as organic halides,[45] heavy metals,[46] halocarbons, CCl_4,[47] *etc.*, which have to be removed or degraded into non-toxic products. Noble metal and oxide nanoparticles are examples of catalysts, which have been used in the degradation of halocarbons.[48] Due to the novel catalytic properties of metal clusters, there is also a recent interest in their use in such degradation reactions. Bootharaju *et al.*[49] used sub-nanometer Ag$_9$ clusters protected with mercaptosuccinic acid (MSA)[50] (both supported on alumina, Al$_2$O$_3$, and unsupported) in the catalytic degradation of CCl$_4$, CHCl$_3$ and C$_6$H$_5$CH$_2$Cl at room temperature in IPA (isopropylalcohol). They observed the presence of AgCl (indicating the removal of a MSA monolayer from the clusters) and also a decrease of the pH during the reaction, which was ascribed to IPA transformation into acetone. They concluded that the ligand monolayer of the cluster was chemically transformed during the reaction and assumed that the global reaction proceeds as follows (monolayer changes are not included):

$$Ag_9MSA_7(aq) + (CH_3)_2CHOH + CCl_4 \rightarrow C + AgCl \Downarrow + (CH_3)_2CO + 2H^+$$

AgCLs not only catalyse the oxidation of the IPA to acetone but also activate the halocarbon (C–Cl bond). The proposed mechanism proceeds through the removal of the e$^-$ released in the oxidation of IPA by activated CCl$_4$, leading to the formation of Cl$^-$ and other active Cl species, which may act as oxidizing agents. Then Cl$^-$ replaces MSA (which is degraded to sulfite compounds) forming AgCl. The authors observed that the catalytic efficiency of the clusters was considerably higher than the corresponding nanoparticles, but the non-reusability was their main limitation. In spite of the great efforts made by the authors, the mechanism of this reaction still remains unclear. Because of this we propose, in line with previous discussions, the following alternative mechanism:

As can be seen in Figure 8.11, the mechanism consists of e$^-$ transfer from the MSA ligands to the LUMO orbital of the Ag$_9$ metal core. The e$^-$ can be

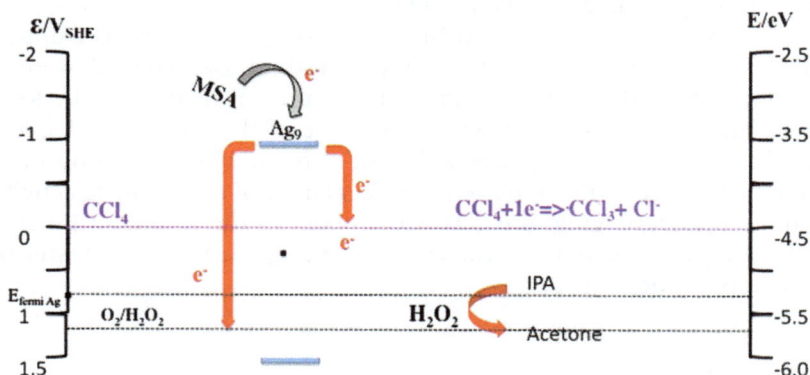

Figure 8.11 Alternative mechanism for CCl$_4$ degradation showing the e$^-$ transfer from MSA to the LUMO orbital of AgCLs. The e$^-$ can be further transferred to the O$_2$ forming the corresponding superoxide or to the CCl$_4$ with consequent degradation.

further transferred to the CCl_4 forming radicals ($^{\cdot}CCl_3$), which can also act as oxidants of the Ag_9 clusters, once the protecting ligands are removed. The mechanism would also explain the oxidation of the IPA to acetone by e^- transfer from the metal core to the O_2 forming H_2O_2, which would be the oxidant species.

8.2.1.5 Other Reactions

Recently, Oliver-Meseguer *et al.*[51] reported the size-dependent specific catalytic activity of very small gold clusters (Au_nCLs, $n \leq 9$) for the ester-assisted intermolecular hydration of alkynes. They showed that Au_nCLs ($3 \leq n \leq 5$) could catalyse the reaction with efficiencies up to 5 orders greater than those previously reported. On the other hand, the Au_nCLs ($7 \leq n \leq 9$) were found to be catalytic active for the bromination of arenes. The catalytic activity mechanism of these small AuCLs was ascribed to their Lewis character, with an available LUMO orbital for the nucleophilic interaction. At the same time, they checked the catalytic activity using exogenous Au_nCLs ($n = 5$ and $n = 8$) stabilized with the dendrimer poly(amineamide-ethanol) (PAMAM-OH). They observed the specificity of Au_5CLs for the ester-assisted intermolecular hydration of alkynes and the corresponding specificity of Au_8CLs for the bromination of arenes, despite being of a lower catalytic efficiency than that observed with their *in situ* counterparts, which could be attributed to cluster shielding by the relatively voluminous dendrimers used as capping agents.

8.2.2 Heterogeneous Catalysis

Heterogeneous catalysis refers to a reaction carried out using supported catalysts in a different thermodynamic state than the reaction itself. Because recyclability is one of the main issues in catalysis, the use of supported catalysts can avoid such a problem. Different metal clusters adsorbed on metal oxides such as: silica (SiO_2), alumina (Al_2O_3), titanium dioxide (TiO_2), cerium oxide (CeO_2) have shown exceptional properties in different reaction catalysis.[52] The development of new heterogeneous catalysts is still under investigation, because the controlling factors of the catalysis have not yet been fully understood. Here, we will summarize the most important recent studies about the use of metal clusters in heterogeneous catalysis, making special emphasis on the influence of the complex interactions between cluster and support on the catalytic properties.

8.2.2.1 Oxidation Reactions

8.2.2.1.1 Carbon Monoxide Oxidation. Gold clusters on oxide supports are being used to catalyse an increasing number of reactions. The origin of the catalysis has different interpretations. Some authors have emphasized the important role of the size and morphology of the clusters on their catalytic properties. Others, however, postulate the metal oxidation state and the supporting material as the main important parameters.

There is continuous interest in the relatively simple low temperature oxidation of CO for industrial application as CO removal from H_2 in fuel cells.[53] CO oxidation using AuNPs on oxide supports has been extensively investigated since the initial studies developed by Haruta *et al.*[7b,54] Herzing *et al.*[52a] carried out a study of the influence of Au particle size on this reaction. For this purpose they used two samples with mean particle sizes of 5.4 nm (1) and 7.0 nm (2). The catalytic test revealed an unexpected behaviour: inactivity of sample 1 and high activity of sample 2. The reason for this behaviour was attributed to the presence of sub-nanometer clusters on sample 2, opening a new investigation field where Au clusters can act as catalysts for CO oxidation. More recently, Nie *et al.*[55] reported the activity of well monodispersed $Au_{25}(SCH_2CH_2Ph)_{18}$ clusters (1 nm metal core) supported by impregnation on TiO_2, Fe_2O_3 and CeO_2 for carbon monoxide (CO) oxidation (Figure 8.12). One of the advantages of using such monodisperse clusters is that their structure[21] is well known, allowing investigation into the connection between the catalytically active sites and the reaction mechanism. The catalytic activity of these supported AuCLs was studied at different temperatures (from 20 to 200 °C) and different pre-treatment conditions (N_2 or O_2). Unexpectedly, the $Au_{25}(SR)_{18}/TiO_2$ catalyst shows almost no catalytic activity, even up to 200 °C, which is strikingly different from conventional Au/TiO_2 catalysts that exhibit extraordinary activity as reported by many groups.[56] Among the three types of supported catalysts, $Au_{25}(SR)_{18}/CeO_2$ gave the highest CO oxidation conversion, achieving 50% conversion at ~ 150 °C [Figure 8.12(A)]. Besides this interesting result, it was also observed that the catalytic activity was improved remarkably by thermal pre-treatment of the catalysts with O_2 (≈ 150 °C without removal of the surface thiolate ligands), reaching a conversion of 92.4% under mild conditions (80 °C) (Figure 8.12(C).

Although they do not offer a deep insight into the catalytic mechanism (neither identification of the exact form of the active oxygen species, nor the mechanism for CO and O_2 activation), they suggest that the interface of $Au_{25}(SR)_{18}/CeO_2$ is critical for the catalytic reaction. The mechanism they proposed is as follows: O_2 adsorption at the $Au_{25}(SR)_{18}/CeO_2$ interface (due to the capability of CeO_2 to activate O_2 molecules by the rich oxygen vacancies[57]) with conversion to peroxyl or hydroperoxyl species. The small activity of the other substrates allows them to conclude that the perimeter sites between the substrate (CeO_2) and the cluster (Au_{25}) are the actives sites, instead of the external surface of the Au_{25}. The little activity of $Au_{25}(SR)_{18}/TiO_2$ (even after 250 °C pre-treatment in O_2 atmosphere) was a surprise and it remains to be elucidated in the future. In line with previous cases, we also propose an alternative mechanism here (similar to that presented in section 8.1.1.2).

According to this mechanism, the substrate can display active sites which are responsible for its activity/inactivity depending on the corresponding

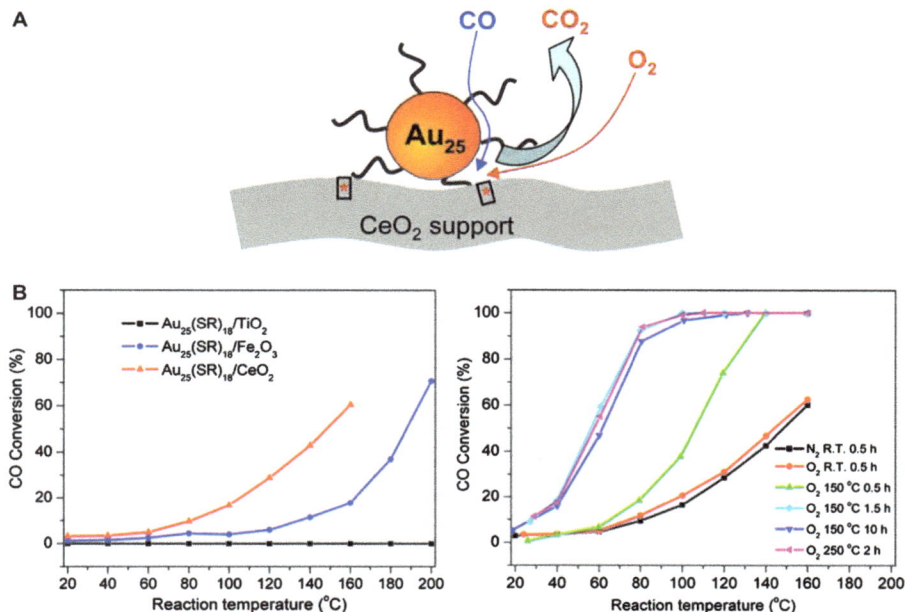

Figure 8.12 Proposed model for CO oxidation at the perimeter sites of $Au_{25}(SR)_{18}/$ CeO_2 catalyst. (A) Reaction temperature dependence of CO conversion using $Au_{25}(SR)_{18}/MO_x$ catalyst. Pre-treatment condition: N_2 at room temperature (RT) for 0.5 h. (B) Reaction temperature dependence of CO conversion using $Au_{25}(SR)_{18}/CeO_2$ catalyst after different pre-treatments.
(Reprinted with permission from ref. 55. Copyright © 2012 American Chemical Society.)

redox potential. For example, the redox potential of Ce^{4+}/Ce^{3+} located around ≈ 1.5 V, near to the HOMO orbital for the Au_{25} cluster, can be one key factor for its high catalytic activity. As can be seen in Figure 8.13 the greater the proximity of the redox potential to the LUMO orbital the more catalytic activity is observed: $TiO_2 < Fe_2O_3 < CeO_2$.

8.2.2.1.2 Alcohol Oxidation. Various nanocatalysts immobilized by different inorganic supports have been used in alcohol oxidation reactions up to now.[58] Recently, the possibility of using synthetic methods to prepare monodisperse sub-nanometer metal clusters, together with the possibility of their immobilization inside solid supports opened a new way to carry out heterogeneous catalysis with clusters. Due to the small size of sub-nanometer metal clusters, this kind of immobilization remains a great challenge in terms of catalytic recyclability. In this sense, Liu *et al.*[59] reported the immobilization of ≈ 1 nm Au clusters within mesoporous channels of pure silica supports (SBA-15, MCF, HMS) using triphenylphosphine-protected Au_{11} (Au_{11}:TPP) clusters as precursors (Figure 8.14). They firstly

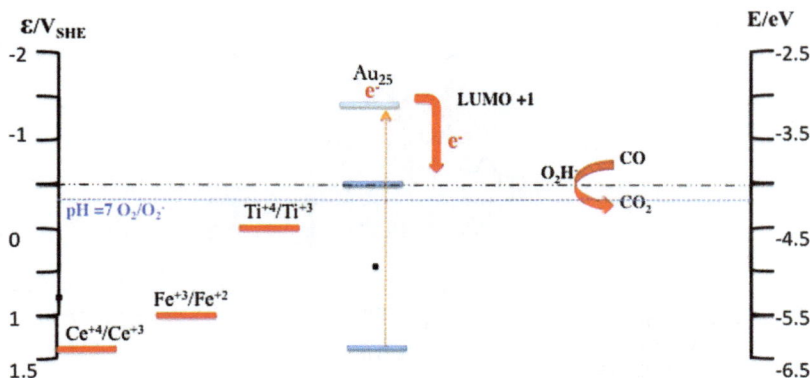

Figure 8.13 Alternative mechanism for CO oxidation with supported Au_{25} clusters. e^- transfer from the HOMO to the LUMO + 1 orbital of the Au_{25}CL by photoactivation is shown; then, this e^- can be transferred to O_2 forming the corresponding superoxide, which would be responsible for the CO oxidation.

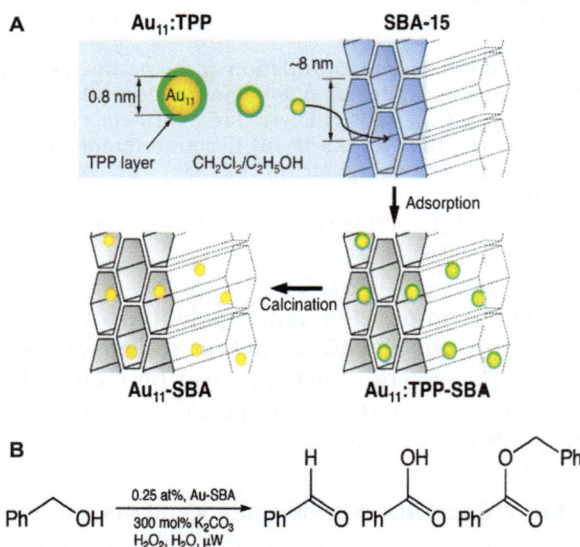

Figure 8.14 Synthesis of sub-nanometer sized AuCLs within SBA-15 using Au_{11}:TPP as precursor (A); oxidation of benzyl alcohol catalysed by clusters (B).
(Reprinted with permission from ref. 59. Copyright © 2009 American Chemical Society.)

prepared Au_{11} clusters, most likely in the form of $[Au_{11}(TPP)_8Cl_2]^+$ (≈ 0.8 nm), following reported procedures.[60] Then, clusters were deposited on SBA-15 in organic media optimizing the solvent-mediated interaction. Finally, they carried out the calcination of the composite to remove the organic ligands suppressing the aggregation of the resulting Au clusters (Figure 8.14).

The authors tested the catalytic performance of the composite Au_{11}-SBA for different alcohol oxidations (water–H_2O_2 as an oxidant) under microwave heating (small activity under aerobic conditions at ambient temperature). As an example, they carried out benzyl alcohol oxidation obtaining benzoic acid as the major product [Figure 8.14(B)]. The catalyst could be recycled at least four times without any loss of activity. Although the authors did not investigate the oxidation mechanism in detail, they observed that the presence of H_2O_2 and microwave irradiation were the key aspects affecting the catalytic performance. Although the reaction cannot be defined as a typical O_2 oxidation catalysis, it can be considered as a first approach to alcohol oxidation by clusters used as a heterogeneous catalyst.

8.2.2.1.3 Alkene Oxidation. The selective oxidation of styrene has been chosen by many researchers to study the catalytic properties of supported metal clusters. Turner and co-workers[61] observed that Au_{55} clusters supported on silica could catalytically activate O_2 for styrene epoxidation (conversion yield $\sim 25.8\%$ and epoxide selectivity $\sim 12\%$). Their results showed that AuCLs could adsorb and activate O_2 for selective oxidation (by dissociating O_2 in O ad-atoms, which initiate the partial oxidation), whereas larger particles are totally inactive. The catalyst could be recovered and reused for 2 runs without any deactivation, this behaviour being attributed to the oxidation resistance of "magic-number" clusters such as Au_{55}. Recently, Liu *et al.*[62] thought that a critical cluster size must exist for the efficient and selective oxidation of alkenes. Therefore, they studied the cluster size effect on styrene epoxidation (Figure 8.15). They obtained an excellent catalytic performance by preparing $Au_{25}(SG)_{18}$ (G = Gluthatione) clusters immobilized on hydroxyapatite (HAP; $Ca_{10}(PO_4)_6(OH)_2$) using *tert*-butyl hydroperoxide (TBHP) as oxidant (more effective than O_2 for alkene oxidations).[63] Epoxide was obtained with 92% selectivity and 100% conversion (higher conversion yield than larger analogues), but with the difference that another oxidant was used, rather than the common O_2.

Zhu *et al.*[20] further investigated the effect of cluster size immobilized in different inorganic supports, focused on selective styrene oxidation. They explored two types of supported $Au_n(SR)_m$ (R = CH_2CH_2Ph) catalyst using HAP and SiO_2 supports. The sizes of the three clusters used were: $Au_{25}(SR)_{18}$ (0.98 nm), $Au_{38}(SR)_{24}$ (1.3 nm) and $Au_{144}(SR)_{60}$ (1.6 nm). Two different oxidants were also used in order to check the activity: TBHP (*tert*-butyl hydroperoxide) and O_2. With the SiO_2-supported Au clusters they observed conversions of styrene larger than 90% with highest selectivity for benzaldehyde ($\sim 100\%$). The trends observed with the HAP-supported catalysts are similar to the SiO_2-system, including that the systems involving TBHP show higher conversions of styrene than the sole O_2 system, and that O_2 activation needs smaller clusters. TBHP can be activated even with large Au nanoparticles. As was mentioned in section 8.1.1.2., the electronic properties of

A

B

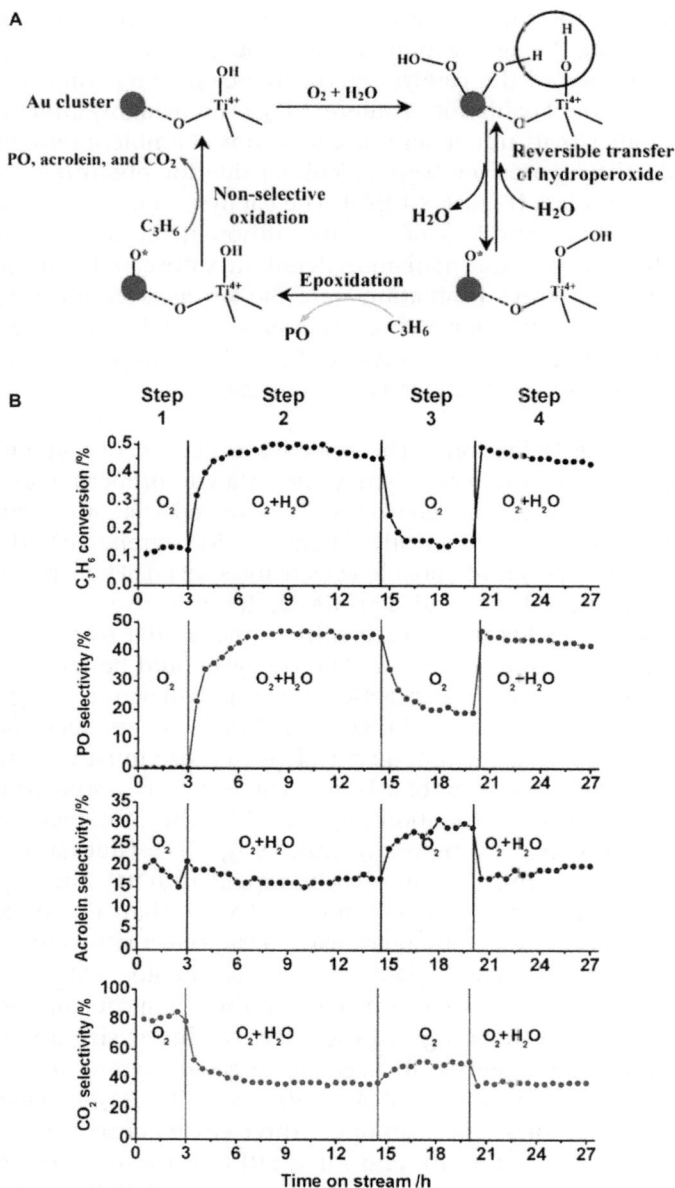

Figure 8.15 (A) A possible route for C_3H_6 epoxidation with $O_2 + H_2O$ using Au/TS-1-K1. PO = propene epoxide. (B) Enhancing effect of H_2O on propene epoxidation with O_2 using 0.10 wt% Au/TS-1-K1. Reaction conditions: catalyst 0.30 g; reaction temperature 473 K; feed gas $C_3H_6:O_2:$ $Ar = 10:10:80$ or $C_3H_6:O_2:H_2O:Ar = 10:10:2:78$; space velocity 4000 mL g^{-1} cat h^{-1}. Step 1: without H_2O; Step 2: H_2O was added to the feed gas; Step 3: the addition of H_2O was stopped; Step 4: H_2O was added to the feed gas again.

(Reprinted with permission from ref. 71. Copyright Angew. Chem., Int. Ed., 2009. Copyright © 2010 WILEY-VCH Verlag GmbH & Co. KGaA, Weinheim.)

clusters are believed to have a strong impact on their catalytic activity. Assuming that the Au_{25} structure is formed by an Au_{13} icosahedral core and an Au_{12} (positive charged) shell, they attributed the catalytic activity to O_2 activation by e^- transfer from the Au_{13} core, followed by oxidative conversion of $[Au_{25}(SR)_{18}]^-$ into neutral $[Au_{25}(SR)_{18}]^0$.

Continuing the above work, recently, Quian *et al.*[52d] studied the relationship between catalytic activity and the presence of some doping on the metal cluster for the selective epoxidation of styrene. $Au_{25}(SR)_{18}/TiO_2$ and $Pt_1Au_{24}(SR)_{18}/TiO_2$ ($R = C_2H_4Ph$) clusters supported on TiO_2 were tested for the selective oxidation of styrene with $PhI(OAc)_2$ as oxidant. Results showed an increase of the styrene conversion for the $PtAu_{24}CLs$ compared to the $Au_{25}CLs$ (from 58.9 to 90.8%). The selectivity of $PtAu_{24}CLs$ was also higher obtaining 89.9% for benzaldehyde, compared to 54% using $Au_{25}CLs$. This interesting behaviour can be explained by the drastic effects caused by the platinum doping on the cluster electronic properties.[64] Different reaction mechanisms can take place for Au clusters and monodoping Pt clusters, which must be investigated more deeply.

In spite of gold and silver being the favourite metals for metal cluster synthesis, nowadays, copper clusters[5a,65] are opening a wide range of new affordable possibilities, due to their different electronic behaviour compared to that of gold and silver. On the basis of previous results obtained by Rongchao *et al.*[52d] about how metal cluster composition can lead to drastic changes of the catalytic properties, recently, Biswas *et al.*[66] have used copper clusters as oxidation catalysts. They studied styrene oxidation using small Cu_nCLs ($n \approx 13$) protected with *O*-[2-(3-mercaptopropionylamino)ethyl]-*O'*-methylpolyethylene glycol [MPEG] supported on SiO_2, using *tert*-butyl hydroperoxide TBHP as oxidant (under similar conditions to Liu *et al.*[62]). They observed that, similarly to the Au_NCLs, the selectivity of Cu_NCLs goes preferentially to benzaldehyde (67%) with $\sim 72\%$ overall conversion of styrene (similar values to the ones obtained using AuCLs).

8.2.2.1.4 Epoxidation Reactions.

The epoxidation of hydrocarbons is very important due to its use in the production of different chemicals. Concretely, propylene epoxidation is one of the most studied due to its importance in polyurethane and polyol production, and also due to the high cost of the employed oxidants (environmentally unfriendly). Propylene one-step epoxidation using O_2 and H_2 in the gas phase[67] and in the liquid phase[68] is an attractive process due to its simplicity; it is green (H_2O as the only by-product) and has affordable characteristics (O_2) for a sustainable industry. However, the relatively low efficiency using H_2 (<50%) and the risk of explosion (using O_2 and H_2) still limits the application of this process. In this context, the use of metal clusters trying to apply soft reaction conditions is a very appealing novel approach. Considerable attention has been paid since Haruta *et al.*[69] discovered the catalytic properties of small gold nanoparticles (AuNPs) for direct propylene oxidation by an O_2–H_2 mixture. The selectivity of these AuNPs supported onto different substrates is extremely sensitive to their size and shape, obtaining mainly propane with AuNPs ≤ 2 nm and CO_2 and H_2O with AuNPs ≥ 5 nm. From an economical point of view, the use of H_2 should be as low as possible. Focused on this objective, Huang *et al.*[70] employed AuCLs (size ≤ 2 nm) supported on alkaline-treated titanosilicalite-1 (TS-1) for propylene epoxidation with $O_2 + H_2O$. They found that Ti sites are indispensable for AuCLs to catalyse the reaction,[71] obtaining a propylene conversion near 1.0% with a selectivity through the epoxide $\approx 50\%$. The possible route [Figure 8.15(A)] consists of a first reaction between O_2 with H_2O over the gold cluster surfaces to produce –OOH species. Then, the –OOH species are reversibly transferred from the gold cluster surfaces to neighbouring Ti sites to form Ti–OOH species, which would be responsible for subsequent C_3H_6 epoxidation.

Almost at the same time Lee *et al.*[72] reported the great catalytic properties of sub-nanometer gold clusters (Au_6–Au_{10}), supported onto amorphous Al_2O_3 by laser ablation, for propylene oxidation in absence of H_2, being replaced by safe H_2O vapour. Their experimental and theoretical study revealed the partial oxidation of propylene using alumina (instead of TiO_2) as support, and the importance of using water vapour as an efficient method to maintain the hydroxyl equilibrium at the surface. Theoretical simulations assign a more active character to the Au_n/Al_2O_3 interface than to the Au/TiO_2. Lately, they also reported the catalytic activity of Ag_n^+ ($n = 2$ to 4) clusters (with dominant trimer contribution) supported on amorphous alumina films for the propylene epoxidation reaction at low temperatures.[73] They suggested that addition of H_2O to the feed stream could reduce the acrolein production and enhance the epoxidation. Based on theoretical calculations, the reaction mechanism was assumed to go through an Ag–O intermediate, which was confirmed by an *in situ* XPS analysis during the progress of the reaction. A positive shift of the Ag 3d core levels, with values lower than the AgCLs was reported. Therefore, two different oxygen sites were suggested for the O_2 dissociation: (1) the three-fold site on Ag_3, and (2) the alumina at the interface with Ag_3. More recently, Haruta *et al.* continued the studies about the effect of different parameters on propylene epoxidation, paying attention to the deposition method,[74] as well as the role of the basic salts used.[70,71] The experiments pointed out the higher catalytic activity through the AuCLs deposited onto the TS-1 (microporous titanosilicate) than the one inside their channels, and also the higher selectivity to the formation of the propylene oxide using basic potassium salts, like CH_3COO^-, CO_3^{2-}, *etc.*, instead of using common salts, like K_2NO_3 or KSO_4, which cannot produce propylene oxide. In conclusion, many factors have a deep influence on the mechanism of the catalytic reaction, such as the support, the reaction medium and the cluster size. Therefore, deeper studies would be necessary to clarify the mechanisms involved in this very important reaction.

8.2.2.1.5 Cyclohexane Oxidation. One of the challenges nowadays in the chemical industry is the aerobic oxidation of cyclohexane due to the large use of its oxidation products (cyclohexanone and cyclohexanol (Ketone/Alcohol or KAoil)), as key intermediates in the nylon industry. Many studies have successfully reported the employment of supported/gold clusters for this reaction: Au/ZSM-5,[75] Au/Al$_2$O$_3$, Au/TiO$_2$ and Au/SB-15,[76] the last one being the system which exhibits the best performance, with 71% selectivity and 4% conversion. Recently, Liu *et al.*[77] used this reaction as a model to determine the catalytic activity of size-selected Au$_n$ clusters supported on HAP (hydroxyapatite). They used different glutathionate (GS)-protected Au$_n$ clusters Au$_n$(SG)$_m$ with $(n,m) = (10,10)$, $(18,14)$, $(25,18)$, and $(39,24)$, supported onto HAP. The catalytic properties were investigated under solvent-free conditions using O$_2$ as the oxidant and TBHP (*tert*-butylhydroperoxide) as initiator. They observed an expected cluster size activity dependence. Initially, the turnover frequency increased with increasing the cluster size obtaining the same selectivity for both products with the maximum conversion value for the Au$_{39}$ cluster ($\approx 15\%$). Then, the activity decreased dramatically when increasing the size. The highest catalytic activity observed in the case of Au$_{39}$ was associated with its unique electronic structure.

8.2.2.1.6 Cyclization Reactions.

One important cyclization example is the Ullmann-type reaction. This is an aryl halide homocoupling reaction, normally used for the preparation of pharmaceuticals and polymers, which are normally catalysed by Pd, Ni and Cu complexes. Nowadays, due to increasing attention on the catalytic properties of metal clusters, Li *et al.*[78] studied the catalytic activity of Au$_{25}$(SR)$_{18}$ clusters (R = CH$_2$CH$_2$Ph) supported on oxides (CeO$_2$, TiO$_2$, SiO$_2$, Al$_2$O$_3$) for such a reaction. No noticeable differences between the supports were observed in this case. Besides, unsupported Au$_{25}$CLs showed a much lower catalytic activity, which was ascribed to the gradual decomposition of such clusters at the high temperatures used for this reaction (130 °C). Different studies carried out with AuCLs/CeO$_2$ reported their recyclability and also versatility for different homocoupling reactions. An important result in this study has to been pointed out: the use of a CO$_3$$^{2-}$ instead of PO$_4$$^-$ as counterion clearly increases the conversion yield. This behaviour agrees with that reported for Haruta for the propylene epoxidation reaction. The above catalytic results demonstrate the good e$^-$ transfer capability of the Au$_{25}$(SR)$_{18}$ cluster. The mechanism can be described as follows: first, the substrate is adsorbed on the surface of the Au$_{25}$(SR)$_{18}$/CeO$_2$ catalyst to form [Au$_{25}$](R$_0$I)$_2$ (R$_0$ = Ph). Then, the C–I bond becomes activated on the cluster Au$_{12}$ (positive charged Au$^{\delta+}$, $0 < \delta < 1$), giving rise to Au$_{25}$Ph$_2$I$_2$ intermediates (active sites for activating iodobenzene). Finally, the reductive elimination of [Au$_{25}$Ph$_2$]I$_2$ generates the final product: biphenyl.

8.2.2.1.7 Other types of oxidation reactions.

Very recently, Corma *et al.*[79] studied the aerobic oxidation of thiophenol (Ph–SH) to benzene disulfide using AuCLs supported onto multiwalled carbon nanotubes. They showed that whereas single isolated gold atoms and gold nanoparticles display negligible catalytic activity, gold clusters (from 5 up to 10 atoms) show activities similar to sulfhydril oxidase enzymes.[80] The proposed reaction mechanism again assumes O_2 activation by the small AuCLs. A time resolved study of the reaction showed that the reaction begins with an induction period, ascribed to the majority population of inactive gold atoms. Then, the reaction proceeds with a high reaction rate, observing the formation of small AuCLs (4–13 atoms). Finally, in the last stage of the reaction, the activity of the catalyst stops, which is associated with the aggregation of metal clusters forming inactive nanoparticles. To confirm this mechanism they performed the same experiment using *ex situ* synthesized AuCLs of different sizes, observing that the active ones were Au_nCLs with $n = 5$–10 atoms. Again a similar mechanism to that presented in section 8.2.1.1.2 can be used to explain the existence of an optimal cluster size for this reaction taking into account that the AuCL LUMO has to be located between the Ph–SH HOMO (≈ -0.5 V *vs.* SHE) and the O_2 LUMO (≈ 0.6 V *vs.* SHE), which requires a minimum size for the Au_nCLs ($n = 5$ atoms). But, a further increase of the AuCL size above this limit involves an increase of the bonding strength between the AuCLs and the Ph–SH (due to an increase of the energy difference between the AuCL LUMO and Ph–SH HOMO), giving rise to an inactivation of the cluster for attracting the O_2 molecules when $n >\approx 10$ atoms.

8.2.2.2 Hydrogenation Reactions

Recyclability of selective hydrogenation catalysts remains a challenging objective nowadays. Zhu *et al.*[19b] explored this field after observing complete selectivity to the formation of the unsaturated alcohol, which was possible using unsupported $Au_{25}(SR)_{18}$ clusters ($R = CH_2CH_2Ph$). They prepared oxide-supported catalysts (Fe_2O_3, TiO_2, SiO_2) simply by soaking the oxide powder into a solution of $Au_{25}(SR)_{18}$ clusters. They found that the supported clusters show significantly higher conversion yields (43, 40 and 19% respectively) than the unsupported clusters (22%). The authors also observed that the type of support has an influence on the catalytic activity: $Fe_2O_3 > TiO_2 > SiO_2$. Gold sub-nanometer clusters supported on various metal oxides (such as: γ-Al_2O_3, La_2O_3 and CeO_2) have also been recently reported by Ohyama *et al.*[81] as potential catalysts for the hydrogenation of 2-hydroxymethyl-5-furfural (HMF). The selective hydrogenation of the aldehyde produces 2,5-bis (hydroxymethyl)-furfuran (BHF), which is very useful for the production of resin additives, polymers, *etc.*[82] The results showed that the acid–base properties of the different metal oxide supports have strong effects on the selective-acid sites, leading to ring opening. They obtained the best results for Au/Al_2O_3, showing that the basic properties of Al_2O_3 can display an important role on the selectivity mechanism. Besides, it was also observed that the catalytic activity drastically increases when the cluster size

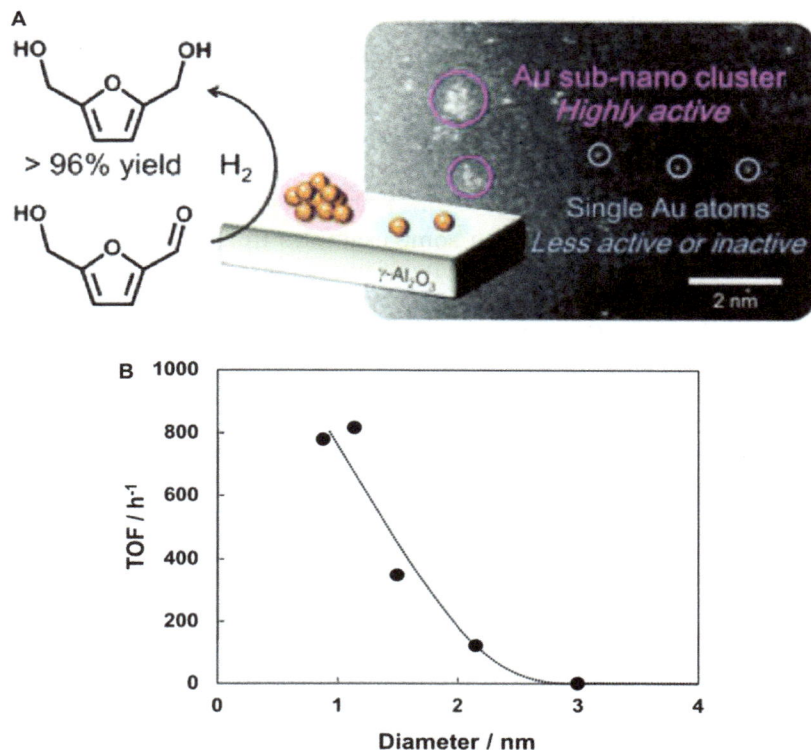

Figure 8.16 HMF hydrogenation reaction and HAADF STEM images of Au/Al$_2$O$_3$ prepared clusters. Some single gold atoms are indicated by blue circles and clusters by magenta circles (A). TOF dependence on Au particle size of Au/Al$_2$O$_3$ for HMF hydrogenation (B). (Reprinted from ref. 81.)

decreases (Figure 8.16B), and that no activity was observed for gold ions themselves. Different cluster metals have also been studied supported onto Al$_2$O$_3$ (such as Pt, Pd, Cu and Ag, although cluster sizes were not reported), but no good results were obtained in such cases. It is assumed that the Au$_{13}$ cluster is the most active species because it provides highly efficient adsorption sites for the C=O group of HMF and also to the H$_2$ to react selectively.

8.2.2.3 Reduction Reactions

Leelavathi *et al.*[83] reported the catalytic activity of Ag$_7$ and Ag$_8$ clusters protected with MSA, synthesized by interfacial etching and supported on different substrates (SiO$_2$, Fe$_2$O$_3$, Al$_2$O$_3$ and TiO$_2$), for *p*-nitrophenol reduction with NaBH$_4$. It was observed that the substrate and temperature display a strong effect on the reaction rate. The catalyst efficiency follows the order $SiO_2 > TiO_2 > Fe_2O_3 > Al_2O_3$. They did not report any reaction mechanism, but a similar mechanism to the one proposed for homogeneous catalysis could be applied (see section 8.1.3), *i.e.* e^- transfer mediated by the LUMO orbital of the metal clusters.

8.3   Conclusions and Remarks

This chapter has commented on some examples from the recent literature that clearly show the intrinsic activity of unsupported and supported metal clusters for some typical reactions. In the case of oxidation reactions, a common pathway for O_2 activation clearly appears. Moreover, it is found that there is a critical size of metal clusters to activate the O_2. The critical size of the cluster seems to be related to the position of its LUMO orbital, which would be located above the O_2 redox potential in order to allow superoxide formation, as could be deduced from an alternative mechanism used as a common framework to explain different type of reactions. The solvent and the pH are also very important parameters in the oxidation reactions due to the change of the O_2 redox potential with such parameters. For example, for some types of oxidations the reaction cannot take place under acidic conditions. With regard to the activity of clusters, an issue that has not been sufficiently studied is the role of the protecting ligand. Some studies reveal that the ligand does not affect the activity, but if one compares different reactions, one can clearly see that bulky ligands, such as dendrimers or polymers, display a big steric effect, hindering the accessibility of the reactants to the cluster surface. Another interesting parameter is the supporting material. Although, in some cases, an increase of the activity for supported clusters has been reported, before achieving any conclusion it should be important to carry out more studies of the effect of the support in a wide number of reactions. Until now the temperature has not been paid too much attention, but it was observed that some clusters can be degraded at high temperatures (due to the instability of their ligands), and also that, in some cases, clusters clearly increase their catalytic activity at low temperatures. Therefore, an exhaustive study in this direction would also be very desirable.

Finally, we have to point out that three important findings could be deduced from this review:

(1) Activity yields in the alkene oxidation reaction using Au_{25} clusters and Cu_{13} clusters (with different protecting ligands) were similar. This fact supports the hypothesis explored here using an alternative mechanism to explain different kinds of reactions, namely, that the catalytic activity relies on the band gap of the clusters and on their LUMO orbital position. Therefore, clusters of different metals can be useful for the same reaction by only changing their size. On the other hand, the selectivity cannot be explained with such simple alternative mechanisms, which would need the introduction of a new hypothesis like

the existence of some spatial restrictions[37,39] (from the clusters and also from the substrate), the preference of interaction of clusters with oxygen,[19,40] *etc.*, but this would need a deeper study.

(2) Different results were obtained for reactions catalysed by metal clusters supported onto different substrates indicating the additional effect of such a support on the catalytic activity of the cluster. Consequently, the best support has to be found for each specific reaction.

(3) Recent major advances in the monodisperse synthesis of ultrasmall metal clusters open a new possibility of using exogeneous well defined clusters with specific functionalities as heterogeneous catalysts. This can optimize the catalytic properties for a particular reaction in terms of activity, selectivity and recyclability.

In conclusion, we envisage that the research area of metal clusters as catalysts will grow rapidly in the next few years, and will become a very important issue for fast application in many industrial processes in the near future, improving their performance, and being more green and sustainable.

Acknowledgements

This work was supported by the MCI, Spain (MAT2010-20442; MAT2011-28673-C02-01); MINECO, Spain (MAT2012-36754-C02-01); Xunta de Galicia (Grupos Ref. Comp. GRC2013-044, FEDER Funds); and Obra Social Fundación La Caixa (OSLC-2012-007)).

References

1. J. Calvo, J. Rivas and M. A. López-Quintela, in *Synthesis of Subnanometric Nanoparticles, Encyclopedia of Nanotechnology*, ed. B. Bharat., Springer Verlag, Dordrecht, 2012, 2639–2648.
2. J. Li, X. Li, H. J. Zhai and L. S. Wang, *Science*, 2003, **299**, 864.
3. (a) N. Vilar-Vidal, J. Rivas and M. A. López-Quintela, *ACS Catal.*, 2012, **2**, 1693; (b) N. Vilar-Vidal, PhD Thesis: Synthesis, Characterization and Properties of Copper Clusters, University of Santiago de Compostela, 2012 – results to be published.
4. H. Tsunoyama, Y. Liu, T. Akita, N. Ichikuni, H. Sakurai, S. Xie and T. Tsukuda, *Catal. Surv. Asia*, 2011, **15**, 230.
5. (a) N. Vilar-Vidal, M. C. Blanco, M. A. López-Quintela, J. Rivas and C. Serra, *J. Phys. Chem. C*, 2010, **114**, 15924; (b) B. Santiago González, M. J. Rodríguez, C. Blanco, J. Rivas, M. A. López-Quintela and J. M. G. Martinho, *Nano Lett.*, 2010, **10**, 4217; (c) J. Selva, S. E. Martinez, D. Buceta, M. J. Rodriguez-Vazquez, M. C. Blanco, M. A. Lopez-Quintela and G. Egea, *J. Am. Chem. Soc.*, 2010, **132**, 6947.

6. M. B. Smith and J. March, *March's Advanced Organic Chemistry: Reactions, Mechanism and Structures*, John Wiley and Sons, Hoboken, 6th edn, 2007.

7. (a) M. Haruta, N. Yamada, T. Kobayashi and S. Iijima, *J. Catal.*, 1989, **115**, 301; (b) M. Haruta, T. Kobayashi, H. Sano and N. Yamada, *Chem. Lett.*, 1987, **2**, 405.

8. M. Comotti, C. Della Pina, R. Matarrese and M. Rossi, *Angew. Chem.*, 2004, **116**, 5936; *Angew. Chem., Int. Ed.*, 2004, **43**, 5812.

9. H. Tsunoyama, H. Sakurai, Y. Negishi and T. Tsukuda, *J. Am. Chem. Soc.*, 2005, **127**, 9374.

10. H. Tsunoyama, H. Sakurai and T. Tsukuda, *Chem. Phys. Lett.*, 2006, **429**, 528.

11. N. K. Chaki, H. Tsunoyama, Y. Negishi, H. Sakurai and T. Tsukuda, *J. Phys. Chem. C*, 2007, **111**, 4885.

12. H. Tsunoyama, N. Ichikuni, H. Sakurai and T. Tsukuda, *J. Am. Chem. Soc.*, 2009, **131**, 7086.

13. (a) A. Fielicke, G. von Helden, G. Meijer, D. B. Pedersen, B. Simard and D. M. Rayner, *J. Am. Chem. Soc.*, 2006, **128**, 6341; (b) M. Chen, Y. Cai, Z. Yan and D. W. Goodman, *J. Am. Chem. Soc.*, 2006, **128**, 6341; (c) H.-J. Freund, *Catal. Today*, 2006, **117**, 6.

14. M. Okumura, Y. Kitagawa, T. Kawakami and M. Haruta, *Chem. Phys. Lett.*, 2008, **459**, 133.

15. (a) A. Sanchez, S. Abbet, U. Heiz, W. D. Schneider, H. Häkkinen, R. N. Barnett and U. Landman, *J. Phys. Chem. A*, 1999, **103**, 9573; (b) M. Okumura, Y. Kitagawa, M. Haruta and K. Yamaguchi, *Chem. Phys. Lett.*, 2001, **346**, 163; (c) B. Yoon, H. Häkkinen and U. Landman, *J. Phys. Chem. A*, 2003, **107**, 4066; (d) B. Yoon, H. Häkkinen, U. Landman, A. S. Wörz, J.-M. Antonietti, S. Abbet, K. Judai and U. Heiz, *Science*, 2005, **307**, 403; (e) M. Okumura, Y. Kitagawa, M. Haruta and K. Yamaguchi, *Appl. Catal. A*, 2005, **291**, 37; (f) B. Yoon, P. Koskinen, B. Huber, O. Kostko, B. Issendorff, H. Häkkinen, M. Moseler and U. Landman, *Chem. Phys. Chem.*, 2007, **8**, 157.

16. T. Tsukuda, H. Tsunoyama and H. Sakurai, *Chem. – Asian J.*, 2011, **6**, 736.

17. H. Tsunoyama, T. Tsukuda and H. Sakurai, *Chem. Lett.*, 2007, **36**, 212.

18. (a) A. Corma and H. García, *Chem. Rev.*, 2002, **102**, 3837; (b) M. Alvaro, C. Aprile, A. Corma, B. Ferrer and H. Garcia, *J. Catal.*, 2007, **245**, 249.

19. (a) Y. Zhu, H. Qian, M. Zhu and R. Jin, *Adv. Mater.*, 2010, **22**, 1915; (b) Y. Zhu, H. Qian, B. A. Drake and R. Jin, *Angew. Chem., Int. Ed.*, 2010, **49**, 1295.

20. Y. Zhu, H. Qian and R. Jin, *Chem. – Eur. J.*, 2010, **16**, 11455.

21. M. Zhu, C. M. Aikens, F. J. Hollander, G. C. Schatz and R. Jin, *J. Am. Chem. Soc.*, 2008, **130**, 5883.

22. Y. Zhu, H. Qian and R. Jin, *J. Mater. Chem.*, 2011, **21**, 6793.

23. D. R. Kauffman, D. Alfonso, C. Matranga, G. Li and R. Jin, *J. Phys. Chem. Lett.*, 2013, **4**, 195.

24. (a) M. Zhu, C. M. Aikens, M. P. Hendrich, R. Gupta, H. Qian, G. C. Schatz and R. Jin, *J. Am. Chem. Soc.*, 2009, **131**, 2490; (b) M. Zhu, W. T. Eckenhoff, T. Pintauer and R. Jin, *J. Phys. Chem. C*, 2008, **112**, 14221.

25. H. Qian, Y. Zhu and R. Jin, *ACS Nano*, 2009, **11**, 3795.

26. The uncertainty of the position of the orbitals could explain why the reaction could still proceed in the case of small clusters through the LUMO orbital. In such a case, the presence of small amounts of smaller clusters in the large cluster samples could also explain the lower activity observed with the larger cluster samples.

27. D. T. Sawyer and J. S. Valentine, *Acc. Chem. Res.*, 1981, **14**, 393.

28. H. Hata, H. Shinokubo and A. Osuka, *J. Am. Chem. Soc.*, 2005, **127**, 8264.

29. A. S. Demir, O. Reis and M. Emrullahoglu, *J. Org. Chem.*, 2003, **68**, 10130.

30. A. S. Demir, O. Reis and M. Emrullahoglu, *J. Org. Chem.*, 2003, **68**, 578.

31. H. Sakurai, H. Tsunoyama and T. Tsukuda, *J. Organomet. Chem.*, 2007, **692**, 368.

32. I. Kamiya, H. Tsunoyama, T. Tsukuda and H. Sakurai, *Chem. Lett.*, 2007, **36**, 646.

33. H. Kitahara and H. Sakurai, *J. Organomet. Chem.*, 2011, **696**, 442.

34. (a) S. T. Qi, B. A. Cheney, R. Y. Zheng, W. W. Lonergan, W. T. Yu and J. G. G. Chen, *Appl. Catal., A*, 2011, **393**, 44; (b) K. Pattamakomsan, E. Ehret, F. Morfin, P. Gelin, Y. Jugnet, S. Prakash, J. C. Bertolini, J. Panpranot and F. J. C. S. Aires, *Catal. Today*, 2011, **164**, 28; (c) A. Primo, P. Concepcion and A. Corma, *Chem. Commun.*, 2011, **47**, 3613; (d) J. Fu, X. Y. Lu and P. E. Savage, *ChemSusChem*, 2011, **4**, 481.

35. (a) M. Okumura, T. Akita and M. Haruta, *Catal. Today*, 2002, **74**, 265; (b) A. Hugon, L. Delannoy, J.-M. Krafft and C. Louis, *J. Phys. Chem. C*, 2010, **114**, 10823; (c) H. Shi, N. Xu, D. Zhao and B. Q. Xu, *Catal. Commun.*, 2008, 1949.

36. P. G. N. Mertens, P. Vandezande, X. Ye, H. Poelman, I. F. J. Vankelecon and D. E. DeVos, *Appl. Catal. A*, 2009, **355**, 176.

37. (a) M. Zhu, E. Lanni, N. Garg, M. E. Bier and R. Jin, *J. Am. Chem. Soc.*, 2008, **130**, 1138; (b) Z. Wu, J. Suhan and R. Jin, *J. Mater. Chem.*, 2009, **19**, 622.

38. (a) C. G. Arellano, A. Corma, M. Iglesias and F. Sanchez, *Chem. Commun.*, 2005, 3451; (b) A. Corma, M. Boronat, S. Gonzalez and F. Illas, *Chem. Commun.*, 2007, 3371; (c) C. Mohr, H. Hofmeister, J. Radnik and P. Claus, *J. Am. Chem. Soc.*, 2003, **125**, 1905; (d) S. A. Varganov, R. M. Olson and M. S. Gordon, *J. Chem. Phys.*, 2004, **120**, 5169; (e) L. Andrews, *Chem. Soc. Rev.*, 2004, **33**, 123.

39. Y. Zhu, Z. Wu, C. Gayathri, H. Qian, R. R. Gil and R. Jin, *J. Catal.*, 2010, **271**, 155.

40. P. Maity, S. Yamazoe and T. Tsukuda, *ACS Catal.*, 2013, **3**, 182.

41. (a) G. H. Guvelioglu, P. Ma, X. He, R. C. Forrey and H. Cheng, *Phys. Rev. B*, 2006, **73**, 155436; (b) Y. Yang, J. Evans, J. A. Rodrigues, M. G. White and P. Liu, *Phys. Chem. Chem. Phys.*, 2010, **12**, 9909.

42. X. Crispin, C. Bureau, V. Geskin, R. Lazzaroni and J.-L. Brédas, *Eur. J. Inorg. Chem.*, 1999, **1999**, 349.

43. J. S. Shen, Y. L. Chen, Q. P. Wang, T. Yu, X. Y. Huang, Y. Yanga and H. W. Zhang, *J. Mater. Chem. C*, 2013, **1**, 2092.

44. A. Shivhare, S. J. Ambrose, H. Zhang, R. W. Purves and R. W. J. Scott, *Chem. Commun.*, 2013, **49**, 276.

45. B. Schrick, J. L. Blough, A. D. Jones and T. E. Mallouk, *Chem. Mater.*, 2002, **14**, 5140.

46. E. Sumesh, M. S. Bootharaju, S. Anshup and T. Pradeep, *J. Hazard. Mater.*, 2011, **189**, 450.

47. J. Farrell, M. Kason, N. Melitas and T. Li, *Environ. Sci. Technol.*, 1999, **34**, 514.

48. (a) A. S. Nair and T. Pradeep, *Curr. Sci.*, 2003, **84**, 1560; (b) A. S. Nair and T. Pradeep, *J. Nanosci. Nanotechnol.*, 2007, 7, 1871; (c) M. S. Bootharaju and T. Pradeep, *Langmuir*, 2012, **28**, 2671.

49. M. S. Bootharaju, G. K. Deepesh, T. Udayabhaskararao and T. Pradeep, *J. Mater. Chem. A*, 2013, **1**, 611.

50. T. U. B. Rao, B. Nataraju and T. Pradeep, *J. Am. Chem. Soc.*, 2010, **132**, 16304.

51. J. Oliver-Meseguer, J. R. Cabrero-Antonino, I. Dominguez, A. Leyva-Pérez and A. Corma, *Science*, 2012, **338**, 1452.

52. (a) A. A. Herzing, C. J. Kiel, A. F. Carley, P. Landon and G. J. Hutchings, *Science*, 2008, **321**, 1331; (b) Y. Liu, H. Tsunoyama, T. Akita and T. Tsukuda, *J. Phys. Chem. C*, 2009, **113**, 31; (c) K. Shimizu, R. Sato and A. Satsuma, *Angew. Chem., Int. Ed.*, 2009, **48**, 3982; (d) H. Qian, D. Jiang, G. Li, Ch. Gayathri, A. Das, R. R. Gil and R. Jin, *J. Am. Chem. Soc.*, 2012, **134**, 16159.

53. H. Imagawa, A. Suda, K. Yamamura and S. Sun, *J. Phys. Chem. C*, 2011, **115**, 1740.

54. (a) M. Haruta, N. Yamada, T. Kobayashi and S. Iijima, *Gold J. Catal.*, 1989, **115**, 301; (b) M. Valden, X. Lai and D. W. Goodman, *Science*, 1998, **281**, 1647; (c) C.-J. Jia, Y. Liu, H. Bongard and F. Schuth, *J. Am. Chem. Soc.*, 2010, **132**, 1520.

55. X. Nie, H. Qian, Q. Ge, H. Xu and R. Jin, *ACS Nano*, 2012, **6**, 6014.

56. M. C. Kung, R. J. Davis and H. H. Kung, *J. Phys. Chem. C*, 2007, **111**, 11767.

57. F. Gao, T. Wood and D. Goodman, *Catal. Lett.*, 2010, **134**, 9.

58. (a) A. Abad, C. Almela, A. Corma and H. Garcia, *Chem. Commun.*, 2006, 3178; (b) L. C. Wang, Y. M. Liu, M. Chen, Y. Cao, H. Y. He and K. N. Fan, *J. Phys. Chem. C*, 2008, **112**, 6981; (c) F. Z. Su, M. Chen, L. C. Wang, X. L. Huang, Y. M. Liu, Y. Cao, H. Y. He and K. N. Fan, *Catal. Commun.*, 2008, 1027.

59. Y. Liu, H. Tsunoyama, T. Akita and T. Tsukuda, *J. Phys. Chem. C*, 2009, **113**, 13457.
60. (a) G. H. Woehrle, M. G. Warner and J. E. Hutchison, *J. Phys. Chem. B*, 2002, **106**, 9979; (b) Y. Shichibu, Y. Negishi, T. Tsukuda and T. Teranishi, *J. Am. Chem. Soc.*, 2005, **127**, 13464; (c) Y. Yanagimoto, Y. Negishi, H. Fujihara and T. Tsukuda, *J. Phys. Chem. B*, 2006, **110**, 11611.
61. M. Turner, V. B. Golovko, O. P. H. Vaughan, P. Abdulkin, A. Berenguer-Murcia, M. S. Tikhov, B. F. G. Johnson and R. M. Lambert, *Nature*, 2008, **454**, 981.
62. Y. Liu, H. Tsunoyama, T. Akita and T. Tsukuda, *Chem. Commun.*, 2010, **46**, 550.
63. (a) D. Gajan, K. Guillois, P. Delichere, J.-M. Basset, J.-P. Candy, V. Caps, C. Coperet, A. Lesage and L. Emsley, *J. Am. Chem. Soc.*, 2009, **131**, 14667; (b) L. Luo, N. Yu, R. Tan, Y. Jin, D. Yin and D. Yin, *Catal. Lett.*, 2009, **130**, 489.
64. (a) C. M. Aikens, *J. Phys. Chem. Lett.*, 2011, **2**, 99; (b) J. Jung, S. Kang and Y.-K. Han, *Nanoscale*, 2012, **4**, 4206.
65. N. Goswami, A. Giri, M. S. Bootharaju, P. L. Xavier, T. Pradeep and S. K. Pal, *Anal. Chem.*, 2011, **83**, 9676.
66. S. Biswas, J. T. Miller, Y. Li, K. Nandakumar and C. S. S. R. Kumar, *Small*, 2012, **8**, 688.
67. (a) J. Huang, T. Takei, T. Akita, H. Ohashi and M. Haruta, *Appl. Catal. B*, 2010, **95**, 430; (b) C. Qi, J. Huang, S. Bao, H. Su, T. Akita and M. Haruta, *J. Catal.*, 2011, **281**, 12.
68. G. Jenzer, T. Mallat, M. Maciejewski, F. Eigenmann and A. Baiker, *Appl. Catal. A*, 2001, **208**, 125.
69. T. Hayashi, K. Tanaka and M. Haruta, *J. Catal.*, 1998, **178**, 566.
70. (a) J. Huang, T. Takei, T. Akita, H. Ohashi and M. Haruta, *Appl. Catal., B*, 2010, **95**, 430; (b) J. Huang, T. Takei, H. Ohashi and M. Haruta, *Appl. Catal., A*, 2012, **435**, 115.
71. J. Huang, T. Akita, J. Faye, T. Fujitani, T. Takei and M. Haruta, *Angew. Chem., Int. Ed.*, 2009, **48**, 7862.
72. S. Lee, L. M. Molina, M. J. López, J. A. Alonso, B. Hammer, B. Lee, S. Seifert, R. E. Winans, J. W. Elam, M. J. Pellin and S. Vajda, *Angew. Chem., Int. Ed.*, 2009, **48**, 1467.
73. Y. Lei, F. Mehmood, S. Lee, J. Greeley, B. Lee, S. Seifert, R. E. Winans, J. W. Elam, R. J. Meyer, P. C. Redfern, D. Teschner, R. Schlögl, M. J. Pellin, L. A. Curtiss and S. Vajda, *Science*, 2010, **328**, 224.
74. J. Huang, E. Lima, T. Akita, A. Guzmán, C. Qi, T. Takei and M. Haruta, *J. Catal.*, 2011, **278**, 8.
75. R. Zhao, D. Ji, G. Lv, G. Quian, L. Yan, X. Wang and J. Suo, *Chem. Commun.*, 2004, 904.
76. B. P. C. Hereijgers and B. M. Weckhuysen, *J. Catal.*, 2010, **270**, 16.
77. Y. Liu, H. Tsunoyama, T. Akita, S. Xie and T. Tsukuda, *ACS Catal.*, 2011, **1**, 2.

78. G. Li, C. Liu, Y. Lei and R. Jin, *Chem. Commun.*, 2012, **48**, 12005.
79. A. Corma, P. Concepción, M. Boronat, M. J. Sabater, J. Navas, M. J. Yacaman, E. Larios, A. Posadas, M. A. López-Quintela, D. Buceta, E. Mendoza, G. Guilera and A. Mayoral, *Nat. Chem.*, 2013, **5**, 775.
80. (a) K. L. Hoober and C. Thorpe, *Biochemistry*, 1999, **38**, 3211; (b) J. Jaje, H. N. Wolcott, O. Fadugba, D. Cripps, A. J. Yang, I. H. Mather and C. Thorpe, *Biochemistry*, 2007, **46**, 13031.
81. J. Ohyama, A. Esaki, Y. Yamamoto, S. Arai and A. Satsuma, *RSC Adv.*, 2013, **3**, 1033.
82. J. N. Chheda, G. W. Huber and J. A. Dumesic, *Angew. Chem., Int. Ed.*, 2007, **46**, 7164.
83. A. Leelavathi, T. U. B. Rao and T. Pradeep, *Nanoscale Res. Lett.*, 2011, **6**, 123.

CHAPTER 9

Metal Nanoclusters: Size-Controlled Synthesis and Size-Dependent Catalytic Activity

YIZHONG LU[a,b] AND WEI CHEN[*a]

[a] State Key Laboratory of Electroanalytical Chemistry, Changchun Institute of Applied Chemistry, Chinese Academy of Sciences, Changchun 130022, Jilin, China; [b] University of Chinese Academy of Sciences, Beijing 100039, China
*Email: weichen@ciac.ac.cn

9.1 Introduction

Metal nanoparticles (1–100 nm), including nanocrystals and nanoclusters, have aroused significant scientific research interest in nanoscience and nanotechnology in recent years due to their importance in both fundamental science and potential applications such as catalysis,[1] optics,[2] molecular imaging,[3] chemical sensing,[4] and nanomedicine,[5] *etc.* Among these, nanoclusters, which consist of only several to tens of metal atoms and possess sizes comparable to the level of the Fermi wavelength of electrons, are a fascinating area of widespread interest in nanomaterials. Because of the ultrasmall size (<2 nm) and the resulting quantum confinement effect, such metal nanoclusters exhibit discrete, molecular-like electronic structures and unique properties, such as enhanced photoluminescence, HOMO–LUMO electronic transition, intrinsic magnetism, and high catalytic properties that are fundamentally different from those of their larger nanocrystal

RSC Smart Materials No. 7
Functional Nanometer-Sized Clusters of Transition Metals: Synthesis, Properties and Applications
Edited by Wei Chen and Shaowei Chen
© The Royal Society of Chemistry 2014
Published by the Royal Society of Chemistry, www.rsc.org

counterparts.[6] Up to now, various metal clusters, including Au, Ag, Pt, Pd, and Cu, have been intensively studied, from their synthesis, characterizations, and properties to potential applications, which have been systematically reviewed in our recent critical review.[7] Among various potential applications, the catalytic application of metal nanoclusters has attracted much more attention over past years. Recently, intense efforts have been made to unravel the origin of the catalytic properties of atomically precise metal nanoclusters. Due to their high monodispersity, clear structure, and definite composition, metal nanoclusters can serve as model catalysts to help us easily understand the relationship between the observed catalytic performance and the structure and intrinsic properties of individual nanoclusters. Overall, atomically precise metal nanoclusters are expected to become a promising class of model catalysts for fundamental catalysis research. In this chapter, we will first choose several typical atomically precise gold nanoclusters, including $Au_{25}(SR)_{18}$, $Au_{38}(SR)_{24}$, $Au_{102}(SR)_{44}$, and $Au_{144}(SR)_{60}$, and Pt, Pd, Ag, and Cu nanoclusters to illustrate the size-controlled synthesis. We will then demonstrate the recent progress in size-dependent catalytic applications of these nanoclusters, for instance, CO oxidation, the oxygen reduction reaction, aerobic oxidation, and so on. Finally, a brief conclusion and an outlook on the future research challenges for size-controlled synthesis and size-dependent catalytic applications of metal nanoclusters will be provided.

9.2 Size-Controlled Synthesis of Metal Nanoclusters

The pioneering work by Brust and co-workers[8] in 1994 provides a convenient and efficient method to synthesize monolayer-protected metal nanoclusters. Since then, the Brust–Schiffrin method and the following modified synthetic strategies have been widely applied to the syntheses of various metal nanoclusters with different core size and composition. With such methods, the particle dimension, size dispersity, and surface functionality can be easily manipulated by changing the experimental conditions. Moreover, the resulting thiolate-protected products exhibit extraordinary stability to air, long-term storage, solvents, temperature, and concentration extremes.[9] In the classical method, metal precursors are first dissolved in an aqueous solution and then transferred to an organic phase (mostly toluene) by phase-transferring reagents such as tetraoctylammonium bromide (TOABr). Subsequently, organic protecting ligands and reducing agents are added into the solution to generate metal nanoclusters. The reaction mechanism was suggested as follows:[8]

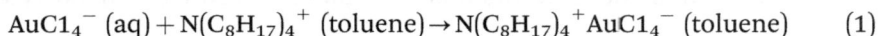

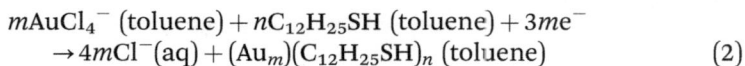

$$AuCl_4^- \text{ (aq)} + N(C_8H_{17})_4^+ \text{ (toluene)} \rightarrow N(C_8H_{17})_4^+ AuCl_4^- \text{ (toluene)} \tag{1}$$

$$
\begin{aligned}
mAuCl_4^- \text{ (toluene)} &+ nC_{12}H_{25}SH \text{ (toluene)} + 3me^- \\
&\rightarrow 4mCl^- \text{(aq)} + (Au_m)(C_{12}H_{25}SH)_n \text{ (toluene)}
\end{aligned}
\tag{2}
$$

where BH_4^- is the reducing agent, m and n are the mole numbers of the metal precursor and the protecting ligand, respectively. Such a preparation process is defined as a two-phase method. Brust *et al.*[10] also developed a one-phase method where the solvents used are usually polar solvents, such as tetrahydrofuran (THF). The Brust–Schiffrin synthesis usually produces metal clusters whose core size, shape, and atomic packing properties are determined in large part by the thiolate ligands[11] used and the thiolate-to-gold ratio.[12] Recently, there have also been other novel and unique methods developed for synthesizing metal nanoclusters with controlled core size,[7] such as template-based synthesis methods, ligand exchange reaction from phosphine-stabilized metal nanoclusters, precursor- or ligand-induced etching of metal nanoparticles, microemulsion methods, electrochemical synthesis, microwave-assisted synthesis, dimethylformamide (DMF)-based reduction methods, solid-state route, radiolytic approach, photoreduction synthesis method, and so on.

However, despite the significant progress that has been achieved in metal cluster syntheses, most reported synthetic procedures still suffer from the production of a mixture of different cluster sizes and, often, a low yield of nanoclusters with the specific size. In most preparation, in order to obtain atomically precise clusters, the crude products have to be separated *via* various techniques, such as fractional crystallization,[13] chromatography,[14,15] solvent extraction,[16] and polyacrylamide gel electrophoresis (PAGE),[17,18] *etc.* The difficulty in purifying clusters has become a major obstacle to the practical applications of the metal cluster materials. Thus, it is of crucial importance to develop novel synthetic strategies that allow for high yield synthesis of monodisperse metal clusters at the atomic level without the need for complicated size separation processes.[19] At the same time, for in-depth studies on the quantum size effects, structure–property correlations, and the novel physical and chemical properties of these metal nanoclusters, it is of paramount importance to first achieve success in synthesizing homogeneous, high-quality, and atomically monodisperse metal nanoclusters with high yield.

Among various metal nanoclusters, gold nanoclusters have drawn significant research interest in past decades due to their extraordinary chemical stability as well as a wide range of potential applications. Recently, there have been major research advances in the synthesis of atomically precise magic Au nanoclusters which show high thermodynamic and chemical stability with high yield, and the geometrical structures and origin of stability of these nanoclusters have also been well understood. Here, we first choose several typical gold nanoclusters with well-defined core size and number of surface protecting ligands to illustrate the size-controlled synthesis developed in recent years. It should be noted that, due to the extremely tiny size of metal nanoclusters and the limited resolution of electron microscopy, traditional techniques for the size characterization of large metal nanoparticles, including SEM, HR-(TEM), XRD are actually not very reliable for evaluating the core diameter of metal clusters. Mass spectra have been widely used to determine the core size and the precious composition of metal nanoclusters.

9.2.1 Au$_{25}$(SR)$_{18}$ Nanoclusters

Among the monolayer-protected gold nanoclusters, Au$_{25}$ is one of the most studied species. Recently, single-crystal X-ray crystallography analysis and theoretical studies have revealed that the structure of Au$_{25}$(SR)$_{18}$ nanoclusters can be viewed as an Au$_{13}$ icosahedral core encapsulated by an incomplete shell consisting of the exterior 12 gold atoms in the form of six –RS–Au–RS–Au–SR– motifs.[20-22] Such a fully protected structure endows the Au$_{25}$ nanoclusters extraordinary stability against core etching and redox reactions compared to pure Au nanoclusters with similar size.[23] Therefore, various research has been focused on Au$_{25}$ nanoclusters. Much progress has been achieved in the studies of Au$_{25}$ nanoclusters, including composition-controlled synthetic strategies in high yield and high purity, crystal structure determination and properties.

The first successful synthesis of atomically precise Au$_{25}$ nanoclusters with high yield was achieved by Shichibu and co-workers[24] *via* ligand exchange reaction of phosphine-stabilized Au$_{11}$ clusters. Although phosphine-stabilized metal clusters have been studied extensively, the phosphine-passivated Au nanoclusters are not stable and tend to decompose even at ambient conditions, which greatly limit their usage in practical applications. Meanwhile, due to the strong interaction between metal atoms and –SH groups, the as-synthesized phosphine Au nanoclusters can be easily transferred into thiol-stabilized ones in high yield with the same or different core sizes. Thus, ligand exchange strategy has been used to synthesize atomically monodisperse metal nanoclusters. In Shichibu's method, an aqueous solution of glutathione (GSH) was added to the chloroform solution of Au$_{11}$(PPh$_3$)$_8$ nanoclusters, which were prepared by reduction of AuCl(PPh$_3$) in ethanol by NaBH$_4$ under N$_2$ atmosphere,[25,26] with vigorous stirring under N$_2$ atmosphere. Under the optimized conditions, the pre-synthesized Au$_{11}$ nanoclusters underwent aggregation and dissociation in reaction with water-soluble GSH, and finally Au$_{25}$(SG)$_{18}$ nanoclusters could be selectively formed. The formation of Au$_{25}$ nanoclusters in high yield was confirmed by UV-Vis, electrospray ionization mass spectrometry (ESI-MS) and PAGE analysis as described in Figure 9.1. The large-scale synthesis of thiolated Au$_{25}$ clusters by the ligand exchange strategy opens up the possibility of synthesis of atomically monodisperse gold nanoclusters and their subsequent applications.

Another successful high yield synthesis of atomically precise Au$_{25}$ nanoclusters was made by Tsukuda and co-workers.[27] In the method, a crude sample of Au:SC$_n$ nanoclusters (n = 6, 10, 12) were first prepared using the modified Brust method. The Au$_{25}$(SC$_{12}$)$_{18}$ nanoclusters were then extracted from the dried samples using pure acetone. In recent years, this method has been widely used for preparation of highly monodisperse Au$_{25}$ nanoclusters capped with various ligands due to its simplicity and versatility.[23,28] Lately, Zhu *et al.*[19] developed a facile, low-temperature method for synthesizing Au$_{25}$ clusters in high yield *via* controlling the kinetics of the

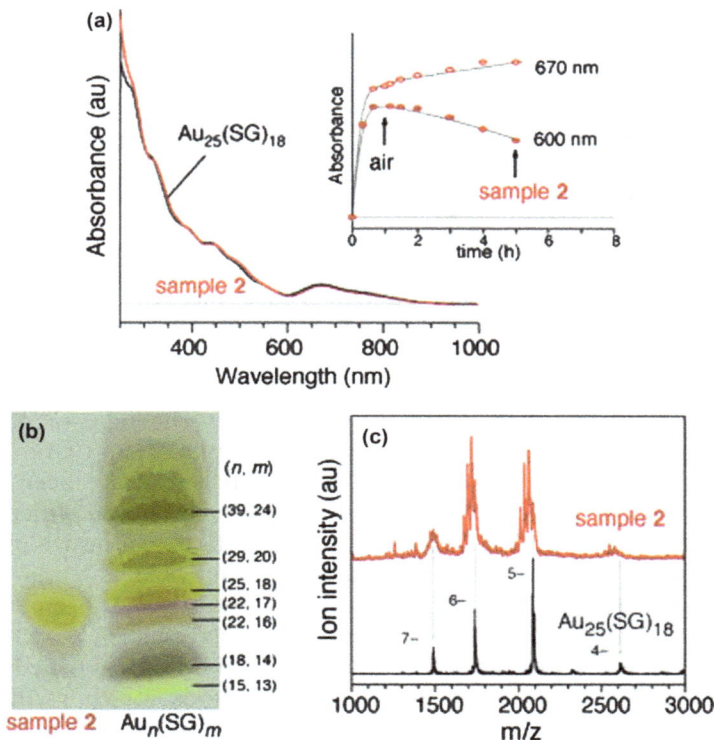

Figure 9.1 (a) Optical absorption spectra of the as-synthesized nanoclusters *via* ligand exchange reaction (sample 2, red) and $Au_{25}(SG)_{18}$ (black). (b) Polyacrylamide gel electrophoresis (PAGE) result and (c) Electrospray ionization (ESI) mass spectra of sample 2 and $Au_{25}(SG)_{18}$. The progression of the mass peaks in the spectrum of sample 2 is due to the partial hydrolysis of the GS ligands.
Reprinted from ref. 24 with permission by the American Chemical Society.

formation of Au(ı):SR intermediates. The authors chose the typical Brust–Schiffrin two-phase method as a model system to study the effect of reaction kinetics on the formation of Au_{25} clusters. The procedure typically involves two steps: (i) reduction of Au(ııı) to Au(ı):SR complexes by $PhCH_2CH_2SH$ at a very low stirring speed (~ 30 rpm) under N_2 atmosphere in an ice bath; (ii) further reduction of Au(ı) to Au(0) by a strong reducing agent ($NaBH_4$) under vigorous magnetic stirring (~ 1100 rpm). The authors pointed out that the kinetics of the formation of the Au(ı):SR intermediate are critical for high yield synthesis of Au_{25} nanoclusters in this method. The exclusive formation of Au_{25} clusters can only be obtained by controlling the reaction temperature at 0 °C and slow stirring conditions. More importantly, the high purity of Au_{25} nanoclusters achieved by this kinetic approach permits facile growth of high-quality single crystals of Au_{25} nanoclusters for X-ray structural determination.[22] This method is promising and has been extended to the synthesis of gold clusters with different core sizes by carefully tuning the reaction kinetics. However, difficulties still exist in attempting to incorporate different types of functionalized thiols into the Au_{25} nanoclusters in the two-phase approach, which have been proven to be of particular importance for practical applications of the Au_{25} nanoclusters.

More recently, Wu *et al.*[29] applied this kinetic control principle to the one-pot approach using tetrahydrofuran (THF) as solvent. This one-pot method eliminates the phase-transfer agent (TOABr) and allows for facile incorporation of thiols with different functionalities into the surface of Au_{25} nanoclusters in high yield. At the same time, the use of THF as a solvent significantly improves the purity and yield of Au_{25} nanoclusters under the kinetically controlled synthetic conditions. By such method, Au_{25} nanoclusters capped with different types of thiol ligands, in particular, those bearing functional groups such as –OH, –COOH, and atom transfer radical polymerization initiator $-OC(O)C(CH_3)_2Br$, *etc.*, have been successfully synthesized. The one-pot method has been further modified by dissolving $HAuCl_4$ and TOABr simultaneously in THF to synthesize molecular purified Au_{25} nanoclusters,[30] doped Au_{25} nanoclusters,[31] as well as $Au_{20}(SCH_2CH_2Ph)_{16}$.[32] Recently, other new approaches have been developed to synthesize atomic monodispersed Au_{25} nanoclusters, such as CO-directed synthesis[33] and the protection–deprotection method[34] established by Xie's group. Overall, almost all of the efficient methods for the synthesis of monodisperse Au_{25} nanoclusters are based on the original or modified Brust–Schiffrin methods.

9.2.2 $Au_{38}(SR)_{24}$ Nanoclusters

While tremendous work has been focused on atomic monodisperse Au_{25} nanoclusters, gold nanoclusters with other compositions have also been prepared and characterized. Chaki *et al.*[27] reported the isolation of alkanethiolate-protected Au_{38} nanoclusters, and the $Au_{38}(SC_nH_{2n+1})_{24}$ formula was unequivocally determined by ESI-MS analysis. Toikkanen *et al.*[35] also reported a simple synthetic strategy to obtain hexanethiolate-protected Au_{38} nanoclusters based on their stability in excess thiols. By changing the reduction time after fast addition of a freshly prepared aqueous solution of $NaBH_4$ in the standard Brust–Schiffrin two-phase synthesis and subsequent thiol etching process, Au_{38} nanoclusters fully passivated by a thiol monolayer were formed as the only product. Unlike the synthesis of Au_{25} in previous reports,[19,23] the formation of atomic monodisperse Au_{38} nanoclusters may be due to the thermodynamic stability of Au_{38} in comparison with the kinetic stability of Au_{25}.[19]

Despite these successes in the synthesis of atomic monodispersed Au_{38} nanoclusters, the nanoclusters are often in low yield and small quantities. In this context, Qian *et al.*[36] recently developed a facile, large-scale method for synthesizing truly monodisperse $Au_{38}(SC_{12}H_{25})_{24}$ nanoclusters in high purity. This new method explores a two-phase ligand exchange process in which glutathione-capped Au_n nanoclusters (a mixture) are utilized as the starting material. Due to the extraordinarily stability of Au_{25}, Au_{38}, and Au_{144}, they are strongly resistant to the thiol etching process.[27,35] When 1-dodecanethiol is added, the ligand exchange process will occur spontaneously at 80 °C under vigorous stirring and ultimately result in the formation of monodisperse Au_{38} nanoclusters in high purity. The monodispersity and purity of the as-synthesized Au_{38} nanoclusters were characterized by size exclusion chromatography (SEC) and UV-Vis spectroscopy. As shown in Figure 9.2(A), the UV-Vis spectrum of the nanoclusters exhibited a stepwise, multiple-absorption band spectrum, which is almost superimposable upon previously reported $Au_{38}(SR)_{24}$ optical spectra.[27,37,38] Figure 9.2(B) shows one single symmetric peak in size-exclusion chromatography. The spectra

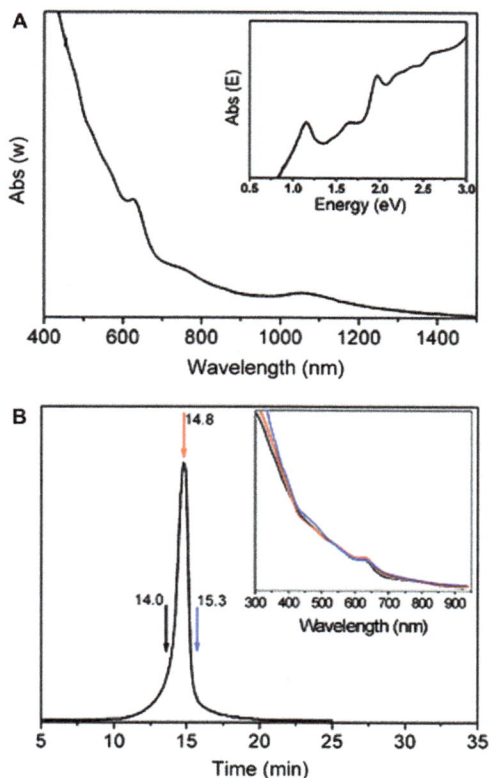

Figure 9.2   (A) UV-Vis spectrum of Au_{38} nanoclusters. The inset shows the absorbance *versus* photo energy. (B) Typical chromatogram of Au_{38} sample detected by diode array detector (DAD) at 630 nm. Inset: UV-Vis spectra obtained by DAD at 14.0 min (black line), 14.8 min (red line), and 15.3 min (blue line).
Reprinted from ref. 36 with permission by the American Chemical Society.

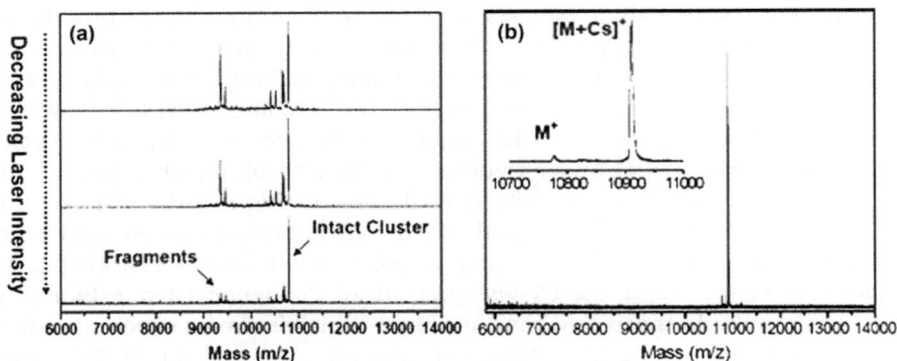

Figure 9.3 (a) Positive MALDI-TOF mass spectra of $Au_{38}(SC_2H_4Ph)_{24}$ nanoclusters corresponding to different laser intensities (decreasing from top to bottom). (b) ESI mass spectrum of $Au_{38}(SC_2H_4Ph)_{24}$; inset shows the zoomed-in spectrum.
Reprinted from ref. 39 with permission by the American Chemical Society.

shown in the inset (300–950 nm) correspond to the eluted components at 14.0 min (left side of the peak), 14.8 min (the peak position), and 15.3 min (right side of the peak), respectively. The three spectra are indeed superimposable, which confirms the high purity of the as-synthesized Au_{38} nanoclusters.

Afterwards, on the basis of this approach, Qian *et al.*[39] developed a size-focusing method to synthesize $Au_{38}(SCH_2CH_2Ph)_{24}$ in a yield of $\sim 25\%$. In the synthesis, a mixture of glutathione-protected $Au_n(SG)_m$ (n ranging from 38 to ~ 102) was first prepared by reducing Au(I)–SG in an acetone solution. Subsequently, excess phenylethylthiol (PhC_2H_4SH) was added and reacted for ~ 40 h at 80 °C. Finally, pure $Au_{38}(SC_2H_4Ph)_{24}$ nanoclusters were formed in high yield. The formula and molecular purity of $Au_{38}(SC_2H_4Ph)_{24}$ nanoclusters were then confirmed by matrix-assisted laser desorption ionization (MALDI) and ESI mass spectrometry, and size-exclusion chromatography. As shown in Figure 9.3(a) (top spectrum), a clean MALDI spectrum with an intense peak at $\sim 10\,780$ Da (assigned to the intact $Au_{38}(SC_2H_4Ph)_{24}$ ion) was observed, indicating their high monodispersity. Moreover, only one intense sharp peak at $10\,910.69$ Da (assigned to $[Au_{38}(SC_2H_4Ph)_{24}Cs]^+$) was observed in the ESI spectrum [Figure 9.3(b)], further indicating the high purity of the as-synthesized Au_{38} nanoclusters. These studies show that the size-focusing process is remarkable, and may be extendable to the synthesis of other stable $Au_n(SR)_m$ nanoclusters since the growth process is primarily driven by cluster stability, thus, making the approach of potentially broad utility.

9.2.3 $Au_{102}(SR)_{44}$ Nanoclusters

In 2007, Au_{102} clusters were firstly synthesized using *p*-mercaptobenzonic acid (*p*-MBA) as protecting ligands, yielding a definite assignment as

$Au_{102}(p\text{-MBA})_{44}$.[40] Single X-ray analysis showed that the $Au_{102}(p\text{-MBA})_{44}$ possesses a core of Au_{79}, which are packed in a Marks decahedron, and a shell of $Au_{23}(p\text{-MBA})_{44}$, which is composed of 19 $Au(SR)_2$ staples and two $Au_2(SR)_3$ staples. Interestingly, the $Au_{102}(p\text{-MBA})_{44}$ exhibits chirality with two alternating enantiomers in the crystal lattice. For the aqueous gold thiolate clusters, Kornberg and co-workers[9] investigated a variety of water-soluble thiols and synthetic conditions (ratios of thiolate : gold, $NaBH_4$: gold, and water : methanol, *et al.*) for the preparation of gold nanoclusters. They finally found the optimal conditions for the synthesis of homogeneous and high-quality Au_{102} nanoclusters. The simple procedure was as follows:[40] first, a mixture of $HAuCl_4$, PMBA, and methanol was allowed to equilibrate for 1 h, adjusted to 10 mM $NaBH_4$ and then agitated on a Vortex mixer for 5 h. The product was then precipitated with a mixture of 10% 2.5 M NaCl and 1 vol methanol at top speed in a microcentrifuge. After being pelleted twice and dried overnight at room temperature, the clusters obtained were resuspended in water. However, the first crystallized Au_{102} nanoclusters for X-ray structure determination were from a mixture in which it was present in a trace amount.

Recently, Kornberg and co-workers[41] reported an improved procedure that yields the $Au_{102}(p\text{-MBA})_{44}$ in abundant, essentially pure form. In the method, p-MBA and $HAuCl_4$ (3 : 1 ratio of p-MBA : gold) were combined in water and 47% methanol at a final gold concentration of 3 mM. $NaBH_4$ with a 2 : 1 ratio of BH_4^- to gold was used and the reaction was allowed to proceed overnight at room temperature. The product is purified by fractional precipitation with methanol. The PAGE shown in Figure 9.4A shows a single band of the same mobility, indicating the high monodispersity of the synthesized nanoclusters. At the same time, the large, black rod-like crystallization shown in Figure 9.4B and C indicate the high monodispersity of the product. Moreover, the yield could reach 50–70% (conversion of Au from $HAuCl_4$ to $Au_{102}(p\text{-MBA})_{44}$) and the as-prepared nanoclusters are stable in aqueous solution at 4 °C for at least 6 months. In order to determine the chemical

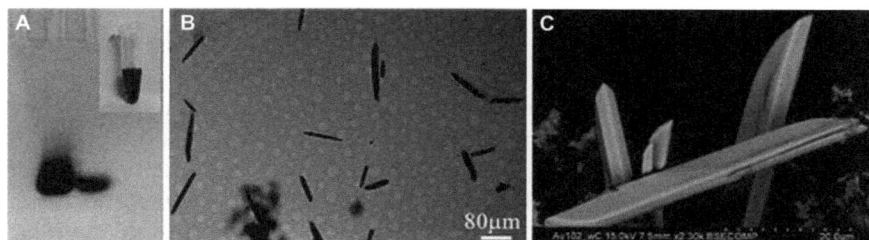

Figure 9.4   (A) Analysis of $Au_{102}(p\text{-MBA})_{44}$ product by 20% PAGE gel before (left lane) and after (right lane) purification by fractional precipitation. Inset: Photograph of the product. Crystal of $Au_{102}(p\text{-MBA})_{44}$ imaged by light microscopy (B) and scanning electron microscopy (C). Reprinted from ref. 41 with permission by the American Chemical Society.

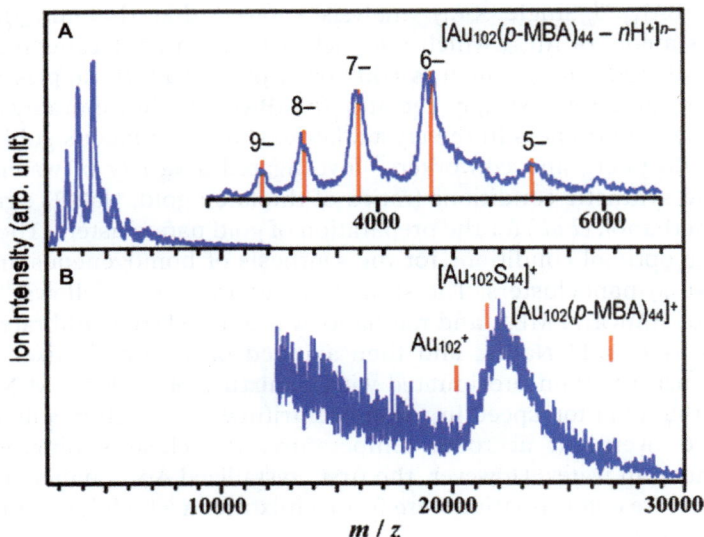

Figure 9.5 ESI (A) and MALDI-TOF (B) mass spectra of the gold clusters. The red bars in panel A indicate the calculated peak positions for $[Au_{102}(p\text{-}MBA)_{44}\text{-}nH^+]^{n-}$.

Reprinted from ref. 41 with permission by the American Chemical Society.

formulas and charge states of the thiolate-protected gold clusters, ESI-MS and MALDI-TOF MS were applied. From the ESI mass spectra of the gold nanoclusters (Figure 9.5A), it can be seen that a series of peaks with mass to charge ratios corresponding to partially deprotonated states of $Au_{102}(p\text{-}MBA)_{44}$ were obtained. The MALDI-TOF mass gave a broad peak centered at 22 kDa (Figure 9.5B), consistent with 102 gold atoms and 44 sulfur atoms. All of the characterizations indicated the successful synthesis of highly monodisperse Au_{102} nanoclusters with high yield and purity.

9.2.4 $Au_{144}(SR)_{60}$ Nanoclusters

Earlier studies have shown that only the 5, 8, and 29 kDa nanoclusters can survive the harsh conditions of etching by free thiols, whereas the others are degraded by core etching,[27,38,42] which indicates that these nanoclusters are thermodynamically and chemically stable. The biggest 29 kDa nanoclusters were determined to be Au_{144} nanoclusters by ESI-MS mass spectra analysis.[43] Recently, Qian and Jin[43] developed a facile, two-step synthetic method for preparing truly monodisperse $Au_{144}(SCH_2CH_2Ph)_{60}$ nanoclusters with a yield of 20–30%. In the first step, polydispersed Au nanoparticles capped by $-SCH_2CH_2Ph$ were prepared *via* a modified Brust–Schiffrin method. The polydispersed Au nanoparticles dissolved in toluene were then etched with excess $PhCH_2CH_2SH$ at 80 °C for ~24 h. Through this procedure, an intriguing size "focusing" process occurred and only the $Au_{144}(SCH_2CH_2Ph)_{60}$ species was finally obtained. The ESI mass spectrometry analysis together with TGA and NMR confirmed the monodisperse $Au_{144}(SCH_2CH_2Ph)_{60}$ composition. A remarkable advantage of this synthetic approach is that it can solely produce Au_{144} nanoclusters, hence eliminating the difficult post-synthetic steps of size separation. These truly monodisperse nanoclusters can serve as a well-defined model system for further study of size-related properties and their practical applications.

More recently, an ambient synthesis of atomically monodisperse $Au_{144}(SR)_{60}$ in methanol has also been developed by Qian and Jin[44] In this method, gold salt precursor was first mixed with excess thiol and TOABr in methanol to form $Au(I)-SR$ polymers. Then, $NaBH_4$ aqueous solution was added to form polydisperse Au nanoclusters, which are size-focused into two monodisperse gold nanoclusters: $Au_{144}(SR)_{60}$ (major product) and $Au_{25}(SR)_{18}$ (side product) over a ~ 5 h period. Through acetone washing, pure $Au_{144}(SR)_{60}$ was obtained with a yield of 10–20%. Importantly, this method proved to be versatile for different thiolate ligands, such as PhC_2H_4SH and various $C_nH_{2n+1}SH$ (where $n = 4$–8).

9.2.5 Other Atomic Monodisperse $Au_n(SR)_m$ Nanoclusters

Except for the magic-number Au nanoclusters discussed above, other monodisperse Au nanoclusters, such as Au_8,[45] Au_{10},[46] Au_{11},[47,48] Au_{13},[49] Au_{18},[50] Au_{19},[51] Au_{20},[52] Au_{36},[53] Au_{55},[54] *etc.*, have also been successfully synthesized over past years by various synthetic approaches. Here, we will not discuss the synthesis of these monodisperse nanoclusters in detail. It should be noted that the successful syntheses of atomically precise metal nanoclusters provide the possibility to investigate the relationship between the structure of nanoclusters, including size, composition, surface chemistry, and their physical and chemical properties. In addition to the gold nanoclusters, significant efforts have also been devoted to the synthesis and property studies of other metal nanoclusters, such as Ag, Pt, Pd, and Cu, *etc.*

9.2.6 Ag Nanoclusters

Although various well-defined monodisperse Au_n nanoclusters have been synthesized and their exact formulas have also been determined by mass spectrometry analysis, fewer truly monodisperse $Ag_n(SR)_m$ nanoclusters have been reported.

Jin's group devised a facile synthetic route for preparing well-defined Ag_7 nanoclusters in high yield using *meso*-2,3-dimercaptosuccinic acid (DMSA) as protecting ligands.[55] The core size and composition of the as-prepared Ag_7 nanoclusters were determined by ESI-MS analysis for the first time. In their synthetic route, $AgNO_3$ dissolved in ethanol was first cooled to 0 °C in an ice bath. Under low stirring speed, DMSA was added to the cold

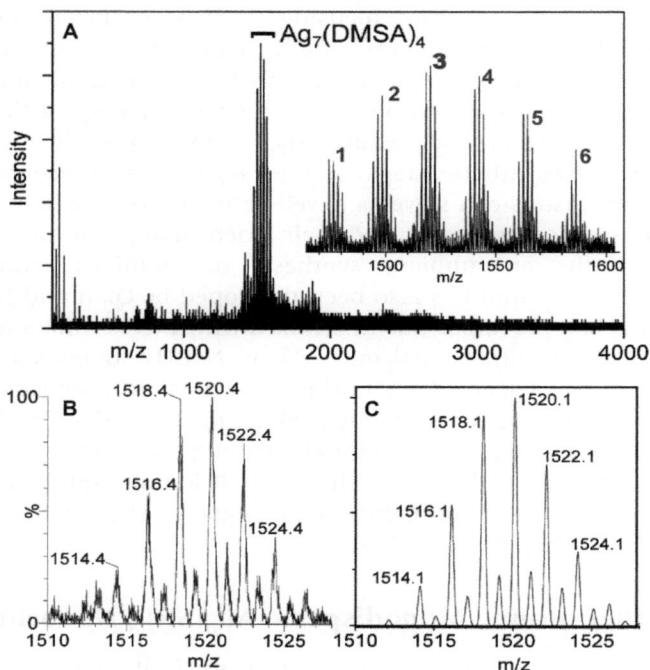

Figure 9.6 (A) ESI spectra of Ag$_7$ clusters (negative ion mode, inset shows the zoomed-in spectrum). (B) and (C) show the experimental and simulated isotopic pattern of Ag$_7$L$_4$–2H + 2Na, respectively.
Reprinted from ref. 55 with permission by the American Chemical Society.

Ag(I) solution. After the complete formation of Ag$_x$(DMSA)$_y$ intermediates, NaBH$_4$ powder was slowly added under vigorous stirring. Finally, Ag$_7$ nanoclusters were formed with a yield of ∼20%. As can be seen from the ESI-MS spectra shown in Figure 9.6, the base peak at *m/z* 1520.4 could be assigned to be Ag$_7$(DMSA)$_4$ nanoclusters with high monodispersity.

Another monodisperse Ag nanocluster (Ag$_9$) was synthesized by a novel solid-state route.[56] As described in Chapter 3, there are three steps in the reported synthesis. The ESI-MS characterization (Figure 9.7) clearly indicated that monodisperse Ag$_9$ nanoclusters were successfully obtained through the solid-state reaction. Interestingly, these Ag$_9$ nanoclusters are highly stable both in the solid state under an inert atmosphere and in solvent mixtures with water as one of the components. More importantly, this solid-state synthesis provides the following advantages: (1) the nanocluster growth could be controlled more easily than that in solvents, where large metallic particles tend to be formed due to the fast reduction process; (2) large scale product can be achieved; (3) compared to most solution-phase syntheses, the solid-state route provides a new synthetic approach for synthesizing tiny metal nanoclusters.

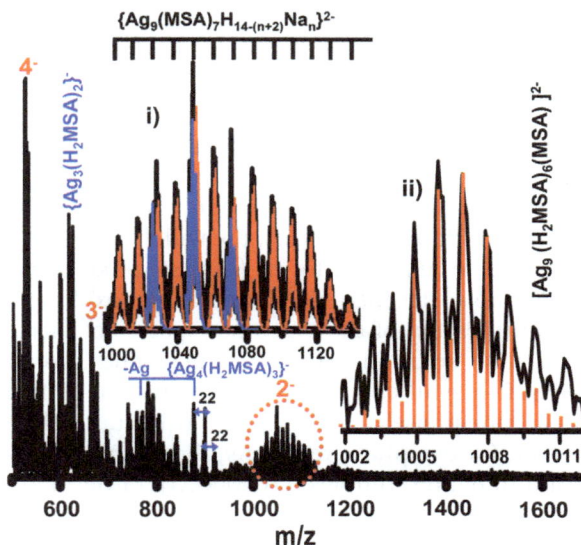

Figure 9.7   ESI-MS spectrum of the separated Ag nanoclusters in a negative mode in the region m/z 500–1700. Inset (i) shows an enlarged spectrum of the $[Ag_9(MSA)_7H_{14-(n+2)}Na_n]^{2-}$ overlapped with peaks due to $[Ag_4(H_2MSA)_3HMSA]^-$ and its sodium adducts, which are shown in blue. Inset (ii) compares the expected and observed MS patterns of $[Ag_9(H_2MSA)_6(MSA)]^{2-}$.
Reprinted from ref. 56 with permission by the American Chemical Society.

Based on this solid-state route, Pradeep and co-workers have also synthesized monodisperse $Ag_{152}(SCH_2CH_2Ph)_{60}$ nanoclusters in high yield.[57] These monodisperse Ag nanoclusters exhibited a single sharp peak near 25 kDa in the MALDI-MS measurement and a well-defined metal core of ~ 2 nm was measured by TEM. It was found that the Ag_{152} nanoclusters displayed a core–shell structure with a 92-atom silver core having icosahedral–dodecahedral symmetry and an encapsulating protective shell containing 60 Ag atoms and 60 thiolates arranged in a network of six-membered rings. Rao and Pradeep also developed an interfacial etching reaction method to synthesize mercaptosuccinic acid (H_2MSA) protected Ag_7 $[Ag_7(H_2MSA)_7]$ and Ag_8 $[Ag_8(H_2MSA)_8]$ nanoclusters in gram quantities.[58]

9.2.7   Cu Nanoclusters

Compared to the extensive studies on gold and silver nanoclusters, reports on copper nanoclusters are still scarce primarily because of their susceptibility to oxidation and the difficulty of preparing extremely tiny particles. Recently, several methods, including template-based synthesis,[59,60] electrochemical synthesis,[61,62] microemulsion technique,[63] and microwave-assisted polyol synthesis[64] have been widely used to synthesize Cu nanoclusters. All the methods mentioned here for copper cluster synthesis have been summarized in our recent mini review.[65]

Although Brust–Schiffrin reactions have been studied extensively by different research groups, there remain significant questions regarding their detailed mechanisms, especially the precursor species present in solution before the addition of $NaBH_4$. In earlier reports on the synthesis of various Au nanoclusters with different core sizes, the widely accepted assumption is that Au(III) is reduced to Au(I) and forms $[Au(I)SR]_n$-like polymers upon the addition of alkanethiols. Recently, Goulet and Lennox[66] reported the identification and quantification of the precursor species of metal nanoclusters based on quantitative solution ^1H NMR study, and proposed a revised view of Brust–Schiffrin syntheses. Based on the proton NMR results, it was proposed that in one-phase reaction, M(I) thiolate is likely to be the precursor. However, for two-phase reactions, the metal precursor is metal(I)-tetraoctylammonium halide complex ([TOA][MX$_2$]) rather than the $[M(I)SR]_n$-like polymers. On the basis of this new finding, our group successfully synthesized stable Cu_n ($n \leq 8$) nanoclusters through a modified one-phase Brust–Schiffrin method.[67] In the synthesis, [TOA]$_2$[CuBr$_4$] precursor was firstly prepared and then dissolved in ethanol. After being cooled to 0 °C in an ice bath, 5 equivalents of MPP (2-mercapto-5-n-propylpyrimidine) *versus* Cu salt were added slowly under an argon atmosphere with vigorous stirring. After addition of $NaBH_4$, the reaction was allowed to proceed for another 7 h. Finally, Cu nanoclusters with high monodispersity were obtained. Positive-ion mode ESI-MS was used to characterize the composition of the as-synthesized Cu nanoclusters. As shown in Figure 9.8, the highest mass peak,

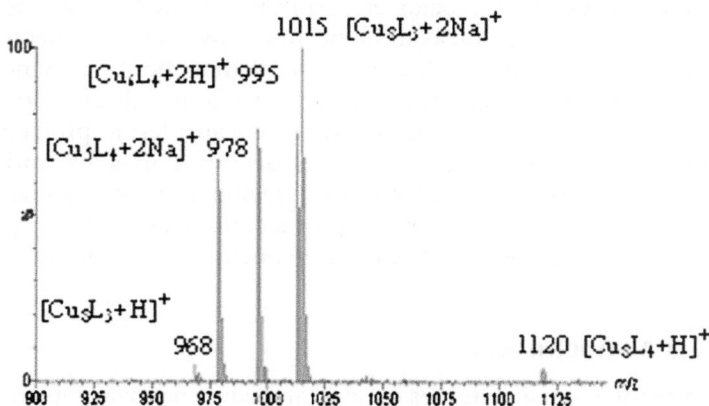

Figure 9.8   Representative ESI mass spectrum of a copper cluster sample detected in the positive–ion mode.
Reprinted from ref. 67 with permission by the American Chemical Society.

$m/z \approx 1120$, could be assigned to the Cu nanoclusters with a composition of Cu_8L_4 ($L = C_7H_9N_2S$) whereas those in the lower range might be assigned to the different fragments as labeled. This result indicates that Cu_8 nanoclusters are dominant in the product. Our approach could provide a facile and versatile strategy to synthesize other metal nanoclusters with different functionalities.

9.2.8 Pt Nanoclusters

As is known to us, colloidal Pt nanoparticles with diameters of 2–5 nm on carbon supports are currently regarded as the best catalysts for electrochemical reactions in fuel cells.[68] However, the high cost and limited supply of Pt largely restrict the wide-spread commercialization of fuel cells. Thus, reduction of the size of Pt particles to the atomic level has become a necessity for efficient use of this precious resource. Sub-nanometer sized Pt nanoclusters have emerged as a promising materials to address this question. Unfortunately, traditional wet chemical reduction methods usually produce Pt nanoparticles larger than 2 nm. Hence, template-based methods have been used to synthesize Pt nanoclusters on the sub-nanometer scale. Yamamoto *et al.*[69] synthesized monodispersed Pt_{12} (0.9 ± 0.1 nm), Pt_{28} (1.0 ± 0.1 nm) and Pt_{60} (1.2 ± 0.1 nm) nanoclusters through stepwise complexation of platinum(IV) chloride ($PtCl_4$) with a spherical macromolecular template (phenylazomethine dendrimer) and a subsequent reduction process. More recently, Tanaka *et al.*[70] reported the synthesis of atomically monodispersed, water-soluble $Pt_5(MAA)_8$ (MAA, mercaptoacetic acid) nanoclusters by using PAMAM (G4-OH, the fourth-generation polyamidoamine dendrimer) as the template. in the synthesis, Pt nanoclusters were first prepared by reducing H_2PtCl_6 with $NaBH_4$ in the presence of PAMAM. And then, MAA was added to the supernatant to replace PAMAM as the ligand. Finally, atomically monodispersed Pt nanoclusters were obtained by using size-exclusion high performance liquid chromatography (HPLC). From the ESI data, a main peak at $m/z = 1712$ could be assigned to $Pt_5(MAA)_8$ nanoclusters.

9.2.9 Pd Nanoclusters

Although monodisperse Au, Ag, Pt, and Cu nanoclusters have been extensively synthesized using various synthetic techniques, there have been few reports on the synthesis of monodisperse Pd nanoclusters except polydisperse Pd nanoclusters. For example, Negishi *et al.*[71] synthesized polydisperse Pd_n ($5 \leq n \leq 60$) nanoclusters with core size around 1.0 nm through the direct mixing of $PdCl_2$ and *n*-alkanethiols (RSH: R = *n*-$C_{18}H_{37}$, *n*-$C_{12}H_{25}$) in toluene. Recently, a new surfactant-free method has been proposed, which can be used to synthesize Pd nanoclusters[72] as well as Au,[73,74] Pt,[75] and Cu[76] nanoclusters. In a typical procedure, a solution of aqueous $PdCl_2$ was added to DMF solution and then heated at 140 °C for 6 h. Here, DMF serves as both the solvent and the reductant. Although monodisperse metal nanoclusters could not be easily obtained under these conditions, the DMF reduction method provides a facile and large-scale synthesis of surfactant-free metal nanoclusters which have been shown to be important in catalysis applications.

9.3 Size-Dependent Catalytic Activity of Metal Nanoclusters

It is well known that at the nanometer scale, materials exhibit very interesting size-dependent optical, electrical, magnetic, and chemical properties that differ drastically from the bulky ones. Much effort has been made in recent years to unravel the origin of the catalytic properties of nanomaterials.[77,78] However, in most catalysis studies, the nanoparticles are polydispersed. Therefore, the observed catalytic properties of nanomaterials reflect only an ensemble average. The polydispersity of nanomaterials and their unknown surface structure preclude the precise correlation of particle structure and electronic properties with their catalytic properties. In order to understand the origin of their catalytic properties, it is critical to first obtain atomically precise nanomaterials. Recently, various metal nanoclusters with high monodispersity and clear structure have been successfully synthesized as discussed above, which will allow for precise measurements of size-dependent catalytic performance. For nanoclusters, owing to their unique electronic structure and high fraction of surface atoms with low coordination numbers, their unusual catalytic activity has attracted more and more attention in catalytic applications over past years. Here, we focus on the size-dependent catalysis of metal nanoclusters, including CO oxidation, oxygen reduction reaction (ORR), as well as aerobic oxidation.

9.3.1 CO Oxidation

Since the seminal work by Haruta *et al.*,[79,80] carbon monoxide oxidation catalyzed by Au nanoparticles supported on oxide supports has been extensively investigated over past years.[81–83] Generally, in such systems Au nanoparticles are dispersed finely on a support with a high surface for the efficient use of catalytically active components. Therefore, the size of Au nanoparticles is one of the most important factors that dictates the performance of a catalyst. Previous studies have shown that the size of gold particles strongly affects their catalytic activity, and gold nanoparticles smaller than 5 nm have high catalytic activity.[83–85] For instance, Goodman and co-workers[83] have investigated the unusual size-dependence of the low-temperature catalytic oxidation of carbon monoxide by using Au clusters supported on single crystalline surfaces of titania (Au/TiO_2) in ultrahigh vacuum. Scanning tunneling microscopy/spectroscopy (STM/STS) and elevated pressure reaction kinetics measurements showed that the structure sensitivity of the CO oxidation reaction on Au/TiO$_2$ is related to a quantum size effect with respect to the thickness of the gold islands. Surprisingly, they found that islands with two layers of gold are most effective for catalyzing the oxidation of CO.

Recently, Au nanoclusters, which consist of only several to tens of metal atoms and possess sizes comparable to the level of Fermi wavelength of electrons, have been extensively used as catalyst for CO oxidation. Due to the ultrasmall size (<2 nm) and the resulting quantum size effect, such metal nanoclusters exhibit discrete, molecular-like electronic structure and unique properties, such as enhanced photoluminescence, HOMO–LUMO electronic transition, intrinsic magnetism, and high catalytic properties that are fundamentally different from those of their larger counterparts. Recent theoretical and experimental results demonstrated that sub-nanometer clusters have shown better catalytic activity and/or selectivity than nanometer sized particles.[86] CO oxidation on Au/rutile TiO$_2$ (110) has been well documented, and activities typically are found to peak for Au$_n$ in the few nanometer size range.[85] Lopez and Norskov[87] used Au$_{10}$ nanoclusters as a model system and predicted that isolated Au$_{10}$ nanoclusters can be highly active for the CO oxidation even below room temperature both in the gas phase and on the rutile TiO$_2$ support *via* a density functional study. It was proposed that, compared to the extended Au surfaces, the extraordinary activity can be attributed to the very low coordinated Au atoms which are able to interact stronger with adsorbates. Lee *et al.*[85] presented a study of room-temperature CO oxidation on Au$_n$/TiO$_2$ catalysts prepared by deposition of size-selected Au$_n^+$, $n = 1$, 2, 3, 4, 7. As seen from Figure 9.9, CO oxidation activity is strongly dependent on the cluster size, with substantial activity for Au$_n$ as small as three atoms, due to the size-dependent electronic and geometric structures. In another report, Lee *et al.*[88] studied the room temperature CO oxidation activity of Au$_n$/TiO$_2$ (110), $n = 1$–7. They found that the activity of Au clusters showed strong size-dependence, first appearing for Au$_3$, declining to Au$_5$, then increasing substantially for Au$_6$ and Au$_7$ nanoclusters. More recently, Gao *et al.*[89] systematically studied the catalytic activities of sub-nanometer sized gold nanoclusters for CO oxidation using *ab initio* calculations. They found that anionic nanoclusters can adsorb CO and O$_2$ more strongly than neutral ones and the co-adsorption energies of both CO and O$_2$ molecules decrease as the cluster size increases, with the exception of Au$_{34}$ (an electronic "magic-number" cluster).

In addition to TiO$_2$ support for dispersion of metal nanoclusters, other metal oxide materials have also been studied as supports for catalytic reactions. For instance, Landman and co-workers[90,91] experimentally and theoretically studied the low-temperature combustion of CO on size-selected gold nanoclusters, Au$_n$ ($n \leq 20$), supported on MgO (100) films. They found that the activity of Au$_n$ ($n \leq 20$) for CO oxidation is size-dependent, with Au$_8$ to be the smallest size to catalyze the reaction. Herzing *et al.*[81] also studied CO oxidation on gold nanoclusters supported on iron oxide. The highest catalytic activity for CO oxidation was observed on Au bilayer clusters that

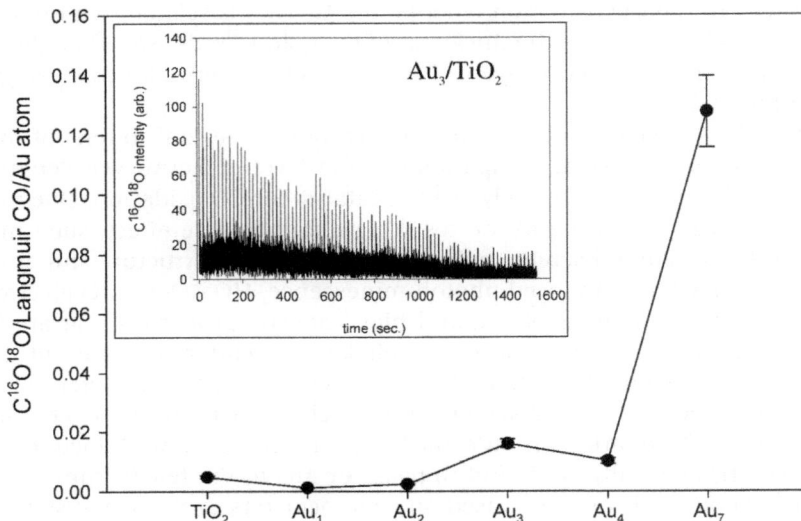

Figure 9.9 Size-dependence of CO oxidation activity. Inset: Effects of 100 CO pulses on Au_3/TiO_2.
Reprinted from ref. 85 with permission by the American Chemical Society.

are ~0.5 nm in diameter and contain only 10 gold atoms. Such results showed that Au nanoclusters supported on metal oxides may represent a unique class of catalysts that deserve further investigation.

Note that, although the activity of gold catalysts depends on the size of the gold particles (quantum size effects), the nature of the supporting materials,[86] charging of the gold particles by interaction with defects in the oxide, availability of low coordinated sites have also been suggested to play key roles in catalysis.[82,90–94] Among these, the role of the support is still a matter of discussion and the metal–support interaction has often been described as charge transfer, *i.e.*, a change in the number of d electrons between the metal nanoclusters and the oxide surface.[95,96] For example, Jin's group studied the application in CO oxidation of atomically monodisperse, thiolate-protected $Au_{25}(SCH_2CH_2Ph)_{18}$ nanoclusters supported on various oxides.[97] In their study, the as-synthesized Au_{25} nanoclusters were directly deposited onto TiO_2, CeO_2, and Fe_2O_3 supports, respectively, and then used as catalysts for CO oxidation reaction in a fixed bed reactor. They found that the supports exhibit a strong effect, and the Au_{25}/CeO_2 catalysts were found to be much more active than the others.

9.3.2 Oxygen Reduction Reaction (ORR)

Due to high power density, high energy-conversion efficiency and zero or low emission of pollutants, fuel cells, devices that efficiently convert the chemical energy stored in fuels into electrical energy through electrochemical reactions, have attracted increasing interest as power sources for portable applications over the past decades. Unfortunately, despite their great promise, there are still many scientific and technological difficulties hampering the widespread commercialization of fuel cells. One of the main obstacles is the high overpotential required for the ORR to proceed at an adequate rate. Nowadays, the state-of-the-art commercial catalysts are usually Pt nanoparticles (2–5 nm) supported on carbon supports (Pt/C). However, the high cost, limited resources, and the crossover and poisoning effects of this precious metal are hindering the commercialization of fuel cells. Moreover, due to the corrosion of the carbon supports and the dissolution–aggregation of Pt nanoparticles, the Pt/C catalysts usually suffer from poor durability caused by the fast and significant loss of electrochemical surface area (ECSA) over time during fuel cell operation. Thus, in order to reduce the cost of the fuel cells, increase the durability and improve the catalytic activity of catalysts, one of the important challenges is the development of Pt-free catalysts or catalysts with a lower content of Pt.

Compared to the Pt-group metals, bulk gold has attracted little attention in electrocatalysis, largely because of its inertness in the bulk state, which is confirmed by both experimental evidence and density functional theory (DFT) calculations. However, when the dimension is diminished to the nanometer scale, gold materials displayed enhanced catalytic activity compared to bulk Au.[79] Recent studies have shown that Au nanoclusters supported on metal oxides displayed enhanced catalytic activity towards CO oxidation[83] and the ORR.[98] The enhanced catalytic activity of Au nanoclusters is always accounted for by the high fraction of low-coordinated surface atoms and higher d-band center predicted by theoretical studies, which result in favorable O_2 adsorption onto these Au cluster surfaces. These studies suggest that sub-nanometer sized gold nanoclusters might represent a unique class of catalysts for the ORR that deserve further investigation.

By modifying Pt nanoparticles supported on carbon with gold nanoclusters, Adzic and co-workers[98] demonstrated that Au nanoclusters could have a stabilizing effect on an underlying Pt metal surface under highly oxidizing conditions and suppress Pt dissolution during potential cycling, without significant changes in the activity and surface area of Au-modified Pt catalysts. Such a stabilizing effect of Au nanoclusters on Pt was determined in an accelerated stability test by continuously applying linear potential sweeps from 0.6 to 1.1 V, which caused surface oxidation–reduction cycles of Pt. From the voltammetric measurements, the catalytic activity of Au/Pt/C showed only 5 mV degradation in half-wave potential over the cycling period, in contrast, a loss of 39 mV was obtained from Pt/C. Moreover, after potential cycling, only ∼55% of the original Pt surface area remained for the Pt/C catalyst, while for Au/Pt/C, there was no recordable loss of Pt surface area. The enhanced stability could be attributed to an efficient spillover of H_2O_2 from Au nanoclusters to the surrounding Pt atoms, where further reduction to H_2O can take place.

With Au nanoclusters as catalysts, much work has been done to explore the size effect on the catalytic activity for the ORR.[1,99] Recently, Chen and Chen[1] systematically studied the electrocatalytic activity of Au nanoclusters with different sizes (from Au_{11} to Au_{140}) for ORR in alkaline media and found that the ORR activity of Au nanoclusters is size-dependent, with Au_{11} exhibiting the best catalytic activity. As can be seen from rotating-disk voltammograms (Figure 9.10A), the Au_{11} nanoclusters exhibit much better defined current density plateaus than those with larger Au nanoparticles.

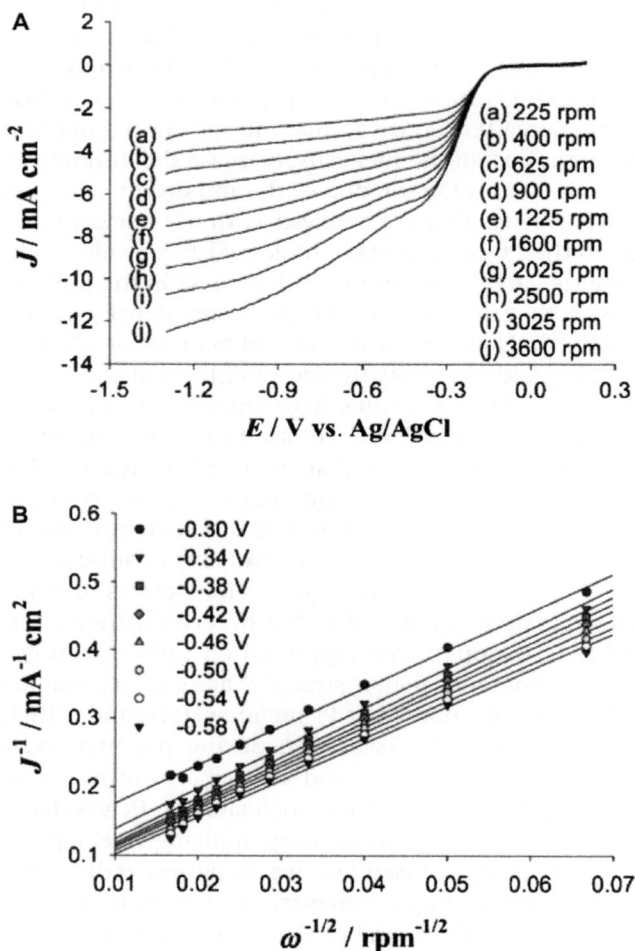

Figure 9.10 (A) Rotating-disk voltammograms recorded for a GC electrode modified by Au_{11} clusters in aqueous 0.1 M KOH saturated with oxygen at different rotation rates. (B) Koutecky–Levich plots (J^{-1} vs. $\omega^{-1/2}$) at different electrode potentials. DC ramp 20 mV s^{-1}. Symbols are experimental data obtained from (A), and lines are linear regressions. Reprinted from ref. 1 with permission by Wiley-VCH.

From the Koutecky–Levich plots shown in Figure 9.10B, the most efficient four-electron ORR process occurs with Au_{11} clusters as the cathodic catalysts. The results strongly suggest the potential application of atomic Au nanoclusters as efficient cathode catalysts in fuel cells.

Although Au nanoclusters are found to be promising cathode catalysts for the ORR, they still suffer big problems during the catalysis reactions, such as the capping agents on the surface which may block the mass transport and electron transfer and easy occurrence of dissolution and aggregation due to their extremely small size and high surface energy. In order to overcome these obstacles, much effort has been devoted to developing surfactant-free metal clusters supported on graphene sheets.[100–102] As shown in Figure 9.11, Tang and co-workers[102] found that in a comparison of reduced graphene oxide sheets (rGO), Au nanoparticle/rGO hybrids, thiol-capped Au clusters with the same sizes, and especially the commercial Pt/C catalyst, the Au nanoclusters supported on rGO (Au/rGO) display excellent electrocatalytic performance toward the ORR, for instance, high onset potential, superior methanol tolerance, and excellent stability in O_2-saturated 0.1 M KOH. Except for alkaline media, Jeyabharathi *et al.*[103] also found that in acidic media Au atomic clusters could electrocatalyze the reduction of oxygen to water through a direct four-electron pathway. Utilizing Au nanoclusters as the ORR electrocatalysts can not only improve the catalytic activity and durability but also considerably decrease the catalyst cost.

Except for Au nanoclusters, studies of size-dependent ORR activity have also been performed on Ag, Cu, and Pt nanoclusters. For instance, Shao *et al.*[104] reported the size-dependent specific and mass activities for the ORR

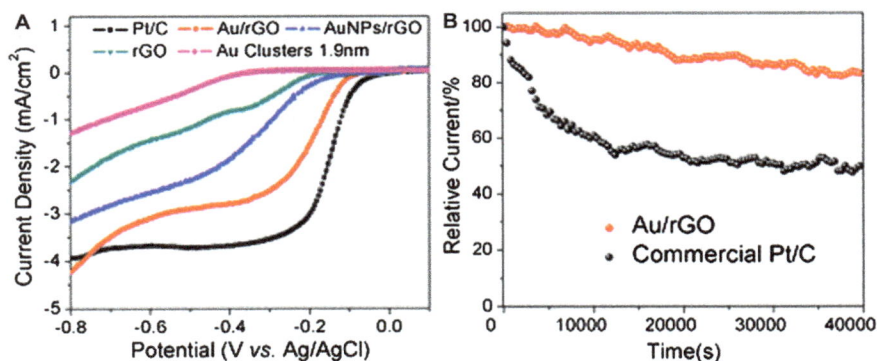

Figure 9.11 (A) RDE curves of commercial Pt/C, Au/rGO hybrids, Au NP/rGO hybrids, rGO sheets, Au clusters in O_2-saturated 0.1 M KOH at a scanning rate of 50 mV s^{-1} at 1600 rpm; (B) Current–time (*i–t*) chronoamperometric responses for the ORR at the Au/rGO hybrids and commercial Pt/C electrodes in an O_2-saturated 0.1 M KOH solution at -0.2 V *vs.* Ag/AgCl.
Reprinted from ref. 102 with permission by the American Chemical Society.

on Pt nanoparticles in the size range of 1–5 nm. In another report, Yamamoto *et al.*[69] successfully synthesized monodisperse Pt_{12} (0.9 ± 0.1 nm), Pt_{28} (1.0 ± 0.1 nm) and Pt_{60} (1.2 ± 0.1 nm) nanoclusters using a spherical macromolecular template (phenylazomethine dendrimer). They found that the as-synthesized Pt nanoclusters exhibit strong size-dependent ORR catalytic activity ($Pt_{12} > Pt_{28} > Pt_{60}$) with Pt_{12} exhibiting the highest catalytic current, which is approximately 13 times higher than that of commercially available platinum nanoparticles supported on carbon black.

As for Ag nanoclusters, we recently synthesized two different sized silver nanoclusters capped with the same ligands and studied their ORR activities.[105] By comparing the onset potential and current densities of the ORR, we found that the 0.7 nm (Ag_7) nanoclusters exhibited higher electrocatalytic activity than that of the 3.3 nm Ag nanoparticles. Size-dependent catalytic activity of Cu nanoclusters for the ORR has also been studied recently.[106] It was found that the core size of copper nanoclusters exhibited a strong effect on their electrocatalytic activity towards the ORR. By comparing the onset potential and current densities of the ORR, the catalytic performance decreased with the core size increasing and the smallest clusters exhibited the best electrocatalytic activity. However, for both of the studied Ag and Cu nanoclusters, a two-electron reduction process of adsorbed oxygen was derived from the kinetic results of RDE, which may be ascribed to the capping ligand-passivated surfaces of the nanoclusters. Further studies about the ORR activity on surfactant-free Ag and Cu nanoclusters with different core sizes are needed in future work.

9.3.3 Aerobic Oxidation

Recently, increased interest has been focused on environmentally benign oxidation processes that utilize molecular oxygen in air as an oxidant.[81,107–110] Extensive studies have also revealed that the size (diameter) of Au NPs has a significant influence on their catalytic properties: the catalytic activity increases with decreasing size.[111] For instance, Turner *et al.*[112] demonstrated that small gold clusters (∼1.4 nm) supported on inert materials are efficient and robust catalysts for the selective oxidation of styrene by dioxygen. At the same time, they found a sharp size threshold in catalytic activity, completely inactive for the particles with diameters of ∼2 nm and above. It was suggested that the catalytic activity arises from the altered electronic structure intrinsic to small gold nanoparticles. Tsukuda and coworkers[113] studied the catalytic activity of small (1–3 nm) monodisperse Au nanoclusters for aerobic oxidation of alcohol. It was found that activity appears at a core size of ∼5 nm and increases rapidly with decrease in the core size, with Au nanoclusters smaller than 1.5 nm showing the highest activity.

In another work, Liu *et al.*[114] synthesized gold nanoclusters, Au_n ($n = 10, 18, 25, 39$), with atomically controlled sizes on hydroxyapatite (HAP) and studied the catalysis for aerobic oxidation of cyclohexane. They found that these Au_n/HAP catalysts could efficiently oxidize cyclohexane to cyclohexanol and

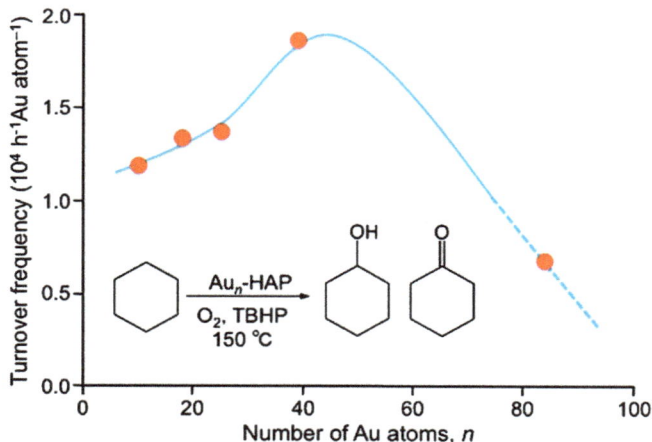

Figure 9.12 TOF values as a function of the gold cluster size n. The curve is a guide for the eye.
Reprinted from ref. 114 with permission by the American Chemical Society.

cyclohexanone. Interestingly, the turnover frequency monotonically increased with an increase in the nanocluster size, reaching values as high as $18\,500$ h^{-1} Au atom^{-1} at $n = 39$, and thereafter decreased with a further increase in n up to $n \sim 85$ (Figure 9.12). This finding further provides a fundamental insight into size-specific catalysts of gold in the clusters regime (<2 nm) and a guiding principle for rational design of Au nanocluster-based catalysts.

In addition, atomically monodisperse Au nanoclusters and doped or modified nanoclusters have also been used as efficient aerobic oxidation catalysts. For example, Xie *et al.*[28] demonstrated that monodisperse single Pd-doped Au$_{25}$ nanoclusters ($Pd_1Au_{24}(SC_{12}H_{25})_{18}$) supported on CNTs significantly improved the aerobic oxidation of benzyl alcohol activity over pure Au$_{25}$ nanoclusters supported on CNTs. This doping effect is proposed to be due to modulation of the electronic structure by intracluster electron transfer from Pd to Au. Recently, Toshima and co-workers[115] reported the synthesis of "crown-jewel-structured" Au/Pd nanocluster (CJ-Au/Pd NC) catalysts with an abundance of top (vertex or corner) Au atoms on Pd nanoparticles (1.8 ± 0.6 nm). As can be seen from Figure 9.13, compared to the Au nanoclusters, Pd mother nanoclusters, and Pd/Au alloy nanoclusters, which possess almost the same particle size, the "crown-jewel-structured" Au/Pd nanoclusters showed the highest catalytic activity towards aerobic glucose oxidation.

9.3.4 Other Catalytic Applications

In addition to the promising catalytic applications of metal nanoclusters mentioned above, other potential applications of various metal nanoclusters

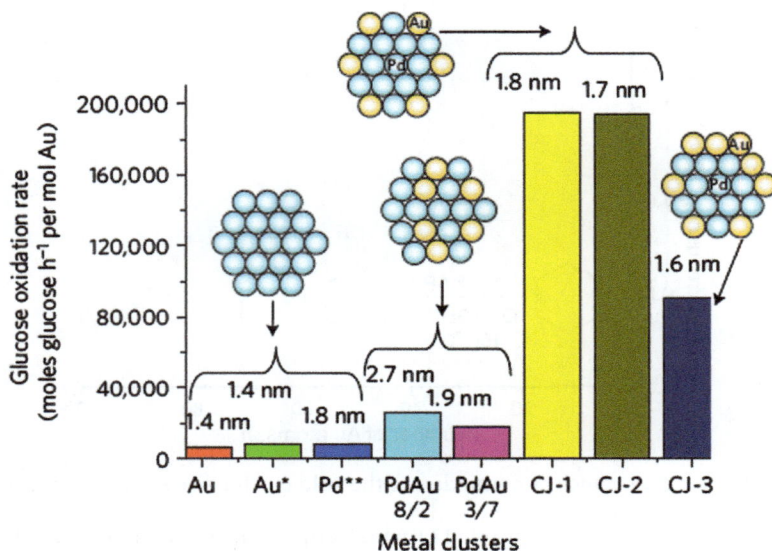

Figure 9.13 Comparison of the catalytic activity of CJ-Au/Pd, Au, Pd and Pd/Au alloy NCs for aerobic glucose oxidation. Schematic insets and numbers shown at the top of each bar indicate the structure models and the average particle sizes, respectively, of the NCs; Au*, the activity was normalized by the number of surface Au atoms in NCs; Pd**, the activity was normalized by the number of surface Pd atoms, activity of 8290 moles glucose h^{-1} per mole surface Pd.
Reprinted from ref. 115 with permission by Nature Publishing Group.

have also been extensively studied over past years. Kawasaki and co-workers[116] investigated the catalytic properties of DMF stabilized Au nanoclusters in a reduction of 4-nitrophenol to 4-aminophenol. It was found that the DMF-stabilized Au nanoclusters showed high catalytic activity even when used in small quantities. In another report, Shivhare *et al.*[117] found that thiol-stabilized $Au_{25}L_{18}$ nanoclusters are also active for the reduction of 4-nitrophenol. More interestingly, they found that these Au_{25} nanoclusters are stable and do not lose their structural integrity during the catalytic process. Recently, Jin's group[118] reported the catalytic performance of Au_{25} nanoclusters for the selective hydrogenation of α,β-unsaturated ketones and aldehydes to unsaturated alcohols. It was found that the Au_{25} nanoclusters can achieve high conversion and 100% chemoselectivity for the unsaturated alcohol. Compared with conventional Au nanoparticles, the Au_{25} nanoclusters are catalytically active for hydrogenation reactions even at low temperatures (*e.g.* 0 °C). The authors attributed the observed catalytic performance of Au_{25} nanoclusters to the core–shell structure (Au_{13} core/Au_{12} shell) and their unique electronic properties (electron-rich Au_{13} core and low-coordinate ($N = 3$) surface gold atoms).

For Pd nanoclusters, Hyotanishi[72] demonstrated that the surfactant-free Pd nanoclusters (1–1.5 nm) display high catalytic activity in the Suzuki–Miyaura cross-coupling and Mizoroki–Heck reactions with a high turnover number, which can be recycled at least five times without loss of catalytic activity. Surfactant-free copper nanoclusters have also been used as an efficient catalyst in the Ullmann-coupling reaction.[76] In addition, Vilar-Vidal *et al.*[119] demonstrated that Cu nanoclusters exhibit strong size-dependent catalytic properties in the clock redox methylene blue–leucomethylene blue (MB–LMB) reaction. It was found that only Cu nanoclusters smaller than $\approx$ 10–13 atoms enhance the rate of the reduction of MB.

9.4   Conclusions and Future Outlook

In this chapter, we summarized the recent research progress on the size-controlled synthesis and size-dependent catalytic activity of metal nanoclusters. Especially, we chose several types of monodisperse Au nanoclusters, such as $Au_{25}(SR)_{18}$, $Au_{38}(SR)_{24}$, $Au_{102}(SR)_{44}$, $Au_{144}(SR)_{60}$, and Ag, Pd, Pt, Cu nanoclusters to illustrate the size-controlled synthesis. Due to the unique properties of metal clusters, various methods have been developed to synthesize monodisperse metal nanoclusters in high yield and high purity over past years. Taking $Au_{25}(SR)_{18}$ as an example, a modified Brust–Schiffrin method *via* controlling the kinetics of the formation of Au(I):SR intermediates has been applied to the synthesis of highly monodisperse Au_{25}.[19] While for Au_{38} nanoclusters, kinetic control does not work due to the thermodynamic stability of Au_{38} *versus* the kinetic stability of Au_{25}. Thus, precise control of the experimental conditions will be of great importance for the synthesis of monodisperse nanoclusters with different core size. Based on the controlled synthesis of atomically monodisperse metal nanoclusters, a lot of potential applications of metal nanoclusters have been explored, especially in the field of catalysis. For example, the metal nanoclusters have been successfully used as catalysts in CO oxidation, the ORR, aerobic oxidation and so on.

Yet despite the substantial progress that has been achieved in the study of size-controlled synthesis and size-dependent catalytic activity of metal nanoclusters, some challenges remain in future work in this field. The emphases of future investigations should mainly include: (1) further developing facile, versatile synthetic techniques to produce highly monodisperse metal nanoclusters in high yield; (2) understanding theoretically the correlation between nanocluster structure, electronic properties and their catalytic properties; (3) exploring more practical applications of these nanoclusters and improving their catalytic performance and achieving cost-effective catalysts.

Acknowledgements

This work was supported by the National Natural Science Foundation of China with the Grant Numbers of 21275136, 21043013 and the Natural Science Foundation of Jilin province, China (No. 201215090).

References

1. W. Chen and S. W. Chen, *Angew. Chem., Int. Ed.*, 2009, **48**, 4386–4389.
2. H. Y. Fan, K. Yang, D. M. Boye, T. Sigmon, K. J. Malloy, H. F. Xu, G. P. Lopez and C. J. Brinker, *Science*, 2004, **304**, 567–571.
3. C. S. Wang, J. Y. Li, C. Amatore, Y. Chen, H. Jiang and X. M. Wang, *Angew. Chem., Int. Ed.*, 2011, **50**, 11644–11648.
4. H. Wohltjen and A. W. Snow, *Anal. Chem.*, 1998, **70**, 2856–2859.
5. N. L. Rosi, D. A. Giljohann, C. S. Thaxton, A. K. R. Lytton-Jean, M. S. Han and C. A. Mirkin, *Science*, 2006, **312**, 1027–1030.
6. R. C. Jin, *Nanoscale*, 2010, **2**, 343–362.
7. Y. Z. Lu and W. Chen, *Chem. Soc. Rev.*, 2012, **41**, 3594–3623.
8. M. Brust, M. Walker, D. Bethell, D. J. Schiffrin and R. Whyman, *J. Chem. Soc., Chem. Commun.*, 1994, 801–802.
9. C. J. Ackerson, P. D. Jadzinsky and R. D. Kornberg, *J. Am. Chem. Soc.*, 2005, **127**, 6550–6551.
10. M. Brust, J. Fink, D. Bethell, D. J. Schiffrin and C. Kiely, *J. Chem. Soc., Chem. Commun.*, 1995, 1655–1656.
11. T. G. Schaaff and R. L. Whetten, *J. Phys. Chem. B*, 2000, **104**, 2630–2641.
12. M. J. Hostetler, J. E. Wingate, C. J. Zhong, J. E. Harris, R. W. Vachet, M. R. Clark, J. D. Londono, S. J. Green, J. J. Stokes, G. D. Wignall, G. L. Glish, M. D. Porter, N. D. Evans and R. W. Murray, *Langmuir*, 1998, **14**, 17–30.
13. R. L. Whetten, J. T. Khoury, M. M. Alvarez, S. Murthy, I. Vezmar, Z. L. Wang, P. W. Stephens, C. L. Cleveland, W. D. Luedtke and U. Landman, *Adv. Mater.*, 1996, **8**, 428–433.
14. A. C. Templeton, D. E. Cliffel and R. W. Murray, *J. Am. Chem. Soc.*, 1999, **121**, 7081–7089.
15. V. L. Jimenez, M. C. Leopold, C. Mazzitelli, J. W. Jorgenson and R. W. Murray, *Anal. Chem.*, 2003, **75**, 199–206.
16. Z. K. Wu, J. Chen and R. C. Jin, *Adv. Funct. Mater.*, 2011, **21**, 177–183.
17. Y. Negishi, K. Nobusada and T. Tsukuda, *J. Am. Chem. Soc.*, 2005, **127**, 5261–5270.
18. S. Kumar, M. D. Bolan and T. P. Bigioni, *J. Am. Chem. Soc.*, 2010, **132**, 13141–13143.
19. M. Zhu, E. Lanni, N. Garg, M. E. Bier and R. Jin, *J. Am. Chem. Soc.*, 2008, **130**, 1138–1139.
20. M. W. Heaven, A. Dass, P. S. White, K. M. Holt and R. W. Murray, *J. Am. Chem. Soc.*, 2008, **130**, 3754–3755.
21. J. Akola, M. Walter, R. L. Whetten, H. Hakkinen and H. Gronbeck, *J. Am. Chem. Soc.*, 2008, **130**, 3756–3757.
22. M. Zhu, C. M. Aikens, F. J. Hollander, G. C. Schatz and R. Jin, *J. Am. Chem. Soc.*, 2008, **130**, 5883–5885.
23. Y. Negishi, N. K. Chaki, Y. Shichibu, R. L. Whetten and T. Tsukuda, *J. Am. Chem. Soc.*, 2007, **129**, 11322–11323.

24. Y. Shichibu, Y. Negishi, T. Tsukuda and T. Teranishi, *J. Am. Chem. Soc.*, 2005, **127**, 13464–13465.
25. G. H. Woehrle, M. G. Warner and J. E. Hutchison, *J. Phys. Chem. B*, 2002, **106**, 9979–9981.
26. Y. Y. Yang and S. W. Chen, *Nano Lett.*, 2003, **3**, 75–79.
27. N. K. Chaki, Y. Negishi, H. Tsunoyama, Y. Shichibu and T. Tsukuda, *J. Am. Chem. Soc.*, 2008, **130**, 8608–8610.
28. S. H. Xie, H. Tsunoyama, W. Kurashige, Y. Negishi and T. Tsukuda, *ACS Catal.*, 2012, **2**, 1519–1523.
29. Z. Wu, J. Suhan and R. C. Jin, *J. Mater. Chem.*, 2009, **19**, 622–626.
30. Z. K. Wu, *Angew. Chem., Int. Ed.*, 2012, **51**, 2934–2938.
31. H. F. Qian, D. E. Jiang, G. Li, C. Gayathri, A. Das, R. R. Gil and R. C. Jin, *J. Am. Chem. Soc.*, 2012, **134**, 16159–16162.
32. M. Z. Zhu, H. F. Qian and R. C. Jin, *J. Am. Chem. Soc.*, 2009, **131**, 7220–7221.
33. Y. Yu, Z. T. Luo, Y. Yu, J. Y. Lee and J. P. Xie, *ACS Nano*, 2012, **6**, 7920–7927.
34. X. Yuan, Y. Yu, Q. F. Yao, Q. B. Zhang and J. P. Xie, *J. Phys. Chem. Lett.*, 2012, **3**, 2310–2314.
35. O. Toikkanen, V. Ruiz, G. Ronholm, N. Kalkkinen, P. Liljeroth and B. M. Quinn, *J. Am. Chem. Soc.*, 2008, **130**, 11049–11055.
36. H. F. Qian, M. Z. Zhu, U. N. Andersen and R. C. Jin, *J. Phys. Chem. A*, 2009, **113**, 4281–4284.
37. T. G. Schaaff, M. N. Shafigullin, J. T. Khoury, I. Vezmar, R. L. Whetten, W. G. Cullen, P. N. First, C. Gutierrez-Wing, J. Ascensio and M. J. Jose-Yacaman, *J. Phys. Chem. B*, 1997, **101**, 7885–7891.
38. H. Tsunoyama, P. Nickut, Y. Negishi, K. Al-Shamery, Y. Matsumoto and T. Tsukuda, *J. Phys. Chem. C*, 2007, **111**, 4153–4158.
39. H. F. Qian, Y. Zhu and R. C. Jin, *ACS Nano*, 2009, **3**, 3795–3803.
40. P. D. Jadzinsky, G. Calero, C. J. Ackerson, D. A. Bushnell and R. D. Kornberg, *Science*, 2007, **318**, 430–433.
41. Y. Levi-Kalisman, P. D. Jadzinsky, N. Kalisman, H. Tsunoyama, T. Tsukuda, D. A. Bushnell and R. D. Kornberg, *J. Am. Chem. Soc.*, 2011, **133**, 2976–2982.
42. T. G. Schaaff and R. L. Whetten, *J. Phys. Chem. B*, 1999, **103**, 9394–9396.
43. H. F. Qian and R. C. Jin, *Nano Lett.*, 2009, **9**, 4083–4087.
44. H. F. Qian and R. C. Jin, *Chem. Mater.*, 2011, **23**, 2209–2217.
45. J. Zheng, J. T. Petty and R. M. Dickson, *J. Am. Chem. Soc.*, 2003, **125**, 7780–7781.
46. P. Yu, X. M. Wen, Y. R. Toh and J. Tang, *J. Phys. Chem. C*, 2012, **116**, 6567–6571.
47. C. D. Grant, A. M. Schwartzberg, Y. Y. Yang, S. W. Chen and J. Z. Zhang, *Chem. Phys. Lett.*, 2004, **383**, 31–34.
48. G. E. Johnson, C. Wang, T. Priest and J. Laskin, *Anal. Chem.*, 2011, **83**, 8069–8072.
49. Y. Shichibu, K. Suzuki and K. Konishi, *Nanoscale*, 2012, **4**, 4125–4129.

50. Q. Xu, S. X. Wang, Z. Liu, G. Y. Xu, X. M. Meng and M. Z. Zhu, *Nanoscale*, 2013, **5**, 1176–1182.
51. Z. K. Wu, M. A. MacDonald, J. Chen, P. Zhang and R. C. Jin, *J. Am. Chem. Soc.*, 2011, **133**, 9670–9673.
52. H. F. Zhang, M. Stender, R. Zhang, C. M. Wang, J. Li and L. S. Wang, *J. Phys. Chem. B*, 2004, **108**, 12259–12263.
53. P. R. Nimmala and A. Dass, *J. Am. Chem. Soc.*, 2011, **133**, 9175–9177.
54. H. Tsunoyama, Y. Negishi and T. Tsukuda, *J. Am. Chem. Soc.*, 2006, **128**, 6036–6037.
55. Z. K. Wu, E. Lanni, W. Q. Chen, M. E. Bier, D. Ly and R. C. Jin, *J. Am. Chem. Soc.*, 2009, **131**, 16672–16674.
56. T. U. B. Rao, B. Nataraju and T. Pradeep, *J. Am. Chem. Soc.*, 2010, **132**, 16304–16307.
57. I. Chakraborty, A. Govindarajan, J. Erusappan, A. Ghosh, T. Pradeep, B. Yoon, R. L. Whetten and U. Landman, *Nano Lett.*, 2012, **12**, 5861–5866.
58. T. U. B. Rao and T. Pradeep, *Angew. Chem., Int. Ed.*, 2010, **49**, 3925–3929.
59. M. Q. Zhao, L. Sun and R. M. Crooks, *J. Am. Chem. Soc.*, 1998, **120**, 4877–4878.
60. L. Balogh and D. A. Tomalia, *J. Am. Chem. Soc.*, 1998, **120**, 7355–7356.
61. N. Vilar-Vidal, M. C. Blanco, M. A. Lopez-Quintela, J. Rivas and C. Serra, *J. Phys. Chem. C*, 2010, **114**, 15924–15930.
62. M. T. Reetz and W. Helbig, *J. Am. Chem. Soc.*, 1994, **116**, 7401–7402.
63. C. Vazquez-Vazquez, M. Banobre-Lopez, A. Mitra, M. A. Lopez-Quintela and J. Rivas, *Langmuir*, 2009, **25**, 8208–8216.
64. H. Kawasaki, Y. Kosaka, Y. Myoujin, T. Narushima, T. Yonezawa and R. Arakawa, *Chem. Commun.*, 2011, **47**, 7740–7742.
65. Y. Z. Lu, W. T. Wei and W. Chen, *Chin. Sci. Bull.*, 2012, **57**, 41–47.
66. P. J. G. Goulet and R. B. Lennox, *J. Am. Chem. Soc.*, 2010, **132**, 9582–9584.
67. W. T. Wei, Y. Z. Lu, W. Chen and S. W. Chen, *J. Am. Chem. Soc.*, 2011, **133**, 2060–2063.
68. B. C. H. Steele and A. Heinzel, *Nature*, 2001, **414**, 345–352.
69. K. Yamamoto, T. Imaoka, W. J. Chun, O. Enoki, H. Katoh, M. Takenaga and A. Sonoi, *Nat. Chem.*, 2009, **1**, 397–402.
70. S. I. Tanaka, J. Miyazaki, D. K. Tiwari, T. Jin and Y. Inouye, *Angew. Chem., Int. Ed.*, 2011, **50**, 431–435.
71. Y. Negishi, H. Murayama and T. Tsukuda, *Chem. Phys. Lett.*, 2002, **366**, 561–566.
72. M. Hyotanishi, Y. Isomura, H. Yamamoto, H. Kawasaki and Y. Obora, *Chem. Commun.*, 2011, **47**, 5750–5752.
73. X. F. Liu, C. H. Li, J. L. Xu, J. Lv, M. Zhu, Y. B. Guo, S. Cui, H. B. Liu, S. Wang and Y. L. Li, *J. Phys. Chem. C*, 2008, **112**, 10778–10783.
74. H. Kawasaki, H. Yamamoto, H. Fujimori, R. Arakawa, Y. Iwasaki and M. Inada, *Langmuir*, 2010, **26**, 5926–5933.

75. H. Kawasaki, H. Yamamoto, H. Fujimori, R. Arakawa, M. Inada and Y. Iwasaki, *Chem. Commun.*, 2010, **46**, 3759–3761.
76. Y. Isomura, T. Narushima, H. Kawasaki, T. Yonezawa and Y. Obora, *Chem. Commun.*, 2012, **48**, 3784–3786.
77. J. C. Fierro-Gonzalez and B. C. Gates, *Chem. Soc. Rev.*, 2008, **37**, 2127–2134.
78. W. L. Xu, J. S. Kong, Y. T. E. Yeh and P. Chen, *Nat. Mater.*, 2008, 7, 992–996.
79. M. Haruta, N. Yamada, T. Kobayashi and S. Iijima, *J. Catal.*, 1989, **115**, 301–309.
80. M. Haruta, S. Tsubota, T. Kobayashi, H. Kageyama, M. J. Genet and B. Delmon, *J. Catal.*, 1993, **144**, 175–192.
81. A. A. Herzing, C. J. Kiely, A. F. Carley, P. Landon and G. J. Hutchings, *Science*, 2008, **321**, 1331–1335.
82. N. Lopez, T. V. W. Janssens, B. S. Clausen, Y. Xu, M. Mavrikakis, T. Bligaard and J. K. Norskov, *J. Catal.*, 2004, **223**, 232–235.
83. M. Valden, X. Lai and D. W. Goodman, *Science*, 1998, **281**, 1647–1650.
84. G. R. Bamwenda, S. Tsubota, T. Nakamura and M. Haruta, *Catal. Lett.*, 1997, **44**, 83–87.
85. S. S. Lee, C. Y. Fan, T. P. Wu and S. L. Anderson, *J. Am. Chem. Soc.*, 2004, **126**, 5682–5683.
86. N. Nikbin, G. Mpourmpakis and D. G. Vlachos, *J. Phys. Chem. C*, 2011, **115**, 20192–20200.
87. N. Lopez and J. K. Norskov, *J. Am. Chem. Soc.*, 2002, **124**, 11262–11263.
88. S. Lee, C. Y. Fan, T. P. Wu and S. L. Anderson, *J. Chem. Phys.*, 2005, **123**, 124710.
89. Y. Gao, N. Shao, Y. Pei, Z. F. Chen and X. C. Zeng, *ACS Nano*, 2011, **5**, 7818–7829.
90. A. Sanchez, S. Abbet, U. Heiz, W. D. Schneider, H. Hakkinen, R. N. Barnett and U. Landman, *J. Phys. Chem. A*, 1999, **103**, 9573–9578.
91. B. Yoon, H. Hakkinen, U. Landman, A. S. Worz, J. M. Antonietti, S. Abbet, K. Judai and U. Heiz, *Science*, 2005, **307**, 403–407.
92. L. M. Molina and B. Hammer, *J. Catal.*, 2005, **233**, 399–404.
93. M. Mavrikakis, P. Stoltze and J. K. Norskov, *Catal. Lett.*, 2000, **64**, 101–106.
94. J. D. Grunwaldt and A. Baiker, *J. Phys. Chem. B*, 1999, **103**, 1002–1012.
95. Z. P. Liu, X. Q. Gong, J. Kohanoff, C. Sanchez and P. Hu, *Phys. Rev. Lett.*, 2003, **91**, 266102.
96. D. Matthey, J. G. Wang, S. Wendt, J. Matthiesen, R. Schaub, E. Laegsgaard, B. Hammer and F. Besenbacher, *Science*, 2007, **315**, 1692–1696.
97. X. T. Nie, H. F. Qian, Q. J. Ge, H. Y. Xu and R. C. Jin, *ACS Nano*, 2012, **6**, 6014–6022.
98. J. Zhang, K. Sasaki, E. Sutter and R. R. Adzic, *Science*, 2007, **315**, 220–222.
99. W. Tang, H. F. Lin, A. Kleiman-Shwarsctein, G. D. Stucky and E. W. McFarland, *J. Phys. Chem. C*, 2008, **112**, 10515–10519.

100. E. Yoo, T. Okata, T. Akita, M. Kohyama, J. Nakamura and I. Honma, *Nano Lett.*, 2009, **9**, 2255–2259.
101. X. M. Chen, G. H. Wu, J. M. Chen, X. Chen, Z. X. Xie and X. R. Wang, *J. Am. Chem. Soc.*, 2011, **133**, 3693–3695.
102. H. J. Yin, H. J. Tang, D. Wang, Y. Gao and Z. Y. Tang, *ACS Nano*, 2012, **6**, 8288–8297.
103. C. Jeyabharathi, S. S. Kumar, G. V. M. Kiruthika and K. L. N. Phani, *Angew. Chem., Int. Ed.*, 2010, **49**, 2925–2928.
104. M. H. Shao, A. Peles and K. Shoemaker, *Nano Lett.*, 2011, **11**, 3714–3719.
105. Y. Z. Lu and W. Chen, *J. Power Sources*, 2012, **197**, 107–110.
106. W. T. Wei and W. Chen, *Int. J. Smart Nano Mater.*, 2013, **4**, 62–71.
107. M. Haruta, *Nature*, 2005, **437**, 1098–1099.
108. T. Ishida and M. Haruta, *Angew. Chem., Int. Ed.*, 2007, **46**, 7154–7156.
109. J. K. Edwards, B. Solsona, E. N. N. A. F. Carley, A. A. Herzing, C. J. Kiely and G. J. Hutchings, *Science*, 2009, **323**, 1037–1041.
110. D. Gajan, K. Guillois, P. Delichere, J. M. Basset, J. P. Candy, V. Caps, C. Coperet, A. Lesage and L. Emsley, *J. Am. Chem. Soc.*, 2009, **131**, 14667–14669.
111. P. P. Edwards and J. M. Thomas, *Angew. Chem., Int. Ed.*, 2007, **46**, 5480–5486.
112. M. Turner, V. B. Golovko, O. P. H. Vaughan, P. Abdulkin, A. Berenguer-Murcia, M. S. Tikhov, B. F. G. Johnson and R. M. Lambert, *Nature*, 2008, **454**, 981–983.
113. H. Tsunoyama, N. Ichikuni, H. Sakurai and T. Tsukuda, *J. Am. Chem. Soc.*, 2009, **131**, 7086–7093.
114. Y. M. Liu, H. Tsunoyama, T. Akita, S. H. Xie and T. Tsukuda, *ACS Catal.*, 2011, **1**, 2–6.
115. H. J. Zhang, T. Watanabe, M. Okumura, M. Haruta and N. Toshima, *Nat. Mater.*, 2012, **11**, 49–52.
116. H. Yamamoto, H. Yano, H. Kouchi, Y. Obora, R. Arakawa and H. Kawasaki, *Nanoscale*, 2012, **4**, 4148–4154.
117. A. Shivhare, S. J. Ambrose, H. X. Zhang, R. W. Purves and R. W. J. Scott, *Chem. Commun.*, 2013, **49**, 276–278.
118. Y. Zhu, H. F. Qian, B. A. Drake and R. C. Jin, *Angew. Chem., Int. Ed.*, 2010, **49**, 1295–1298.
119. N. Vilar-Vidal, J. Rivas and M. A. Lopez-Quintela, *ACS Catal.*, 2012, **2**, 1693–1697.

CHAPTER 10

Metal Clusters in Catalysis

SEIJI YAMAZOE[a,b] AND TATSUYA TSUKUDA*[a,b]

[a] Department of Chemistry, School of Science, The University of Tokyo, Japan; [b] Elements Strategy Initiative for Catalysts and Batteries, Kyoto University, Japan
*Email: tsukuda@chem.s.u-tokyo.ac.jp

10.1 Introduction

10.1.1 Why Metal Clusters?

Metal nanoparticles with diameters in the range of a few to several tens of nanometers have conventionally been used as active sites in supported metal catalysts.[1] The primary motivation for using metal nanoparticles in catalytic applications is to enhance the surface-to-volume ratio and population of low coordination sites, which are expected to show higher activity than flat surfaces of bulk metals. Another motivation is to selectively expose specific facets that are suitable for a given catalytic conversion by controlling the morphology of the nanoparticles. The basic idea behind these applications is that the catalytic properties of nanoparticles can be predicted from those of model surface systems once the dimensions and morphology are determined.

How will the catalytic properties change when the diameter of the metal particles becomes smaller than 2 nm or when the number of constituent atoms is less than ~100? One may argue that the catalytic properties of such small particles, *metal clusters*, will smoothly converge with those of the corresponding single atom with the reduction of size because both clusters

RSC Smart Materials No. 7
Functional Nanometer-Sized Clusters of Transition Metals: Synthesis, Properties and Applications
Edited by Wei Chen and Shaowei Chen

and nanoparticles have similar structural features: large surface-to-volume ratios and quantized electronic structures. However, the catalytic properties of clusters and nanoparticles are essentially different as described below (Scheme 10.1).

Reduction of the surface energy is a major factor determining the geometrical structures of metal clusters and nanoparticles. In the case of nanoparticles, the surface energy is reduced by truncation of the corner sites of nanocrystals, as exemplified by the formation of cuboctahedral nanocrystals.[2] The packing arrangement of the constituent atoms is the same as that of bulk metal, although the metal-to-metal distance tends to shrink slightly so as to minimize the surface area. In other words, metal nanoparticles can be viewed as fragments of the corresponding bulk metal. Hence, the surface structure of metal nanoparticles such as population of low coordination sites (vertex, corner, edge) and plane indices can be estimated once their size and morphology are given. Therefore, we can design their catalysis by controlling these structural parameters of metal nanoparticles. On the other hand, constituent atoms of metal clusters are arranged in a significantly different manner from those of the bulk metal to reduce the surface energy.[3] As a result, metal clusters take unique morphologies, such as icosahedral and decahedral motifs having five-fold

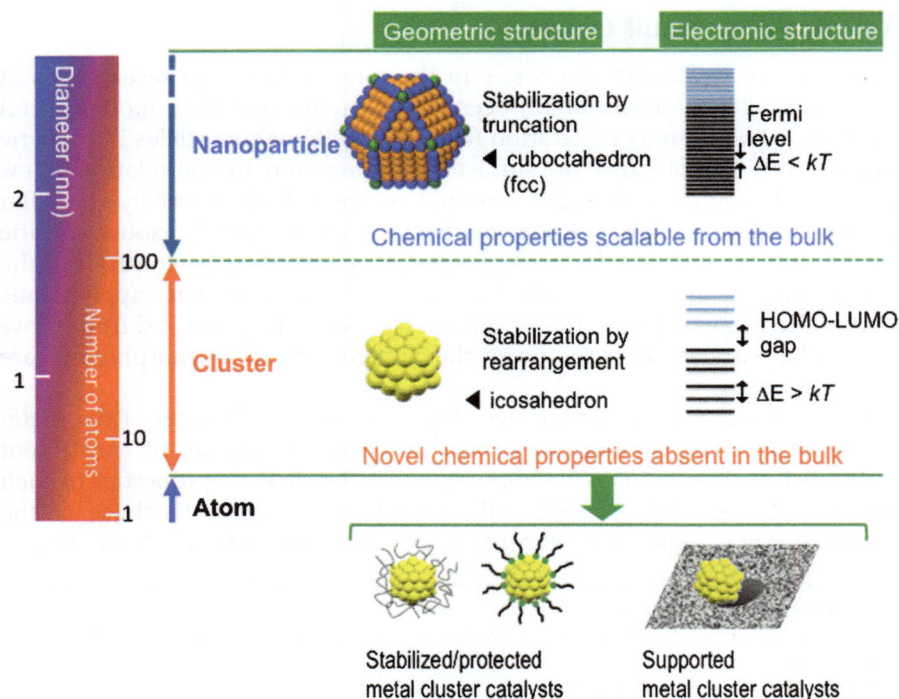

Scheme 10.1 Development of novel catalysts based on metal clusters.

symmetry. Such distinct packing of the atoms in the clusters may provide unique reaction sites.

Regarding the electronic structures, metal nanoparticles have quantized electronic levels with a typical energy gap smaller than the thermal energy (k_BT) as predicted by the theory of Kubo *et al.*[4] Because of the quantization of the electronic structures, metal nanoparticles exhibit novel magnetic and thermal properties at low temperature. In contrast, the catalytic properties of metal nanoparticles may not be affected appreciably by the quantized nature of electronic structures because the electronic structure can be viewed as a continuous band at room temperature. On the other hand, the energy gap between the quantized states of metal clusters (<2 nm) is estimated to exceed the thermal energy.[5] This is evidenced by a clear onset of optical absorption due to the transition from the highest occupied molecular orbital (HOMO) to the lowest unoccupied molecular orbital (LUMO). Activation of reactants on metal clusters by electronic interaction will differ from that by metal nanoparticles, reflecting the quantized electronic structures.

In addition to the above static properties, dynamic properties will significantly affect the catalysis of metal clusters. In general, the melting points of the clusters are much lower than those of the bulk metal[6] and, as a result, their geometrical structures may be readily changed during the catalytic reaction because of exothermicity induced by adsorption and chemical reactions of reactants. Moreover, fluctuation of the electronic structures is accompanied by geometrical isomerization because they are strongly coupled with each other. Such flexibility in geometrical and electronic structures of clusters will open up new reaction pathways having lower activation barriers. Because of these unique static and dynamic properties, metal clusters are expected to show novel chemical properties that are absent in the corresponding nanoparticles and bulk surfaces. Indeed, experimental and theoretical studies on bare metal clusters isolated in the gas phase,[3,7] dispersed in liquids,[8] and anchored on well-defined surfaces,[9] have demonstrated the high potential of metal clusters as unique catalytic sites.

However, it is challenging to use metal clusters as real-world catalysts because of the difficulty of controlled synthesis. The primary requisite for the synthesis of metal clusters is to stabilize them against aggregation. They aggregate easily and irreversibly into larger particles upon contact with each other because of their large surface energy. At the same time, a part of the cluster surface must be exposed for catalytic conversion of chemical substrates, which induces destabilization of the clusters. Thus, we must optimize the trade-off between stability and activity. The second requirement is to precisely control the structural parameters of the clusters, such as size and composition, because the intrinsic chemical properties of the clusters are strongly dependent on these parameters. If the structural parameters have distributions, the correlation between structure and catalysis cannot be established because we cannot exclude the possibility that clusters with a low population or undetectable by conventional experimental probes play an essential role in the catalysis. Therefore, precision synthesis at the atomic level is essential to understand the origin of catalysis and to develop novel cluster-based catalysts by taking advantage of their unique properties.

10.1.2 Classification

There are two main approaches to avoiding physical contact between individual clusters: (1) steric protection of the clusters by organic ligands or polymers and (2) immobilization of the clusters on a solid surface. The advantages and disadvantages of each system are briefly summarized in the next subsection.

10.1.2.1 *Stabilized/Protected Metal Clusters*

Metal clusters can be stabilized by organic polymers *via* multiple weak coordination as in the case of metal nanoparticles.[1,10] The resulting polymer-stabilized metal clusters can be finely dispersed in solvents as in the case of conventional homogeneous catalysts. Benefits of this system are high controllability of the cluster size and high catalytic activity due to the large exposed area of the surface. However, in most cases, the clusters are not robust enough under harsh reaction conditions to avoid aggregation into larger particles. Moreover, it is not easy to reuse the clusters because they have to be recovered by a time-consuming ultrafiltration process. The stability of the clusters can be improved by protection with ligands such as phosphines[11] and thiols.[12] Although it is believed that they show no or low catalytic activity due to high surface coverage of the organic ligands, several exceptions have been reported recently.[13-15] Thus, this chapter also deals with the synthesis and catalysis of ligand-protected clusters.

10.1.2.2 *Supported Metal Clusters*

Metal clusters can be stabilized against sintering by immobilization on various solid materials while exposing the surfaces.[16,17] Metal oxides and carbon materials have conventionally been used as supports, but mesoporous materials with high specific surface areas are promising materials to prevent the aggregation of clusters during synthesis and catalytic reactions. Usually this class of supported clusters cannot be dispersed in solvents as in the case of conventional heterogeneous catalysts. Compared with stabilized/protected clusters, supported metal clusters offer the benefits of robustness and ease of reuse, but their drawback is the difficulty of controlling size by the conventional method.

10.2 Stabilized/Protected Metal Cluster Catalysts

10.2.1 Size-Controlled Synthesis

Chemical reduction of metal ions in the presence of organic polymers or ligands yields stabilized/protected metal clusters. To synthesize metal clusters with a diameter of <2 nm, the growth of the nucleus should be terminated at an early stage by increasing the relative concentration of the protecting agents to the metal ions and by rapid consumption of the atoms in the formation of nuclei by using a strong reducing agent ($NaBH_4$, *etc.*). In the following, controlled synthesis of clusters stabilized by polymers or protected by ligands is overviewed.

10.2.1.1 Polymer-Stabilized Metal Clusters

Linear polymers conventionally used in colloidal metal particles, such as polyvinylpyrrolidone (PVP), polyaminothiophenol (PAT) and polyethylene glycol (PEG), can also be used as stabilizers of metal clusters.[1,10] However, it is very difficult to kinetically control the cluster size at the atomic level by adjusting synthetic conditions such as temperature and concentration. Post-synthetic fractionation by size also cannot be applied because the number of polymers involved in the stabilization of individual clusters cannot be fixed. Tsukuda *et al.* demonstrated that the size distribution was improved remarkably by homogeneous mixing of solutions of metal ions and reductants using a microfluidic mixer.[18] In the microfluidic mixer, the two solutions of $HAuCl_4$ and $NaBH_4$ containing PVP were laminated into thin substreams (70 µm in the case of SIMM-V2, Institute für Mikrotechnik Mainz GmbH) and overlayed. Figure 10.1 compares transmission electron micrograph (TEM) images of PVP-stabilized gold clusters (Au:PVP) prepared by the batch mixing and microfluidic mixing methods.[19] The size distribution was drastically improved as the average diameter was decreased to 1.1 nm using the micro mixer. Moreover, the size distribution of Au:PVP was characterized with atomic resolution by matrix assisted laser desorption/ionization (MALDI) mass spectrometry.[20] Interestingly, populations of $Au_{\sim 35}$ and Au_{58}, which are known as magic clusters in the isolated form,[21] became prominent by increasing the temperature of the mixer. Similarities in the magic numbers indicate that the stability of the Au clusters in PVP is determined by the electronic structures.[20]

Cluster size can be controlled more precisely using molecular templates, such as dendrimers (DEN) (an arborized polymer with a regularly-branched structure from a core)[22–28] and ferritin (a self-aggregate constructed by 24 protein subunits with 8 nm inner diameter).[29,30] Crooks and coworkers proposed atomically-precise synthesis by reducing a given number of metal ions coordinated in the interior of the DENs (Figure 10.2).[25] Figure 10.3(a) shows a TEM image of Pd_{40} encapsulated within a polyamidoamine (PAMAM) DEN.[25] Dickson and coworkers demonstrated the production of a series of size-controlled Au clusters (Au_5, Au_8, Au_{13}, Au_{23}, and Au_{31}) within PAMAM-DEN that are highly photoluminescent.[24] Somorjai and coworkers synthesized Rh_{30} and Pt_{40} in PAMAM-DENs and immobilized them within channels of mesoporous silica [Figure 10.3(b) and (c)] for catalytic applications.[26] Other types of dendrimer have also been used: Yamamoto *et al.* used phenyl azomethine (DPA)-DENs to synthesize Pt_{12}, Pt_{28}, and Pt_{60}[27] and Kaneda and coworkers used polypropylenimine (PPI)-DENs to synthesize

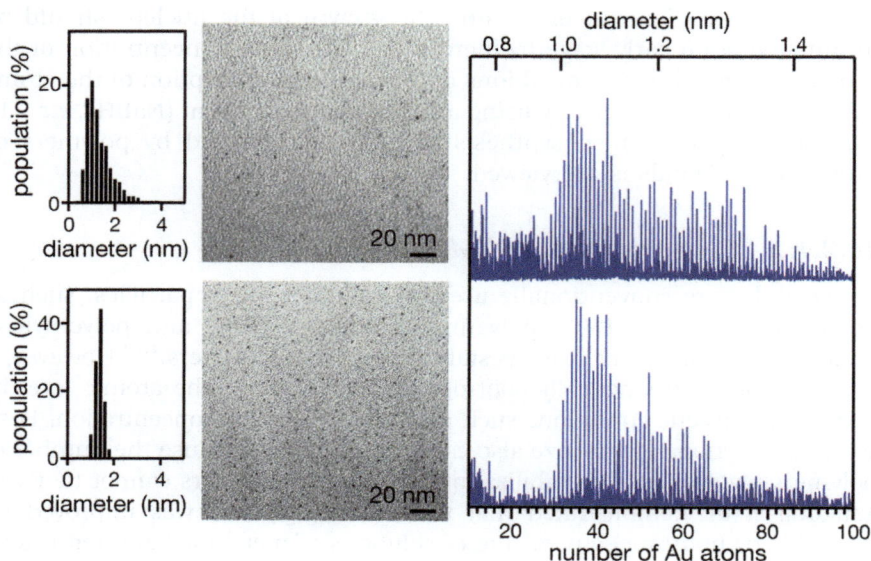

Figure 10.1   TEM images, distributions of diameters and MALDI mass spectra of Au:PVP prepared by batch mixing (top) and microfluidic mixing methods (bottom).[19]
Copyright © 2011 WILEY-VCH Verlag GmbH & Co. KGaA, Weinheim.

Figure 10.2   Synthesis scheme of metal clusters within PAMAM-DENs.[25]
Copyright © 2005 American Chemical Society.

Pd_4, Pd_8, and Pd_{16}.[28] These examples illustrate that the interior of DENs is a promising field to control the cluster size with atomic precision for catalytic applications.

10.2.1.2   Ligand-Protected Metal Clusters

Metal clusters protected by organic ligands, such as phosphine (R_3-P)[11] and thiol $(R-SH)$,[12,31,32] have several advantages in terms of controlled synthesis as compared with those stabilized by polymers. First, it is possible to fractionate the clusters from the polydispersed mixtures using size exclusion chromatography, polyacrylamide gel electrophoresis, and extraction with solvents.[33] Then, the chemical composition of the protected clusters (the numbers of metal atoms and organic ligands) can be determined precisely

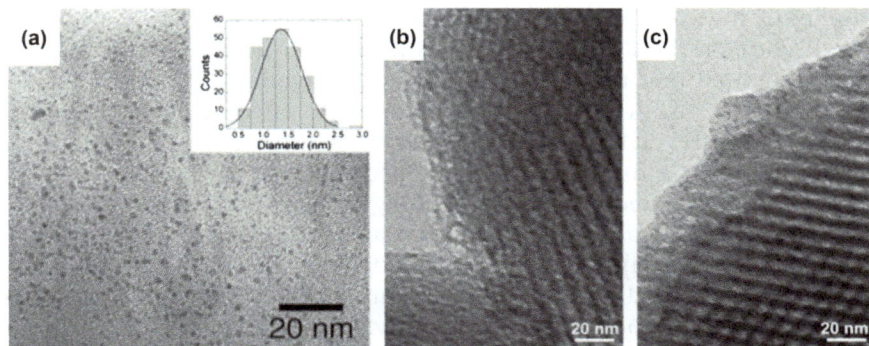

Figure 10.3 (a) High-resolution TEM image and a cluster diameter distribution of $Pd_{40}@PAMAM-DENs.$[25] TEM images of PAMAM-DEN encapsulated (b) Rh_{30} and (c) Pt_{40} immobilized on SBA-15.[26] Copyright © 2005 and 2008 American Chemical Society.

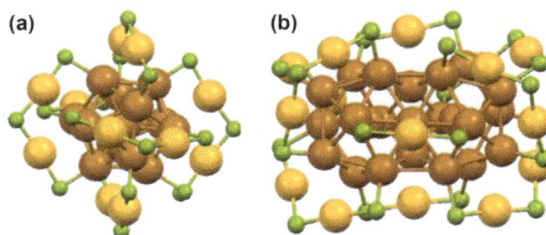

Figure 10.4 Crystallographic structures of the cores of (a) $Au_{25}(SR)_{18}$ and (b) $Au_{38}(SR)_{24}$. The brown, yellow, and green balls show Au(0), Au(I), and S atoms, respectively. The R groups are omitted for clarity.

using non-destructive mass spectrometry (MS), such as MALDI-MS and electrospray ionization ESI-MS.[33] Nevertheless, the key advantage is that the geometric structures can be determined by X-ray diffraction if single crystals are obtained.[34]

Thiolate-protected gold clusters have been studied extensively in the last two decades since the first report by Brust *et al.*[12] A series of $Au_n(SR)_m$ with well-defined sizes ($n = 11$, 12, 18, 22, 25, 29, 33, 36, 38, 39, 55, 68, 102, 130, 144, 187, 333) have been isolated so far by Whetten, Murray, Tsukuda, Negishi, Jin, and Dass.[5,35] After the groundbreaking report on the structure determination of $Au_{102}(SR)_{44}$ by Kornberg and coworkers,[34] the structures of $Au_{25}(SR)_{18}$, $Au_{36}(SR)_{24}$, and $Au_{38}(SR)_{24}$ have been determined by single-crystal X-ray diffraction (XRD).[36-39] These XRD studies revealed that the stable $Au_n(SR)_m$ clusters are comprised of a highly symmetrical Au core whose surface atoms are completely protected by $-SR-(Ag-SR)_n-$ oligomers (called "staples"[34]) of different lengths. Figure 10.4 shows that $Au_{25}(SR)_{18}$ has an icosahedral Au_{13} core surrounded by six staples of $-SR-(Au-SR)_2-$ units,[36,37] whereas $Au_{38}(SR)_{24}$ has a face-sharing bi-icosahedral Au_{23} core surrounded by three $-SR-(Au-SR)-$ and six $-SR-(Au-SR)_2-$ staples.[39] The density functional theory (DFT) calculations not only explained the high stability of the isolated clusters by a superatom concept,[40] but also "pre-dicted" correctly the structures of $Au_{25}(SR)_{18}$ and $Au_{38}(SR)_{24}$.[41–43] In contrast, it was proposed that the formation of Au–SR oligomers at the interface is suppressed in $Au_{41}(S–Eind)_{12}$ due to the steric effect around the sulfur moiety (Eind–SH = 1,1,3,3,5,5,7,7-octaethyl-s-hydrindacene-4-thiol).[44] Thio-late-protected Ag_n clusters with well-defined sizes ($n = 7$, 8, 9, 32, 44, ~75, 152) have been isolated by the groups of Bigioni, Stellacci, and Pradeep.[45–50] At present, there is no report on single-crystal XRD. Theoretical calculations for $Ag_{152}(SCH_2CH_2Ph)_{60}$ by Landman predicted that 60 Ag atoms and 60 thiolates are arranged in a network of six-membered rings on an Au_{92} core having icosahedral-dodecahedral symmetry.[48]

Mono-dentate phosphines, such as triphenyl phosphines, have long been used as protecting ligands of gold clusters. The crystal structures of Au_{11}[51] and Au_{13}[52] showed that the Au core favors icosahedral-based motifs. The cuboctahedral core structure proposed for the most famous $Au_{55}(PPh_3)_{12}Cl_6$[11] has remained a matter of debate. Recently, bi-dentate phosphines, such as $Ph_2P(CH_2)_2PPh_2$ (DPPE), have been used as more effi-cient protecting reagents. Shichibu and Konishi reported the synthesis of icosahedral Au_{13}[53] and an isomeric form of Au_{11}^{3+} having a nonspherical [core + *exo*]-type structure.[54]

Au clusters with icosahedral-based structures have been synthesized using mixed ligands of phosphines and thiolates. Nunokawa *et al.* synthesized an icosahedral Au_{11}^{3+}; interestingly, thiolates are directly bonded to the Au_{11} core through Au–S bonds and act as electron scavengers.[55] Tsukuda and coworkers synthesized a vertex-sharing biicosahedral Au_{25}^{9+}[56] whereas Jin and coworkers reported Au_{24}^{8+}[57] which is composed of two incomplete (*i.e.*, one vertex missing) icosahedral Au_{12} units joined by five thiolate linkages (Figure 10.5). Recently, Zheng and coworkers determined the structures of Ag clusters protected by mixed ligands, such as $Ag_{14}(SC_6H_{13}F_2)_{12}(PPh_3)_8$, $Ag_{16}(SC_6H_3F_2)_{14}(DPPE)_4$ and $[Ag_{32}(DPPE)_5(SC_6H_4CF_3)_{24}]^{2-}$, by single-crystal

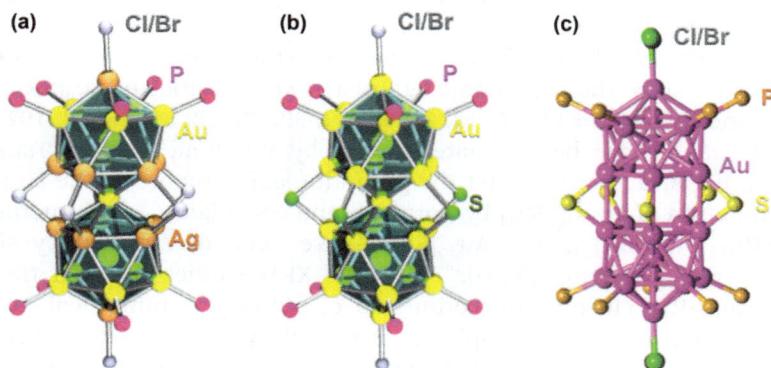

Figure 10.5   Crystallographic structures of the cores of (a) $[Au_{13}Ag_{12}(PR_3)_{10}X_7]^{2+}$,[65] (b) $[Au_{25}(PR_3)_{10}(SR')_5X_2]^{2+}$,[56] (c) $[Au_{24}(PR_3)_{10}(SR')_5X_2]^+$ (X=Cl, Br).[57] The R and R' groups are omitted for clarity.
Copyright © 1990, 2007, and 2012 American Chemical Society.

XRD.[58,59] They have core–shell structures with Ag_6^{4+}, Ag_8^{6+} and Ag_{22}^{12+} as their cores, which are not simply either fragments of face-centered cubic metals or the five-fold twinned counterparts. Mizuno and coworkers also synthesized octahedral Ag_6^{4+} clusters which are protected by two silico-tungstate POM ligands.[60] Recently, alkynes have been used as new protecting ligands of Au clusters *via* Au–C covalent bonding: stable clusters such as $Au_{54}(C\equiv CPh)_{26}$ have been isolated.[61]

10.2.2 Composition-Controlled Synthesis

The catalytic properties of metal clusters can be improved by introducing second elements. In the early 1990s, Toshima *et al.* developed a preparation of PVP-stabilized bimetallic clusters with a core–shell structure by simultaneous reduction of two metal ions. For example, the formation of core–shell clusters with a Pt_{13} core and a shell of 42 Pd atoms was proposed although decisive evidence has not been provided.[62] Recently, Toshima and coworkers replaced Pd atoms at the corner sites of Pd_{147} with Au atoms by galvanic replacement reactions.[63] Crooks and coworkers prepared bimetallic clusters with alloy (or solid solution) and core–shell structures by the co-reduction and sequential reduction of two metal ions coordinated within PAMAM-DENs (Figure 10.6).[25] Yamamoto *et al.* synthesized bimetallic

1. Co-complexation Method

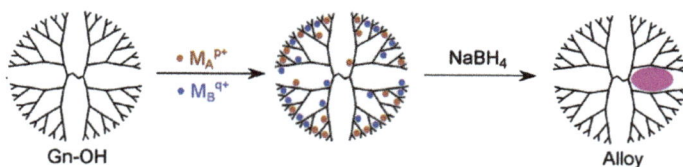

2. Sequential Method

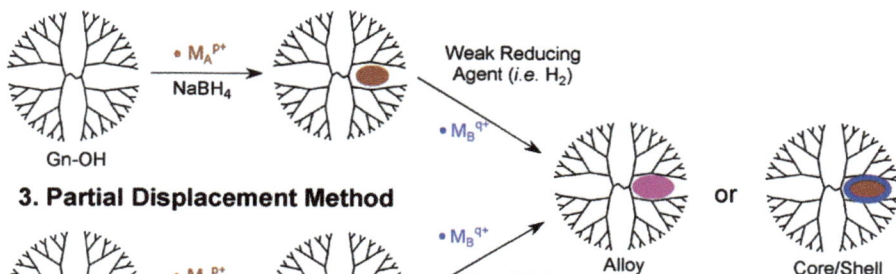

3. Partial Displacement Method

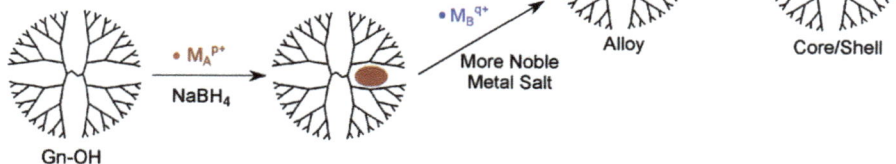

Figure 10.6 Schematic illustration of bimetallic clusters within DENs.[25] Copyright © 2005 American Chemical Society.

clusters with well-controlled compositions, $Pt_{16}Sn_{12}$ and $Pt_{24}Sn_4$ within DPA-DENs.[27]

Although "average" compositions of an ensemble of bimetallic clusters can be controlled by co-reduction of two metal precursor ions at a given molar ratio, it is formidably difficult to make uniform compositions of individual particles with atomic precision. In addition, it is very difficult to keep the average diameters constant while changing the molar ratio of the two elements.[64] Pioneering work on the precise control of compositions has been conducted by Teo and coworkers who synthesized mixed metal clusters with well-defined compositions using mixed ligands of phosphines and halides, such as $[Au_{13}Ag_{12}(PR_3)_{10}Cl_7]^{2+}$[65] and $[Au_{18}Ag_{20}(PR_3)_{14}Cl_{12}]^{2+}$.[66] Recently the groups of Murray, Negishi, Dass, and Jin reported the synthesis of unique sets of ligand-protected bimetallic clusters having the same number of total atoms as the pure Au clusters, $Au_{11}(PPh_3)_8Cl_2$, $Au_{25}(SR)_{18}$, $Au_{38}(SR)_{24}$, and $Au_{144}(SR)_{60}$.

(i) 11-atom clusters: $Pd_1Au_{10}(PPh_3)_8Cl_2$[67]

(ii) 25-atom clusters: $Pt_1Au_{24}(SR)_{18}$,[68] $Pd_1Au_{24}(SR)_{18}$,[69,70] $Ag_xAu_{25-x}(SR)_{18}$ $(1 \leq x \leq 11)$,[71,72] and $Cu_xAu_{25-x}(SR)_{18}$ $(1 \leq x \leq 5)$[72,73]

(iii) 38-atom clusters: $Pd_2Au_{36}(SR)_{24}$[74] and $Ag_xAu_{38-x}(SR)_{24}$ $(1 \leq x \leq 10)$[75]

(iv) 144-atom clusters: $Ag_xAu_{144-x}(SR)_{60}$ $(30 \leq x \leq 54)$[76]

The structures of the bimetallic clusters listed above have not yet been revealed by single-crystal XRD. Recently, Jin and coworkers reported that the doped Pt atom is located at the center of the icosahedral $PtAu_{12}$ core of $Pt_1Au_{24}(SR)_{18}$.[77] Selective introduction of one and two Pd atom(s) into $Au_{25}(SR)_{18}$ and $Au_{38}(SR)_{24}$, respectively, strongly supported the conjecture that the Pd atom(s) are located at the center of the icosahedral Au_{13} unit(s) (Figure 10.4), as theoretically supported by Nobusada and coworkers (Figure 10.7).[69,74] The preferred composition of $Au_{84}Ag_{60}(SC_2H_4Ph)_{60}$ has been ascribed to the formation of a core–shell structure in which all the Ag atoms occupy 60 surface sites of the Au_{84} core.[78]

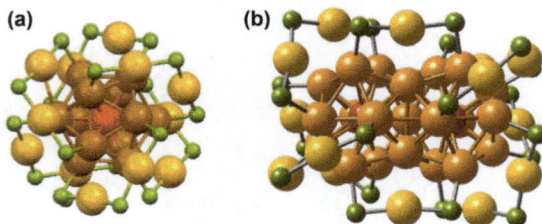

Figure 10.7   Optimized   structures   of $Pd_1Au_{24}(SCH_3)_{18}$ and $[Pd_2Au_{36}(SCH_3)_{18}]^{2-}$.[69,74] The CH_3 groups are omitted for clarity. The brown, yellow, and green balls show Au(0), Au(I), and S atoms, respectively. The red spheres represent Pd atoms.

10.2.3 Catalytic Applications

This section gives an overview of catalytic applications of the stabilized/protected metal clusters for various types of chemical reactions (Table 10.1).

10.2.3.1 Oxidations

An example of cluster-specific catalysis can be found in Au:PVP, which are active against a wide variety of oxidation reactions using oxygen as an oxidant (Table 10.1).[19] The particle size dependence on catalytic activity for the aerobic oxidation of *p*-hydroxybenzyl alcohol was investigated using monodispersed Au:PVP with average diameters in the range of 1–10 nm. Turn-over frequency (TOF; a reaction rate constant normalized by the number of surface atoms of the cluster) was plotted against the particle diameter (Figure 10.8).[19,79] The Au nanoparticles with ≤4 nm diameter showed catalytic activity and the activity monotonically increased with smaller particle size. This behavior clearly demonstrates that the chemical properties of the clusters intrinsically differ from those of the nanoparticles. Various spectroscopic methods and DFT calculations indicated that the Au clusters are negatively charged by electron donation from PVP.[80,81] Namely, PVP not only acts as a stabilizer, but also activates the Au clusters by modulating the electronic structures.[80] On the basis of these results, Tsukuda and coworkers proposed that aerobic alcohol oxidation proceeds by abstraction of the α-hydrogen of the alkoxide by an O_2 anionic species formed by electron transfer from the Au clusters to the LUMO of O_2. In support of this mechanism, the catalytic activity of Au:PVP was further enhanced by doping a small amount of Ag (≤5%), which was ascribed to an increase of electronic charge supplied to the Au site by intracluster electron transfer from the Ag site (Table 10.1).[64] A similar mechanism was proposed by Toshima and

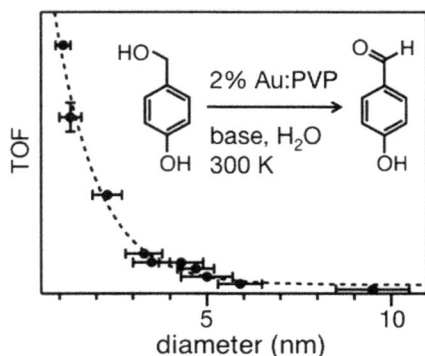

Figure 10.8 Particle size dependence of Au clusters on the catalytic activity for *p*-hydroxybenzyl alcohol oxidation over Au:PVP catalysts.[19]
Copyright © 2011 WILEY-VCH Verlag GmbH & Co. KGaA, Weinheim.

coworkers to explain the enhanced catalytic activity for aerobic glucose oxidation by Au atoms assembled at the top sites of $Pd_{\sim 147}$:PVP (Table 10.1).[63] These results show that Au clusters are catalytically active for the aerobic oxidations when they are small enough and negatively charged, as in the case of Au clusters in the gas phase.

Catalytic properties of ligand-protected small Au clusters have been studied theoretically. Häkkinen and coworkers predicted that phosphine-protected Au_{11} can catalyze CO oxidation when the ligands are partially removed.[82] Zeng and coworkers proposed that molecular oxygen can be activated at low-coordinated Au atoms of $Au_{55}(PPh_3)_{12}Cl_6$ due to negative charge on the Au_{55} core back-donated from the ligands.[83] Surprisingly, Jin and coworkers discovered that $Au_{25}(SR)_{18}$, $Au_{38}(SR)_{24}$, and $Au_{144}(SR)_{60}$ ($R = PhCH_2CH_3$) dispersed in toluene catalyze styrene oxidation (Table 10.1) although all the Au atoms on the cluster surfaces are completely "poisoned" by the thiolates (Figure 10.4).[13] The durability of $Au_n(SC_2H_4Ph)_m$ can be improved by immobilizing them on a solid support. $Au_n(SC_2H_4Ph)_m$ immobilized on silica and hydroxyapatite showed activities for epoxidation of styrene using *tert*-butyl hydroperoxide as an oxidant (Table 10.1).[84] $Au_{25}(C_2H_4Ph)_{18}/CeO_2$ showed high activity for CO oxidation[85] and homo-coupling of aryl iodides (Table 10.1).[86] $Au_{25}(SR)_{18}/TiO_2$ catalyzed selective oxidation of sulfide to sulfoxide by PhIO (Table 10.1).[87] The catalysis of $Pt_1Au_{24}(SR)_{18}/TiO_2$ was found to be superior to $Au_{25}(SR)_{18}/TiO_2$ in styrene oxidation indicating a remarkable effect on catalysis by the doping of a single Pt atom (Table 10.1).[68]

10.2.3.2 Reductions

Zhao and Crooks reported that Pd_{40} and Pt_{40} within PAMAM-DENs showed high catalytic activity for hydrogenation of allyl alcohol and N-isopropylacrylamide (Table 10.1).[22] Tsukuda *et al.* showed that Cu_{10-60} within PAMAM-DENs acted as a chemoselective catalyst for hydrogenation of cinnamaldehyde to cinnamyl alcohol and that Cu_{30} showed the highest activity (Table 10.1).[88] Somorjai and coworkers reported that Pt_{20}, Pt_{40}, Rh_{15}, and Rh_{30} within PAMAM-DENs adsorbed within mesoporous silica (SBA-15) were active against the hydrogenation of ethylene and pyrrole (Table 10.1).[26] Yamamoto *et al.* found that Pt_{12}, Pt_{28}, and Pt_{60} within DPA-DENs catalyzed the electrochemical reduction of oxygen and that catalytic activity was improved as the cluster size decreased (Table 10.1).[27]

Jin and coworkers demonstrated that $Au_{25}(SR)_{18}$ acted as a chemoselective catalyst for hydrogenation of α,β-unsaturated ketones to the corresponding alcohols (Table 10.1).[89] Scott and coworkers[90] and Kawasaki *et al.*[91] reported that $Au_{25}(SR)_{18}$ efficiently reduced 4-nitrophenol to 4-aminophenol using $NaBH_4$ as a reductant (Table 10.1). Chen and Chen studied electrocatalytic activity for oxygen reduction by ligand-protected Au_{11}, Au_{25}, Au_{55} and Au_{144} and found that smaller Au clusters showed higher activity (Table 10.1).[92] Kauffman *et al.* demonstrated that electrocatalytic activity of $Au_{25}(SR)_{18}$ for

Table 10.1 Catalytic reactions over stabilized/protected metal cluster catalysts.

Reactions	Reactants	Products	Catalysts	Conditions	Ref.
Oxidation	4-(hydroxymethyl)phenol	4-hydroxybenzaldehyde	Au (1.1 ± 0.2 nm):PVP Au–Ag (1.3–2.0 nm):PVP	27 °C, 6 h, H_2O 27 °C, 6 h, H_2O	19 64
	glucose	gluconic acid (open-chain polyhydroxy aldehyde)	Top Au atoms decorating Pd $\sim_{147}$ (1.6–1.8 nm):PVP	60 °C, NaOH sol.	63
	CO	CO_2	$Au_{25}(SC_2H_4Ph)_{18}/CeO_2$	80–100 °C, gas phase	85
	styrene	styrene oxide / benzaldehyde / benzoic acid	$Au_{25}(SC_2H_4Ph)_{18}$, $Au_{38}(SC_2H_4Ph)_{24}$, $Au_{144}(SC_2H_4Ph)_{60}$/SiO_2, HAP	80 °C, 12 h, toluene	84
			$Au_{25}(SCH_2CH_2Ph)_{18}$, $Au_{38}(SCH_2CH_2Ph)_{24}$, $Au_{144}(SCH_2CH_2Ph)_{60}$ $Pt_1Au_{24}(SC_2H_4Ph)_{18}/TiO_2$	100 °C, 24 h, toluene	13
	(substituted aryl iodide, R)	(biphenyl, R)		70 °C, 10 h, acetonitrile	68
	sulfide (R_1–S–R_2)	sulfone / sulfoxide	$Au_{25}(SC_2H_4Ph)_{18}/CeO_2$	130 °C, 48 h, DMF	86
			$Au_{25}(SR)_{18}/TiO_2$	40 °C, 12 h, DCM	87

Table 10.1　(*Continued*)

Reactions	Reactants	Products	Catalysts	Conditions	Ref.
Hydrogenation	(N-isopropyl acrylamide)	(N-isopropyl amide)	Pd_{40}@PAMAM	20 °C, H_2O	23
	R–C6H4–CHO	R–C6H4–CH_2OH	Cu_{30}@PAMAM	25 °C, 6 h, H_2O	88
	CH_2=CH (pyrrole)	CH_3CH_3 / H_2N– (propylamine)	Pt_{20}, Pt_{40}, Rh_{15}, Rh_{30}/SBA-15	50–150 °C, gas phase	26
Electrochemical reduction	O_2	H_2O	Pt_{12}, Pt_{28}, Pt_{60}@ DPA-DENs	R.T., H_2O	27
	(4-phenyl-3-buten-2-one)	(4-phenyl-3-buten-2-ol)	$Au_{25}(SC_2H_4Ph)_{18}$, $Au_{25}(SC_2H_4Ph)_{18}$/ TiO_2, Fe_2O_3, SiO_2	0 °C, 3 h, toluene	89
	HO–C6H4–NO_2	HO–C6H4–NH_2	$Au_{25}(SC_nH_{2n+1})_{18}$ (n = 6, 8, 12), $Au_{180}(SC_6H_{13})_{100}$, $Au_{25}(SG)_{18}$	R.T., THF–H_2O	90,91
	O_2	O^{2-}	$Au_{11}(PPh_3)_8Cl_3$, $Au_{25}(SC_2H_4Ph)_{18}$, $Au_{55}(PPh_3)_{12}Cl_6$, $Au_{140}(SC_6H_{13})_{53}$/ glassy carbon	KOH	92
	CO_2	CO	$Au_{25}(SC_2H_4Ph)_{18}$	$KHCO_3$	93

Reaction	Catalyst	Conditions	Ref.
Homocoupling	Au (1.3–1.6 nm): PVP	R.T, 24 h, H_2O	94
Stille coupling	Pd (1.6 ± 0.3 nm)@PAMAM	80 °C, 24 h, H_2O	95
Heck reaction	Pd (2–3 nm)@PPPI(polyether-derivatized poly(propylene imine))	90 °C, 24 h	96
Suzuki–Miyaura coupling	Pd (1.4–3.6 nm)@PAMAM	24 h, EtOH	97
Allylic substitution reaction	Pd (0.5–0.97 nm)@PPI(poly(propylene imine))	80 °C, 24 h, toluene, DMF, DMSO, ethyl acetate	28
Photo-oxidation	$Au_{25}(SG)_{18}/TiO_2$	Photo-irradiation (500–750 nm, 10 mW cm^{-2}), CH_3COONa sol	98

CO_2 reduction was higher than that of larger Au particles (Table 10.1).[93] These studies suggest that the ligand-protected Au clusters can catalytically activate small molecules such as O_2 and CO_2 by interfacial electron transfer from the Au core.

10.2.3.3　Other Reactions

Small Au:PVP catalyzed aerobic homocoupling of phenylboronic acids (Table 10.1).[94] In this reaction, it is proposed that adsorption of molecular oxygen created a Lewis acidic site on the small Au clusters for catalysis.[19] Pd clusters within DENs have been used as catalysts for Heck, Stille and Suzuki–Miyaura coupling reactions (Table 10.1).[95–97] Kaneda and coworkers compared the catalysis of Pd_4, Pd_8, and Pd_{16} within PPI-DENs for allylic substitution reaction and found that the initial TOF increases with the cluster size (Table 10.1).[28] Tatsuma and coworkers reported that $Au_{25}(SG)_{18}/TiO_2$ acted as a photocatalyst under visible light irradiation (Table 10.1).[98] The recombination of electrons and holes is retarded by electron transfer from $Au_{25}(SG)_{18}$ to the TiO_2 support and the holes produced on $Au_{25}(SG)_{18}$ react with electron donors such as hydroquinone.

10.3　Supported Metal Cluster Catalysts

10.3.1　Size-Controlled Synthesis

10.3.1.1　Impregnation Method

Supported metal clusters can be prepared by the impregnation method, a convenient route to prepare conventional supported metal catalysts (Scheme 10.2). In this method, metal precursors are first adsorbed on the support surface by mixing them in solvents. After filtration and drying, the adsorbed metal ion species are reduced by heating or chemical treatment to form metal clusters on the supports. The particle size can be controlled by adjusting synthesis conditions, such as type and specific surface area of the supports, the amount of metal ions adsorbed, and the reduction methods. For example, Satsuma and coworkers prepared Ag clusters with a diameter of ~1.0 nm on Al_2O_3[99] and Fan and coworkers prepared sub-nanometer Ir clusters on TiO_2 (Figure 10.9) by the impregnation method Table 10.2.[100]

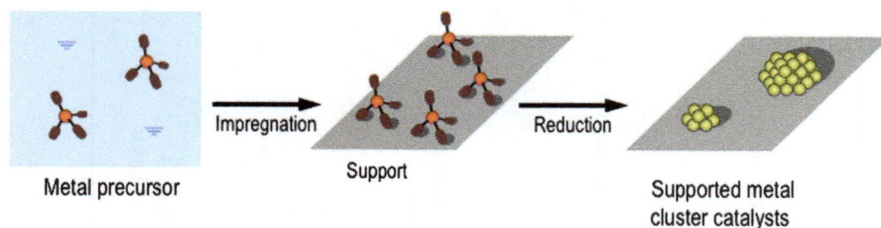

Metal precursor　　Impregnation　　Support　　Reduction　　Supported metal cluster catalysts

Scheme 10.2　Impregnation method.

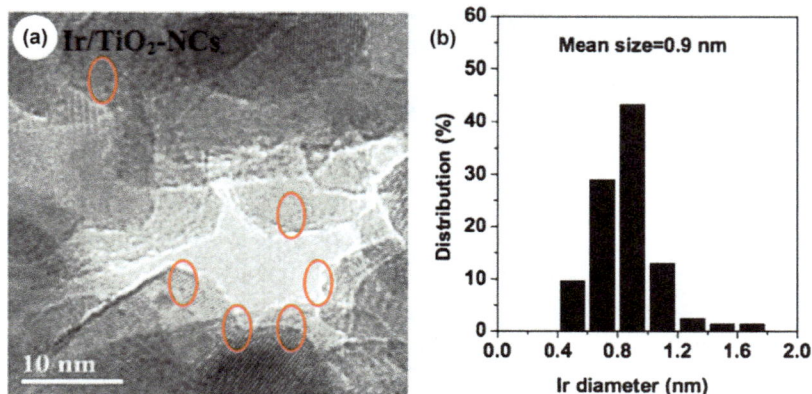

Figure 10.9 HRTEM image of Ir/TiO$_2$.[100]
Copyright © 2011 WILEY-VCH Verlag GmbH & Co. KGaA, Weinheim.

Haruta and coworkers developed a new method of synthesizing supported small Au clusters. Au clusters (diameter <1.8 nm) were loaded on lanthanum oxide (Au/La(OH)$_3$) by calcinating the co-precipitated gold–lanthanum hydroxide, which was obtained by adding a mixed solution of chloroauric acid and lanthanum nitrate to that of sodium hydroxide.[101] Haruta and coworkers also synthesized the supported Au catalyst (diameter <2 nm) by hydrogen reduction after solid grinding of a volatile dimethyl(acetylacetonate)gold complex with titania–silica (TS-1) pretreated with alkaline solution (Figure 10.10).[102,103]

Although the above methods yielded metal clusters smaller than 2 nm, it is impossible to control the size and composition of the clusters with atomic precision. One method to overcome this problem is the ship-in-bottle method in which the cluster size is regulated by well-defined, monodisperse pores.[16] For example, Ir$_6$ and Re$_{10}$ were synthesized using the pores of zeolites.[104,105]

10.3.1.2 Soft-Landing Method

The soft-landing method has been developed to synthesize size-selected metal clusters on a well-defined clean surface.[7] In this method, an intense beam of metal cluster ions is produced by laser ablation or magnetron sputtering in a high vacuum. Then, they are atomically size-selected using a mass filter, decelerated and allowed to collide against the surfaces installed under ultra high vacuum (Scheme 10.3). The kinetic energy of the cluster ions must be reduced sufficiently so as to suppress fragmentation during the deposition. Heiz *et al.* prepared size-selected Pt$_n$ ($8 \leq n \leq 20$) on a clean MgO surface[106] and Yamaguichi and Murakami prepared W$_{2-6}$ on a graphite surface.[107] The model systems prepared by this approach help provide a fundamental understanding of the catalysis of supported metal clusters. The observed size- and composition-dependence on catalysis was discussed in terms of the specific structures of the clusters.

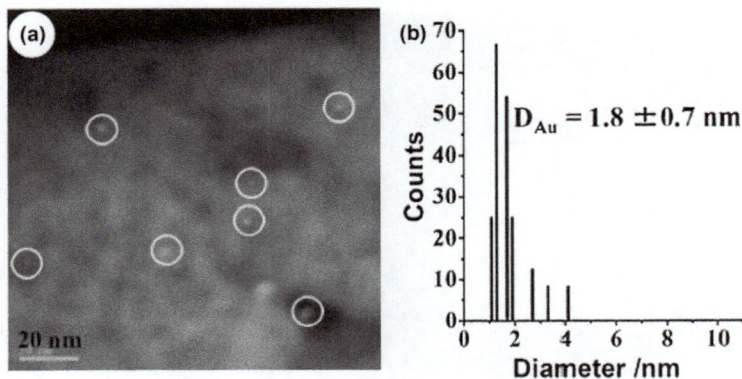

Figure 10.10   (a) HAADF-STEM image and (b) cluster diameter distribution of Au/TS pretreated with alkaline solution.[103]
Copyright © 2012 Elsevier.

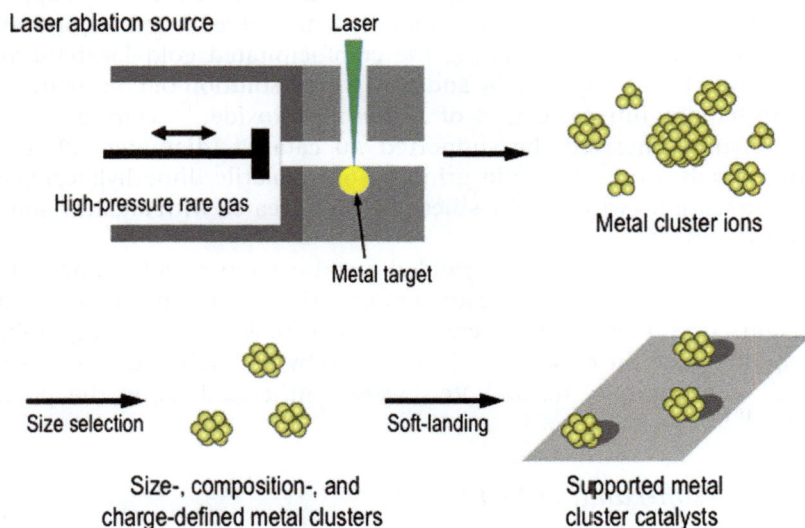

Scheme 10.3   Soft-landing method.

Recently, this approach has been used to fabricate real-world catalysts. Vajda and coworkers studied the catalysis of Pt_{8-10} on Al_2O_3[108] and Au_{6-10} on TiO_2 or Al_2O_3[109] prepared by the soft-landing method. Heiz and co-workers produced $Pt_{\geq 36}$ and Pt_{46} on CdS semiconductor nanorods.[110] The HAADF-STEM images indicate that size-selected Pt clusters were densely loaded on CdS nanorods without aggregation (Figure 10.11).[110] Palmer and coworkers soft-landed and immobilized size-selected Pd_{55}, Pd_{120}, Pd_{200},

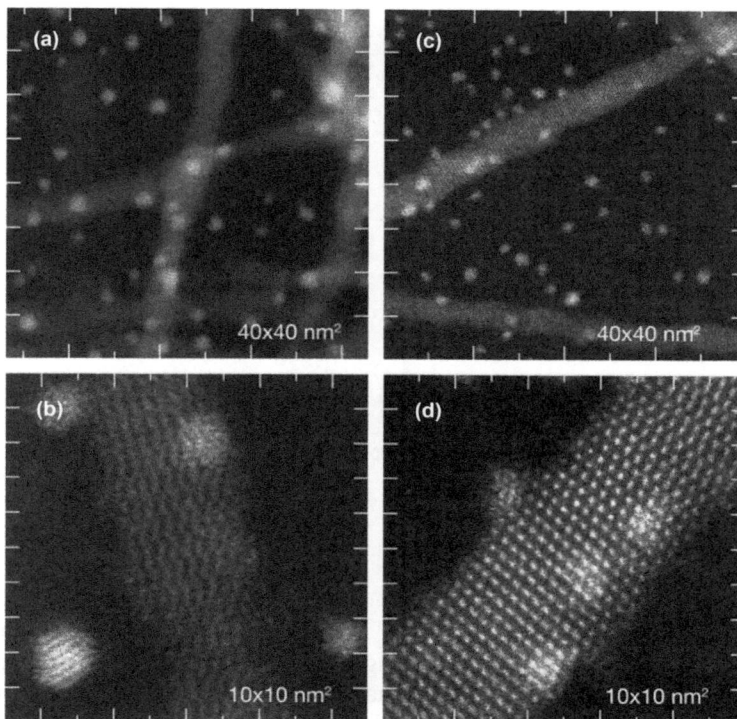

Figure 10.11 HAADF-STEM images of CdS nanorods decorated with (a, b) $Pt_{\geq 36}$ and (c, d) Pt_{46}.[110]
Copyright © 2012 American Chemical Society.

Pd_{300}, and Pd_{400} on graphite tapes having surface defects generated by Ar^+ ion bombardment.[111] These studies have demonstrated that the soft-landing method is powerful for the precise control of size of supported metal clusters. On the other hand, the method requires a special, well-designed apparatus so there is a need to develop a simpler method to control the size of supported metal clusters.

10.3.1.3 Precursor Method

Small metal carbonyl clusters have been used as precursors of supported metal clusters.[17] Ligand-protected metal clusters whose constituent atom number is strictly defined (section 10.2.1.2) could be another potential source of the precursors. If the ligands are completely removed from the ligand-protected clusters without aggregation of the resulting metal clusters on solid surfaces, we can control the size and composition with atomic precision (Scheme 10.4). Obviously, removal of the ligands is the key for

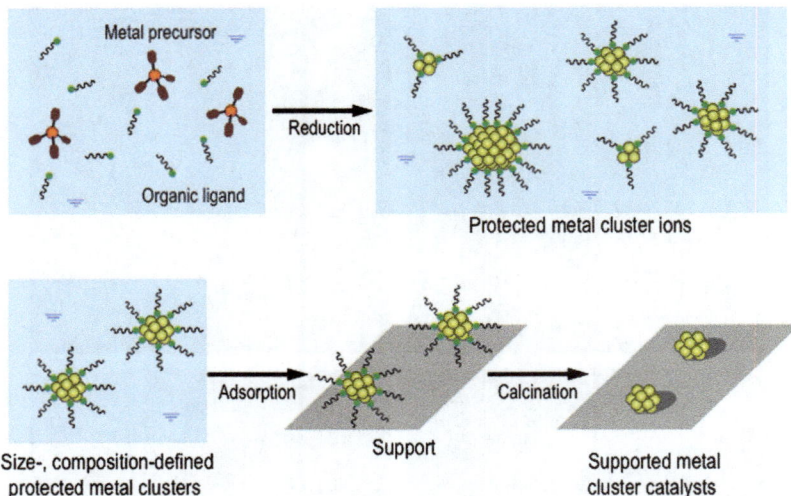

Scheme 10.4 Precursor method.

successful preparation. Calcination is the most simple and often-used method to remove the ligands. Zheng and Stucky[112] and Tai[113] independently prepared monodispersed Au nanoparticles on various metal oxides (TiO$_2$, SiO$_2$, ZnO, Al$_2$O$_3$) using alkanethiolate-protected Au nanoparticles and showed the promise of this approach for application to the synthesis of size-selected metal clusters. Lambert and coworkers synthesized Au clusters (diameter = 1.4 nm) on BN and SiO$_2$ supports by removing PPh$_3$ ligands from Au$_{55}$(PPh$_3$)$_{12}$Cl$_6$ by heat treatment at 200 °C (Figure 10.12a).[114] Tsukuda and coworkers synthesized Au$_{11}$ within a mesoporous silica (SBA-15) having a high surface area (specific surface area ~900 m^2 g^{-1}) by calcination of [Au$_{11}$(PPh$_3$)$_8$Cl$_2$]$^+$ at 200 °C (Figure 10.12b).[115] They also synthesized a series of Au$_n$ (n = 10, 18, 25, and 39) on hydroxyapatite (HAP) using Au$_n$(SG)$_m$ as a starting material (Figure 10.12c).[116] Recently, Dai and coworkers proposed a new method in which the thiolates are removed from Au$_n$(SR)$_m$ on heterostructured mesoporous supports *via* thermal treatment (Figure 10.12d).[117] The Au clusters immobilized on the heterostructured mesoporous supports preserved their size even under thermal treatment at 500 °C in air because of stronger stabilizing interaction with the metal oxide layer than pure SiO$_2$.

10.3.2 Composition-Controlled Synthesis

Catalytic activity will be changed drastically by introducing heterogeneous metal atoms into the clusters. Especially, in the case of small clusters, even a single atom dopant can modify the chemical properties of the

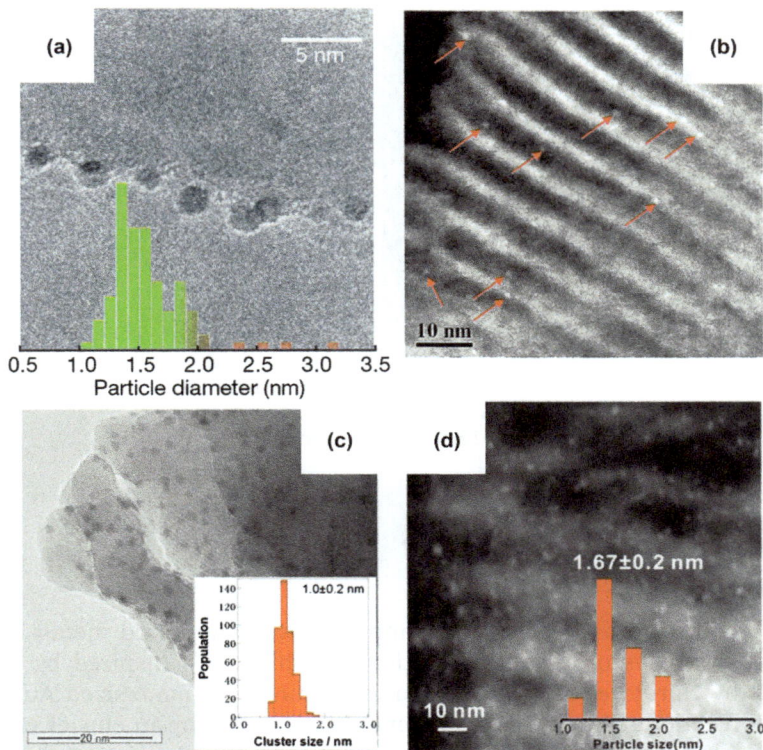

Figure 10.12 (a) HRTEM image and cluster diameter distribution of Au_{55}/SiO_2.[114] (b) HAADF-STEM image of Au_{11}/SBA.[115] (c) TEM image and size distribution of Au_{25}/HAP.[116] (d) STEM image of $Au_{144}/CuO-EP-FDU-12$.[117] Copyright © 2009/2011 American Chemical Society and 2008 Rights Managed by Nature Publishing Group.

whole system. One can precisely control the composition of supported metal clusters by using the soft-landing method. For example, Heiz and coworkers produced $Au_{n-1}Sr$ on a clean MgO surface and showed that single atom doping of Sr into Au_n clusters enhances the CO oxidation activity.[118]

However, it is a challenge to control precisely the composition of the supported clusters *via* a chemical approach while maintaining the total number of constituent atoms. This problem can be overcome using the ligand-protected bimetallic clusters introduced in section 10.2.2 whose composition is definitely and systematically controlled while keeping their constituent atom number constant. Recently, Tsukuda and coworkers synthesized Au_{25} and Pd_1Au_{24} on carbon nanotubes by calcinating $Au_{25}(SC_{12}H_{25})_{18}$ and $Pd_1Au_{24}(SC_{12}H_{25})_{18}$ at 450 °C in a vacuum (Figure 10.13).[119]

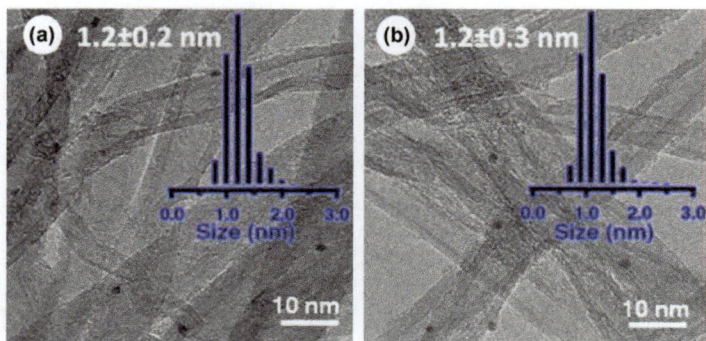

Figure 10.13   TEM images and cluster diameter distributions of (a) Au_{25}/CNT and (b) Pd_1Au_{24}/CNT.[119]
Copyright © 2012 American Chemical Society.

10.3.3   Catalytic Applications

10.3.3.1   *Oxidations*

This section summarizes catalytic application of supported metals clusters for various types of chemical reactions (Table 10.2). Inspired by the discovery by Haruta *et al.* of CO oxidation catalysis of nano-sized Au,[120] oxidation reactions have been studied most extensively for Au cluster catalysts. Haruta and co-workers reported that Au clusters smaller than 2 nm immobilized on $La(OH)_3$ showed high activity for CO oxidation at low temperature (Table 10.2).[101] Dai and coworkers demonstrated that Au_{25} and Au_{144} on heterostructured transition metal-oxide mesoporous silica are robust and active catalysts for CO oxidation (Table 10.2).[117] Tsukuda and coworkers reported the Au_{11}/SBA-15 catalyzed aerobic oxidations of primary and secondary alcohols using hydrogen peroxide as an oxidant under microwave irradiation (Table 10.2).[115] It was reported that the cluster size increased homogeneously from 0.8 to 1.9 nm by prolonging the duration of maintaining the calcination temperature and that the catalytic activity decreased with increase in size.[121] The catalytic activity for aerobic oxidation of Pd_1Au_{24}/CNT was much higher than that of Au_{25}/CNT. This is the first demonstration of a single atom doping effect on catalysis although the role of the Pd atom is still under investigation (Table 10.2).[119]

There are several challenging targets in oxidation reactions: selective oxidation of alkenes (C=C bonds), alkanes (C–H bonds), and benzene. Haruta and coworkers reported that supported Au clusters smaller than 2 nm showed high and selective epoxidation of propene to propene epoxide using O_2 as an oxidant (Table 10.2).[102,103,122] Vajda and coworkers reported that the selective epoxidation of propene was catalyzed by Au_{6-10}/TiO_2 and /Al_2O_3 (Table 10.2).[109] Lambert and coworkers demonstrated that only Au clusters smaller than Au_{55} oxidize styrene (Table 10.2).[114] Au_{25}/HAP

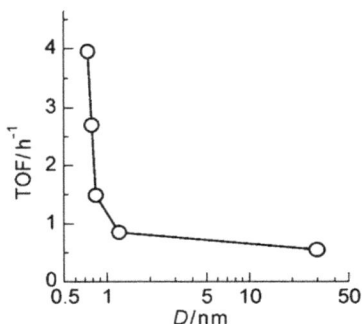

Figure 10.14 TOF for the coupling reaction of 1-phenylethanol and benzyl alcohol at 115 °C as a function of the average particle size of Ag clusters on Al_2O_3.[99]
Copyright © 2009 WILEY-VCH Verlag GmbH & Co. KGaA, Weinheim.

showed much higher selectivity (>90%) toward styrene oxide from styrene (Table 10.2)[123] than conventional supported Au catalysts (*ca.* 50%). Tsukuda and coworkers found that small Au clusters on HAP are highly active and selective toward oxidation of cyclohexane into cyclohexanone and cyclo-hexanol (Table 10.2).[116] In addition, it was revealed that Au_{39}/HAP showed the highest activity among other catalysts, Au_n ($n = 10, 18, 25$, and ~ 85)/HAP (Table 10.2).[116] Vajda *et al.* reported that Pt_{8-10}/Al_2O_3 acted as a highly active and selective catalyst for the oxidative dehydrogenation of propane to pro-pylene under atmospheric conditions (Table 10.2).[108] The DFT calculation suggests that the C–H bond is activated on Pt clusters by electron transfer from the C–H bonding orbital to Pt clusters. It was demonstrated that benzene was directly oxidized to phenol with high selectivity using O_2 as an oxidant over Re_{10} supported on zeolite (Table 10.2).[105]

10.3.3.2 Other Reactions

Small clusters are known to show size-specific catalysis for other types of reactions. López-Quintela and coworkers reported that $Cu_{\leq 10-13}$ clusters on mica catalyzed the reduction of methylene blue to leucomethylene by hy-drazine (Table 10.2).[124] Heiz and coworkers reported that $Pt_{\geq 36}$ and Pt_{46} on CdS nanorods are active for the water splitting reaction to hydrogen under photoirradiation (Table 10.2).[110] Satsuma and coworkers demonstrated that Ag clusters on Al_2O_3 catalyzed the C–C coupling between primary and sec-ondary alcohols *via* oxidative dehydrogenation (Table 10.2).[99] The catalytic activity was enhanced remarkably with the reduction of size (Figure 10.14). Fan and coworkers reported that sub-nanometer Ir clusters on TiO_2 showed high catalytic activity for quinoline synthesis from alcohol and nitroarenes (Table 10.2).[100]

Table 10.2 Catalytic reactions over metal cluster catalysts.

Reactions	Reactants	Products	Catalysts	Methods	Conditions	Refs.
Oxidation	CO	CO_2	Au/Fe_2O_3, Co_3O_4, NiO	IP[a]	-70–30 °C, gas phase	120
	(propene)	(propylene oxide)	Au (<3 nm)/La(OH)$_3$	IP[a]	-80 °C, gas phase	101
			Au$_{25}$, Au$_{144}$, Co_3O_4–mesoporous silica	PR[b]	30–120 °C, gas phase	117
	(styrene)	(benzaldehyde, styrene oxide, benzoic acid)	Au (1.0–2.0 nm)/TS-1	IP[a]	200 °C, gas phase	102, 103, 122
			Au$_{6-10}$/Al_2O_3, TiO_2	SL[c]	200 °C, 30 s, gas phase	109
			Au$_{55}$/BN, SiO_2	PR[b]	100 °C, 15 h, toluene	114
	(cyclohexane)	(cyclohexanone, cyclohexanol)	Au$_{25}$/HAP	PR[b]	80 °C, 4 h, toluene	123
	(benzyl alcohol)	(benzaldehyde, benzoic acid)	Au$_{10}$, Au$_{18}$, Au$_{25}$, Au$_{39}$, Au$_{\sim 85}$/HAP	PR[b]	150 °C, 4 h, solvent free	116
			Au$_{11}$/SBA-15	PR[b]	80 °C, 1 h, H_2O	115
		(benzyl benzoate)	Au$_{25}$, PdAu$_{24}$/CNT	PR[b]	30 °C, 6 h, H_2O	119

Reaction	Substrate	Product	Catalyst	Method	Conditions	Ref.
Oxidative dehydrogenation	(benzene/cyclohexene)	(phenol, —OH)	Re/ZSM-5	PR[b]	280 °C, gas phase	105
	(isobutane)	(isobutene)	Pt$_{8-10}$/Al$_2$O$_3$	SL[c]	550 °C, gas phase	108
Reduction	(methylene blue, S^+)	(leuco form, H–N, S)	Cu (0.4 ± 0.2 nm)	ED[d]	R.T., H$_2$O	123
Photo-reduction	H$_2$O	H$_2$	Pt$_{>36}$, Pt$_{46}$/CdS	SL[c]	R.T., H$_2$O	110
C–C coupling	(1-phenylethanol + 2-phenylethanol, OH/OH)	(1,3-diphenylpropan-1-one, O)	Ag (0.8 nm)/Al$_2$O$_3$	IP[a]	115 °C, 48 h, toluene	99
Quinoline synthesis	(R-nitrobenzene, —NO$_2$ + amino alcohol, —OH)	(quinoline, R)	Ir/TiO$_2$	IP[a]	120 °C, 8 h, solvent free	100

[a]Impregnation method.
[b]Precursor method.
[c]Soft-landing method.
[d]Electrodeposition method.

10.4   Summary and Prospects

Metal clusters have high potential as unique reaction sites and will show novel catalysis under optimized structural parameters beyond that expected based on bulk metals. This chapter examined the current status of precision synthesis and catalytic applications of metal clusters stabilized by polymers, protected by ligands, and immobilized on supports. These efforts will not only provide a deeper understanding of the correlation between structural parameters (size and composition) and catalysis, but also a guiding principle when designing a required catalysis. The understanding will help replace the platinum group metals conventionally used in catalysis with ubiquitous metals and thus reduce the use of precious metals of which there are limited natural resources. However, the following issues must be overcome to develop cluster-based catalysts.

Ligand-protected metal clusters will provide an ideal and unique platform to study the correlation between structures and catalysis at the molecular level since catalysts with a well-established structure can be used. Although catalysis was observed in fully protected Au clusters as shown in section 10.2.3, their catalytic activity will be significantly enhanced by lowering the ligand coverage and exposing the surface area. One way to achieve this is to reduce the number of ligands by taking advantage of the steric repulsion between adjacent ligands having dendritic structures. Katz and coworkers demonstrated that 25% of the surface of Au clusters with a diameter of 0.9 nm was exposed by using calixarene derivatives as ligands.[125]

The precursor method described in sections 10.3.1.3 and 10.3.2 is promising to precisely control the size and composition of supported metal clusters. However, several challenges remain and there is room for further development of this method. Firstly, the systems currently available are limited to Au and Ag-based monometallic and bimetallic clusters protected by phosphines and/or thiolates as shown in sections 10.2.1.2 and 10.2.2. It is necessary to expand the variations of the systems to achieve on-demand catalysis. Secondly, removal of the ligands from the protected clusters has been conducted in most cases by heating, which often induces aggregation and/or dissociation of clusters. A method of removing the ligands without heating should be developed, including chemical methods such as oxidative or reductive desorption of the ligands and physical methods such as plasma etching.[126,127]

One of the most difficult tasks for the research of supported metal cluster catalysis is to determine their structures because of their extremely small size and density. Aberration-corrected high-angle annular dark-field scanning transmission electron microscopy (HAADF-STEM) allows us to count the number of atoms contained in individual clusters based on the contrast of the image and can be viewed as "a mass spectrometry of clusters on a surface".[128] By using size-selected Au clusters soft-landed on supports as mass standards, this approach has been successfully applied to $Au_{38}(SR)_{24}$[129] and $Au_{144}(SR)_{60}$[130] as well as size-selected Au_{25} and Au_{39} on

HAP.[131] However, it is not trivial to reveal the origin of catalysis of supported metal clusters in terms of their structures because the geometrical structures of the clusters prepared by soft-landing or precursor methods are poly-dispersed even if the size and compositions are uniform. Thus, further challenges must be tackled to synthesize clusters with uniform structures.

Another difficulty in structural characterization of clusters is associated with their flexibility intrinsic to small clusters and dynamical change of structures induced by interaction with reactants under catalytic conditions. An insight into the atomic arrangement of small clusters can be obtained by HAADF-STEM.[132] However, the structures change dramatically with time under electron beam exposure in contrast to larger nanoparticles.[132–135] Isomerization dynamics of the clusters can be assessed by combination with theoretical results and statistical analysis. The intrinsic structure of the clusters under reaction conditions was studied by Takeda and coworkers using environmental TEM.[136,137] *In situ* spectroscopic methods also allow us to probe the structural changes of catalysts during catalytic conversion.[138]

References

1. L. N. Lewis, *Chem. Rev.*, 1993, **93**, 2693.
2. T. P. Martin, *Phys. Rep.*, 1996, **273**, 199.
3. A. W. Castleman, Jr. and K. H. Bowen, Jr., *J. Phys. Chem.*, 1996, **100**, 12911.
4. R. Kubo, A. Kawabata and S. Kobayashi, *Ann. Rev. Mater. Sci.*, 1984, **14**, 49.
5. R. Jin, *Nanoscale*, 2010, **2**, 343.
6. A. Aguado and M. F. Jarrold, *Annu. Rev. Phys. Chem.*, 2011, **62**, 151.
7. M. B. Nickelbein, *Annu. Rev. Phys. Chem.*, 1999, **50**, 79.
8. J. Oliver-Meseguer, J. R. Cabrero-Antonino, I. Domínguez, A. Leyva-Pérez and A. Corma, *Science*, 2012, **338**, 1452.
9. T. M. Bernhardt, U. Heiz and U. Landman, in *Nanocatalysis*, ed. U. Heiz and U. Landman, *NanoScience and Technology*, Springer, Berlin, Heidelberg, New York, 2006, ch. 1, p. 1.
10. *Clusters and Colloids*, ed. G. Schmid, VCH, Weinheim, 1994.
11. G. Schmid, *Chem. Rev.*, 1992, **92**, 1709.
12. M. Brust, M. Walker, D. Bethell, D. J. Schiffrin and R. Whyman, *J. Chem. Soc., Chem. Commun.*, 1994, 801.
13. Y. Zhu, H. F. Qian, M. Z. Zhu and R. Jin, *Adv. Mater.*, 2010, **22**, 1915.
14. Y. Zhu, H. F. Qian and R. Jin, *J. Mater. Chem.*, 2011, **21**, 6793.
15. G. Li and R. Jin, *Acc. Chem. Res.*, 2013, **46**, 1749.
16. B. C. Gates, *Chem. Rev.*, 1995, **99**, 511.
17. A. Kulkarni, R. J. Lobo-Lapidus and B. C. Gates, *Chem. Commun.*, 2010, **46**, 5997.
18. H. Tsunoyama, N. Ichikuni and T. Tsukuda, *Langmuir*, 2008, **24**, 11327.
19. T. Tsukuda, H. Tsunoyama and H. Sakurai, *Chem. – Asian J.*, 2011, **6**, 736.
20. H. Tsunoyama and T. Tsukuda, *J. Am. Chem. Soc.*, 2009, **131**, 18216.

21. L.-M. Wang and L.-S. Wang, *Nanoscale*, 2012, **4**, 4038.
22. M. Q. Zhao, L. Sun and R. M. Crooks, *J. Am. Chem. Soc.*, 1998, **120**, 4877.
23. M. Q. Zhao and R. M. Crooks, *Angew. Chem., Int. Ed.*, 1999, **38**, 364.
24. J. Zheng, C. W. Zhang and R. M. Dickson, *Phys. Rev. Lett.*, 2004, **93**, 077402.
25. R. W. J. Scott, O. M. Wilson and R. M. Crooks, *J. Phys. Chem. B*, 2005, **109**, 692.
26. W. Huang, J. N. Kuhn, C. K. Tsung, Y. Zhang, S. E. Habas, P. Yang and G. A. Somorjai, *Nano Lett.*, 2008, **8**, 2027.
27. K. Yamamoto, T. Imaoka, W. J. Chun, O. Enoki, H. Katoh, M. Takenaga and A. Sonoi, *Nat. Chem.*, 2009, **1**, 397.
28. T. Kibata, T. Mitsudome, T. Mizugaki, K. Jitsukawa and K. Kaneda, *Chem. Commun.*, 2013, **49**, 167.
29. T. Ueno, M. Suzuki, T. Goto, T. Matsumoto, K. Nagayama and Y. Watanabe, *Angew. Chem., Int. Ed.*, 2004, **43**, 2527.
30. C. J. Sun, H. Yang, Y. Yuan, X. Tian, L. M. Wang, Y. Guo, L. Xu, J. L. Lei, N. Gao, G. J. Anderson, X. J. Liang, C. Y. Chen, Y. L. Zhao and G. J. Nie, *J. Am. Chem. Soc.*, 2011, **133**, 8617.
31. R. L. Whetten, J. T. Khoury, M. M. Alvarez, S. Murthy, I. Vezmar, Z. L. Wang, P. W. Stephens, C. L. Cleveland, W. D. Luedtke and U. Landman, *Adv. Mater.*, 1996, **8**, 428.
32. A. C. Templeton, W. P. Wuelfing and R. W. Murray, *Acc. Chem. Res.*, 2000, **33**, 27.
33. T. Tsukuda, *Bull. Chem. Soc. Jpn.*, 2012, **85**, 151.
34. P. D. Jadzinsky, G. Calero, C. J. Ackerson, D. A. Bushnell and R. D. Kornberg, *Science*, 2007, **318**, 430.
35. P. Maity, S. Xie, M. Yamauchi and T. Tsukuda, *Nanoscale*, 2012, **4**, 4027.
36. M. W. Heaven, A. Dass, P. S. White, K. M. Holt and R. W. Murray, *J. Am. Chem. Soc.*, 2008, **130**, 3754.
37. M. Zhu, C. M. Aikens, F. J. Hollander, G. C. Schatz and R. Jin, *J. Am. Chem. Soc.*, 2008, **130**, 5883.
38. C. Zeng, H. Qian, T. Li, G. Li, N. L. Rosi, B. Yoon, R. N. Barnett, R. L. Whetthen, U. Landman and R. Jin, *Angew. Chem., Int. Ed.*, 2012, **51**, 13114.
39. H. F. Qian, W. T. Eckenhoff, Y. Zhu, T. Pintauer and R. Jin, *J. Am. Chem. Soc.*, 2010, **132**, 8280.
40. M. Walter, J. Akola, O. Lopez-Acevedo, P. D. Jadzinsky, G. Calero, C. J. Ackerson, R. L. Whetten, H. Grönbeck and H. Häkkinen, *Proc. Natl. Acad. Sci. U.S.A.*, 2008, **105**, 9157.
41. J. Akola, M. Walter, R. L. Whetten, H. Häkkinen and H. Grönbeck, *J. Am. Chem. Soc.*, 2008, **130**, 3756.
42. O. Lopez-Acevedo, H. Tsunoyama, T. Tsukuda, H. Häkkinen and C. M. Aikens, *J. Am. Chem. Soc.*, 2010, **132**, 8210.
43. Y. Pei and X. C. Zeng, *Nanoscale*, 2012, **4**, 4054.

44. J. Nishigaki, R. Tsunoyama, H. Tsunoyama, N. Ichikuni, S. Yamazoe, Y. Negishi, M. Ito, T. Matsuo, K. Tamao and T. Tsukuda, *J. Am. Chem. Soc.*, 2012, **134**, 14295.

45. T. U. B. Rao and T. Pradeep, *Angew. Chem., Int. Ed.*, 2010, **49**, 3925.

46. T. U. B. Rao, B. Nataraju and T. Pradeep, *J. Am. Chem. Soc.*, 2010, **132**, 16304.

47. K. M. Harkness, Y. Tang, A. Dass, J. Pan, N. Kothalawala, V. J. Reddy, D. E. Cliffel, B. Demeler, F. Stellacci, O. M. Bakr and J. A. McLean, *Nanoscale*, 2012, **4**, 4269.

48. I. Chakraborty, A. Govindarajan, J. Erusappan, A. Ghosh, T. Pradeep, B. Yoon, R. L. Whetten and U. Landman, *Nano Lett.*, 2012, **12**, 5861.

49. J. S. Guo, S. Kumar, M. Bolan, A. Desireddy, T. P. Bigioni and W. P. Griffith, *Anal. Chem.*, 2012, **84**, 5304.

50. I. Chakraborty, T. Udayabhaskararao and T. Pradeep, *Chem. Commun.*, 2012, **48**, 6788.

51. R. C. B. Copley and D. M. P. Mingos, *J. Chem. Soc., Dalton Trans.*, 1996, 479.

52. C. E. Briant, B. R. C. Theobald, J. W. White, L. K. Bell and D. M. P. Mingos, *J. Chem. Soc., Chem. Commun.*, 1981, 201.

53. Y. Shichibu and K. Konishi, *Small*, 2010, **6**, 1216.

54. Y. Shichibu, Y. Kamei and K. Konishi, *Chem. Commun.*, 2012, **48**, 7559.

55. K. Nunokawa, S. Onaka, M. Ito, M. Horibe, T. Yonezawa, H. Nishihara, T. Ozeki, H. Chiba, S. Watase and M. Nakamoto, *J. Organomet. Chem.*, 2006, **691**, 638.

56. Y. Shichibu, Y. Negishi, T. Watanabe, N. K. Chaki, H. Kawaguchi and T. Tsukuda, *J. Phys. Chem. C*, 2007, **111**, 7845.

57. A. Das, T. Li, K. Nobusada, Q. Zeng, N. L. Rosi and R. Jin, *J. Am. Chem. Soc.*, 2012, **134**, 20286.

58. H. Y. Yang, J. Lei, B. H. Wu, Y. Wang, M. Zhou, A. D. Xia, L. S. Zheng and N. Zheng, *Chem. Commun.*, 2013, **49**, 300.

59. H. Yang, Y. Wang and N. Zheng, *Nanoscale*, 2013, **5**, 2674.

60. Y. Kikukawa, Y. Kuroda, K. Suzuki, M. Hibino, K. Yamaguchi and N. Mizuno, *Chem. Commun.*, 2013, **49**, 376.

61. P. Maity, T. Wakabayashi, N. Ichikuni, H. Tsunoyama, S. Xie, M. Yamauchi and T. Tsukuda, *Chem. Commun.*, 2012, **48**, 6085.

62. N. Toshima, M. Harada, T. Yonezawa, K. Kushihashi and K. Asakura, *J. Phys. Chem.*, 1991, **95**, 7448.

63. H. Zhang, T. Watanabe, M. Okumura, M. Haruta and N. Toshima, *Nat. Mater.*, 2012, **11**, 49.

64. N. K. Chaki, H. Tsunoyama, Y. Negishi, H. Sakurai and T. Tsukuda, *J. Phys. Chem. C*, 2007, **111**, 4885.

65. B. K. Teo, H. Zhang and X. Shi, *Inorg. Chem.*, 1990, **29**, 2083.

66. B. K. Teo, H. Zhang and X. Shi, *J. Am. Chem. Soc.*, 1990, **112**, 8552.

67. W. Kurashige and Y. Negishi, *J. Cluster Sci.*, 2012, **23**, 365.

68. H. F. Qian, D. E. Jiang, G. Li, C. Gayathri, A. Das, R. R. Gil and R. Jin, *J. Am. Chem. Soc.*, 2012, **134**, 16159.

69. Y. Negishi, W. Kurashige, Y. Niihori, T. Iwasa and K. Nobusada, *Phys. Chem. Chem. Phys.*, 2010, **12**, 6219.
70. S. A. Miller, C. A. Fields-Zinna, R. W. Murray and A. M. Moran, *J. Phys. Chem. Lett.*, 2010, **1**, 1383.
71. Y. Negishi, T. Iwai and M. Ide, *Chem. Commun.*, 2010, **46**, 4713.
72. E. Gottlieb, H. Qian and R. Jin, *Chem. Eur. J.*, 2013, **19**, 4238.
73. Y. Negishi, K. Munakata, W. Ohgake and K. Nobusada, *J. Phys. Chem. Lett.*, 2012, **3**, 2209.
74. Y. Negishi, K. Igarashi, K. Munakata, W. Ohgake and K. Nobusada, *Chem. Commun.*, 2012, **48**, 660.
75. C. Kumara and A. Dass, *Nanoscale*, 2012, **4**, 4084.
76. C. Kumara and A. Dass, *Nanoscale*, 2011, **3**, 3064.
77. S. L. Christensen, M. A. MacDonald, A. Chatt, P. Zhang, H. Qian and R. Jin, *J. Phys. Chem. C*, 2012, **116**, 26932.
78. S. Malola and H. Häkkinen, *J. Phys. Chem. Lett.*, 2011, **2**, 2316.
79. H. Tsunoyama, H. Sakurai and T. Tsukuda, *Chem. Phys. Lett.*, 2006, **429**, 528.
80. H. Tsunoyama, N. Ichikuni, H. Sakurai and T. Tsukuda, *J. Am. Chem. Soc.*, 2009, **131**, 7086.
81. M. Okumura, Y. Kitagawa, T. Kawakami and M. Haruta, *Chem. Phys. Lett.*, 2008, **459**, 133.
82. O. Lopez-Acevedo, K. A. Kacprzak, J. Akola and H. Häkkinen, *Nat. Chem.*, 2010, **2**, 329.
83. Y. Pei, N. Shao, Y. Gao and X. C. Zeng, *ACS Nano*, 2010, **4**, 2009.
84. Y. Zhu, H. F. Qian and R. Jin, *Chem. – Eur. J.*, 2010, **16**, 11455.
85. X. Nie, H. Qian, Q. Ge, H. Xu and R. Jin, *ACS Nano*, 2012, **6**, 6014.
86. G. Li, C. Liu, Y. Lei and R. Jin, *Chem. Commun.*, 2012, **48**, 12005.
87. G. Li, H. Qian and R. Jin, *Nanoscale*, 2012, **4**, 6714.
88. P. Maity, S. Yamazoe and T. Tsukuda, *ACS Catal*, 2013, **3**, 182.
89. Y. Zhu, H. Qian, B. A. Drake and R. Jin, *Angew. Chem., Int. Ed.*, 2010, **49**, 1295.
90. A. Shivhare, S. J. Ambrose, H. Zhang, R. W. Purves and R. W. J. Scott, *Chem. Commun.*, 2013, **49**, 276.
91. H. Yamamoto, H. Yano, H. Kouchi, Y. Obora, R. Arakawa and H. Kawasaki, *Nanoscale*, 2012, **4**, 4148.
92. W. Chen and S. Chen, *Angew. Chem., Int. Ed.*, 2009, **48**, 4386.
93. D. R. Kauffman, D. Alfonso, C. Matranga, H. Qian and R. Jin, *J. Am. Chem. Soc.*, 2012, **134**, 10237.
94. H. Tsunoyama, H. Sakurai, N. Ichikuni, Y. Negishi and T. Tsukuda, *Langmuir*, 2004, **20**, 11293.
95. M. Bernechea, E. de Jesus, C. Lopez-Mardomingo and P. Terreros, *Inorg. Chem.*, 2009, **48**, 4491.
96. L. K. Yeung and R. M. Crooks, *Nano Lett.*, 2001, **1**, 14.
97. Y. Li and M. A. El-Sayed, *J. Phys. Chem. B*, 2001, **105**, 8938.
98. A. Kogo, N. Sakai and T. Tatsuma, *Electrochem. Commun.*, 2010, **12**, 996.

99. K. Shimizu, R. Sato and A. Satsuma, *Angew. Chem., Int. Ed.*, 2009, **48**, 3982.

100. L. He, J. Q. Wang, Y. Gong, Y. M. Liu, Y. Cao, H. Y. He and K. N. Fan, *Angew. Chem., Int. Ed.*, 2011, **50**, 10216.

101. T. Takei, I. Okuda, K. K. Bando, T. Akita and M. Haruta, *Chem. Phys. Lett.*, 2010, **493**, 207.

102. J. H. Huang, T. Takei, T. Akita, H. Ohashi and M. Haruta, *Appl. Catal., B*, 2010, **95**, 430.

103. J. H. Huang, T. Takei, H. Ohashi and M. Haruta, *Appl. Catal., A*, 2012, **435**, 115.

104. C. Aydin, J. Lu, M. Shirai, N. D. Browning and B. C. Gates, *ACS Catal.*, 2011, **1**, 1613.

105. R. Bal, M. Tada, T. Sasaki and Y. Iwasawa, *Angew. Chem., Int. Ed.*, 2006, **45**, 448.

106. U. Heiz, A. Sanchez, S. Abbet and W. D. Schneider, *J. Am. Chem. Soc.*, 1999, **121**, 3214.

107. W. Yamaguchi and J. Murakami, *J. Am. Chem. Soc.*, 2007, **129**, 6102.

108. S. Vajda, M. J. Pellin, J. P. Greeley, C. L. Marshall, L. A. Curtiss, G. A. Ballentine, J. W. Elam, S. Catillon-Mucherie, P. C. Redfern, F. Mehmood and P. Zapol, *Nat. Mater.*, 2009, **8**, 213.

109. S. Lee, L. M. Molina, M. J. Lopez, J. A. Alonso, B. Hammer, B. Lee, S. Seifert, R. E. Winans, J. W. Elam, M. J. Pellin and S. Vajda, *Angew. Chem., Int. Ed.*, 2009, **48**, 1467.

110. M. J. Berr, F. F. Schweinberger, M. Döblinger, K. E. Sanwald, C. Wolff, J. Breimeier, A. S. Crampton, C. J. Ridge, M. Tschurl, U. Heiz, F. Jäckel and J. Feldmann, *Nano Lett.*, 2012, **12**, 5903.

111. V. Habibpour, M. Y. Song, Z. W. Wang, J. Cookson, C. M. Brown, P. T. Bishop and R. E. Palmer, *J. Phys. Chem. C*, 2012, **116**, 26295.

112. N. Zheng and G. D. Stucky, *J. Am. Chem. Soc.*, 2006, **128**, 14278.

113. Y. Tai, J. Murakami, K. Tajiri, F. Ohashi, M. Daté and S. Tsubota, *Appl. Catal. A: Gen.*, 2004, **268**, 183.

114. M. Turner, V. B. Golovko, O. P. H. Vaughan, P. Abdulkin, A. Berenguer-Murcia, M. S. Tikhov, B. F. G. Johnson and R. M. Lambert, *Nature*, 2008, **454**, 981.

115. Y. Liu, H. Tsunoyama, T. Akita and T. Tsukuda, *J. Phys. Chem. C*, 2009, **113**, 13457.

116. Y. Liu, H. Tsunoyama, T. Akita, S. Xie and T. Tsukuda, *ACS Catal.*, 2011, **1**, 2.

117. G. Ma, A. Binder, M. Chi, C. Liu, R. Jin, D. Jiang, J. Fan and S. Dai, *Chem. Commun.*, 2012, **48**, 11413.

118. H. Häkkinen, W. Abbet, A. Sanchez, U. Heiz and U. Landman, *Angew. Chem., Int. Ed.*, 2003, **42**, 1297.

119. S. Xie, H. Tsunoyama, W. Kurashige, Y. Negishi and T. Tsukuda, *ACS Catal.*, 2012, **2**, 1519.

120. M. Haruta, T. Kobayashi, H. Sano and N. Yamada, *Chem. Lett.*, 1987, 405.

121. Y. Liu, H. Tsunoyama, T. Akita and T. Tsukuda, *Chem. Lett.*, 2010, **39**, 159.

122. J. H. Huang, T. Akita, J. Faye, T. Fujitani, T. Takei and M. Haruta, *Angew. Chem., Int. Ed.*, 2009, **48**, 7862.

123. Y. Liu, H. Tsunoyama, T. Akita and T. Tsukuda, *Chem. Commun.*, 2010, **46**, 550.

124. N. Vilar-Vidal, J. Rivas and M. A. López-Quintela, *ACS Catal.*, 2012, **2**, 1693.

125. N. de Silva, J. M. Ha, A. Solovyov, M. M. Nigra, I. Ogino, S. W. Yeh, K. A. Durkin and A. Katz, *Nat. Chem.*, 2010, **2**, 1062.

126. H.-G. Boyen, G. Kästle, F. Weigl, B. Koslowski, C. Dietrich, P. Ziemann, J. P. Spatz, S. Riethmüller, C. Hartmann, M. Möller, G. Schmid, M. G. Garnier and P. Oelhafen, *Science*, 2002, **297**, 1533.

127. R. E. Palmer, S. Pratontep and H.-G. Boyen, *Nat. Mater.*, 2003, **2**, 443.

128. Z. W. Wang and R. E. Palmer, *Nano Lett.*, 2012, **12**, 91.

129. Z. W. Wang, O. Toikkanen, F. Yin, Z. Y. Li, B. M. Quinn and R. E. Palmer, *J. Am. Chem. Soc.*, 2010, **132**, 2854.

130. D. Bahena, N. Bhattarai, U. Santiago, A. Tlahuice, A. Ponce, S. B. H. Bach, B. Yoon, R. L. Whetten, U. Landman and M. Jose-Yacaman, *J. Phys. Chem. Lett.*, 2013, **4**, 975.

131. Y. Han, D. S. He, Y. Liu, S. Xie, T. Tsukuda and Z. Li, *Small*, 2012, **8**, 2361.

132. Z. W. Wang, O. Toikkanen, B. M. Quinn and R. E. Palmer, *Small*, 2011, **7**, 1542.

133. Z. W. Wang and R. E. Palmer, *Nanoscale*, 2012, **4**, 4947.

134. Z. W. Wang and R. E. Palmer, *Nano Lett.*, 2012, **12**, 5510.

135. S. Bals, S. Van Aert, C. P. Romero, K. Lauwaet, M. J. Van Bael, B. Schoeters, B. Partoens, E. Yücelen, P. Lievens and G. Van Tendeloo, *Nat. Comm.*, 2012, **3**, 897.

136. Y. Kuwauchi, H. Yoshida, T. Akita, M. Haruta and S. Takeda, *Angew. Chem., Int. Ed.*, 2012, **51**, 7729.

137. H. Yoshida, Y. Kuwauchi, J. R. Jinschek, K. Sun, S. Tanaka, M. Kohyama, S. Shimada, M. Haruta and S. Takeda, *Science*, 2012, **335**, 317.

138. Special Issue on Operando and In Situ Studies of Catalysis, *ACS Catal.*, 2012, **2**, 2216–2445.

CHAPTER 11

In Silico *Studies of Functional Transition Metal Nanoclusters*

LICHANG WANG* AND PAMELA C. UBALDO

Department of Chemistry and Biochemistry, Southern Illinois University, Carbondale, IL 62901, USA
*Email: lwang@chem.siu.edu

11.1 *In Silico* Synthesis and Characterization of Functional Transition Metal Nanoclusters

In silico study of functional transition metal nanoclusters refers to the investigation of the structural, electronic, optical, and catalytic properties of transition metal nanoclusters using molecular modeling techniques, often denoted as Computational Chemistry. Great progress has been made in Computational Chemistry towards its application as a practical research tool for all areas of Chemistry and Biochemistry since the early 20th century. With breakthroughs in developing practical theories to deal with various Chemistry problems, coupled with the rapid development of faster and more powerful computers, software was developed and employed to study various properties of materials. The first milestone in the history of Computational Chemistry was marked by the appearance of commercial software, which allowed researchers to perform *in silico* studies of primitive problems, such as geometry optimization. The most widely used commercial software include *Gaussian* and *Materials Studio*. The climax of this stage of Computational Chemistry was the awarding of the 1998 Nobel Prize in Chemistry to Walter Kohn and John A. Pople for their contributions to

RSC Smart Materials No. 7
Functional Nanometer-Sized Clusters of Transition Metals: Synthesis, Properties and Applications
Edited by Wei Chen and Shaowei Chen
© The Royal Society of Chemistry 2014
Published by the Royal Society of Chemistry, www.rsc.org

electronic structure calculations; the development of the Density Functional Theory by Walter Kohn and computational methods in Quantum Chemistry by John A. Pople.

Since 1990, publications of *in silico* studies on various research topics have been springing up like mushrooms, starting from only theoretical research groups and extending to many non-theoretical groups, including some industrial settings. The scope of the calculations have also extended from geometrical and stability properties of materials to IR, Raman, electronic and optical properties. The exponential growth of Computational Chemistry applications in research laboratories and industrial research activities is also due to the nature of modern research. New material-based technologies, such as nanotechnology, are frequently developed at the nanoscale and often require the understanding of materials' properties at atomic and molecular level. Thus, Computational Chemistry has become an indispensable tool in the assistance of basic research development and has begun to play an important role in the development of technology in industry.

The second milestone of Computational Chemistry was marked by the appearance of modeling companies, such as Materials Design and MASIS, which further extended the role of Computational Chemistry and its impact on technology development and society.

In order to provide a better understanding of the results of the *in silico* studies of transition metal nanoclusters as catalysts, chiral regulators, sensors, and conductive inks to be discussed in this chapter, we first describe general concepts and procedures for conducting *in silico* studies in this section.

11.1.1 Synthesis

In silico synthesis of a given system of interest deals with performing optimizations of electrons and nuclei by solving the electronic Schrödinger equation for electrons, often denoted as electronic structure calculations. Alternatively, Molecular Dynamics (MD) simulations or on-the-fly calculations can also be used. In the latter two types of calculations, the optimizations are made by solving the Schrödinger or Newtonian equation for nuclei, which is not necessarily the case in electronic structure calculations. A flow chart of the *in silico* synthesis process is illustrated in Figure 11.1.

An *in silico* synthesis consists of three steps. The first step is the preparation of the system that is to be synthesized, namely optimized. One of the first things is to know what types of and how many atoms are in the system. Finding out the type of atoms and composition, in the case of an alloy, is not difficult. However, the number of atoms in a particular cluster size can be challenging. In most wet experimental studies, the size of nanoclusters is determined rather than the number of atoms. This sometimes presents challenges when directly comparing results between wet laboratory measurements and *in silico* studies. Figure 11.2 provides a general idea on how many atoms there are in a nanocluster of up to 11 nm.

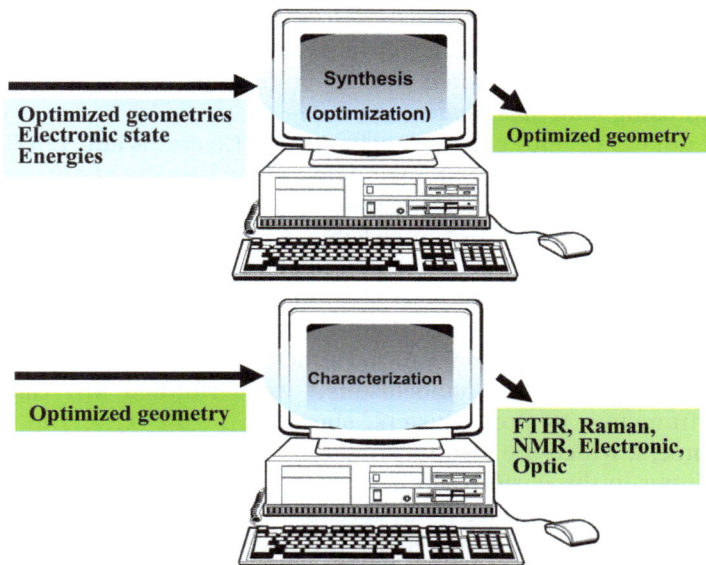

Figure 11.1   An illustration of an In Silico synthesis (top) or characterization (bottom) set-up.

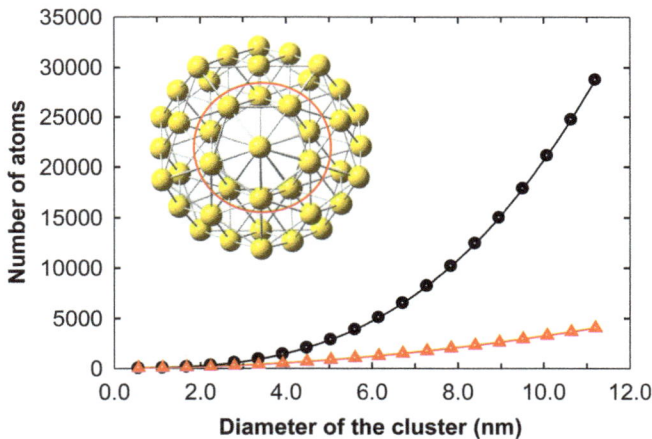

Figure 11.2   The number of atoms as a function of cluster diameter for Au clusters. An icosahedron growth is assumed. The filled circle is the number of total atoms in the cluster and the triangles are the number of surface atoms. Inset: A Au_{55} cluster with a 13-atom inside the circle.

Once the type and number of atoms are known, one needs to provide initial input data of the system to be optimized to the computer. In electronic structure calculations, this means that the coordinates of each atom will need to be provided. If MD calculations are used, the momentum of each atom is also required.[1] Generation of the initial coordinates becomes feasible using various interface software. For instance, a 2 nm Co–Cr alloy cluster with radially distributed 33% Cr atoms, shown in Figure 11.3, was generated using *Materials Studio*.

This 2 nm Co–Cr alloy cluster consists of 256 Co atoms and 126 Cr atoms. Therefore, it is not difficult to imagine that many isomers of different shape and electronic spin multiplicity are possible. The attempt to find all isomers can be a daunting task, if not impossible. Construction of initial inputs to the problem of interest are still experience-based, even though efforts have been made to develop methods to generate initial input more robustly and thoroughly. For instance, one of the practices we employed was to use MD simulations to generate as many initial structures as possible.[1–3] In the study of the coalescence of Ag nanoclusters,[2] we performed MD calculations starting from two 19-atom clusters, shown in Figure 11.4, and obtained many isomers, which are also illustrated in Figure 11.4. These isomers are used as input in the *in silico* synthesis.

A Potential Energy Surface (PES) is required in MD simulations, in which electron optimization is replaced by the use of PES. For the purpose of generating initial structures to be used in *in silico* synthesis, an approximate PES is sufficient. However, if the results (*i.e.* geometries) obtained from MD simulations are to be used as the final optimized structures, an accurate PES will have to be used. Therefore, a lot of effort has been made to obtain an accurate PES.[3,4]

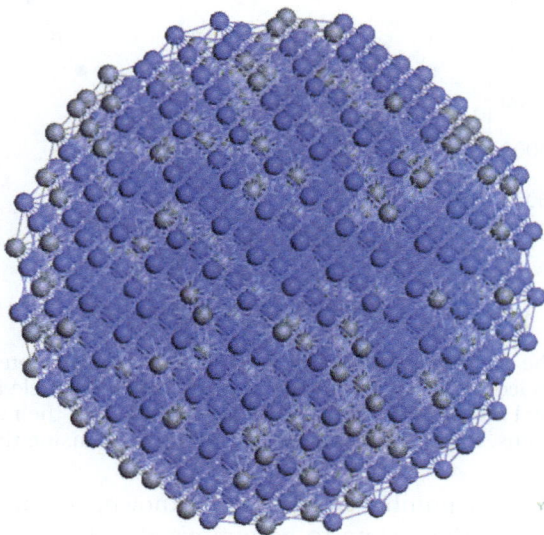

Figure 11.3   A 2 nm Cr–Co nanocluster that contains 382 atoms and was generated using *Materials Studio*.

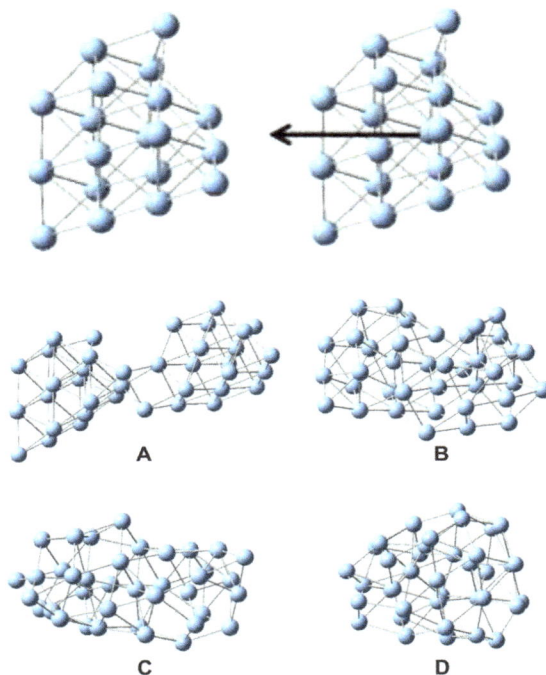

Figure 11.4 The initial configuration of $Ag_{19} + Ag_{19}$ collision (top) and the four products (A)–(D) after MD simulations.
Pictures were taken from ref. 2.

If electronic structure calculations are used for optimization, coordinates of each atom are the only input data required. In the case of on-the-fly or MD simulations, the momentum of each atom is also required.

The second step of the *in silico* synthesis is the choice of theory and determination of the accuracy of the numerical calculations. The most widely used theory in the study of transition metal nanoclusters is DFT. Within the DFT framework, the choice of functional and basis set are required and determine the accuracy and expense of the calculation. The most commonly used functionals in the study of transition metal clusters are the B3LYP, PBE, and PW91. The most popular basis sets used are the Gaussian split valence basis sets, double-numerical basis set with polarization function, and the plane wave basis sets.

Once the method (*i.e.* the functional of DFT) and basis set are chosen, calculations can be performed by computers with frequent and close monitoring by the researchers on the calculations for a successful run. Geometry optimization calculations will terminate when an extreme value, either maximum or minimum, is obtained. To find a minimum value of the system, one needs to perform frequency calculations. If all frequencies (*i.e.* 3N-6 for a nonlinear or 3N-5 for a linear arrangement of atoms in a cluster with N being the number of atoms) are positive, a minimum is obtained. If only one frequency out of 3N-6 or 3N-5 is negative, a transition state is obtained. Therefore, before proceeding to the *in silico* characterization of the calculated material, frequencies of the optimized geometry will need to be determined and probed for any negative values. Any negative frequency means that the stable state optimized geometry is not yet achieved and should be further optimized by adjusting the structure. Only an optimized geometry without any negative frequencies (excluding transition state search) can proceed to the following calculations that aim to characterize the system of interest.

Optimization calculations can be done in solution or in vacuum. It is important to choose an environment that is very similar to the one used in experiments performed in the wet laboratory. The closer the *in silico* model is to the experimental model, the more valuable a comparison between the *in silico* and the wet lab will be.

The third step of the synthesis is to perform analysis of the results. Each geometry optimization calculation should be probed for complete convergence before proceeding to *in silico* characterization techniques. Important information can be directly extracted from the optimized geometry, including bond lengths, bond angles, and atomic charge. The total energy of the system is also obtained, which can be used to calculate heat of formation.

11.1.2 Characterization

In silico synthesized materials can be characterized using electronic structure calculations to obtain FTIR, Raman, NMR, UV-Vis, and other electronic properties. Additionally, heat of formation can be obtained. A flow chart of *in silico* characterization is shown in Figure 11.1.

The initial input of the characterization is the optimized geometry. Single point calculations are performed, often with higher convergence accuracy than those during optimization stage. This means no geometry optimization is involved in the *in silico* characterization process, *i.e.* nuclear positions are fixed in the calculations.

Selection of *in silico* characterization methods can be guided by existing experimental results or the designed characterization methods for wet lab experiments. It also depends on what kind of electronic information is desired. The goal of *in silico* characterization is to gain deeper molecular information about the system and to confirm or predict results that can be compared with the wet lab experiments. Some properties that can be determined using *in silico* studies are the system's stability, chemical reactivity, catalytic, optic, magnetic and other physicochemical properties. Several characterization methods that can be done in *in silico* studies will be discussed in this section.

The Highest Occupied Molecular Orbital (HOMO), the Lowest Unoccupied Molecular Orbital (LUMO), and the HOMO–LUMO energy gap, often denoted as band gap, are the first three results that can be easily extracted from the optimized geometry. The HOMO gives information about the specific location and energy of the last molecular orbital filled with electrons as well as how many molecular orbitals were filled. The LUMO is the next orbital after the HOMO and is the first possible location where an electron will go from HOMO when it gains energy, such as absorption of a photon. The HOMO–LUMO energy gap is the energy difference between HOMO and LUMO, which is the minimum energy required to excite an electron and make it go from HOMO to LUMO.

Several characterization methods that are usually done in wet laboratories can be done *in silico* as well. These methods include IR, Raman, UV-Vis, NMR, and calorimetry. IR spectroscopy is a detection technique where a molecule absorbs infrared radiation and converts it into molecular vibrational energies. Experimentally, IR results are often used for identifying certain functional groups present in the system. The current IR calculation in the *in silico* studies relies on harmonic approximation and often overestimates the frequencies. The IR spectrum can be directly extracted from the calculated frequencies of the ground state optimized geometry and can be visualized by using the *Gaussview* software, for example.

Raman spectroscopy is another technique that is based on molecular frequencies that result from inelastic scattering of photons when exposed to monochromatic light. Similarly, Raman intensities can be easily extracted from calculated frequencies of the optimized geometry and can be visualized using the *Gaussview* software.

UV-Vis spectroscopy is often used to gain information on the excitation of electrons in the system upon absorption of a photon. Fluorescence, surface plasmon, and transition peaks can also be generated in the *in silico* studies. In the *in silico* study of UV-Vis spectroscopy, the chosen method should be applicable for a dynamic system such as Configuration Interaction Singles (CIS), Time Dependent Hartree–Fock (TDHF), and Time Dependent DFT (TDDFT). The UV-Vis spectrum is generated using visualization software, such as *Gaussview*, *GaussSum* and *SWizard*.

Nuclear Magnetic Resonance (NMR) spectroscopy is a characterization technique based on the chemical shifts of atoms present in the system when exposed to a magnetic field. A good level of theory for the ground state optimized structure is needed before it can proceed to NMR calculations to achieve accurate results. NMR shielding tensors and magnetic susceptibilities are calculated to generate the NMR spectrum. Using *Gaussian09* software, NMR shielding tensors and magnetic susceptibilities can be computed using the Hartree–Fock, DFT and MP2 methods together with the Continuous Set of Gauge Transformation (CSGT) or the Gauge Independent Atomic Orbital (GIAO) method. The NMR spectrum can be visualized using *Gaussview* software.

Aside from spectroscopic results, calorimetry, which involves measurement of heat of chemical reactions or any physical change, can also be done by *in silico* studies. The heat of reaction, binding energies, and adsorption energies are just some of the properties that can be predicted. This is done by generating several optimized geometries. For example, to calculate the heat of reaction of the combustion reaction: $2CH_4 + 4O_2 \rightarrow 2CO_2 + 4H_2O$, the optimized geometries of each molecule in the reaction will need to be generated. The total energy of each molecule can be directly extracted from the optimized geometries and is used to calculate the heat of reaction by $\Delta H_{reaction} = (2E_{CO_2} + 4E_{H_2O}) - (2E_{CH_4} + 4E_{O_2})$, where E represents the total energy of a system labelled by the subscript.

In silico characterization results are often compared with wet lab measurements. Some of the results, such as FTIR, Raman, UV-Vis, and NMR can be directly compared with experiments, while others will need further but simple calculations, such as heat of formation. The calculated results will aid in the understanding of the materials' properties or reaction mechanisms, such as what occurs in catalysis.

11.2   Catalysis of Transition Metal Nanoclusters

More than 80% of the chemical reactions that play key roles in both industrially and environmentally important processes occur catalytically under heterogeneous conditions. The VIIIB 4d and 5d transition metals, such as Ru, Rh, Pd, Os, Ir, Pt, and Ag have been widely used as catalysts in many heterogeneous chemical reactions. Over the past twenty years, many studies have indicated that materials at the nanometer scale are not only important for current industrial and environmental interests, but may also revolutionize the design and fabrication of new generation catalysts and other new functional materials. Nanoscale materials hold great promise for providing breakthroughs in areas where technological innovation is needed. For example, developing highly efficient nanoscale cathode catalysts is considered a key step towards the widespread utilization of hydrogen fuel cells.

Nanoparticles (also often denoted specifically as clusters or nanoclusters when they are in the subnanometer size) of transition metals have been recognized for their high catalytic activities in comparison to their bulk counterparts.[5–11] The key challenges in the practical use of nanoparticles as catalysts are consistent activity and robustness, which require fundamental understanding of the nanoclusters and their catalytic activities. Additionally, important issues include sintering and dissolution of catalysts. Small nanoparticles are highly active and may aggregate[12] if they are placed close to each other under suitable conditions, a process termed as sintering in catalysis. Although the formation of large particles due to the aggregation of small clusters is largely responsible for the complete loss of catalytic activities, no effective control has been available due to the lack of understanding of the coalescence processes.

Computational studies have been widely used to investigate catalytic activities of nanomaterials to provide insight into various catalytic processes. The current *in silico* studies of catalysis are largely confined in the size range of less than 1 nm or the bulk surface. For a system consisting of hundreds of atoms, simulations are still prohibited by the cost of vast computing times. In this section, *in silico* studies of two reactions, O_2 reduction and CH_4 dehydrogenation, will be discussed, in addition to the results of Ir nanoclusters.

11.2.1 Sinter-Resistant Ir Nanoclusters

In order to investigate the catalytic activities of nanoclusters, the first step of research is to address issues such as the stability and structure of the nanoclusters, the resemblance of the structure of the nanoclusters to the bulk structure, the correlation of the electromagnetic properties to the size and structure of the nanoclusters, the extra degrees of intrinsic differences of the nanoclusters from their bulk counterparts in addition to the well-known fact of the substantially large surface/volume ratio of nanoclusters. It is also important to study how the above mentioned issues are related to the choice of metal. For instance, is the most stable structure of a 30-atom 4d metal cluster Rh_{30} the same as a 30-atom 5d metal cluster Ir_{30}?

To answer these questions, we systematically investigated about 100 clusters consisting of up to 60 atoms (about 1.5 nm) for each of the elements: Pd, Pt, Ir, Rh, Cu, Ag, Au, and Ru clusters using DFT calculations and reported the results for some of the clusters, such as Rh,[13] Pd,[14] Pt,[15] Ru,[16] Au,[17] and Ir clusters[18,19] as well as bimetallic Pt–Au clusters.[20,21]

One of our interesting DFT results of the Ru, Rh, Pd, Os, Ir, and Pt clusters showed that simple cubic is the most stable structure for the small Os, Ru, Ir, Rh, and Pt clusters.[18] The icosahedron becomes the most stable structure for the Rh clusters and the cubo-icosahedron for the Pt clusters when the cluster size becomes larger than 13 atoms. The structural phase transition from simple cubic to icosahedron occurs at around 40 atoms for the Ru clusters[16] but at 48 atoms for the Ir clusters.[19] Our comparative studies show that the trend of forming the simple cubic structure increases from right to left across a period and from the top to bottom in a group in the periodic table. The trend for the icosahedron structure is opposite. Figure 11.5 depicts structures of four pairs of Ir isomers with their relative stability plotted in Figure 11.6.

The structural phase transition of Ir clusters from simple cubic to the bulk structure at this large size indicates that it is energetically more difficult to change from simple cubic to face-centered cubic (fcc) when two cubic isomers coalesce to form a bigger cluster and therefore these nanoclusters can be resistant to aggregation.

Our DFT prediction was observed experimentally six years later. The cubic shape of the Ir clusters was found in 2012 using aberration-corrected scanning transmission electron microscopy (STEM) by Gates' group.[22] These 1 nm cubic Ir clusters were found to bounce off each other. Their study also found that the critical size and the resistant properties of the Ir clusters are independent of substrates, indicating that this sinter-resistance is the intrinsic character of the 1 nm Ir clusters.

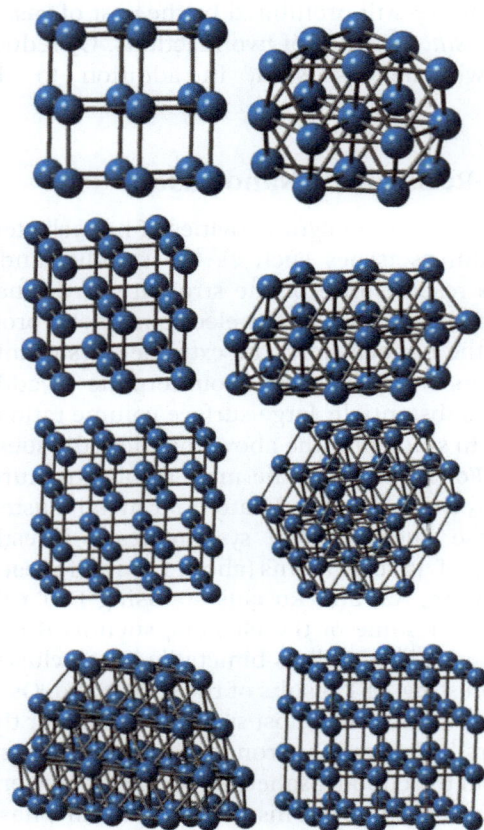

Figure 11.5   Four pairs of Ir isomers of 18, 27, 48, and 64 atoms. The left structures
are more stable than their corresponding isomers on the right.
All the structures were taken from ref. 19.

An extensive review of many synthesized Ir clusters was made by Gates'
group subsequently.[23] Interestingly, their investigation showed results from
various treatments of iridium complexes on different supports, even under
harsh reductive conditions (*e.g.*, 873 K in H_2), still led to the formation of
only uniform iridium clusters limited to a critical diameter of approximately
1 nm. As predicted by our calculations, the cubic structures of this critical
size are resistant to aggregation because coalescence of two such clusters
would require energetically unfavourable rearrangements of atoms.

These results illustrate that the atomic packing of nanoclusters can be
intrinsically different from that of the corresponding bulk metal. These
structural differences could be one of the reasons for the high catalytic activities of nanoclusters. In general, small clusters are highly active and have a
high tendency to aggregate.[12] When small clusters aggregate to form large
particles, they may completely lose their catalytic properties. It is therefore

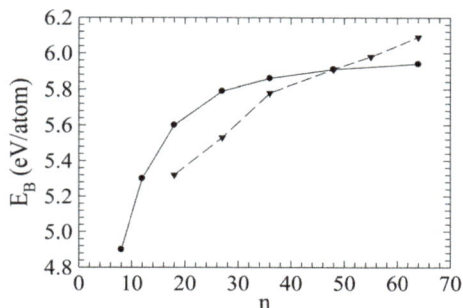

Figure 11.6 Transition from simple cubic to fcc. The solid and dashed lines represent the binding energy (E_B) in eV/atom of simple cubic structures (●) and those cut from the fcc bulk (▼), respectively, as a function of cluster size, n.
The figure was taken from ref. 19.

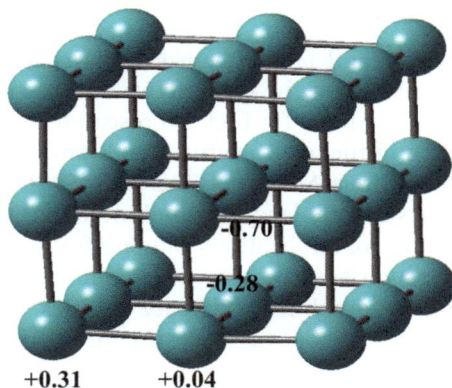

Figure 11.7 A 27-atom Ru cluster. Among the 27 atoms, there are four types of atoms: the corner, the edge center, the face center, and the center atom, whose charge is labeled.
The figure was taken from ref. 16.

important to understand the sintering mechanism, such as how and in what conditions these nanoparticles will coalesce. The effect of temperature on the structure and coalescence of clusters will also be studied here. The coalescence mechanism can be studied using MD simulations.[2,24,25]

Furthermore, catalytic activity of a cluster depends not only on its shape but also on its local environment, *i.e.* local structure. The local structure is partly reflected by the charge distribution. The local charge distribution of a cluster is also important to predict adsorption sites for a molecule and to predict the strength of the bond between adsorbates and nanoparticles. In Figure 11.7, we provided the Mulliken charge distribution of individual atoms in a 27-atom Ru cluster that was obtained from our DFT calculation.[16]

The results show that the central and the face-centered atoms are all negatively charged, while the edge and the corner atoms in the cluster are positively charged, indicating that the low coordinated atoms are positively charged and the atoms with high coordination numbers are negatively charged. The existence of the positively charged atoms at the corner or the edge suggests that electron-rich molecules, such as CO or benzene, will be easily adsorbed on these sites. On the other hand, the face-centered atoms will be open for the absorption of nucleophilic molecular species.

With the increase of computer power and interface software for constructing clusters, we can routinely study clusters of about 1.5 nm, which have around 150 atoms. *In silico* structures of small metal clusters that consist of a few atoms can be confirmed by photoelectron spectroscopy measurements. Structures of larger clusters of about 1–2 nm are still a challenge for experimental confirmation.

11.2.2 Multicomponent Pt Alloy Nanoclusters for O_2 Reduction

Development of efficient and cost effective catalysts for breaking the O–O bond of O_2 molecules is one of the most active and key areas in fuel cell research. The market driving force for fuel cell research is the reality that fossil fuels are running out and fuel cells such as the Proton Exchange Membrane Fuel Cell (PEMFC) and Direct Methanol Fuel Cell (DMFC) become attractive because of high conversion efficiency, low pollution, lightweight, high power density, and a wide range of applications ranging from small power supplies for electronic devices such as PCs, notebooks, and cellular phones to large power sources used in automobiles and space shuttles.

Catalyst accounts for $\sim 30\%$ of the cost of manufacturing fuel cells. Currently, low activity, poor durability and high cost of the platinum-based anode and cathode catalysts in PEMFCs and DMFCs constitute some of the major barriers to commercialization of fuel cells, which has promoted intense research activities towards both fundamental and technological innovations in developing multimetallic nanocluster catalysts with a combination of activity-enhancing, stability-optimizing, and cost-reducing components. Zhong's group has been working on developing nano binary/ternary alloy catalyst technologies and explored the synergistic multifunctional activities and the compromised balance of activity and stability of the catalysts.[26]

In development of robust and cost effective catalysts, efforts have been focused on the design, synthesis, characterization, modeling, and optimization of multimetallic nanoclusters, including binary ($M1_nM2_{100-n}$), ternary ($M1_nM2_mM3_{100-n-m}$) alloys, and core@shell (M1@M2, MO_x@M, and MO_x@M1@M2) nanoparticles (M (1 or 2) = Pt, Co, Ni, V, Fe, Cu, Pd, W, Ag, Au, *etc.*, and $MO_x = Fe_2O_3$, Fe_3O_4, CeO_2, *etc.*). The goal of our efforts is to establish the fundamental correlation between the nanostructural parameters (size, composition and morphology) and the catalytic properties (activity and stability). Many studies focused on understanding the mechanism of oxygen reduction on Pt–Fe, Pt–Ni, and Pt–Co, including CO- or methanol-tolerant. Bulk-melted PtBi, PtIn, and PtPb intermetallic phases and Ru nanoparticles modified with Pt showed some promise for fuel cell applications. The application of high throughput, combinatorial screening methods to screen a large number of catalysts using an optical fluorescence technique by detecting proton production on an array of catalyst inks with different compositions has been successfully demonstrated.[27] Individually addressable array electrodes have also been investigated for rapid screening. Simple thermodynamic principles are proposed as guidelines that assume that one metal breaks the oxygen–oxygen bond of O_2 and the other metal acts to reduce the resulting adsorbed atomic oxygen. The high throughput combinatorial screening of catalysts is very useful for rapid screening, including the work by Zhong's group,[28] but its drawback is the lack of direct insights into the nanoscale correlation. *In silico* study also begins to demonstrate the importance of binary or ternary alloy catalysts, including computational chemistry (*e.g.*, DFT) to determine optimal structures and adsorption energies and to predict synergetic effects.[29]

DFT studies of Pt–Au nanoclusters of about 1.2 nm showed their lattice constant is smaller than the corresponding bulk alloys, depicted in Figure 11.8.[30] This lattice constant difference may be used as a potential

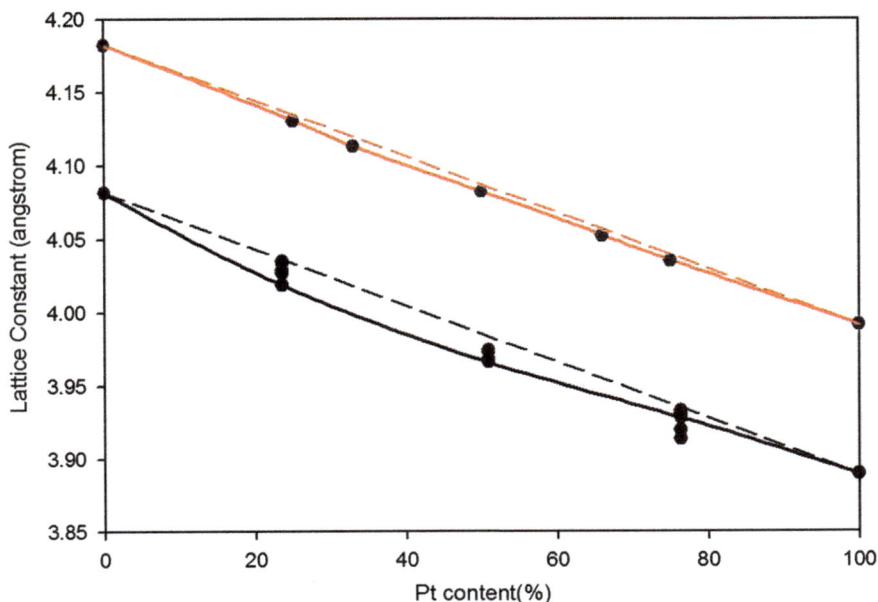

Figure 11.8   The lattice constants as a function of Pt composition for 55-atom PtAu clusters (bottom line) and for bulk PtAu alloys (top line). Figure was taken from ref. 30.

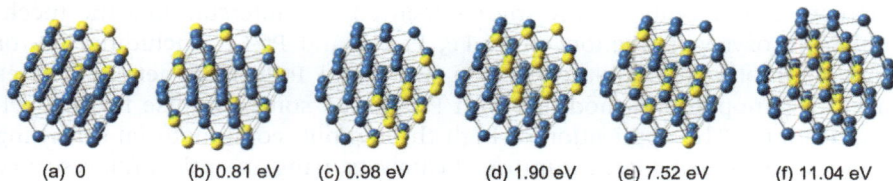

(a) 0	(b) 0.81 eV	(c) 0.98 eV	(d) 1.90 eV	(e) 7.52 eV	(f) 11.04 eV

Figure 11.9   Structures and relative energies of six $Pt_{42}Au_{13}$ clusters. A positive number means that the corresponding structure is less stable than structure (a).
Structures were taken from ref. 30.

correlation to the relative catalytic activity with respect to the bulk alloy. Furthermore, DFT studies on the homogeneities of a $Pt_{42}Au_{13}$ cluster are shown in Figure 11.9. The DFT results indicate that phase segregation is preferred under this composition.

Furthermore, *in silico* studies of interaction of O_2 with various metals were performed and summarized in Figure 11.10, in which the O_2 frequency, electron transferred to O_2, and adsorption energy were plotted for comparison.[29] The results indicated that the changes in the properties of O_2 are caused more by the type of O_2 configuration with respect to the metal than the type of metal.

New concepts and strategies have been explored for the creation of size-, composition-, and morphology-controlled multimetallic nanocatalysts. Because nanoscale phenomena differ from bulk counterparts in many significant ways, including atomic–metallic transition, possible phase-reconstitution, different melting points due to size or alloying effects, and synergistic effects due to modified electronic band structure, answers to how the catalytic activity and stability are influenced by size, composition, and morphology could hold the key to achieving durable and active catalysts.

At the fundamental level there are many important issues to be addressed in view of the fact that multimetallic nanoclusters, given the right combination of metals, often display unique properties and could be used to catalyse new types of reactions not accessible to traditional homogeneous and heterogeneous catalysts, such as whether certain metals form single-phase whereas others segregate or even form core–shell nanostructures or the origin of synergistic catalytic properties. Most existing approaches to multi-component catalysts are limited by either complications in controlling the alloying or phase-segregation properties or the inherent lack of well-defined nanoscale dimension. The phase segregation or core–shell type redistribution of metals on the nanoparticle surface for different bimetallic systems is a major challenge encountered in bimetallic or trimetallic catalyst systems.

In silico studies of O_2 reduction in the presence of model PtVFe nanocatalysts were performed using the PBE functional with a plane wave basis set. The model catalysts consisted of trimers and a 0.6 nm particle. The O_2

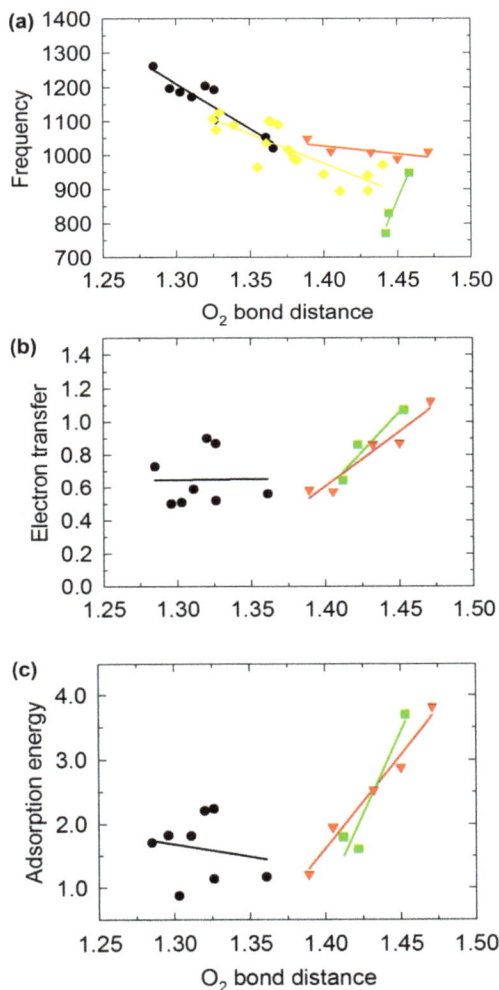

Figure 11.10 O_2 frequency (in cm^{-1}) (a), the number of electrons transferred to O_2 (b), and the adsorption energy (in eV) (c) as a function of O_2 bond distance (in Å) among different metal complexes. The circles, triangles, and squares represent the data obtained from the current calculations for the Pauling, the Griffith, and the Yeager structures, respectively.

Figure was taken from ref. 29.

dissociation pathways were explored using Pt_3 and PtVFe. Interestingly, the *in silico* results suggest that the superior catalytic activity of PtVFe nanocatalysts in the oxygen reduction may be due to effectiveness in charge transfer and the presence of a direct (spontaneous) dissociation pathway. The rate of O_2 reduction is provided in Table 11.1, in which the PtVFe catalyst seems to be 3 times faster than the pure Pt catalysts.

Table 11.1 The O_2 dissociation rate constant as a function of temperature for the reactions on PtVFe and Pt_3. Table was taken from ref. 29.

T (°C)	k_1 (PtVFe), s^{-1}	k_2 (Pt_3), s^{-1}	k_1/k_2
25	4.29×10^9	9.71×10^8	4.4
50	9.39×10^9	2.32×10^9	4.0
75	1.84×10^{10}	4.91×10^9	3.7
100	3.28×10^{10}	9.39×10^9	3.5
125	5.45×10^{10}	1.65×10^{10}	3.3

11.2.3 Pt Nanoclusters for Activation of C–H bonds in CH_4

Shale gas has become an increasingly important source of fuel. It is found to be plentiful and affordable and has the potential to become a dominant fuel in the near future. The world reliance of fuel shifted from coal to petroleum at the beginning of the twentieth century. It seems that another shift, a shift from petroleum to natural gas, is inevitable due to the discovery of plentiful shale gas and their affordability. Permission for fracking is granted in many places and the pros and cons of fracking are being debated and discussed in others before granting the permission. Research in methane, the main component of natural gas, can be expected to intensify in the very near future.

Methane has been widely used in chemical synthesis, hydrogen production, and energy production. As such, methane activation has been studied rather extensively. Activation of the C–H bond in methane over Pt clusters of 2–24 atoms was studied using time-of-flight mass spectroscopy by Trevor *et al.* in 1990.[31] Their experiment showed that the most reactive neutral clusters are Pt_2–Pt_5, while other clusters Pt_6–Pt_{24} are less reactive and the Pt atom is less reactive by at least 1 order of magnitude. Furthermore, among the Pt_2–Pt_5 clusters, the reactivity oscillated with the trimer and pentermer being slightly more active. Studies have also been carried out on the activation of methane molecules on cationic and anionic Pt clusters.[32–36]

In the *in silico* studies of methane dissociation on Pt clusters, we performed DFT calculations using B3LYP and PW91.[37] Catalysis is largely a local phenomenon. We therefore classified the atoms in a catalyst into three categories based on their involvement in the catalytic process: primary, secondary, and tertiary. In Figure 11.11, the top Pt atom in Pt_4 and Pt_{10} can be classified as the primary atom, which is directly involved in the bond making and breaking. The secondary atoms are defined as the atoms that connect directly to the primary atoms in a catalyst. In Figure 11.11, the three atoms below the primary atom are the secondary atoms. The tertiary atoms are those connected directly to the secondary atoms, which are illustrated as the six atoms in the bottom layer of the Pt_{10} cluster in Figure 11.11.

The above classification of atoms in a catalyst needs to be validated by investigating a number of issues that are critical and important to understanding catalysis. As a first step towards validating the above hypothesis on the classification of atoms in a catalyst, we studied CH_4 dehydrogenation on a Pt atom (a cluster with one primary atom only) and a Pt_4 cluster (a catalyst

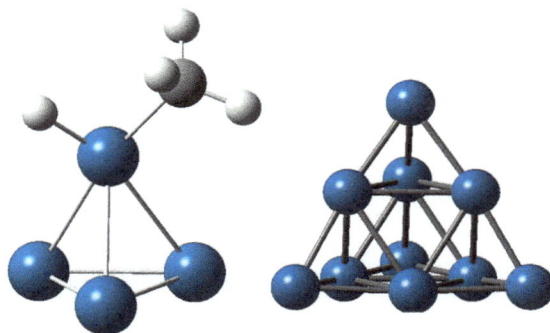

Figure 11.11   An illustration of CH_4–Pt_4 (left) and a 10-atom Pt cluster. Images taken from ref. 37 and ref. 15, respectively.

with one primary and three secondary atoms).[37] Reaction pathways towards dehydrogenation were identified and depicted in Figure 11.12.

For the CH_4–Pt system, the first two C–H bond activations undergo a spin crossing and a three-centred transition state (labelled as TS2 in the top picture in Figure 11.12). The reaction pathway for the methane dehydrogenation on Pt_4 consists of five single reaction steps. The first step is the cleavage of the first C–H bond with an energy barrier of 4 kcal mol^{-1}. Unlike the CH_4–Pt system, there is no spin crossing and the transition state (TS1) is below the reactant asymptote. The second step is the H transfer from one Pt atom to the other before breaking the second C–H bond. The transition state in the H transfer process (TS3) is also below the reactant asymptote. The third step involves breaking up the second C–H bond and is the rate limiting step in dehydrogenation with an energy barrier of 28 kcal mol^{-1}. The fourth and fifth steps are the formation and dissociation of a hydrogen molecule, respectively. Based on our work, the CH_4 dehydrogenation takes place faster on Pt_4 kinetically than on a Pt atom, but thermodynamically, the dehydrogenation in CH_4–Pt is more favoured than that in CH_4–Pt_4 by about 3 kcal mol^{-1}, although both reactions are shown to be endothermic.

Within the context of primary, secondary, and tertiary atoms in a catalyst, the *in silico* studies of CH_4–Pt and CH_4–Pt_4 systems illustrated the importance of secondary atoms. In the first C–H bond activation, the addition of three secondary Pt atoms lowered the energy barrier considerably, lowering the barrier from about 30 kcal mol^{-1} on Pt to about 4 kcal mol^{-1} on Pt_4. We pointed out that another pathway, *i.e.* spin crossing, also opens up to the Pt case. The second C–H bond activation is also interesting. In the case of CH_4–Pt, a barrier of $\sim$36 kcal mol^{-1} has to be overcome to form $(H)_2$–Pt–CH_2, while a H transfer in the case of CH_4–Pt_4 takes place prior to breaking the second C–H bond. Although it undergoes two steps in CH_4–Pt_4, the energy barriers are only 7 and 10 kcal mol^{-1}. Therefore, secondary atoms in the cluster play an important role in the CH_4 hydrogenation. In the hydrogen recombination and desorption steps, the presence of a single Pt atom is

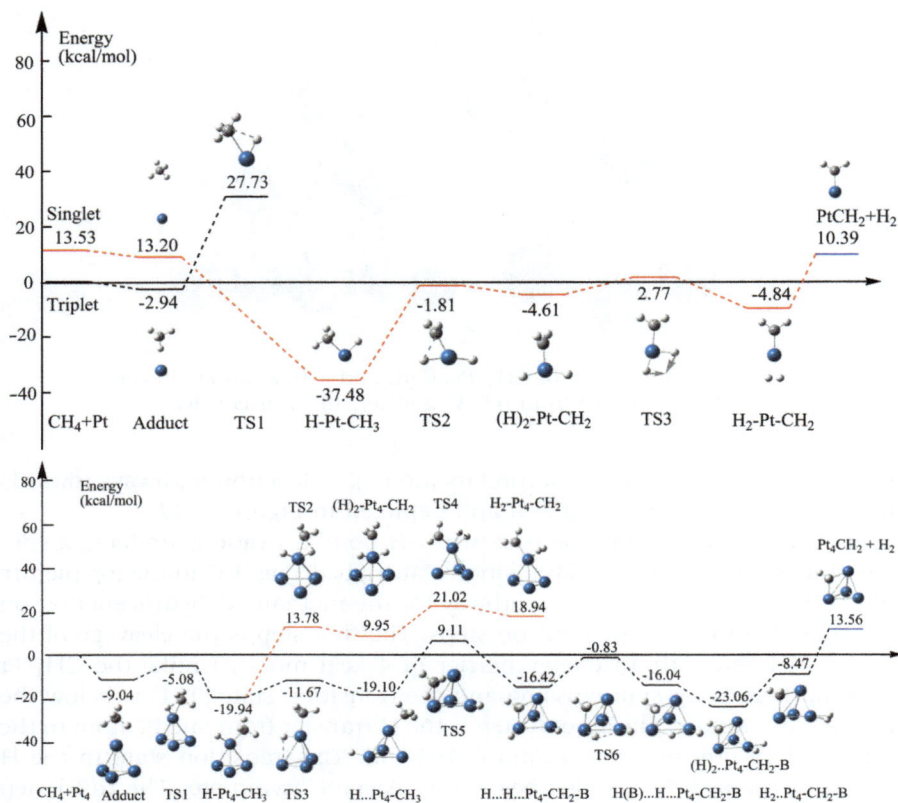

Figure 11.12 The CH_4 activation pathway to form $PtCH_2 + H_2$ (top) and $Pt_4CH_2 + H_2$ (bottom). The energy of the reactants, *e.g.* Pt_4 (triplet state) and CH_4, is set as reference. The data and structures were obtained from the B3LYP calculations.
Figure was taken from ref. 37.

better than that of Pt_4 as the steps required in the case of Pt are fewer and the energy barrier is lower. This role reversal of secondary atoms in a catalyst also demonstrates the importance of secondary atoms.

Our ongoing efforts using more studied Pt clusters[17] in the *in silico* studies of CH_4 dehydrogenation by including tertiary and different types of secondary and more than one primary atom will shield more light on the classification of atoms in a catalyst and on our understanding of catalysis of CH_4 dehydrogenation.

11.3 Other Functionalities of Transition Metal Nanoclusters

Transition metal clusters have been widely explored as functional materials for applications outside of catalysis. In this section, we describe three of those applications: sensing materials, chiral recognition regulators, and conductive ink.

11.3.1 Pd Nanoclusters for Sensing CH_4

The use of a methane (CH_4) gas detector is the best way of ensuring safety at home and work places, such as landfills and mining sites. An ideal CH_4 detector should be simple, cheap, have a high detection efficiency, and have a low detection limit. Understanding the mechanism of CH_4 detection is essential in determining the best CH_4 sensing materials and sensors.

Palladium (Pd) has been a popular doping material for CH_4 sensors. Base materials such as Single Walled Carbon Nanotubes (SWCNTs), Multiple Walled Carbon Nanotubes (MWCNTs), and SnO_2 are often doped with Pd for improved sensitivity and detection. In a study by Mishra and Agarwal, three doping materials, Pt, Au, and Pd, were used for SnO_2-based thick film sensors. Among the three, Pd was found to be the best in terms of sensitivity. A short response and recovery time were also observed, which are usually dependent on the type of gas and concentration up to 400 ppm.[38] Palladium has a lower cost than Pt and Au and can provide better sensitivity, therefore, Pd is the best doping material for methane sensors.

The spill-over effect is the suspected adsorption mechanism when the surrounding conditions favor non-oxidized Pd(0). Palladium can be oxidized under usual operating conditions and in the presence of O_2 to form PdO. The presence of CH_4 will reduce the PdO to Pd(0) and generate CO_2. Part of the Pd(0) will then be re-oxidized by oxygen in the surroundings.[39,40] The above mechanism for Pd clusters doped to CNTs is assumed to be the cause of declining sensitivity for many methane sensors which are not exposed to the target gas for a long time.

When Pd clusters were loaded to SWCNTs, Lu *et al.* demonstrated that these Pd doped SWCNTs can be used as a sensor to detect methane ranging from 6 to 100 ppm in the air at room temperature.[41] Furthermore, the experiment showed that a SWCNT alone does not have any electronic response to CH_4 adsorption at room temperature due to the weak interaction between CH_4 and SWCNTs. This indicates that the SWCNTs alone do not work as a sensor for CH_4. The effects of loading Pd nanoparticles on SWCNTs on the detection of CH_4 were explored. When loaded or doped with Pd clusters, the researchers found that the Pd doped sensors have 100 times smaller power consumption and 10 times higher sensitivity than the conventional metal oxide and catalytic bead sensor for CH_4. DFT studies by Zhou *et al.* investigated the adsorption of CH_4 to one Pd atom doped SWCNTs. The results showed that the interaction between CH_4 and the Pd–SWCNT is electrostatic through the formation of a bound complex of (Pd–SWCNT)–(CH_4) with partial charge transfer from CH_4.[42]

Understanding the binding strength between CH_4 and Pd clusters and how much the CH_4 adsorption can perturb the electromagnetic properties of Pd clusters is an important first step to gaining insight into the superior detection ability exhibited by the Pd–SWCNT sensors. As such, we performed

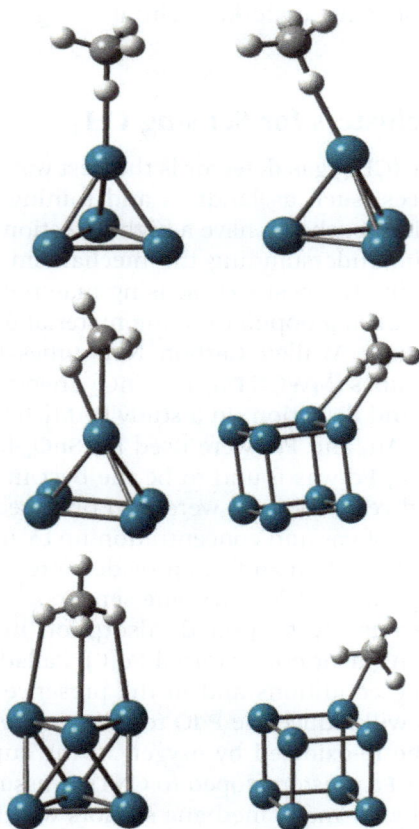

Figure 11.13   The coordination modes η^1 (top), η^2 (middle) and η^3 (bottom) of the
$CH_4 \cdots Pd$ complexes.
Images were taken from ref. 43.

DFT studies on various $CH_4 \cdots Pd$ cluster adducts formed by fifteen Pd clusters with up to 13 atoms at different cluster thickness, adsorption site, and coordination modes, *i.e.* η^1, η^2–C,H, η^2–H,H, and η^3, of the $CH_4 \cdots Pd$ complexes that are illustrated in Figure 11.13.[43]

Our study showed that the binding strength between CH_4 and a Pd cluster is strongly dependent on the cluster thickness, adsorption site, and co-ordination modes of the $CH_4 \cdots Pd$ complexes. Three plots are given in Figure 11.14 summarizing the DFT results according to different classifications. It is clear that single Pd atom doping has the strongest interaction between CH_4 and Pd. Adsorption strength does not change significantly with the increase of cluster size, however, the arrangement of Pd atoms in the cluster affects the adsorption strength. Our DFT calculations indicate that two-atom-layer thickness of Pd clusters can absorb CH_4 more strongly than the monolayer or tri-layer isomers.[43]

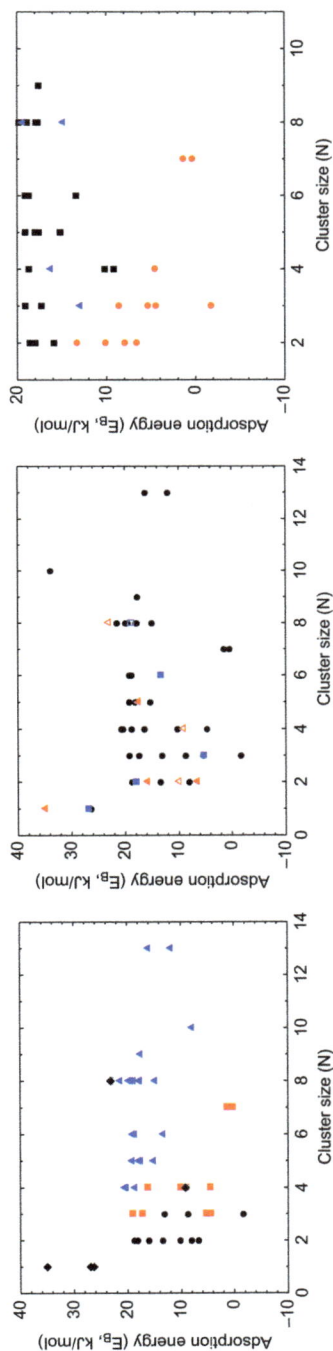

Figure 11.14 Left picture: Plot of the adsorption energy of methane on various Pd clusters as a function of cluster size. The diamonds, circles, squares, and triangles represent the adsorption on a single Pd atom, the linear Pd clusters, the planar clusters, and the three-dimensional clusters, respectively. Middle picture: Plot of the adsorption energy of methane on various Pd clusters as a function of cluster size. The filled circles represent the η^1 complexes, the filled triangles represent the η^2 complexes, and the filled squares represent the η^3 complexes. The open triangles and squares represent the η^2-H,H, and η^3-C,H,H complexes, respectively. Right picture: Plot of the adsorption energy of methane on various Pd clusters as a function of cluster size. The circles, squares, and triangles represent adsorption complexes with a cluster thickness of a monolayer, 2-atom layer, and 3-atom layer, respectively. Figures were taken from ref. 43.

In silico studies of weak interaction systems, such as CH_4–Pd clusters, are challenging due to the accuracy of DFT calculations. With the possibility of including the dispersion energy in the DFT calculations, we expect more *in silico* studies will be possible.

11.3.2 Au Nanoclusters for Chiral Recognition

Separation of enantiomers is often a necessary step in producing homochiral drugs when one of the enantiomers is the active drug ingredient, but the other displays a toxic effect.[44] Recognition of chirality is an important step in the enantiomer separation. Conventional methods used in detecting chirality include fluorescence, Circular Dichroism (CD), and Raman spectroscopy.[45] Characterization methods used in chiral recognition include changes in the Surface Plasmon (SP) band through UV-Vis spectroscopy, in optical rotation through polarimetry, and of ΔE using Cyclic Voltammometry (CV).

Several studies using Au nanoclusters with different chiral thiol containing organic ligands were conducted for enantioselective recognition of other enantiomers. Shukla *et al.* used 5 nm Au nanoclusters with racemic or homochiral (L or D) cysteines in solution to selectively adsorb one enantiomer of propylene oxide from a solution of racemic propylene oxide (PO). The enantioselective properties of the Ag nanoclusters arise from the local interaction between the chirally modified cysteine surface and the chiral propylene oxide adsorbates. These interactions were characterized by polarimetry measurements performed at 436 nm and room temperature.[46]

Work by Kang *et al.* used 2 nm penicillamine (Pen) modified Au clusters (Pen–Au) as an electrochemical sensor for the enantioselective recognition of 3,4-dihydroxyphenylalanine (DOPA). The ligands (D- or L-Pen and D- or L-Pen–Au) were immobilized on gold electrodes and analysed using CV. The peak of DOPA in CV curves is known to appear around 0.6 V due to two-electron–two-proton oxidation and reduction of the DOPA–dopaquinone couple. ΔE values for D- and L-Pen immobilized on the gold electrodes were found to be very small. However, the ΔE values for D- and L-Pen–Au immobilized on the gold electrodes were increased to 119 and 78 mV, respectively. Furthermore, the CV curves of DOPA on D-Pen-Au showed that the heterochiral interaction between D- (or L-) DOPA and L- (or D-) Pen–Au was more favourable than the homochiral interaction.

The effects of cluster size on enantioselective recognition of DOPA were also investigated by CV measurements using gold electrodes by examining 2, 3, and 4 nm Pen–Au systems. The increase in size resulted in a decrease in ΔE values for both 3 nm (42 and 46 mV) and 4 nm (~ 25 mV).[47] This indicates that smaller Pen–Au clusters are better for high resolution and better detection.

The use of Au clusters with thiol ligands has been proven to be effective in chiral recognition of enantiomers. The smaller the clusters, the better the enantioselectivity, detection property, and resolution. Adsorption of enantiomers on the Au cluster surface is through the thiol group by forming Au–S bond. The presence of chiral ligands adsorbed on the surface of the Au cluster can cause assembly of Au clusters or can be further used to selectively identify other enantiomers not necessarily containing a thiol group.

Zhong's group recently introduced a nanoparticle-based strategy in the chiral recognition of enantiomers through regulating interparticle assembly.[48] The model system in their study consists of enantiomeric (L and D) cysteines and Au clusters. As expected, the enantiomeric cysteines adsorb to the Au cluster surface through the thiol group by forming strong binding between Au and S atoms. About 1800–3000 cysteines were adsorbed to a 13 nm Au cluster. After adsorption of cysteines, the Au clusters assemble *via* the pairwise zwitterionic dimerization of enantiomeric cysteines, which were characterized by the change in SP resonance band of the Au clusters in the visible region. When cysteines were introduced in the solution, the 13 nm Au SP band showed a decrease in absorbance at 520 nm and an increase at ~630 nm, together with the appearance of an isosbestic point at 540 nm, reflecting cysteine mediated assembly of the Au cluster.

A cluster size effect was observed on the SP band. As the cluster size increases, the SP band shifts to higher wavelength. The rate of AuNP assembly mediated by homocysteines (LL or DD) adsorbed on the surface was found to be higher than the one mediated by heterocysteines (DL), as depicted in Figure 11.15.

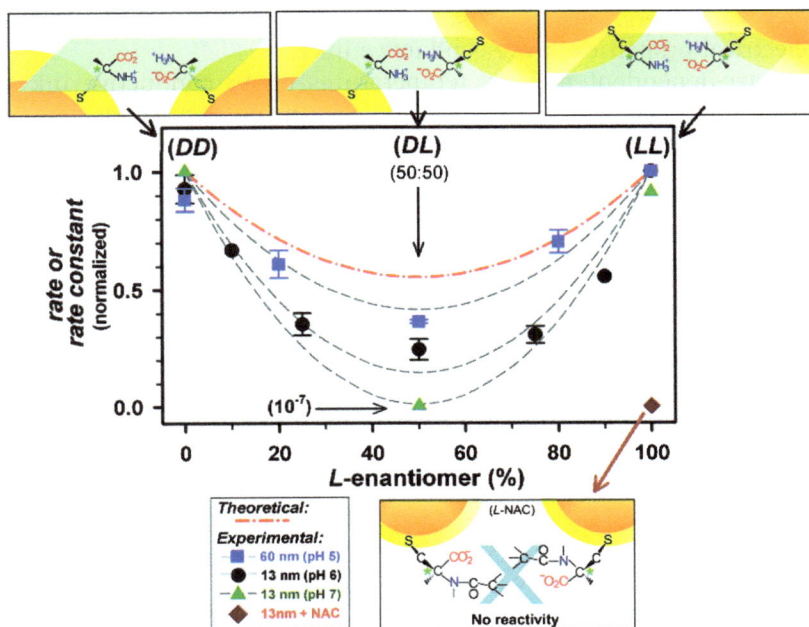

Figure 11.15 Kinetics of D- and L-cysteine adsorption on Au clusters. Figure was taken from ref. 48.

The drastic difference in kinetics was attributed to the pairwise zwitterionic interaction regulated by cluster assembly. The DL pairwise zwitterionic interaction causes Au clusters aligned at a different side of the molecular plane, which is shown in Figure 11.15. DFT studies of cysteine dimerization on the Au cluster surface confirmed the geometry presented in Figure 11.15. The calculated binding energies of a LL (and DD) dimer and DL dimer are 11.2 kcal mol^{-1} and 8.5 kcal mol^{-1}, respectively. This indicates that LL and DD dimers are more stable.[48]

Other factors that affect the kinetics of the interparticle chiral reactivity are pH and temperature. An increase in pH resulted in an enlarged difference between the heterochiral and homochiral cysteines and a decrease in pH decreased the rate of assembly.[48] It would be interesting to perform *in silico* studies to understand these observations.

11.3.3 Ag Nanoclusters as Conductive Ink

Inkjet printing using metal nanoparticle conductive inks has attracted much attention due to its simplicity, high efficiency, and low cost. This advanced technique is carried out by directly printing conductive inks onto non-conductive substrates for application in electronics.[49] Desirable characteristics for an ideal conductive ink include desirable electrical properties, good stability, and low thermal treatment temperature.

Silver clusters are one of the commonly used conductive inks. Protective agents (PAs) are often used for stability and minimization of agglomeration of these Ag clusters. The cluster size, shape, morphology, and PAs all affect the electrical properties of the conductive inks.

The size-dependent melting temperatures of Ag conductive ink were found to influence conductive properties. As the cluster size decreases, so does the melting point, which was explained by the increasing fraction of surface atoms containing weaker bonds.[50] Our *in silico* studies of various small transition metal clusters showed evidently that the binding among atoms decreases with decrease in cluster size and results in lower melting temperature.[13,15,19,21,51,52] Decreasing melting temperature is a desired quality of a conductive ink, however, as the Ag cluster size decreases, the resistivity increases as well due to an increase in the Kubo gap, making the material less conductive.[53] In this case, thermal treatment is then introduced to convert Ag clusters into conductive materials through sintering.

Other factors that influence the conductivity of Ag conductive inks are the shapes and type of exposed surface planes. Yang *et al.* looked into the effect of different shapes of Ag clusters on the electrical performance and microstructure of the conductive tracks. Three different shapes of silver particles were used as conductive inks, Ag spherical particles (64 nm), Ag nanorods (100 nm × 5 μm), and Ag nanoplates (30 nm × 600 nm). The thermal treatment was performed at 160 °C for 15 min. The XRD showed that the (111), (200), (220), and (311) planes of the face-centred Ag were present in all three shaped Ag particles. The electrical properties were measured through resistivity of conductive inks prepared using different shapes of Ag clusters as fillers. Resistivity was found to significantly decrease as the Ag cluster content of the conductive ink increased. Conductive inks containing both Ag nanorods and Ag spherical particles resulted in the lowest resistivity of $\sim 3.2 \times 10^{-5}$ Ω cm at $\sim 54\%$ weight, which is twenty times larger than the resistivity of bulk silver. This result was explained by the well dispersed Ag spherical particles and Ag nanorods. The Ag spherical particles filled the voids caused by volume shrinkage, resulting in a better interconnected conduction network.[54]

There is a wide variety of available PAs but the commonly used ones can be categorized into two functionalized polymers and smaller compounds with long alkyl chains and a polar head. The choice of PA affects the overall conductive property of Ag conductive inks. A list of the resistivity of Ag conductive inks with different PAs is shown in Table 11.2.

In the study of the effects of DDA (dodecylamine) and DDT (dodecanethiol) PAs, we specifically investigated their impact on the post thermal treatment of nano-Ag films. The results showed that Ag–DDA films exhibited lower resistivity and lower temperature in thermal treatment than Ag–DDT films (see Table 11.2).[49] We also found that aggregation of Ag–DDA took place a day after synthesis while Ag–DDT remained essentially the same, as shown by the TEM images in Figure 11.16.

These properties, thermal treatment temperature and aggregation, are all associated with the bonding energy of Ag to the PAs. Our *in silico* studies used a simplified model to represent the Ag cluster and PAs. The models used were Ag–SCH_3 for DDT and Ag–$NHCH_3$ for DDA. This simplification was made to differentiate the binding between Ag–DDT and Ag–DDA. Although a longer alkyl chain increases the interaction between the alkyl chains further increasing the binding between the Ag cluster and PAs, the binding increases in both PAs are the same as they have the same length of alkyl chain. The DFT calculations showed that the binding energy between Ag and SCH_3 is 2.19 eV and the binding energy between Ag and $NHCH_3$ is 1.78 eV. This indicates that a stronger Ag–S bond is present in DDT protected Ag conductive inks, thus Ag–DDT requires a higher thermal treatment temperature to convert to a conductive Ag–DDT.[55] Future *in silico* studies

Table 11.2   Thermal treatment conditions and resistivity of Ag nanoparticles with different PAs and sizes. Table taken from ref. 55.

PA	size (nm)	Thermal Treatment Conditions	$\rho(\mu\Omega\text{cm})$
DDA ($C_{12}H_{27}N$)	5	140 °C/60 min	2.9
DDT ($C_{12}H_{26}S$)	5	280 °C/60 min	18
PVP (($C_6H_9NO)_n$)	50	260 °C/3 min	16
PVP (($C_6H_9NO)_n$)	21 or 47	200 °C/30 min	3.2
Dodecanoic acid ($C_{12}H_{24}O_2$)	7	250 °C/60 min	6

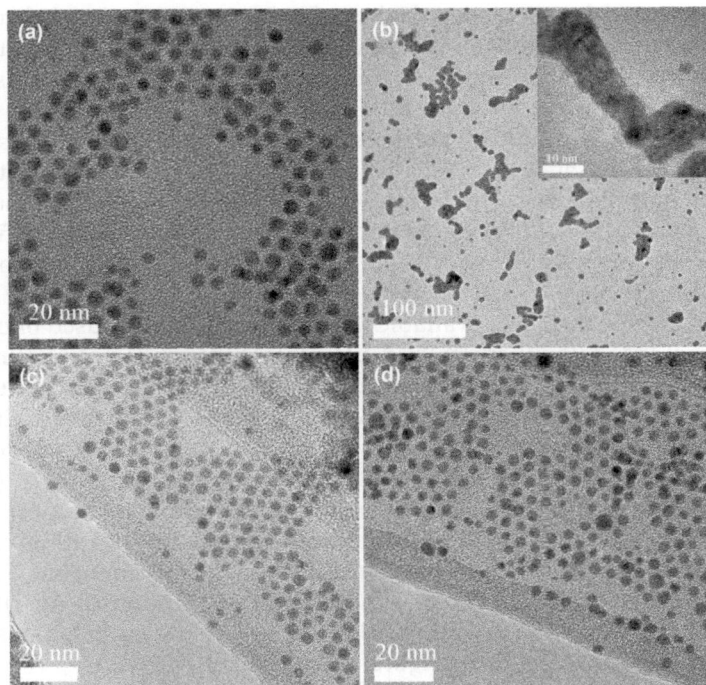

Figure 11.16   TEM images of Ag–DDA nanoclusters taken upon synthesis (a) and
after 24 h air storage (b) and of Ag–DDT nanoclusters upon synthesis
(c) and after 24 h air storage (d).
Figure was taken from ref. 55.

include the entire molecules and Ag clusters to quantify the binding and
predict the resistivity.

11.4   Conclusions

This chapter summarizes our *in silico* studies of transition metal nanoclusters in their role as catalysts and other functional materials. Our *in silico*
studies of structural properties of various transition metal clusters predicted
a new structural phase, *i.e.* simple cubic, which can exist at a size range of
about 1 nm for Ir, Rh, and Os. In 2012, Gates' group observed cubic Ir
nanoclusters that can resist aggregation using Aberration-corrected STEM.
The calculated lattice constant for small clusters, such as Pt–Au clusters, is
smaller than that of bulk materials. This phenomenon may be used as one
factor to draw comparisons of the catalytic activity between the clusters and
their bulk counterpart. Furthermore, by classifying atoms in a cluster into
three categories based on their degrees of participation in the bond breaking
and forming process, systematic *in slico* studies of catalytic activities may
shed more light on our understanding of heterogeneous catalysis.

The work described in this chapter is only a few of the many *in silico* studies that can be found in the literature. Presently, *in silico* studies provide understanding and interpretation of experimental results. Occasionally, with increasing frequency, *in silico* studies are used as a selection tool in designing materials to help eliminate unnecessary bench-top experimental trials. With the development of more accurate theory, combined with faster and more powerful computers, we envision the materials research field will shift more towards using *in silico* studies to predict and evaluate the initial design strategies and only then perform bench-top experiments on evaluation of the refined designs. One such example can perhaps be best witnessed in the research and development of metamaterials. Another significant contribution of *in silico* studies in the near future will be in the Materials Genome Initiative.

References

1. *Molecular Dynamics – Theoretical Developments and Applications in Nanotechnology and Energy*, ed. L. Wang, In Tech, Rijeka, 2012.
2. G. A. Hudson, J. Li and L. Wang, *Chem. Phys. Lett.*, 2010, **498**, 151.
3. T. Pawluk, L. Xiao, J. Yukna and L. Wang, *J. Chem. Theory Comput.*, 2007, **3**, 328.
4. Z. Xu, S. Lu, J. Li and L. Wang, IEEE Proceedings of 2010 Sixth International Conference on Natural Computation, ed. S. Yue, H. L. Wei, L. Wang, and Y. Song, 2010, vol. 3, p. 1586.
5. M. Valden, X. Lai and D. W. Goodman, *Science*, 1998, **281**, 1647.
6. Y. B. Lou, M. M. Maye, L. Han, J. Luo and C. J. Zhong, *Chem. Commun.*, 2001, 473.
7. A. M. Molenbroek, J. K. Norskov and B. S. Clausen, *J. Phys. Chem. B*, 2001, **105**, 5450.
8. R. B. Thompson, V. V. Ginzburg, M. W. Matsen and A. C. Balazs, *Science*, 2001, **292**, 2469.
9. A. S. U. Heiz, S. Abbet and W.-D. Schneider, *J. Am. Chem. Soc.*, 1999, **121**, 3214.
10. T. F. Jaramillo, S. H. Baeck, B. R. Cuenya and E. W. McFarland, *J. Am. Chem. Soc.*, 2003, **125**, 7148.
11. P. Fayet, A. Kaldor and D. M. Cox, *J. Chem. Phys.*, 1990, **92**, 254.
12. C. T. Campbell, S. C. Parker and D. E. Starr, *Science*, 2002, **298**, 811.
13. L. Wang and Q. Ge, *Chem. Phys. Lett.*, 2002, **366**, 368.
14. W. Zhang, Q. Ge and L. Wang, *J. Chem. Phys.*, 2003, **118**, 5793.
15. L. Xiao and L. Wang, *J. Phys. Chem. A*, 2004, **108**, 8605.
16. W. Zhang, H. Zhao and L. Wang, *J. Phys. Chem. B*, 2004, **108**, 2140.
17. L. Xiao and L. Wang, *Chem. Phys. Lett.*, 2004, **392**, 452.
18. W. Zhang, L. Xiao, Y. Hirata, T. Pawluk and L. Wang, *Chem. Phys. Lett.*, 2004, **383**, 67.
19. T. Pawluk and L. Wang, *J. Phys. Chem. B*, 2005, **109**, 20817.
20. Q. Ge, C. Song and L. Wang, *Comput. Mater. Sci.*, 2006, **35**, 247.

21. C. Song, Q. Ge and L. Wang, *J. Phys. Chem. B*, 2005, **109**, 22341.
22. C. Aydin, J. Lu, N. D. Browning and B. C. Gates, *Angew. Chem., Int. Ed.*, 2012, **51**, 5929.
23. C. A. J. Lu, N. D. Browning, L. Wang and B. C. Gates, *Catal. Lett.*, 2012, **142**, 1445.
24. J. Yukna and L. Wang, *J. Phys. Chem. C*, 2007, **111**, 13337.
25. T. Pawluk and L. Wang, *J. Phys. Chem. C*, 2007, **111**, 6713.
26. B. Wanjala, B. Fang, J. Luo, Y. Chen, J. Yin, M. Engelhard, R. Loukrakpam and C. J. Zhong, *J. Am. Chem. Soc.*, 2011, **133**, 12714.
27. T. E. Mallouk and E. S. Smotkin, in *Handbook of Fuel Cells - Fundamentals, Technology and Applications*, ed. W. Vielstich, A. Lamm and H. A. Gasteiger, John Wiley & Sons, Weinheim, 2003, vol. 2, part 3, p. 334.
28. T. He, E. Kreidler, L. Xiong, J. Luo and C. J. Zhong, *J. Electrochem. Soc.*, 2006, **153**, A1637.
29. L. Wang, J. I. Williams, T. Lin and C. J. Zhong, *Catal. Today*, 2011, **165**, 150.
30. B. Wanjala, J. Lou, R. Loukrakpam, D. Mott, P. Njoki, B. Fang, M. Engelhard, H. R. Naslund, J. K. Wu, L. Wang, O. Malis and C. J. Zhong, *Chem. Mater.*, 2010, **22**, 4282.
31. D. J. Trevor, D. M. Cox and A. Kaldor, *J. Am. Chem. Soc.*, 1990, **112**, 3742.
32. C. Adlhart and E. Uggerud, *Chem. Commun.*, 2006, 2581.
33. J. S. Owen, J. A. Labinger and J. E. Bercaw, *J. Am. Chem. Soc.*, 2006, **128**, 2005.
34. X. H. Gu, J. Zhang, J. H. Dong and T. M. Nenoff, *Catal. Lett.*, 2005, **102**, 9.
35. R. L. Martins, M. A. Baldanza, M. Souza and M. Schmal, *Stud. Surf. Sci. Catal.*, 2004, **147**, 643.
36. F. Xia and Z. Cao, *J. Phys. Chem. A*, 2006, **110**, 10078.
37. L. Xiao and L. Wang, *J. Phys. Chem. B*, 2007, **111**, 1657.
38. V. N. Mishra and R. P. Agarwal, *Microelectron. J.*, 1998, **29**, 861.
39. S. C. Su, J. N. Carstens and A. T. Bell, *J. Catal.*, 1998, **176**, 125.
40. T. Wagner, M. Bauer, T. Sauerwald, C. D. Kohl and M. Tiemann, *Thin Solid Films*, 2011, **520**, 909.
41. Y. Lu, J. Li, J. Han, H.-T. Ng, C. Binder, C. Partridge and M. Meyyappan, *Chem. Phys. Lett.*, 2004, **391**, 344.
42. X. Zhou, W. Q. Tian and X. L. Wang, *Sens. Actuator, B*, 2010, **151**, 56.
43. W. Zhang and L. Wang, *Comput. Theor. Chem*, 2011, **963**, 236.
44. N. M. Davies and X. W. Teng, *Adv. Pharm*, 2003, **1**, 242.
45. M. Tatarkovic, Z. Fisar, J. Raboch, R. Jirak and V. Setnicka, *Chirality*, 2012, **24**, 951.
46. N. Shukla, M. A. Bartel and A. J. Gellman, *J. Am. Chem. Soc.*, 2010, **132**, 8575.
47. Y. J. Kang, J. W. Oh, Y. R. Kim, J. S. Kim and H. Kim, *Chem. Commun.*, 2010, **46**, 5665.
48. I. S. Lim, D. Mott, M. H. Engelhard, Y. Pan, S. Kamodia, J. Luo, P. N. Njoki, S. Zhou, L. Wang and C. J. Zhong, *Anal. Chem.*, 2009, **81**, 689.
49. C. L. Lee, K. C. Chang and C. M. Syu, *Colloids Surf., A*, 2011, **381**, 85.

50. *Nanocatalysis*, ed. U. Heiz and U. Landman, Springer-Verlag, Berlin, Heidelberg, 1st edn, 2007.
51. L. Xiao, B. Tollberg, X. Hu and L. Wang, *J. Chem. Phys.*, 2006, **124**, 114309.
52. W. Zhang, X. Ran, H. Zhao and L. Wang, *J. Chem. Phys.*, 2004, **121**, 7717.
53. E. Roduner, *Chem. Soc. Rev.*, 2006, **35**, 583.
54. X. J. Yang, W. He, S. X. Wang, G. Y. Zhou, Y. Tang and J. H. Yang, *J. Mater. Sci.: Mater. Electron.*, 2012, **23**, 1980.
55. L. Mo, D. Liu, W. Li, L. Li, L. Wang and X. Zhou, *Appl. Surf. Sci.*, 2011, **257**, 5746.

CHAPTER 12

DNA-Templated Metal Nanoclusters and Their Applications

ZHIXUE ZHOU[a,b] AND SHAOJUN DONG*[a,b]

[a] State Key Laboratory of Electroanalytical Chemistry, Changchun Institute of Applied Chemistry, Chinese Academy of Sciences, Changchun, Jilin 130022, P. R. China; [b] Graduate School of the Chinese Academy of Sciences, Beijing 100039, P. R. China
*Email: dongsj@ciac.jl.cn

12.1 Introduction

Fluorescent metal nanoclusters (NCs), composed of a few to a hundred atoms, are a class of novel materials whose properties are defined by their sub-micrometer dimensions. The physicochemical properties of NCs are size-dependent, and differ significantly from those of the corresponding bulk material. As a consequence of this tight correlation, the dimension and dispersity of NCs must be controlled during their preparation in order to obtain populations with the desired characteristics. Therefore, synthetic protocols for the production of monodisperse NCs whereby the size can be intimately controlled is an area of research that is intensely studied. In past decades, various methods have been developed to synthesize metal NCs on a sub-nanometer scale, such as precursor- or ligand-inducing etching of metal manoparticles,[1-4] microemulsion methods,[5,6] and electrochemical approaches.[7,8] Among the present strategies, template-based methods have

RSC Smart Materials No. 7
Functional Nanometer-Sized Clusters of Transition Metals: Synthesis, Properties and Applications
Edited by Wei Chen and Shaowei Chen
© The Royal Society of Chemistry 2014
Published by the Royal Society of Chemistry, www.rsc.org

proved to be efficient synthetic techniques for the preparation of fluorescent metal NCs.[9] The role of the templating molecule in the formation of sub-nanometer sized metal NCs is to provide a synthetic microenvironment in which the inorganic phase morphology is tightly controlled by a range of interactions. Up to now, polymers,[10,11] polyelectrolytes,[12] proteins,[13] dendrimers[14,15] and DNA,[16] have been extensively exploited as templates for the synthesis of metal NCs.

DNA is composed of nucleotides, which are linked together by phosphodiester bonds. Further, nucleotides are made up by nucleobases, a 2'-deoxy-ribosesugar ring and phosphate (Figure 12.1A). The nucleobases most commonly found in DNA are shown in Figure 12.1B. The attractiveness of DNA-based metal NCs processes stems from a set of desirable features: (1) DNA interacts with metal ions *via* several different and well-established binding modes;[17–19] (2) DNA in solution adopts a well-defined three-dimensional shape, termed conformation, which is strongly dependent on the nucleotide sequence and which might provide the necessary microenvironment to confine and control NC events; (3) compared with other templates and capping agents, DNA presents a major advantage since an exquisite control over its length and sequence can be easily achieved by chemical and biochemical methods. Length and sequence are important properties of the template that could be exploited to fine-tune the shape and size of the resulting stabilized nano-particles; (4) DNA is chemically stable and its structure is thermally renaturable, even if it is denatured at high temperature; (5) DNA can be generated more efficiently and cost effectively than polypeptide sequences *via* the use of powerful synthetic, enzymatic and *in vitro* evolutionary techniques; (6) DNA

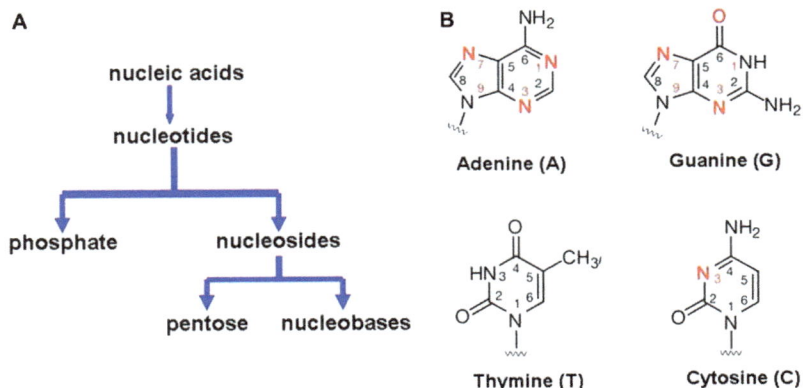

Figure 12.1 (A) Nucleotides are the building blocks of DNA. They consist of a nucleobase, a 2'-deoxyribose sugar ring and phosphate. (B) The four nucleobases most commonly found in DNA. Nucleotides comprise many different chemical groups that are able to bind to metal ions. The atoms in nucleobases responsible for these interactions are highlighted in red and numbered for clarity.
Reprinted from ref. 20 with permission by the Nature Publishing Group.

libraries consist of millions of different sequences, and researchers can *in vitro* identify the optimized DNA for metal NC encapsulation.

In this chapter, we will discuss the main topics of DNA-based metal NCs, including the synthesis, optical and electronic properties, and their applications *etc.* First, we will summarize the interactions between DNA and metal ions. Second, we will focus on the synthesis, characterization and unique properties of DNA-templated metal NCs. Third, we will discuss Ag NCs stabilized by DNA with different sequences and structures. The potential applications of DNA–Ag NCs in biosensing, fluorescent imaging and logic gates will be demonstrated. Fourth, DNA-stabilized Au NCs will be generalized. Finally, we will give a brief outlook on the future development of metal NCs and indicate some challenges.

12.2   The Interactions Between DNA and Metal Ions

The initiating event of the NC synthesis process is the formation of a precursor complex *via* specific interactions between a metal or metal ion and the biotemplate. The phosphate group, as well as a number of other functional groups present in DNA, interact extensively with metal ions (Figure 12.1B). Two different modes of interaction are possible and the extent of each is determined by the type of metal (Table 12.1).[20] At physiological pH, the negatively charged phosphate backbone interacts extensively with positively charged metal cations[18-26] through electrostatic attraction. Alternatively, empty orbitals of the metal cations can accept electrons from the phosphate oxygen or from the nucleobase moieties forming coordination complexes.[27] These are the primary types of interaction for alkali, alkali earth metals and some transition metal ions (Table 12.1). For nucleobases guanine (G) and adenine (A), N1 and N7 are the favoured binding sites, with O6 of G playing a secondary role in metal coordination. As for nucleobases cytosine (C) and thymine (T), metal ion binding occurs primarily at N3, which is highly pH dependent.[21] At neutral pH, the order of stability for nucleobase complexes with transition metal ions has been

Table 12.1   Nucleic acids can bind metal ions through different types of interactions. Often, more than one interaction is responsible for metal binding. The ions are listed in increasing order of binding strength or in increasing order of affinity for the base (adapted from ref. 17). Reprinted from ref. 20 with permission by the Nature Publishing Group.

Binding Site	Metal Ion
Phosphate	$Li^+, Na^+, K^+, Rb^+, Cs^+, Mg^{2+}, Ca^{2+}, Sr^{2+}, Ba^{2+}$, trivalent lanthanides
Phosphate and base	$Co^{2+}=Ni^{2+}, Fe^{2+a}, Mn^{2+}, Zn^{2+}, Cd^{2+}, Pb^{2+}, Cu^{2+}$
Base	Ag^+, Hg^{2+}, Pt^{2+}

^{a}Approximate position of Fe^{2+} is extrapolated from ref. 25.

determined as $N7/O6(G) > N3(C) > N7(A) > N1(A) > N3(A,G)$,[22,28] with T giving no significant complexation through base coordination. According to previous studies, only Ag^+, Hg^{2+} and Pt^{2+} combine exclusively to nucleobases, while most metals form mixed complexes, where the metal ions interact simultaneously with both the nucleobase and the nucleic acid backbone (Table 12.1).[20] All the mentioned interactions between DNA and various metal ions provide the possibility for the *in situ* synthesis of DNA-stabilized metal NCs through reduction of the located metal ions.

12.3 The Synthesis, Characterization and Unique Properties of DNA-Templated Metal NCs

12.3.1 The Synthesis of DNA/Metal NCs

Although a general model for DNA-templated NC synthesis has not been unequivocally identified, the general procedures can be summarized according to previous studies:

Step 1: Localization. DNA ligands bind with metal cations to form a NC precursor, and provide the necessary microenvironment for triggering the nucleation event. Both metal–phosphate and metal–nucleoside interactions are crucial in this initial step. In addition, the concentration of DNA, DNA sequences and the ratio of metal ions to DNA also play important roles.

Step 2: Reduction. The localized metal ions are then reduced to form metal NCs, and sodium borohydride is the commonly used reducing agent. Small clusters produced during the initial step tend to aggregate into larger clusters, and this process needs to overcome the electrostatic barrier formed by the DNA capping. Eventually, upon reaching a critical size, the electrostatic barrier is too large to be overcome, preventing further aggregation.

Step 3: Solubilization. Nucleic acid ligands ensure that NCs remain dispersed in aqueous solution.

In addition, new strategies for the synthesis of DNA–metal NCs have been developed in recent years. For example, Yan and coworkers have described a new DNA-based method for the synthesis of water-soluble fluorescent Ag NCs with a narrow size distribution by use of the well-known Tollens reaction: that is, one aldehyde sugar molecule can reduce two Ag^+ to Ag^0_2.[29] In short, the Tollens reagent ($[Ag(NH_3)^{2+}]$) was added to sugar-modified DNA, and the mixture was incubated overnight in the dark at room temperature. Incorporation of sugar moieties into a DNA sequence enabled the synthesis of AgNCs with the Tollens reaction. This strategy has been successfully used for the site-specific synthesis and *in situ* immobilization of Ag NCs at predefined positions on the DNA nanoscaffold.

12.3.2 Characterization

At present, the reported techniques for NC characterization mainly include ultraviolet and visible (UV-Vis) spectroscopy, electron microscopy (TEM,

HRTEM), mass spectrometry (MS), nuclear magnetic resonance (NMR) spectroscopy, extended X-ray absorption fine structure (EXAFS) and circular dichroism (CD) *etc.*

12.3.2.1 UV-Vis Spectroscopy

The UV-Vis absorption of sub-nanometer sized metal NCs exhibit molecular-like optical transitions with absorbance bands.[3,30–32] Thus, UV-Vis absorption spectroscopy has been used as a powerful and convenient diagnostic tool to study the electronic structures of sub-nanometer sized metal clusters. In addition, DNA has its characteristic absorption peak located at about 260 nm, and the observation of UV-Vis spectral changes during the NC synthesis process can provide quite important information to analyse the interactions between DNA and metal ions or NCs.

For example, Dickson and coworkers have reported the synthesis of Ag NCs with a 12-mer DNA as template.[16] They found changes in the electronic absorption spectra which indicated Ag^+ and NCs associating with the bases. The absorption maximum of the DNA template (λ_{max}) shifts from 257 to 267 nm upon Ag^+ complexation (Figure 12.2A). Following reduction of the bound ions, further spectral changes occur. Initially, λ_{max} shifts from 267 to 256 nm, and the molar absorptivity increases. This effect may be attributed to new, overlapping electronic bands for small silver clusters, which are known to absorb in this spectral region.[33] Eventually, as Ag NCs grow and

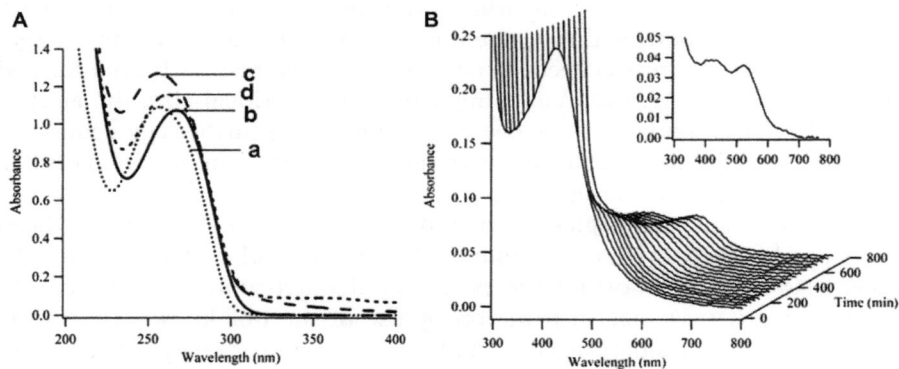

Figure 12.2 (A) Response of the electronic transition of DNA bases to association with Ag^+ and Ag NCs. (a) dotted line, 10 μM DNA; (b) solid line, DNA with 60 μM Ag^+; (c) coarse dashed line, 2 min after adding BH_4^- to the DNA–Ag^+ solution; (d) fine dashed line, 1100 min after adding BH_4^- to the DNA–Ag^+ solution. (B) Absorption spectra associated with DNA–Ag NCs. The foremost spectrum in the time series was acquired 9 min after adding the BH_4^-, and it has $\lambda_{max} = 426$ nm. Subsequent spectra were acquired approximately every 30 min. The inset spectrum shows the last spectrum in the series (692 min), and peaks are observed at 424 and 520 nm.
Reprinted from ref. 16 with permission by the American Chemical Society.

visible absorptions evolve, the λ_{max} shifts to 262 nm and the molar absorptivity decreases. In the long wavelength range, reduction of Ag^+ bound to DNA results in new species with electronic transitions in the visible region of the spectrum (Figure 12.2B). The prominent transition initially has a $\lambda_{max} = 426$ nm at 9 min after adding BH_4^-. Over a period of 12 h, the absorbance of this band decreases and a broad absorption band with peaks at 424 and 520 nm develops. As determined through theoretical and low-temperature spectroscopic studies, electronic transitions for small silver clusters, in particular Ag_2 and Ag_3, are expected in this spectral region.[33-35] In general, DNA–Ag NCs often demonstrate a characteristic absorption peak with a wavelength and bandwidth identical to the excitation peak that appeared in the fluorescence spectrum.

Similar to DNA–Ag NCs, Shao and coworkers have reported the characteristic absorption of DNA-templated Au NCs, which can assist in identifying the formation of NCs.[36]

12.3.2.2 Transmission Electron Microscopy (TEM and HRTEM)

TEM has traditionally been used to obtain information about nanoparticle size and shape. However, due to limited resolution, TEM is actually not a reliable method to obtain accurate sizes (atom numbers) of sub-nanometer sized metal NCs. High resolution TEM (HRTEM) is a primary tool in the study of size and crystalline structure of metal NCs. Even though HRTEM can provide structural information with a highest resolution of ~0.1 nm, the experimental errors, including systematic error, random error and human error, render it not an ideal tool to determine the size of NCs with diameters less than 1 nm.[37] In practical investigations, HRTEM is usually performed as a complementary characterization to evaluate the size and dispersity of the obtained NCs. For example, Wang and coworkers utilized TEM to characterize DNA-stabilized Ag colloid (Figure 12.3). It was found that different sizes (Ø 1–2 nm) of Ag particles are formed in a one-pot synthesis, which can further confirm the formation of Ag NCs.[38]

12.3.2.3 Mass Spectrometry (MS)

At present, MS is widely used in the structural studies of sub-nanometer sized metal clusters. Various types of MS techniques, such as laser desorption ionization mass spectrometry (LDI-MS),[39,40] laser desorption ionization time-of-flight (LDI-TOF),[41] matrix-assisted laser desorption ionization mass spectrometry (MALDI-MS),[42] fast atom bombardment ionization mass spectrometry (FAB-MS),[43] and electrospray ionization mass spectrometry (ESI-MS) in positive- or negative-ion modes[30,44-46] have been applied to the study of metal NCs. Among them, ESI-MS is the frequently used method for the characterization of DNA-templated metal NCs.

Petty and coworkers have used ESI-MS to characterize NCs with C_{12} DNA as template.[47] Electrospray mass spectra demonstrate that small numbers of silver atoms are bound to C_{12} DNA (Figure 12.4). The silver atoms exhibit a

Figure 12.3   TEM image of as-prepared HP26 DNA-stabilized Ag NCs.
Reprinted from ref. 38 with permission by the American Chemical
Society.

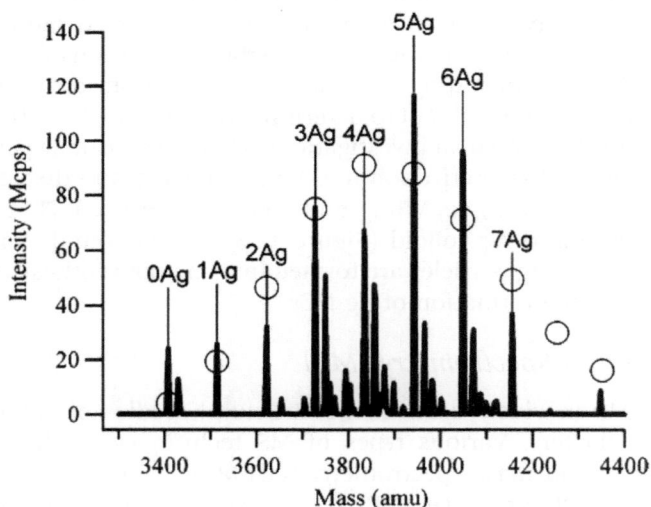

Figure 12.4   Electrospray ionization mass spectra of the C_{12}:Ag complexes 4 h after
adding the BH_4^-. The peaks are labeled with the number of silver
atoms attached to C_{12}. The oligonucleotide has the expected mass of
3408 amu. The peaks to the right of the marked peaks correspond to
Na^+ adducts. The open circles associated with the peaks represent the
predicted stoichiometric distribution based on Poisson statistics. The
average number of bound silvers from this fit is 4.8 ± 0.4.
Reprinted from ref. 47 with permission by the American Chemical
Society.

distribution of stoichiometries that are modeled using Poisson statistics. The average number of bound silvers from this fit is 4.8 ± 0.4.

12.3.2.4 Nuclear Magnetic Resonance (NMR) Spectroscopy

NMR, including proton (^{1}H) and carbon (^{13}C) NMR, is a powerful tool to study the structure and chemical environment of molecules. For DNA-stabilized metal NCs, NMR can be used to probe the protecting ligands surrounding the metal core, accompanied by spectral broadening.[37] Dickson *et al.* researched the interactions between Ag NCs and DNA bases according to the corresponding NMR spectra.[16] The aromatic proton resonances in the ^{1}H NMR spectra of the DNA template are well-resolved prior to the addition of Ag$^+$ (Figure 12.5). However, the proton resonances are essentially broadened to baseline upon addition of Ag$^+$ (spectrum not shown). In contrast, most of the aromatic proton resonances in the ^{1}H spectra with Ag NCs are almost as narrow as those of the free DNA (Figure 12.5). The cytosine H6 proton resonances exhibit the largest change in chemical shift in the

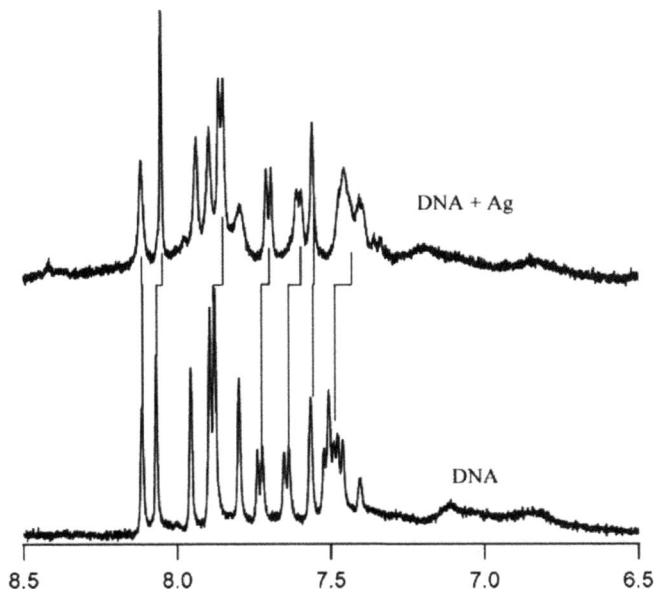

Figure 12.5 The aromatic proton region from the ^{1}H NMR spectra of DNA with and without the Ag NCs. Vertical lines indicate aromatic proton resonances with chemical shifts that change after nanocluster formation. Cytosine H6 resonances, which exhibit the largest changes in chemical shift, can be identified by their splitting due to coupling to cytosine H5. [DNA] = 0.93 mM, [Ag$^+$] = 5.6 mM, and [BH$_4^-$] = 5.6 mM in a solution of 90% 1 mM phosphate buffer and 10% D$_2$O at 25 °C. Reprinted from ref. 16 with permission by the American Chemical Society.

presence of Ag NCs (Figure 12.5). Similar upfield chemical shift changes were also observed for the H5 protons of cytosine (spectrum not shown). These observations indicate that cytosine bases are the most favored interaction sites for Ag NCs.

12.3.2.5   Extended X-ray Absorption Fine Structure (EXAFS) Technique

Extended X-ray absorption fine structure (EXAFS) analysis is a powerful method for obtaining information on atomic-scale bonding in clusters, including metal–metal and metal–ligand bonding, and estimating cluster size.[48–51] Neidig and coworkers reported the use of Ag K-edge EXAFS to identify the effects of DNA sequence modification on Ag NC structure and bonding, and demonstrated correlations between structure and emission properties.[52] Figure 12.6 shows the χ (R) k^3-weighted EXAFS data and fits for the three DNA-templated Ag NCs (Ag 1–3), as well as the individual waves from the curve-fitting results using the FEFF code[53] to calculate the amplitudes and phases. While the spectra of Ag2 and Ag3 exhibit distinct similarities in the nearest-neighbor region, the spectrum of Ag1 is markedly different on both sides of the principal $R = 2.5$ Å feature. Focusing on Ag–Ag interactions, the presence of first nearest-neighbor Ag is observed for all three DNA-templated Ag NCs. The observed Ag–Ag bond distances are significantly contracted from the value of 2.89 Å for silver metal (Ag1, $R = 2.75 \pm 0.01$ Å; Ag2, $R = 2.74 \pm 0.02$ Å; and Ag3, $R = 2.77 \pm 0.02$ Å), consistent with results previously observed for small gold clusters due to particle surface tension and other effects.[54] The χ (R) EXAFS data (Figure 12.6) also indicate the presence of first-shell N/O neighbors at distances (<2.3 Å) consistent with ligation to silver. Overall, the EXAFS results could demonstrate the *in situ* formation of Ag–Ag bonded Ag NCs directly coordinated to DNA, with variations in cluster structure and ligation for the different DNA sequences used. Furthermore, EXAFS also could provide an estimate of the average cluster size due to the correlation between the average number of nearest metal neighbors in the first metal–metal shell and the cluster size.

12.3.2.6   Circular Dichroism (CD)

Circular dichroism (CD), the unequal absorption of right and left circularly polarized light, is a manifestation of optical activity in the vicinity of absorption bands. CD offers a relatively new and powerful means to understand the environment of chromophoric residues. At present, CD is widely used to infer the configuration of asymmetric moleculaes, such as DNA and protein. For DNA, the CD technique is one of the primary means to identify the secondary structure, and a typical DNA structure (such as ssDNA, dsDNA, G-quadruplex, triplex or i-motif) has its characteristic CD spectra. Dickson and coworkers have used CD spectra to indicate that Ag^+ and NCs associate

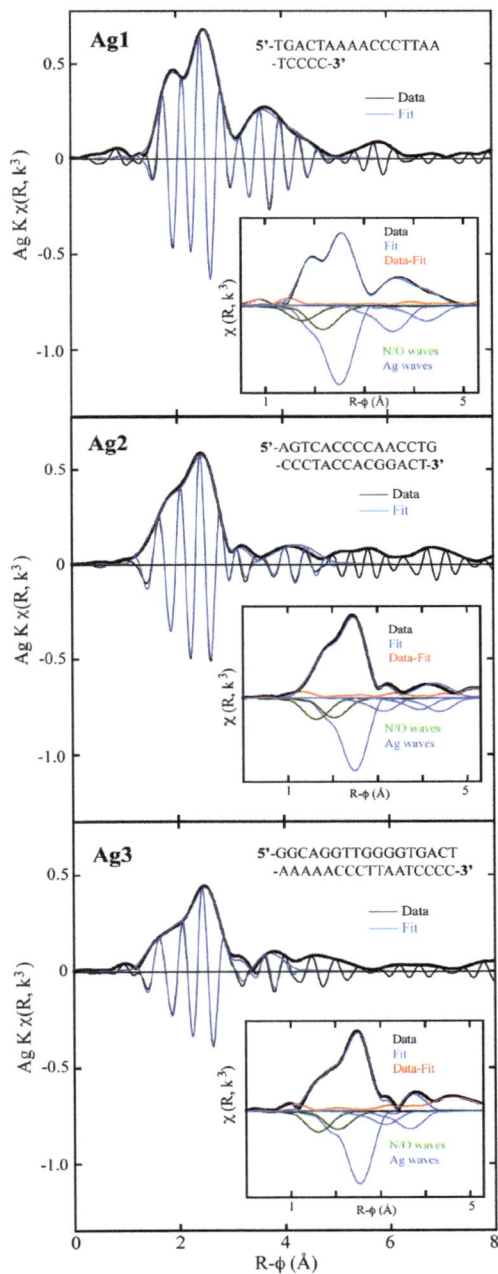

Figure 12.6 $k^3\chi (R)$ of the EXAFS of DNA-templated Ag NCs. The modulus and real part of the transform of both the data and fit are shown using the same vertical scale. The insets show the $k^3\chi$ spectra of the moduli of the data, fit, difference between the data and fit, and the individual contributions to the fit (inverted for clarity).

Reprinted from ref. 52 with permission by the American Chemical Society.

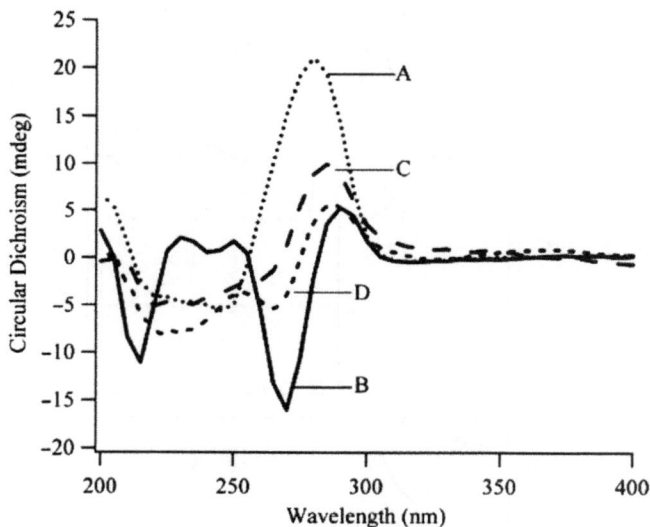

Figure 12.7   Response of the CD associated with electronic transition of DNA bases to association with Ag$^+$ and Ag NCs. A (dotted line), 10 μM DNA template; B (solid line), DNA with 60 μM Ag$^+$ (1 Ag$^+$: 2 bases); C (coarse dashed line), 120 min after adding BH$_4^-$ to the DNA–Ag$^+$; D (fine dashed line), 4300 min after adding BH$_4^-$ to the DNA–Ag$^+$. Reprinted from ref. 16 with permission by the American Chemical Society.

with the bases.[16] Circular and linear dichroism studies show that Ag$^+$ induces nonplanar and tilted orientations of the bases, suggesting that Ag$^+$ may cause perturbation of the bases in ssDNA. Besides, the CD spectra also change upon reduction of the Ag$^+$. The differences between the spectra in Figure 12.7 indicate that Ag NCs and Ag$^+$ induce different structural changes in DNA.

In addition, the CD method can be used to study the relationship between DNA structure and the properties of DNA-stabilized metal NCs. For example, Li and coworkers applied CD measurements to illustrate the aggregation behaviors of silver atoms modulated by polymorphic DNA templates involving i-motif, G-quadruplex, and duplex.[55]

12.3.3   The Unique Properties of DNA–Metal NCs

While the highly polarizable electrons of bulk metal result in many advantageous electronic and optical properties,[56,57] clusters consisting of only a few atoms exhibit discrete electronic transitions, properties more akin to molecular behavior.[58] For example, small metal NCs often show fluorescence, which is size-dependent. Such unique electronic and optical properties of sub-nanometer sized metal NCs render them promising for applications in novel materials or nanodevices.

12.3.3.1 Absorption and Fluorescence

Along the atom to bulk transition, discrete atomic energy levels merge into highly polarizable, continuous, plasmon-supporting bands. When sufficiently small, metal NCs exhibit very strong absorption and emission. For example, as long as the highly polarizable Ag cluster electronic states are discrete and continuous band energies are not exhibited, fluorescence can be a major relaxation pathway for cluster excitation. Luminescence of metals is normally extremely weak, owing to efficient nonradiative decay and the absence of an energy gap.[57] However, small metal NCs behave as "artificial atoms" and show high molecular brightness due to fluorescence quantum yields of 10–70% and extinction coefficients from 100 000 to 350 000 M^{-1} cm^{-1}.[59–61]

12.3.3.2 Solvatochromic Effect

The solvatochromic effect is well known for larger metal nanoparticles and is mainly related to their surface plasmon phenomenon.[61] Various DNA–Ag NCs were found to exhibit strong, discrete fluorescence with solvent polarity-dependent absorption and emission.[58]

12.3.3.3 Two-Photon Absorption

Compared with one-photon excitation, using two near-infrared (NIR) photons of half the transition energy for the excitation of fluorophores can offer substantial advantages. For example, two-photon excited fluorescence offers high spatial resolution owing to its inherent quadratic intensity dependence, while also providing spectral selectivity and improved sensitivity resulting from decreased NIR absorption, scattering, and fluorescence in tissue. Dickson and coworkers first reported the two-photon property of DNA–Ag NCs.[62] They used DNA microarray strategy to identify three distinct red/NIR-emitting DNA–Ag NCs (Figure 12.8). These NCs yield spectrally pure solutions emitting at 660, 680, or 710 nm. Two-photon excitation (150 fs, 80 MHz, 680–1040 nm, Coherent Mira) yields emission indistinguishable from one-photon excitation (OPE) (Figure 12.8). These DNA–Ag NCs exhibit two-photon excitation (TPE) cross sections in buffer far surpassing all known water-soluble fluorophores and are comparable to much larger quantum dot cross sections. In addition, Goodson and coworkers further investigated the two-photon absorption (TPA) efficiencies of DNA–Ag NCs at 800 nm excitation.[63] The log of pump power dependence against the log of the fluorescence at 630 nm for the Ag NCs on dsDNA is shown in Figure 12.9. These data are fitted well by a line of slope two, indicating a two-photon excitation. The method for calculating the TPA cross-section is the comparative two-photon excited fluorescence method with standard reference H2TPP in toluene. The TPA action cross-section at 800 nm for the Ag NC on dsDNA was determined to be ∼3000 GM (based upon the measured one-photon 30% quantum yield for activated clusters). This is a large TPA

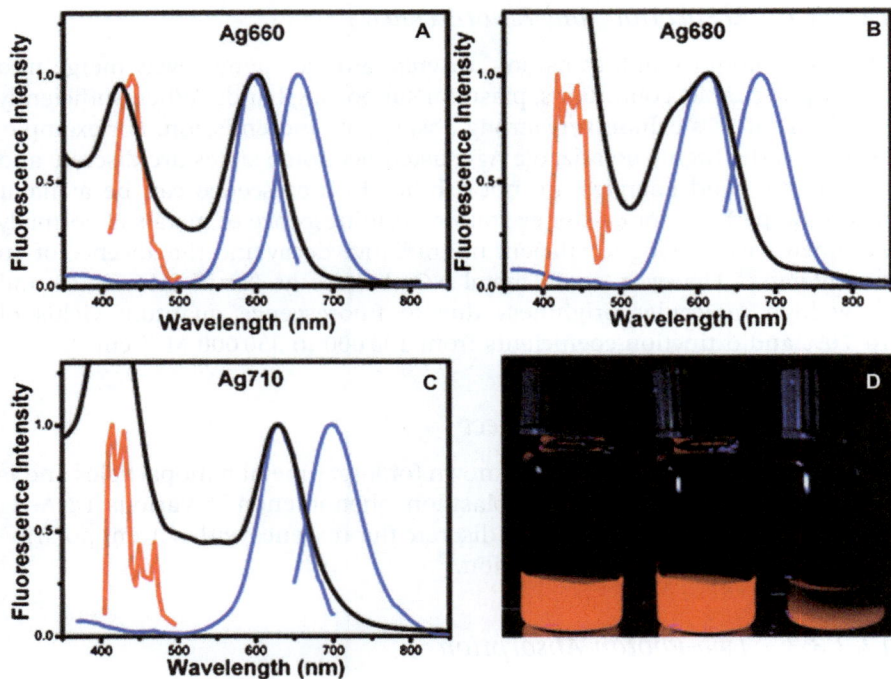

Figure 12.8   OPE and fluorescence spectra (blue), TPE spectra (red), and absorption spectra (black) of (A) 660 nm emitting species created in 5'-CCCATATTCCCC-3'; (B) 680 nm emitting species created in 5'-CCCTATAACCCC-3'; (C) 710 nm emitting species created in 5'-CCCTAACTCCCC-3'. (D) Picture showing from left to right the OPE emission of the 660, 680, and 710 nm species.
Reprinted from ref. 62 with permission by the American Chemical Society.

cross-section value, which would promote the possibility of using the DNA–Ag NCs for multi-photon imaging.

12.3.3.4   Chirality Properties

The issue of chirality in metal clusters protected by a chiral molecular coating has gained much research interest since the first report by Schaaf and Whetten.[64] This chirality was detected by the appearance of a CD signal at wavelengths which were typically on the red side of the absorption of the molecules capping the clusters and corresponded to transitions between metal core electronic states. Dickson and coworkers reported the appearance of a chiroptical response in Ag_{1-4} clusters capped with 12-mer ssDNA.[16] Later on, Kotlyar and coworkers showed that Ag NCs produced by reduction of Ag^+ bound to 1000 base-pair poly(dA)–poly(dT) are chiral and characterized by a negative CD band around 400 nm.[65] Cutting of the DNA template with

Two Photon Power Dependence for dsDNA

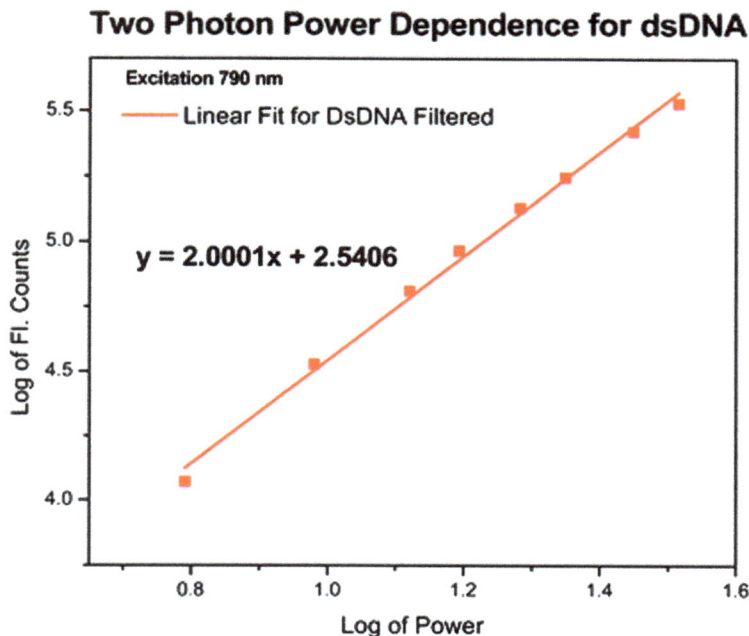

Excitation 790 nm

—— Linear Fit for DsDNA Filtered

$y = 2.0001x + 2.5406$

Log of Fl. Counts (y-axis)

Log of Power (x-axis)

Figure 12.9 Two-photon power dependence for DNA–Ag NC systems. Reprinted from ref. 63 with permission by the Royal Society of Chemistry.

DNase, the enzyme that cleaves double-stranded DNA and, therefore, separates the DNA coated NPs, does not abolish the CD signal of the particles.

12.4 DNA-Templated Ag NCs and Applications

12.4.1 DNA-Templated Ag NCs

In 1962 Yamane and Davidson reported the strong interaction between DNA and Ag^+.[66] They found that the absorption of DNA at 260 nm was shifted in the presence of Ag^+, and thus suggested that the electronic structure of DNA was perturbed. As the absorption of DNA results from heterocyclic bases (pyrimidine and purine) but not from sugar–phosphate backbones, they concluded that Ag^+ was likely to bind with the bases, rather than the negatively charged sugar–phosphate backbone. Eichhorn *et al.* further investigated the interactions between individual nucleobases (*i.e.*, A, C and G) and Ag^+ under different pH (4, 6 and 9) conditions.[67] For A and C, addition of Ag^+ can cause a significant absorbance change at pH 6 and 9, while the absorbance change for G occurred at pH 4 and 6. They extended the studies to polyadenine (poly A) and polycytosine (poly C). Similar results as the nucleosides were observed from the polynucleotide silver complexes. Up to now, the reported interactive sites of Ag^+ are mainly the electron-rich nitrogen and oxygen atoms of the nucleobases, such as the N3, N1, and N7 atoms of A, the N7, N3, the deprotonated N1 and O6 of G, the N3 of C[47,68] and the deprotonated N3 of T.[69] By taking advantage of the strong affinities of Ag^+ toward bases, Ag NCs can be obtained through the reduction of localized Ag^+.

In a pioneering work, Dickson and coworkers[16] reported the synthesis of DNA-encapsulated Ag NCs with $NaBH_4$ as the reductant. Mass spectroscopy demonstrated the formation of small Ag NCs with 1–4 atoms bound to the 12-mer DNA (5'-AGGTCGCCGCCC-3'), and distinct absorption and fluorescence spectral features were observed. NMR spectroscopy further revealed that cytosine experienced the largest chemical shifts with Ag NCs. Base-specific interactions could be a significant feature of these NCs, as suggested from the chemical shifts in the NMR spectra for the cytosine bases and the truncated distribution in the mass spectra. Together, these results suggest that it may be feasible to control the formation of NCs using DNA strands with specific sequences, and cytosine is the most favored base for the formation of Ag NCs. Therefore, Petty *et al.* sequentially discussed the extremely bright and photostable emission of C_{12} DNA-encapsulated Ag NCs, which are characterized by their red and blue–green emission bands.[47] The metal binding was confirmed to occur at the N3 of cytosine in a pH-dependent fashion. Through the use of oxidizing and reducing agents, the red emission was attributed to fully reduced silver clusters, while the blue–green emission was attributed to oxidized clusters. In addition, the photophysics of C_{12} DNA-templated Ag NCs have been elucidated by intensity-dependent correlation analysis and suggested a heavy atom effect of silver that rapidly depopulates an excited dark level before quenching by oxygen, thereby conferring great photostability, very high single-molecule emission rates, and essentially no blinking on experimentally relevant time scales (0.1 to >1000 ms). Strong antibunching and significant dark-state quantum yield are observed from these biocompatible species, which emit $>10^9$ photons before photobleaching. Thus, DNA–Ag NCs exhibit great improvements over existing dyes, and the nonpower law blinking kinetics suggest that these very small species may be alternatives to much larger and strongly intermittent semiconductor quantum dots.[70]

Inspired by the above mentioned excellent properties of few-atom Ag NCs, several studies were carried out to emphasize the key role of DNA sequences and secondary structures on the formation of DNA-stabilized Ag NCs. The main results are summarized as follows.

12.4.1.1   C-rich ssDNA, G-rich ssDNA, duplex and hairpin structures as templates for the formation of Ag NCs

Gwinn and coworkers[71] used six 19-mer DNA oligomers as templates for the synthesis of Ag NCs. They found that both C-rich and G-rich single-strand DNA (ssDNA) produced brightly fluorescent Ag NCs, while the duplex formed by C-rich and G-rich DNA was an invalid template for the formation of fluorescent Ag NCs, accompanied by negligible visible fluorescence. Thus, the spectral properties of DNA-attached few-atom Ag clusters are sensitive to the secondary structure of DNA, and double-strand DNA (dsDNA) does not host fluorescent Ag species. Furthermore, a set of hairpin structures named C-loop, G-loop, A-loop and T-loop were designed to compare the relative capacities of C, G, A, and T homopolymers to host fluorescent species. C-loop and G-loop solutions exhibit fluorescence of similar brightness, while A-loop solutions fluoresce weakly. T-loop solutions even produce no fluorescence for excitation at visible wavelengths. These observations indicate that the spectral properties of few-atom Ag NCs attached to ssDNA are sensitive to the constituted bases. In addition, the apparent similarity in binding affinities of Ag for single-stranded G and C in this work differs from the interpretation of previous studies[16] of Ag bound to a 12-mer DNA (5'-AGGTCGCCGCCC-3'), which used NMR to identify C as the main site for Ag attachment; however, because the solutions studied had roughly 40 times higher concentrations than used here, the dominant mode of Ag binding may be different.[72]

In addition, O'Neill and coworkers further explored the templates of C-loop DNA with 3 to 12 cytosines in the loop, and investigated loop-dependent fluorescence of the formed Ag NCs.[73] All of these hairpins support fluorescence, and the number of cytosines in the loop tunes the stability and fluorescence of Ag NCs. These poly-C loops stabilize four different types of fluorescent Ag NCs based on their wavelengths and chemical stability. For the case of the 9C hairpin, fluorescence intensity and mass abundance correlate well to red emission and Ag_{13}:DNA, and for green emission and Ag_{11}:DNA, consistent with the possibility that the grouping represents clusters with different numbers of atoms.

12.4.1.2 Using T-containing oligonucleotides dT_{12}, $dT_4C_4T_4$, and $dC_4T_4C_4$ to investigate the influence of bases

Petty and coworkers discussed blue–green-emitting NCs with T-containing oligonucleotides dT_{12}, $dT_4C_4T_4$, and $dC_4T_4C_4$ as templates,[69] which can further be used to understand how the bases and sequence influence cluster formation. Although T-loop solutions can not support the formation of Ag NCs as indicated by the interpretation of previous studies,[71] when dT_{12} DNA is used as the template under alkaline conditions, green emission is dominant with $\lambda_{ex} = 350$ nm/$\lambda_{em} = 540$ nm and is stable for several days. Through variations in the pH, the green emitting NCs are shown to bind with the N3 of T. The clusters that form with dT_{12} and $dT_4C_4T_4$ have similar excitation maxima in the fluorescent spectra, but the Stokes shift is smaller for $dT_4C_4T_4$-templated clusters, which suggests that the sequence-specific interactions influence the environment and hence the spectral properties of the clusters. In the case of $dC_4T_4C_4$, formation of NCs depends on the concentration of DNA, and higher concentrations favor a red-emitting species. A blue–green emissive species dominates at lower concentrations of dC$_4$T$_4$C$_4$, which has spectroscopic, physical, and chemical properties that are similar to those of the clusters that form with dT$_{12}$ and dT$_4$C$_4$T$_4$. Thus, the bases have a dominant influence on the formation and stabilization of Ag NCs. Another important finding is that nonfluorescent clusters also form following the reduction of the Ag$^+$–DNA conjugates, and formation of fluorescent NCs is favored with oxygen, thus indicating that DNA-templated NCs are partially oxidized.

12.4.1.3   Color tunability of Ag NCs with five microarray-screened ssDNA as templates

Dickson and coworkers reported on five distinct Ag NC emitters encapsulated in distinct 12-mer ssDNA. The highlight of the present work is that they used DNA microarrays to identify optimized sequences for Ag NC encapsulation.[74] The arrays were imaged at various excitation wavelengths, and emission intensities *versus* pH were used to determine the best sequences for stabilizing fluorescent Ag NCs (Figure 12.10). In the present work, yellow, red, blue, green and NIR emitting Ag NCs were yielded with the selected five distinct ssDNA as templates. Among them, the yellow, red, and NIR emitting species show photophysical parameters comparable or better than those of cyanine dyes. Besides, their greatly reduced bleaching and blinking lead to significant advantages over existing dyes for single molecule studies. Thus, the DNA microarray strategy shown here would allow researchers to create new Ag NC fluorophores with outstanding spectral and photophysical properties, and promote the applications of Ag NC emitters for single molecule spectroscopy and imaging.

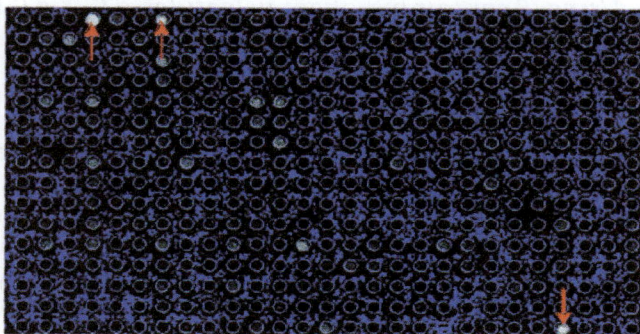

Figure 12.10   Emission from a DNA microarray excited with 543 nm light showing the specificity of particular sequences for the formation of the red emitting Ag NCs. These sequences were then used in bulk synthesis to identify the oligonucleotide with the highest affinity for the formation of the target cluster.
Reprinted from ref. 74 with permission by the American Chemical Society.

12.4.1.4 DNA templates with different secondary structures, including i-motif, G-quadruplex, duplex and triplex structures

Petty and coworkers first explored pH-dependent i-motif DNAs as templates for the synthesis of Ag NCs.[75] Two C-rich sequences $(dTA_2C_4)_4$ and $(dC_4A_2)_3C_4$ with a common C_4 i-motif core are chosen as model systems. Both DNAs adopt i-motif structures that are dependent on the pH, and the role of this particular secondary structure on silver cluster formation is considered. Results demonstrate that the red and green emissive species are most prominent. The red emission is highest for slightly acidic solution, at which the i-motif forms of DNA templates are also stable. While the green emission is highest for basic solution, at which the oligonucleotide alone is unfolded. In addition, they used size exclusion chromatography to study the compact shapes of cluster-oligonucleotide conjugates under different pH conditions. Results indicate that protons dominate DNA folding for the red emissive species, while the green emissive clusters themselves determine the shape of their DNA matrix. These studies provide the basis for understanding how specific base arrangements and environmental factors influence the formation of fluorescent NCs.

Afterwards, Li and coworkers further investigated the effect of DNA templates with different secondary structures, such as i-motif, G-quadruplex and Watson–Crick duplex.[55] First, the affinity sites of Ag^+ on different DNA templates were analyzed by using density functional theory (DFT) calculations, and the conformational variations of DNA templates caused by Ag^+ and atoms were disclosed by means of CD spectra. Second, nanosilvers templated by polymorphic DNA templates involving the G-quadruplex, i-motif and duplex show distinct fluorescent properties associated with the molar ratio of $[BH_4^-]/[Ag^+]$. With DNA template adopting i-motif or duplex structure, silver atoms tend to aggregate inside the encapsulated spaces of nucleobases, and the formed Ag NCs are positively charged with high fluorescent spectral features; whereas with G-quadruplex DNA as template, silver atoms favor aggregation outside of the G-tetrad, which results in the formation of larger silver crystals without fluorescence properties. These results are useful to explore the nucleation and growth mechanism of silver nanomaterials regulated by structure-specific DNA templates.

The DNA triple helix is one of the most useful recognition motifs in the design of sequence-specific labeling, regulation of gene expression, and construction of DNA-based nanostructures.[76] It can form in two main ways: an oligopyrimidine strand binds to the major groove of the duplex parallel to the strand carrying the purine tract using Hoogsteen base-pairing, or alternatively, an oligopurine binds to the purine strand in an antiparallel orientation using reversed Hoogsteen base-pairing.[77] In the parallel motif, cytosines in the third strand need to be protonated at their N3 positions $(pK_\alpha = 4.5)$ to form $CG.C^+$ base triplets, which stabilize the triplex by favorable electrostatic effects. Therefore, the stability of triplexes containing

CG.C$^+$ largely depends on the pH of the solution. In 2009, Ihara and cow-orkers showed that Ag$^+$ remarkably stabilizes the structure of a parallel-motif DNA triplex even at neutral pH.[77] Inspired by this phenomenon, Qu and coworkers have demonstrated site-specific, homogeneous, and bright Ag NCs with triplex DNA as template.[78] By reasonable design of the DNA sequence, a homogeneous Ag$_2$ cluster was obtained in the predefined pos-ition of the CG.C$^+$ site of triplex DNA. This strategy was also explored for controlled alignment of Ag NCs on the DNA nanoscaffold.

12.4.1.5   *DNA abasic or gap site-directed formation of Ag NCs*

Shao and coworkers previously found the preferential binding of Ag$^+$ to the unpaired bases opposite an abasic site (AP site) in dsDNA duplexes.[79] Based on this result, they exploited the potential of an AP site in duplexes as a new scaffold for the synthesis of Ag NCs.[80] In another work, they in-depth studied the effect of bases near the AP site on the properties of formed Ag NCs (Figure 12.11A).[81] Although varying the DNA sequence one base away from the AP site (*i.e.*, the Ag NC growth site) does not alter the size of the fluorescent Ag NCs, the emissions of the formed Ag NCs are still gradually

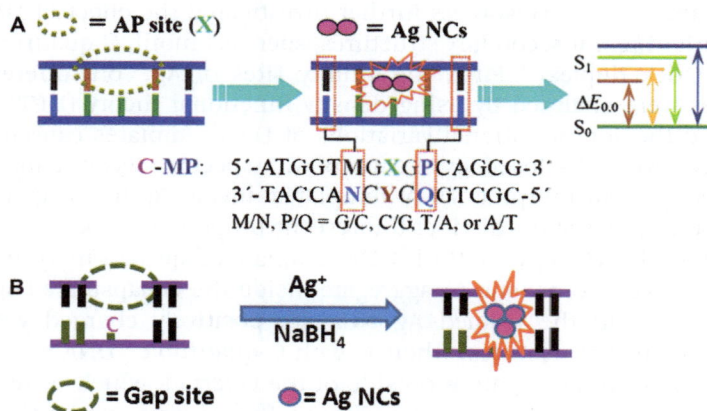

Figure 12.11   (A) The AP site was flanked by two guanines and opposed by a base Y (where Y = C, A, G, or T). Sequences M/N and P/Q (M/N, P/Q = G/C, C/G, T/A, or A/T) one base away from the AP site were systematically changed in order to investigate their roles in the electronic prop-erties of the guanines directly flanking the AP site. The two flanking guanines were thus different in sequence direction. The guanine flanked by M was located on the 3' side (3'-G), and the guanine flanked by P was located on the 5' side (5'-G). (B) Schematic represen-tation of the gap site-directed formation of fluorescent Ag NCs. Hybridizing the target strand with the two probe strands that are one base shorter in length than the former produces a gap site that is thus surrounded by an unpaired target base and two flanking bases. Reprinted from ref. 81 and 82 with permission by the American Chemical Society and Elsevier.

red shifted as the sequence changes from T to C, A, and G. Furthermore, the 3′-side sequences of the 5′-G stack directly flanking the AP site induced larger emission shifts than the 5′-side sequences of the 3′-G stack. Additionally, the size of the formed Ag NCs seems to be dependent on the consecutive AP site number. Thus, the AP site design in this work provides an easy way to shed light on the role of DNA base stacking in the optical properties of Ag NCs.

Similarly to the AP site, the gap site can be expected to provide a binding pocket for Ag^+.[82] The gap site is produced by eliminating one nucleotide from a DNA duplex. Therefore, one long ssDNA strand (target) and two short ssDNA strands (probes) complementary to the target strand can offer a gap site when these three DNA strands form a duplex (Figure 12.11B). The formation of Ag NCs in the gap site-containing DNA duplex is very rapid and also highly dependent on the bases opposite the gap site.

Overall, the properties of Ag NCs are largely dependent on the sequence and structure of the DNA stabilizers. In addition, many other factors, such as the concentration of DNA templates,[69,71] pH[67,75] and solvent conditions,[58] all play important roles. For example, thymine is not beneficial for the formation of fluorescent Ag NCs under neutral conditions.[71] However, T_{12} DNA is an efficient template for the formation of Ag NCs under alkaline conditions.[69] All these aspects would endow DNA-directed Ag NCs many unique properties and applications.

12.4.2 Application of Fluorescent DNA–Ag NCs

12.4.2.1 Detection of Metal Ions, Sulfide (S^{2-}) Ion and Small Biomolecules

Hg^{2+} is an example of a highly toxic and widespread pollutant, and its damage to the brain, nervous system, endocrine system and even the kidneys is well known.[83] Wang and coworkers have developed a new strategy for the selective detection of Hg^{2+} using DNA-stabilized Ag NCs as fluorescent probes.[84] The water-soluble C_{12} DNA-stabilized Ag NCs were found to be good indicators for the selective identification of Hg^{2+} based on the quenching effect of Hg^{2+} on the fluorescence of Ag NCs. The proposed detection protocol is environmentally-friendly, sensitive (with a detection limit of 5 nM) and selective (with no interference from other metal ions). It should be pointed out that this Hg^{2+} sensor is a "turn-off" mode, which is not preferred in practice. The development of a "turn-on" sensor may be more desirable than the reported method. Dong and coworkers have developed a label-free and "turn-on" method for highly selective and sensitive detection of Hg^{2+}, using DNA duplex stabilized Ag NCs as fluorescent probes.[85] A DNA duplex with an inserted cytosine loop could serve as a scaffold for the generation of fluorescent Ag NCs. With T–T mismatches introduced into this DNA duplex, the configuration of Ag NCs forming sequences would be disturbed, resulting in less formation of Ag NCs. In the presence of Hg^{2+}, the

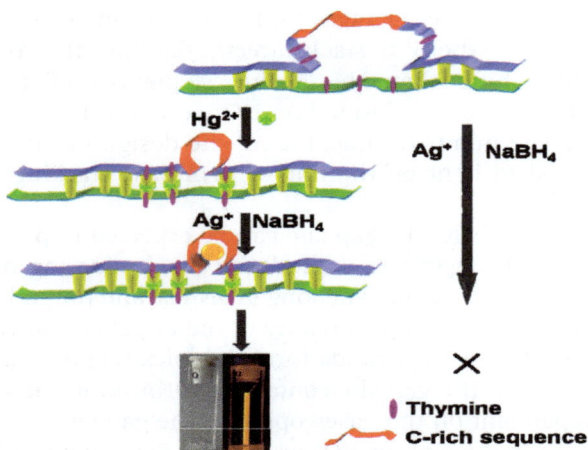

Figure 12.12   DNA duplexes with inserted cytosine loop working as synthetic scaffolds to generate fluorescent Ag NCs for the detection of Hg^{2+}. Reprinted from ref. 85 with permission by the Royal Society of Chemistry.

DNA duplex was strengthened and highly sequence-dependent fluorescent silver species could be formed, since Hg^{2+} can stabilize a T–T mismatch by forming a T–Hg^{2+}–T base pair.[86,87] The fluorescent Ag NCs formed in these duplex scaffolds could be used as a "turn-on" fluorescent probe (Figure 12.12). This detection protocol provided high selectivity and sensitivity for the detection of Hg^{2+}. Its application potential was demonstrated by analyzing the practical sample.

Cu^{2+} is another significant environmental pollutant.[88] Short-term exposure to high levels of Cu can cause gastrointestinal disturbance, and long-term exposure can cause damage to the liver and kidneys.[89] Recently, considerable efforts have been directed at the detection of Cu^{2+} with fluorescent Ag NCs as probes.

Chang and coworkers have developed a novel, simple and "turn-on" method for the detection of Cu^{2+} using water-soluble DNA–Ag NCs.[90] They prepared fluorescent Ag NCs with a reported DNA template (5′-CCCTTAATCCCC-3′). The resulting DNA–Ag NCs featured an emission band centered at 564 nm and a quantum yield of 11.5% that was determined through comparison with that (quantum yield = 94%) of Rhodamine 6G. Interestingly, the introduction of Cu^{2+} induced a quite significant fluorescence enhancement, which could be developed for selective and sensitive detection of Cu^{2+}. The DNA–Ag NC probe is sensitive (limit of detection 8 nM), selective (by at least 350-fold over other metal ions), and simple for the detection of Cu^{2+} in pond water and soil samples. To further improve the sensitivity, they developed another simple and homogeneous DNA–Cu/Ag NC-based fluorescent assay of Cu^{2+}.[91] The fluorescence of DNA–Cu/Ag

NCs was quenched by 3-mercaptopropionic acid (MPA), which was recovered in the presence of Cu^{2+}. The DNA–Cu/Ag NC probes provided a detection limit of 2.7 nM for Cu^{2+}, with high selectivity (by at least 2300-fold over other tested metal ions). The practicality of the present strategy for the detection of Cu^{2+} in environmental samples was validated through analysis of soil and pond water samples.

Another DNA–Ag NC-based method for "turn-on" detection of Cu^{2+} was developed by Li *et al.*[92] The C-rich ssDNA (5′-(CCCTAA)₃CCCTA-3′) was employed as the scaffold for the synthesis of Ag NCs, and the obtained DNA–Ag NCs exhibit outstanding spectral properties. Upon the introduction of cysteine, the intensity of the emission peak maximum of DNA–Ag NCs decreased significantly. However, after the addition of Cu^{2+}, the weak light-scattering signal was remarkably enhanced. Such an interesting phenomenon allows "turn-on" assay of Cu^{2+}.

S^{2-} is widely distributed in natural water and wastewater, and the concentration of S^{2-} is an important environmental index.[93,94] Chang and coworkers presented a one-pot synthesis of fluorescent DNA–Au/Ag NCs with DNA 5′-CCCTTATCCCC-3′ as template, and then employed the as-prepared DNA–Au/Ag NCs to selectively detect S^{2-} on the basis of its specific interactions with Au^+/Ag^+ ions and/or Au/Ag atoms.[95] Relative to DNA–Ag NCs, DNA–Au/Ag NCs are much more stable in high ionic strength media (*e.g.*, 200 mM NaCl). The fluorescence of DNA–Au/Ag NCs was found to be quenched by S^{2-}, and the changes in fluorescence intensity allowed sensitive detection of S^{2-} as low as 0.83 nM. They validated the practicality of this probe for the detection of S^{2-} in hot spring and seawater samples, demonstrating its advantages of simplicity, sensitivity, selectivity, and low cost.

Thiol compounds such as cysteine (Cys), homocysteine (Hcy) and glutathione (GSH) play important roles in many biological processes, and monitoring of these thiol compounds is of great importance for disease diagnosis, such as cancer and cardiovascular disease.[96–98] Inspired by the template-dependent properties of DNA–AgNCs, Qu and coworkers developed a novel fluorescence "turn-on" assay of thiol compounds through modulating DNA-templated Ag NCs.[99] A series of DNA templates with different length and base composition were employed for Ag NC synthesis, and then used to select the optimal fluorescent response patterns toward thiol compounds. It was found that the fluorescence of some DNA–Ag NCs (such as C_{12}–Ag NCs) could be enhanced in the presence of thiol compounds, while no obvious enhancement was induced by the other nineteen amino acids or biologically relevant analytes. Thus, a new type of sensitive, specific and "turn-on" assay of thiol compounds was developed with DNA–Ag NCs as output signal. Significantly, the present work showed that fluorescent "turn-on" detection of a specific analyte could be realized simply by choosing appropriate DNA templates in the NC synthesis process. Given the diversity of DNA, this template-directed response offers a potential approach to the detection of a wide spectrum of analytes.

Metallic NPs and semiconductor QDs find ongoing interest for probing biocatalytic transformations and biologically relevant small molecules.[100–102] Recently, Willner and coworkers have reported on the novel application of fluorescent DNA–AgNCs for monitoring biocatalytic transformations.[103] They demonstrated that H_2O_2 quenched the luminescence of DNA–AgNCs, which enabled the probe of biocatalytic oxidase-stimulated H_2O_2-generating biotransformations. This was exemplified with the analysis of glucose oxidase (GOx)-mediated oxidation of glucose that yields gluconic acid and H_2O_2. Also, quinone derivatives quenched the luminescence of Ag NCs, and this function enabled the probe of tyrosinase (TYR), a melanoma cancer cell biomarker. TYR-stimulated oxidation of tyrosine, dopamine, or tyramine yields quinone derivatives, which quench the fluorescence of AgNCs. This enabled the sensitive detection of tyrosinase (detection limit 1×10^{-4} units mL^{-1}). The successful analysis of tyrosinase or glucose was further implemented to probe bienzyme biocatalytic cascades, including the alkaline phosphatase–tyrosinase coupled hydrolysis and oxidation of *o*-phospho-L-tyrosine, the acetylcholine esterase–choline oxidase-catalyzed hydrolysis of acetylcholine and subsequent oxidation of choline. These results pave the way to implementing AgNCs as optical labels to follow the activity of other oxidases and other cascaded biocatalytic transformations.

Adenosine triphosphate (ATP) is generally acknowledged as "energy currency" in most animate beings, and plays an important role in most enzymatic activities. The concentration and dissipative rate of ATP have been found to be closely related to many diseases such as hypoxia, hypoglycemia, ischemia, Parkinson's disease and some malignant tumors. Wang and coworkers have reported a novel DNA–Ag NC-based photoinduced electron transfer (PET) system that enables the specific detection of ATP with high sensitivity.[104] They successfully prepared fluorescent Ag NCs with A-G4 DNA (5'-CCTCCTTCCTCC*TTGGGTAGGGCGGGTTGGG*-3') as template. The formed DNA–Ag NCs display strong fluorescence, whereas the G-rich DNA sequence (in italic) can fold into a G-quadruplex structure. When hemin is introduced, the G-quadruplex–hemin complex is formed and acts as the electron acceptor. PET from the DNA–Ag NCs to the G-quadruplex–hemin complex then leads to quenching of the fluorescence of the DNA–Ag NCs (Figure 12.13). Interestingly, this novel PET system could be used to develop a new sensing platform for the detection of ATP. As shown in Figure 12.13, the designed A-G4-ATP DNA assembles into a hairpin structure consisting of the G-quadruplex sequence (black) and the aptamer part against ATP (blue). This hairpin composition ensures that parts of the aptamer and G-quadruplex sequences are caged in the stem of the hairpin. Formation of the ATP–aptamer complex opens the hairpin structure and deprotects the G-quadruplex sequence, accompanied by self-assembling of the G-quadruplex–hemin complex and quenched fluorescence of DNA–Ag NCs. This principle allowed a detection limit of 8.0 nM for ATP based on the signal-to-noise ratio. Besides, this approach exhibits excellent selectivity for ATP detection over other ATP analogues GTP, CTP, and UTP.

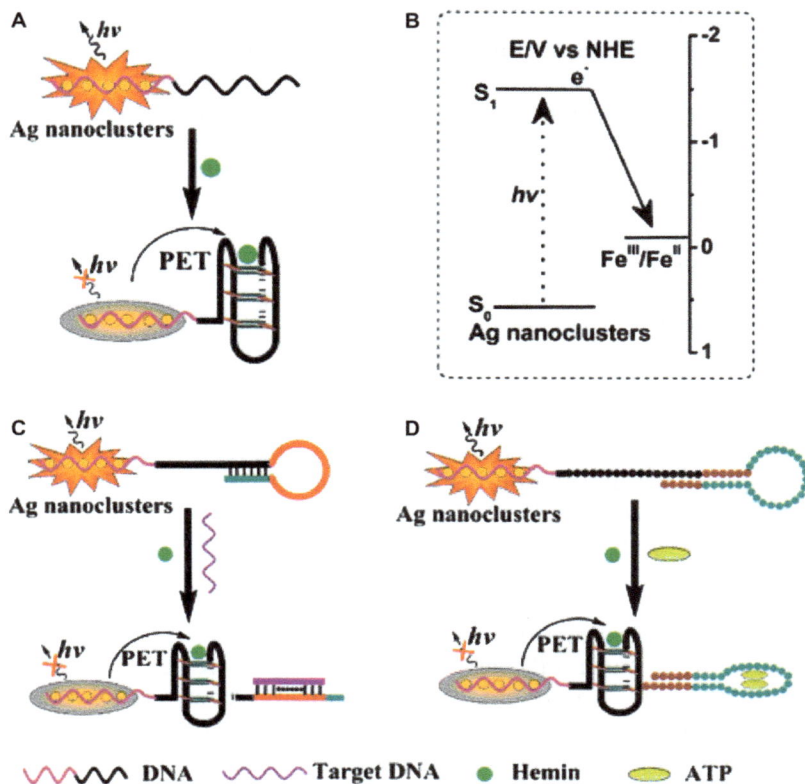

Figure 12.13   (A) Schematic illustration of PET between DNA–Ag NCs and G-quad-ruplex–hemin complex; (B) Comparison of the redox state of G-quad-ruplex–hemin complex and the energy levels of DNA–Ag NCs; (C), (D) Schematic illustrations of the analysis of DNA and ATP using DNA–Ag NC-G-quadruplex–hemin conjugates.
Reprinted from ref. 104 with permission by the American Chemical Society.

12.4.2.2   Analysis of Proteins

The selective detection of proteins with DNA-templated fluorescent Ag NCs as probes often needs to combine selective recognition molecules (*e.g.* antibodies and aptamers). For example, Martinez and coworkers developed a new strategy for the detection of thrombin that combines bright DNA-templated Ag NCs with the specificity and strong binding affinity of DNA aptamers for thrombin.[105] They created an aptamer–nanocluster template chimera through extending a thrombin-binding aptamer with a cytosine-rich DNA sequence that generates fluorescent AgNCs. With this DNA chimera as template, the formed Ag NCs showed strong red fluorescence, and are highly photostable, bright

(~60% quantum yield), and physically stable over a range of bioassay conditions. It was found that the fluorescence of Ag NCs was quenched considerably upon addition, and subsequent binding, of thrombin to the aptamer. While little fluorescence quenching was observed upon addition of denatured thrombin or other control proteins. This strategy integrates the advantages of bright Ag NCs and an aptamer with high selectivity for the cognate protein, thus ensuring high sensitivity (detection limit 1 nM) and selectivity.

SsDNA-binding protein (SSB) is an important protein in cells of all living organisms. It binds selectively and cooperatively to the single-strand region of DNA to prevent premature annealing that is essential for DNA replication, recombination, and repair.[106–109] Chang and coworkers have reported a DNA–Cu/Ag NC-based approach for the highly sensitive and selective detection of SSB.[110] SSB can strongly and specifically interact with the ssDNA template, and thus change the conformation of the template, leading to the decreased fluorescence of ssDNA-stabilized Cu/Ag NCs. The DNA–Cu/Ag NC probes responded selectively toward SSB over the other tested proteins (such as trypsin, lysozyme, myoglobin and thrombin) by factors of at least 25-fold based on high DNA–SSB interaction.[111]

It should be noted that the above mentioned strategies for the detection of protein are a "turn-off" pattern. Therefore, a series of "turn-on" ways have emerged in subsequent studies. Zhu and coworkers presented a binding-induced fluorescence "turn-on" assay of protein using aptamer-functionalized DNA–AgNCs as probes (Figure 12.14).[112] Key features of this assay include aptamer binding-induced DNA hybridization and fluorescent enhancement of Ag NCs with guanine-rich DNA sequences. Taking human α-thrombin as a model system, two aptamers (Apt15 and Apt29) were modified by extending with additional sequence elements. The first aptamer was extended with a 12-mer sequence and a nanocluster nucleation sequence at the 5'-end. The second aptamer was linked through a complementary 12-mer sequence to a G-rich overhang at the 3'-end (Figure 12.14). When thrombin binds with its two aptamer, a hybridization process would be initiated between the complementary sequences attached to each aptamer, thereby making the G-rich overhang in proximity with Ag NCs and resulting in a significant fluorescence enhancement. With this approach, a detection limit of 1 nM and a linear dynamic range of 5 nM–2 μM were achieved for human α-thrombin. The principle of the binding-induced DNA hybridization and fluorescence enhancement of Ag NCs can be extended to other homogeneous assay applications provided that two appropriate probes are available to bind with the same target molecule.

Yang and coworkers also reported a "turn-on" and homogeneous apta-sensor for platelet-derived growth factor B-chain homodimer (PDGF-BB), which relies on target induced formation of Ag NCs.[113]

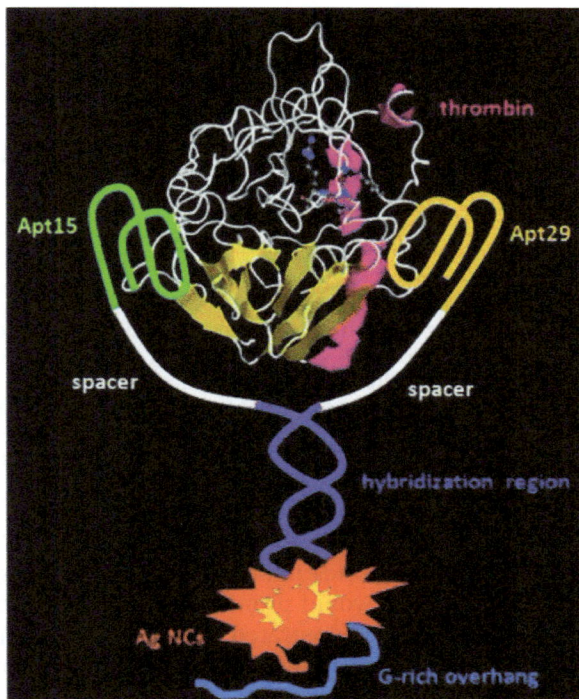

Figure 12.14 Schematic diagram showing the principle of a binding-induced fluorescence "turn-on" assay of thrombin.
Reprinted from ref. 112 with permission by the American Chemical Society.

12.4.2.3 Detection of DNA and miRNA

The detection and quantification of DNA are important for *in vivo* real-time monitoring of cellular processes and for *in vitro* biosensing and clinical diagnosis. Werner and coworkers showed that DNA-templated Ag NCs could be used to detect specific nucleic acid targets on the basis of the phenomenon that the red fluorescence of DNA–Ag NCs can be enhanced 500-fold when placed in proximity to guanine-rich DNA sequences.[114] In the study, dark DNA–Ag NCs can be transformed into bright red-emitting clusters when placed in proximity to guanine bases, which enables the specific detection of DNA targets in a separation-free format. This study is the first example of using spectral conversions of DNA–Ag NCs for DNA detection, and exhibits advantages of high signal-to-background ratio and sensitivity.

By virtue of the highly sequence-dependent generation of fluorescent DNA–Ag NCs, Wang and coworkers[115] developed a Ag NC-based assay that can specifically identify a single-nucleotide mutation (Figure 12.15). In this study, DNA duplexes with an inserted cytosine loop have been developed as

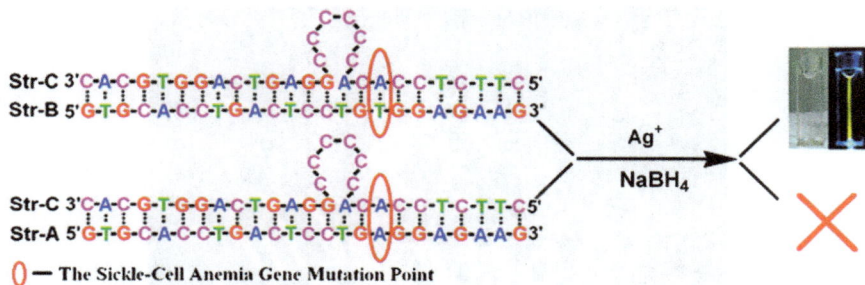

Figure 12.15   Two different DNA duplexes with inserted cytosine loops working as synthetic scaffolds to generate fluorescent Ag NCs for the identification of the sickle cell anemia gene mutation (black dots represent hydrogen bonds formed in base pairing and black dashed lines the sugar–phosphate backbone).
Reprinted from ref. 115 with permission by the American Chemical Society.

capping scaffolds for the generation of fluorescent Ag NCs. Their studies showed that even a single-nucleotide mismatch located two bases away from the nanocluster formation site would prohibit the generation of fluorescent Ag NCs. As an example, they demonstrated the capability of this strategy to identify the sickle cell anemia mutation in the hemoglobin beta chain (HBB) gene.

Ren and coworkers reported the site-specific growth of fluorescent Ag NCs by using a mismatched dsDNA template.[116] Few-atom, molecular-scale Ag NCs are found to localized at the mismatched site, and these DNA-encapsulated NCs can be utilized as functional biological probes to identify single-nucleotide polymorphisms.

MicroRNAs (miRNAs) are regulatory small RNAs that have important roles in numerous developmental, metabolic, and disease processes of plants and animals. The individual levels of miRNAs can be useful biomarkers for cellular events or disease diagnosis. Yang and Vosch have designed a DNA–Ag NC probe that can monitor the presence of target miRNA.[117] The red fluorescence of the DNA–Ag NC probe is diminished in the presence of target miRNA, and high specificity toward detecting specific miRNA sequences is performed. (Figure 12.16) Also, when adding whole plant endogenous RNA to the DNA–Ag NC probe, the emission was significantly higher for the mutant where miRNA was deficient.

Vosch and coworkers investigated the influence of nucleic acid secondary structure on the fast (1 h) formation of bright red emissive Ag NCs in a DNA sequence (DNA-12nt-RED-160), designed for the detection of a microRNA sequence (RNA-miR160).[118] They found that a reduction in secondary structures led to a decreased amount of red emissive Ag NCs. To enhance the emission of the DNA-12nt-RED-172 probe, they rearranged the sequence of the low-emissive DNA-12nt-RED-172 probe to increase formation of secondary structures. The redesigned DNA-GG172-12nt-RED showed a dramatic

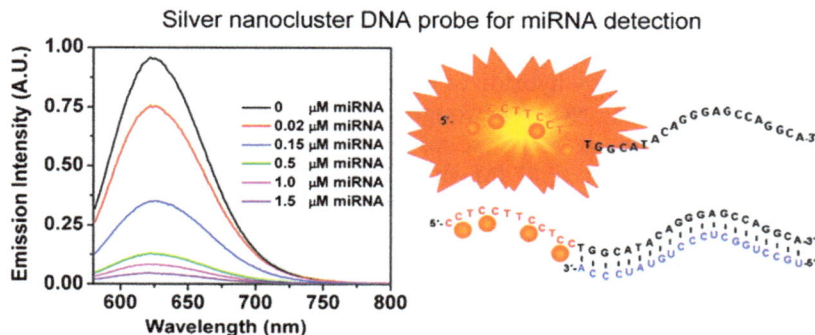

Figure 12.16 Schematic showing the detection of miRNA using a red emitted DNA–Ag NC probe.
Reprinted from ref. 117 with permission by the American Chemical Society.

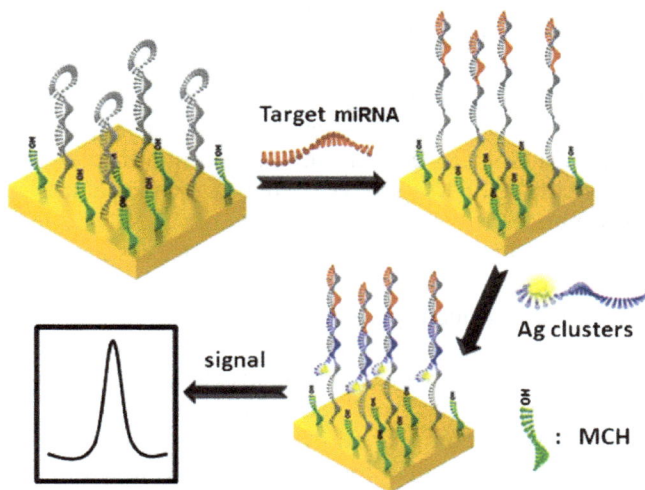

Figure 12.17 Illustration of electrochemical detection of miRNA using DNA-encapsulated Ag NCs.
Reprinted from ref. 119 with permission by the American Chemical Society.

increase in red emission, and the presence of target RNA-miR172 could efficiently quench its fluorescence, providing an approach for RNA-miR172 detection.

Zhang and coworkers discussed a simple, sensitive, and label-free method for miRNA biosensing using DNA–Ag NCs as effective electrochemical probes.[119] (Figure 12.17) The functional DNA probe integrates both a recognition sequence for hybridization and a template sequence for *in situ* synthesis of Ag NCs, which appears to possess exceptional metal mimic enzyme properties for catalyzing H_2O_2 reduction. The miRNA assay employs gold electrodes to immobilize the molecular beacon (MB) probe. After the MB probe subsequently hybridizes with the target and functional probe, the DNA-encapsulated Ag NCs are brought to the electrode surface and produce a detection signal, in response to H_2O_2 reduction. An electrochemical miRNA biosensor down to 67 fM with a linear range of 5 orders of magnitude was obtained. Meanwhile, the MB probe ensures high selectivity or miRNA assay. This is the first application of the electrocatalytic activity of Ag NCs in bioanalysis, which would be attractive for genetic analysis and clinic bio-medical application.

12.4.2.4   Logic Gates Based on DNA-Regulated Ag NCs

In recent years, much research interest has been directed to unconventional chemical computing, which has resulted in the remarkable progress of various Boolean logic systems, such as AND, OR, XOR, NAND, NOR, IN-HIBIT, half-adder, and half-subtractor.[120–122] DNA structure-tuned fluorescent Ag NCs were utilized to construct a conceptually new class of molecular-scale logic gates by Wang *et al.* in 2011.[38] A well-chosen hairpin DNA (HP26 DNA) with a poly-C loop serves as the template for synthesizing two species of Ag NCs. Several G-tracts and C-tracts on its two terminals enable the hairpin DNA to convert into the G-quadruplex and/or i-motif structures upon input of K^+ and H^+. Such a dramatic structural change remarkably influences the spectral behaviors of Ag NCs. In particular, different species of Ag NCs have a distinct fluorescence response to the input of K^+ and H^+. These unique features of DNA-tuned Ag NCs enable multiple logic operations *via* multichannel fluorescence output, indicating their versatility as molecular logic devices. By altering the specific sequence of the hairpin DNA, more logic gates can be constructed utilizing Ag NCs.

The fluorescence (FL) "off–on" switching of designed DNA duplex-stabilized Ag NCs was accomplished by the same group.[123] As illustrated in Figure 12.18, three DNA strands (S-1 to S-3) are designed. S-1 (30 bases) is partially complementary to S-2 (34 bases), while S-3 (34 bases) is totally complementary to S-2. An S-1/S-2 duplex with an inserted six base cytosine-loop can work as an efficient template to generate the red-emitting silver species. The unpaired 10 bases of S-2 in the S-1/S-2 duplex can serve as a toehold to initiate the strand exchange reaction in which S-3 can displace S-1 from the S-1/S-2 duplex to "turn off" the Ag NC emitter. Further addition of S-2 will regenerate the S-1/S-2 duplex and "turn on" the FL of the emitter. The implementation of consecutive FL off–on switching cycles with DNA–Ag NCs as responsers demonstrates that DNA–Ag NCs are promising materials in DNA nanoswitch construction.

Qu and coworkers used Ag NCs as signal transducers to construct various DNA-based logic gates (AND, OR, INHIBIT, XOR, NOR, XNOR, NAND, and a sequential logic gate).[124] Moreover, a biomolecular keypad lock that was capable of constructing crossword puzzles was also fabricated. However, this

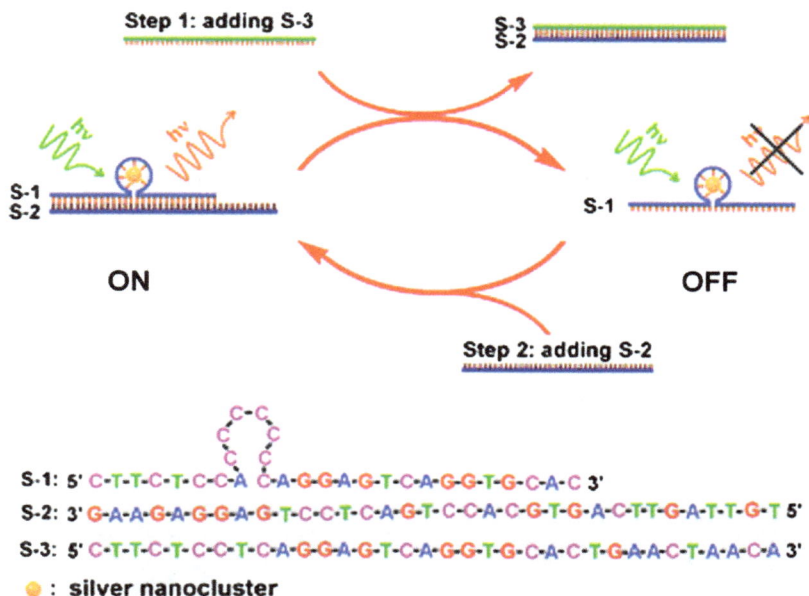

Figure 12.18 The FL off–on switching cycle of hybridized DNA duplex-stabilized Ag NCs modulated by strand exchange reaction.
Reprinted from ref. 123 with permission by the Royal Society of Chemistry.

keypad system without RESET function is two-input mode, and could only perform the function once. Dong and coworkers further presented a resettable molecular keypad lock that utilizes DNA-modulated Ag NCs as signal responser. In their work, an exonuclease-catalyzed DNA hydrolysis reaction was used to achieve the RESET function of a molecular keypad.[125]

12.4.2.5 Application of DNA–Ag NCs in Biological Imaging and Labeling

Most molecular or cellular labeling utilizes organic dyes that are conjugated to bioactive molecules through general organic or inorganic reactions. However, organic dyes often suffer from poor photostability[126] and low brightness,[127] which would limit their applications in live-cell imaging and single-molecule studies. The newly emerging Ag NCs possess an attractive set of features, such as ultrasmall size, good biocompatibility, brightness and photostability, which offers excellent potential as molecular labeling agents. Dickson and coworkers[128] first synthesized Ag NCs with the linear poly(acrylic acid) (PA) as stabilizer and sodium borohydride as reductant. They found that the PA–Ag NCs formed readily transfer NCs to high-affinity ssDNA sequences, which results in the loss of PA–Ag NC fluorescence and generation of characteristic DNA-encapsulated Ag NC emission.

Concomitant with the spectral shifts, cluster transfer from low-quantum yield PA–Ag NCs to high-quantum yield ssDNA–NC results in a 10-fold increase in cluster brightness. Using this fluorogenic NC shuttle, they fluorescently labeled cellular components by first conjugating C_{12} DNA to antibodies (Abs), staining with these nonfluorescent DNA-conjugated antibodies, and transferring NCs as the final step. Results demonstrate that the emission image from the NCs [Figure 12.19(b)] colocalized well with that of the FITC [Figure 12.19(c)], thus revealing antibody location and indicating actin staining with NCs by NC transfer [Figure 12.19(d)]. Furthermore, they have successfully stained microtubules and labeled live cells in the same manner. Vosch and coworkers used C_{24}-encapsulated fluorescent Ag NCs to monitor the transfection of DNA–Ag NCs inside living HeLa cells.[129] The C_{24}–Ag NCs were complexed with a commercially available transfection reagent Lipofectamine, and the internalization of C_{24}–Ag NCs was followed with

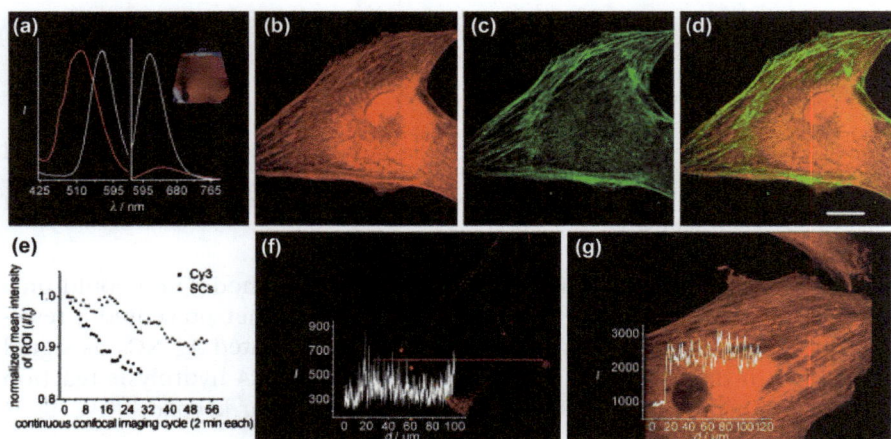

Figure 12.19   BPAECs (bovine pulmonary artery endothelial cells), stained with anti-actin Ab/C_{12} (actinC_{12}) and visualized by NC transfer. (a) NCs transferred from PA to C_{12}. Left panel, normalized excitation spectra of reaction mixture detected at 640 nm before (red) and after (white) the addition of C_{12}, which corresponds to individual excitation spectra of PA–NCs and C_{12}–NCs, respectively. Right panel, 10-fold increase in the fluorescence intensity after the addition of C_{12} (white); excited at 545 nm. The inset shows PA–NCs (100 mL) prepared in a Pyrex conical flask, under 354 nm excitation. (b) Fluorescence images of BPAEC from NCs and (c) FITC tagged to actin C_{12} to indicate the location of actin, with colocalization in (d). (e) Photostability and (f), (g) brightness comparison between Cy3-labeled (f) and NC-labeled (g) cells under identical staining and imaging conditions. The insets show the intensity profiles *versus* distance d along the indicated line within the selected images. NCs and Cy3 were excited at 543 nm (20 W cm^{-2}); FITC was excited at 488 nm. The scale bar is 25 μm.
Reprinted from ref. 128 with permission by Wiley-VCH.

confocal fluorescence microscopy. Bright near-infrared fluorescence was observed from inside the transfected HeLa cells, when exciting with 633 nm excitation, opening up the possibility for the use of Ag NCs for biological labeling and imaging of living cells and for monitoring the transfection process with limited harm to the living cells.

Combining the specific recognition of aptamer towards CCRF-CEM live cells, Gao and coworkers have successfully achieved specific marking of the nucleus of live cells by DNA–Ag NCs.[130] Sgc8c aptamer, selected by cell-SELEX, has been shown to be a specific aptamer for the endosomes of CCRF-CEM cells.[131] Eight cytosine bases were inserted at the 5′ end of the sgc8c aptamer to form Msgc8, which was used for the synthesis of Ag NCs. The cluster–aptamer hybrids formed could specially target the nucleus of CCRF-CEM cells after incubation with the cells for 2 hours at room temperature.

AS1411 is an antiproliferative G-rich phosphodiester DNA and currently an anticancer agent under phase II clinical trials. Its function as an aptamer to nucleolin has led to the identification of nucleolin as a new molecular target for cancer therapy.[132,133] Zhu and coworkers rationally connected AS1411 with poly(cytosine) *via* a –TTTTT– loop (called a T5 loop) and employed it as the scaffold to synthesize Ag NCs through a one-step process.[134] The Ag NCs obtained emitted red color and the fluorescence quantum yield could reach 40.1%, which was beneficial to the applications of biological imaging. AS1411 retained its anticancer nature within the complex, and unexpectedly, functionalized Ag NCs demonstrated enhanced efficiency of growth inhibition compared with naked AS1411. In addition, such functionalized Ag NCs could be internalized into the cells and stain the nucleus *via* receptor-mediated endocytosis. It could improve the permeability of DNA protection groups for live cell imaging and help understand the therapeutic mechanism of AS1411.[135,136]

12.5 DNA-Templated Au NCs

Fluorescent Au NCs have been widely synthesized in a bottom-up manner by reduction of gold precursors that are associated with various biotemplates[137] such as bovine serum albumin,[138] horseradish peroxidase,[139] lysozyme,[140] and transferrin protein.[141] Nevertheless, synthesis of Au NCs templated by DNA has rarely been reported, maybe because of the weak association between the negatively charged DNA and commonly used precursor $AuCl_4^-$. Chen and coworkers synthesized atomically monodispersed Au NCs by etching gold nanocrystals (particles and rods) with the assistance of biomolecules (amino acids, peptides, proteins, and DNA) under sonication in water.[142] Specifically, Au nanorods were gradually etched by dsDNA under sonication, leading to the formation of fluorescent Au_8 NCs. The involvement of violent sonication during synthesis, however, may lead to the disruption of the DNA structures.

Shao and coworkers have reported water-soluble and red-emitting Au NCs that were synthesized with ssDNA as a promising biotemplate and dimethylamine borane as a mild reductant.[36] The fluorescent Au NCs can be formed in a weakly acidic aqueous solution that is free from the simultaneous formation of large nanoparticles. The cluster feature of the formed Au species has been revealed by fluorescence spectra, absorption spectra, and TEM. Additionally, DNA sequences could be used to tune the emissions of Au NCs. The as-prepared Au NCs display high stability at physiological pH conditions, and thus, wide potential applications are anticipated for the biocompatible fluorescent Au NCs serving as nanoprobes in bioimaging and related fields.

12.6 Conclusions and Future Outlook

DNA-templated metal NC synthesis is a young and rapidly evolving field. The ability to tune the size and physicochemical properties of NCs according to DNA structure, composition and sequence holds enormous potential for the rational design of NCs. In addition, as the Dickson group have shown,[74] a unique advantage of DNA as templates is that DNA microarrays can be used to identify suitable templating sequences. Thus, it can be envisaged that *in vitro* evolutionary techniques (such as the so-called SELEX process[143]) can be developed to select efficient template sequences for specific metal NC synthesis. Moreover, one can anticipate that the use of unnatural nucleotides incorporating reactive chemical groups would allow further modulation of metal–nucleic acid interactions and augmentation of applications. Despite this great potential and promising future for DNA–metal NCs, a variety of challenges remain and need to be addressed.

First, it is particularly challenging to produce NCs that are both biocompatible and are outfitted with a defined set of recognition motifs for bioconjugation. The ideal situation is to generate bifunctional DNA modules where one end templates the synthesis of NCs while the other interacts with the surrounding environment without interfering with the NC surface.

Second, the lack of a fundamental understanding of the mechanisms for DNA-templated NCs is currently an issue. It is both desirable and foreseeable that a great deal of effort will be oriented to the templating mechanism. For instance, a systematic investigation of the role that each component of the template (sequence, composition, length, concentration, and so on) plays in NC synthesis would be an important undertaking in order to program the synthesis of NCs. Other important questions that need to be addressed are the role of metal ion–ligand bond strength in defining the outcome of the templated NC synthesis and the effect of the nucleobase on the size and photophysical properties of templated NCs.

Third, previous studies concentrated much effort on Ag NCs, and the synthesis of other metal NCs (such as Au, Cu, Pt, Pb, *etc.*) with DNA as the capping agent is scarce. It is well known that some noble metals (such as Pt and Pb) exhibit excellent catalytic properties. The application of DNA–metal NCs in catalysts would become an important step forward, and the development of DNA-templated noble and alloy metal clusters deserves much focus.

Finally, in addition to fluorescence, more property studies of DNA–metal NCs should be carried out. For example, the catalytic properties of metal NCs synthesized with other modes has been exhaustively investigated, while fewer studies on DNA–metal NCs as active catalysts are reported. In addition, it would be useful to simultaneously endow metal NCs with other properties as bi- or multifunctional probes bear many promising applications. For example, probes comprising fluorescence and magnetism could be advantageous for *in vivo* imaging. Particularly, one can expect that these properties of DNA–metal NCs would be tuned by the DNA template (sequence, composition, length, concentration, and so on) like fluorescent properties.

In conclusion, the synthesis and application of water-soluble fluorescent DNA–metal NCs have achieved significant progress. At the same time, more challenges need to be addressed in future, and a concerted multidisciplinary effort involving chemistry, biology and physics is required.

References

1. H. W. Duan and S. M. Nie, *J. Am. Chem. Soc.*, 2007, **129**, 2412.
2. R. C. Jin, S. Egusa and N. F. Scherer, *J. Am. Chem. Soc.*, 2004, **126**, 9900.
3. T. U. B. Rao and T. Pradeep, *Angew. Chem., Int. Ed.*, 2010, **49**, 3925.
4. C. N. R. Rao and K. P. Kalyanikutty, *Acc. Chem. Res.*, 2008, **41**, 489.
5. C. Vazquez-Vazquez, M. Banobre-Lopez, A. Mitra, M. A. Lopez-Quintela and J. Rivas, *Langmuir*, 2009, **25**, 8208.
6. S. Q. Qiu, J. X. Dong and G. X. Chen, *J. Colloid Interface Sci.*, 1999, **216**, 230.
7. N. Vilar-Vidal, M. C. Blanco, M. A. Lopez-Quintela, J. Rivas and C. Serra, *J. Phys. Chem. C*, 2010, **114**, 15924.
8. B. S. Gonzalez, M. J. Rodriguez, C. Blanco, J. Rivas, M. A. Lopez-Quintela and J. M. G. Martinho, *Nano Lett.*, 2010, **10**, 4217.
9. V. S. Myers, M. G. Weir, E. V. Carino, D. F. Yancey, S. Pande and R. M. Crooks, *Chem. Sci*, 2011, **2**, 1632.
10. J. G. Zhang, S. Q. Xu and E. Kumacheva, *Adv. Mater.*, 2005, **17**, 2336.
11. Z. Shen, H. W. Duan and H. Frey, *Adv. Mater.*, 2007, **19**, 349.
12. L. Shang and S. J. Dong, *Chem. Commun.*, 2008, 1088.
13. J. P. Xie, Y. G. Zheng and J. Y. Ying, *J. Am. Chem. Soc.*, 2009, **131**, 888.
14. J. Zheng, J. T. Petty and R. M. Dickson, *J. Am. Chem. Soc.*, 2003, **125**, 7780.
15. J. Zheng and R. M. Dickson, *J. Am. Chem. Soc.*, 2002, **124**, 13982.
16. J. T. Petty, J. Zheng, N. V. Hud and R. M. Dickson, *J. Am. Chem. Soc.*, 2004, **126**, 5207.
17. R. M. Izatt, J. J. Christensen and J. H. Rytting, *Chem. Rev.*, 1971, **71**, 440.
18. R. B. Martin, *Acc. Chem. Res.*, 1985, **18**, 32.
19. H. Sigel, *Chem. Soc. Rev.*, 1993, **22**, 255.
20. L. Berti and G. A. Burley, *Nat. Nanotechnol*, 2008, **3**, 81.

21. S. J. Klug and M. Famulok, *Mol. Biol. Rep.*, 1994, **20**, 97.
22. C. Wilson and A. D. Keefe, *Curr. Opin. Chem. Biol.*, 2006, **10**, 607.
23. R. K. O. Sigel, E. Freisinger and B. Lippert, *J. Biol. Inorg. Chem.*, 2000, **5**, 287.
24. J. Anastassopoulou, *J. Mol. Struct.*, 2003, **651**, 19.
25. H. Sigel and R. Griesser, *Chem. Soc. Rev.*, 2005, **34**, 875.
26. H. Sigel, S. S. Massoud and N. A. Corfu, *J. Am. Chem. Soc.*, 1994, **116**, 2958.
27. M. de la Fuente, A. Hernanz and R. Navarro, *J. Biol. Inorg. Chem.*, 2004, **9**, 973.
28. J. V. Burda, J. Sponer, J. Leszczynski and P. Hobza, *J. Phys. Chem. B*, 1997, **101**, 9670.
29. S. Pal, R. Varghese, Z. Deng, Z. Zhao, A. Kumar, H. Yan and Y. Liu, *Angew. Chem., Int. Ed.*, 2011, **50**, 1.
30. W. T. Wei, Y. Z. Lu, W. Chen and S. W. Chen, *J. Am. Chem. Soc.*, 2011, **133**, 2060.
31. R. C. Jin, J. E. Jureller, H. Y. Kim and N. F. Scherer, *J. Am. Chem. Soc.*, 2005, **127**, 12482.
32. K. L. Kelly, E. Coronado, L. L. Zhao and G. C. Schatz, *J. Phys. Chem. B*, 2003, **107**, 668.
33. W. Harbich, S. Fedrigo, F. Meyer, D. M. Lindsay, J. Lignieres, J. C. Rivoal and D. Kreisle, *J. Chem. Phys.*, 1990, **93**, 8535.
34. V. Bonacic-Koutecky, J. Pittner, M. Boiron and P. Fantucci, *J. Chem. Phys.*, 1999, **110**, 3876.
35. A. P. Marchetti, A. A. Muenter, R. C. Baetzold and R. T. McCleary, *J. Phys. Chem. B*, 1998, **102**, 5287.
36. G. Liu, Y. Shao, K. Ma, Q. Cui, F. Wu and S. Xu, *Gold Bull.*, 2012, **45**, 69.
37. Y. Lu and W. Chen, *Chem. Soc. Rev.*, 2012, **41**, 3594.
38. T. Li, L. Zhang, J. Ai, S. Dong and E. Wang, *ACS Nano*, 2011, **5**, 6334.
39. T. G. Schaaff, M. N. Shafigullin, J. T. Khoury, I. Vezmar, R. L. Whetten, W. G. Cullen, P. N. First, C. Gutie'rrez-Wing, J. Ascensio and M. J. Jose-Yacama' n, *J. Phys. Chem. B*, 1997, **101**, 7885.
40. M. M. Alvarez, J. T. Khoury, T. G. Schaaff, M. N. Shafigullin, I. Vezmar and R. L. Whetten, *J. Phys. Chem. B*, 1997, **101**, 3706.
41. C. A. Waters, A. J. Mills, K. A. Johnson and D. J. Schiffrin, *Chem. Commun.*, 2003, 540.
42. A. Dass, A. Stevenson, G. R. Dubay, J. B. Tracy and R. W. Murray, *J. Am. Chem. Soc.*, 2008, **130**, 5940.
43. A. Dass, G. R. Dubay, C. A. Fields-Zinna and R. W. Murray, *Anal. Chem.*, 2008, **80**, 6845.
44. J. B. Tracy, G. Kalyuzhny, M. C. Crowe, R. Balasubramanian, J. P. Choi and R. W. Murray, *J. Am. Chem. Soc.*, 2007, **129**, 6706.
45. Y. Shichibu, Y. Negishi, T. Tsukuda and T. Teranishi, *J. Am. Chem. Soc.*, 2005, **127**, 13464.
46. J. B. Tracy, M. C. Crowe, J. F. Parker, O. Hampe, C. A. Fields-Zinna, A. Dass and R. W. Murray, *J. Am. Chem. Soc.*, 2007, **129**, 16209.

47. C. M. Ritchie, K. R. Johnsen, J. R. Kiser, Y. Antoku, R. M. Dickson and J. T. Petty, *J. Phys. Chem. C*, 2007, **111**, 175.
48. D. C. Koningsberger and R. Prins, *X-Ray Absorption: Principles, Applications, Techniques of EXAFS, SEXAFS and XANES*, John Wiley & Sons Inc., New York, 1988.
49. B. J. Kip, F. B. M. Duivenvoorden, D. C. Konigsberger and R. Prins, *J. Catal.*, 1987, **105**, 26.
50. A. Jentys, *Phys. Chem. Chem. Phys.*, 1999, **1**, 4059.
51. B. K. Teo, *EXAFS: basic principles and data analysis*, Springer-Verlag, Berlin, 1986.
52. M. L. Neidig, J. Sharma, H. Yeh, J. S. Martinez, S. D. Conradson and A. P. Shreve, *J. Am. Chem. Soc.*, 2011, **133**, 11837.
53. A. L. Ankudinov, B. Ravel and S. D. Conradson, *Phys. Rev. B*, 1998, **58**, 7565.
54. L. D. Menard, H. Xu, S.-P. Gao, R. D. Twesten, A. S. Harper, Y. Song, G. Wang, A. D. Douglas, J. C. Yang, A. I. Frenkel, R. W. Murray and R. G. Nuzzo, *J. Phys. Chem. B*, 2006, **110**, 14564.
55. J. Wu, Y. Fu, Z. He, Y. Han, L. Zheng, J. Zhang and W. Li, *J. Phys. Chem. B*, 2012, **116**, 1655.
56. U. Kreibig and M. Vollmer, *Optical Properties of Metal Clusters*, Springer, Berlin, 1995, 25.
57. R. L. Johnston, *Atomic and Molecular Clusters*, Taylor & Francis, London, 2002.
58. S. A. Patel, M. Cozzuol, J. M. Hales, C. I. Richards, M. Sartin, J.-C. Hsiang, T. Vosch, J. W. Perry and R. M. Dickson, *J. Phys. Chem. C*, 2009, **113**, 20264.
59. C. I. Richards, S. Choi, J.-C. Hsiang, Y. Antoku, T. Vosch, A. Bongiorno, Y.-L. Tzeng and R. M. Dickson, *J. Am. Chem. Soc.*, 2008, **130**, 5038.
60. J. Sharma, H. C. Yeh, H. Yoo, J. H. Werner and J. S. Martinez, *Chem. Commun.*, 2010, **46**, 3280.
61. L. D. Lavis and R. T. Raines, *ACS Chem. Biol.*, 2008, **3**, 142.
62. S. A. Patel, C. I. Richards, J.-C. Hsiang and R. M. Dickson, *J. Am. Chem. Soc.*, 2008, **130**, 11602.
63. S. H. Yau, N. Abeyasinghe, M. Orr, L. Upton, O. Varnavski, J. H. Werner, H.-C. Yeh, J. Sharma, A. P. Shreve, J. S. Martinez and T. Goodson, III, *Nanoscale*, 2012, **4**, 4247.
64. T. G. Schaaff and R. L. Whetten, *J. Phys. Chem. B*, 2000, **104**, 2630.
65. T. Molotsky, T. Tamarin, A. Ben Moshe, G. Markovich and A. B. Kotlyar, *J. Phys. Chem. C*, 2010, **114**, 15951.
66. T. Yamane and N. Davidson, *Biochim. Biophys. Acta*, 1962, **55**, 609.
67. G. L. Eichhorn, J. J. Butzow, P. Clark and E. Tarien, *Biopolymers*, 1967, **5**, 283.
68. A. Ono, S. Cao, H. Togashi, M. Tashiro, T. Fujimoto, T. Machinami, S. Oda, Y. Miyake, I. Okamoto and Y. Tanaka, *Chem. Commun.*, 2008, **39**, 4825.

69. B. Sengupta, C. M. Ritchie, J. G. Buckman, K. R. Johnsen, P. M. Goodwin and J. T. Petty, *J. Phys. Chem. C*, 2008, **112**, 18776.

70. T. Vosch, Y. Antoku, J.-C. Hsiang, C. I. Richards, J. I. Gonzalez and R. M. Dickson, *Proc. Natl. Acad. Sci. U.S.A.*, 2007, **104**, 12616.

71. E. G. Gwinn, P. O'Neill, A. J. Guerrero, D. Bouwmeester and D. K. Fygenson, *Adv. Mater.*, 2008, **20**, 279.

72. H. Arakawa, J. F. Neault and H. A. Tajmir-Riahl, *Biophys. J.*, 2001, **81**, 1580.

73. P. R. O'Neill, L. R. Velazquez, D. G. Dunn, E. G. Gwinn and D. K. Fygenson, *J. Phys. Chem. C*, 2009, **113**, 4229.

74. C. I. Richards, S. Choi, J.-C. Hsiang, Y. Antoku, T. Vosch, A. Bongiorno, Y.-L. Tzeng and R. M. Dickson, *J. Am. Chem. Soc.*, 2008, **130**, 5038.

75. B. Sengupta, K. Springer, J. G. Buckman, S. P. Story, O. Henry Abe, Z. W. Hasan, Z. D. Prudowsky, S. E. Rudisill, N. N. Degtyareva and J. T. Petty, *J. Phys. Chem. C*, 2009, **113**, 19518.

76. V. N. Soyfer and V. N. Potaman, *Triple-Helical Nucleic Acids*, Springer, New York, 1995.

77. T. Ihara, T. Ishii, N. Araki, A. W. Wilson and A. Jyo, *J. Am. Chem. Soc.*, 2009, **131**, 3826.

78. L. Feng, Z. Huang, J. Ren and X. Qu, *Nucleic Acids Res.*, 2012, **3**, 1.

79. Y. Shao, Z. J. Niu and S. Y. Zou, *Electrochem. Commun.*, 2009, **11**, 417.

80. K. Ma, Q. Cui, G. Liu, F. Wu, S. Xu and Y. Shao, *Nanotechnology*, 2011, **22**, 305502.

81. K. Ma, Y. Shao, Q. Cui, F. Wu, S. Xu and G. Liu, *Langmuir*, 2012, **28**, 15313.

82. Q. Cui, K. Ma, Y. Shao, F. Wu, G. Liu, N. Teramae and H. Bao, *Anal. Chim. Acta*, 2012, **724**, 86.

83. E. M. Nolan and S. J. Lippard, *Chem. Rev.*, 2008, **108**, 3443.

84. W. Guo, J. Yuan and E. Wang, *Chem. Commun.*, 2009, 3395.

85. L. Deng, Z. Zhou, J. Li, T. Li and S. Dong, *Chem. Commun.*, 2011, **47**, 11065.

86. X. F. Liu, Y. L. Tang, L. H. Wang, J. Zhang, S. P. Song, C. Fan and S. Wang, *Adv. Mater.*, 2007, **19**, 1471.

87. C. K. Chiang, C. C. Huang, C. W. Liu and H. T. Chang, *Anal. Chem.*, 2008, **80**, 3716.

88. M. M. Yu, Z. X. Li, L. H. Wei, D. H. Wei and M. S. Tang, *Org. Lett.*, 2008, **10**, 5115.

89. P. G. Georgopoulos, A. Roy, M. J. Yonone-Lioy, R. E. Opiekun and P. J. Lioy, *J. Toxicol. Environ. Health, Part B*, 2001, **4**, 341.

90. G.-Y. Lan, C.-C. Huang and H.-T. Chang, *Chem. Commun.*, 2010, **46**, 1257.

91. Y.-T. Su, G.-Y. Lan, W.-Y. Chen and H.-T. Chang, *Anal. Chem.*, 2010, **82**, 8566.

92. G. Liu, D.-Q. Feng, T. Chen, D. Li and W. Zheng, *J. Mater. Chem.*, 2012, **22**, 20885.

93. J. W. Morse, F. J. Millero, J. C. Cornwell and D. Rickard, *Earth Sci. Rev*, 1987, **24**, 1.
94. T. Bagarinao, *Aquat. Toxicol.*, 1992, **24**, 21.
95. W.-Y. Chen, G.-Y. Lan and H.-T. Chang, *Anal. Chem.*, 2011, **83**, 9450.
96. F. Pu, Z. Huang, J. Ren and X. Qu, *Anal. Chem.*, 2010, **82**, 8211.
97. W. G. Christen, U. A. Ajani, R. J. Glynn and C. H. Hennekens, *Arch. Intern. Med.*, 2000, **160**, 422.
98. J. B. Schulz, J. Lindenau, J. Seyfried and J. Dichgans, *Eur. J. Biochem.*, 2000, **267**, 4904.
99. Z. Huang, F. Pu, Y. Lin, J. Ren and X. Qu, *Chem. Commun.*, 2011, **47**, 3487.
100. M. B. Jr., M. Moronne, P. Gin, S. Weiss and A. P. Alivisatos, 2, *Science*, 1998, **81**, 2013.
101. I. L. Medintz, H. T. Uyeda, E. R. Goldman and H. Mattoussi, *Nat. Mater.*, 2005, **4**, 435.
102. R. Freeman, B. Willner and I. Willner, *J. Phys. Chem. Lett.*, 2011, **2**, 2667.
103. X. Liu, F. Wang, A. Niazov-Elkan, W. Guo and I. Willner, *Nano Lett.*, 2013, **13**, 309.
104. L. Zhang, J. Zhu, S. Guo, T. Li, J. Li and E. Wang, *J. Am. Chem. Soc.*, 2013, **135**, 2403.
105. J. Sharma, H.-C. Yeh, H. Yoo, J. H. Werner and J. S. Martinez, *Chem. Commun.*, 2011, **47**, 2294.
106. R. R. Mayer and P. S. Laine, *Microbiol. Rev.*, 1990, **54**, 342.
107. J. W. Chase and K. R. Williams, *Annu. Rev. Biochem.*, 1986, **55**, 103.
108. J. H. Weiner, L. L. Bertsch and A. Kornberg, *J. Biol. Chem.*, 1975, **250**, 1972.
109. S. C. Kowalczykowski, D. G. Bear and P. H. V. Hippel, *The Enzymes*, Academic Press, New York, 3rd edn, 1981, vol. 14, 373.
110. G.-Y. Lan, W.-Y. Chen and H.-T. Chang, *Analyst*, 2011, **136**, 3623.
111. C.-W. Liu, Y.-W. Lin, C.-C. Huang and H.-T. Chang, *Biosens. Bioelectron.*, 2009, **24**, 2541.
112. J. Li, X. Zhong, H. Zhang, X. C. Le and J.-J. Zhu, *Anal. Chem.*, 2012, **84**, 5170.
113. J.-J. Liu, X.-R. Song, Y.-W. Wang, A.-X. Zheng, G.-N. Chen and H.-H. Yang, *Anal. Chim. Acta*, 2012, **749**, 70.
114. H.-C. Yeh, J. Sharma, J. J. Han, J. S. Martinez and J. H. Werner, *Nano Lett.*, 2010, **10**, 3106.
115. W. Guo, J. Yuan, Q. Dong and E. Wang, *J. Am. Chem. Soc.*, 2010, **132**, 932.
116. Z. Huang, F. Pu, D. Hu, C. Wang, J. Ren and X. Qu, *Chem. – Eur. J*, 2011, **17**, 3774.
117. S. W. Yang and T. Vosch, *Anal. Chem.*, 2011, **83**, 6935.
118. P. Shah, A. Rorvig-Lund, S. Ben Chaabane, P. W. Thulstrup, H. G. Kjaergaard, E. Fron, J. Hofkens, S. W. Yang and T. Vosch, *ACS Nano*, 2012, **6**, 8803.

119. H. Dong, S. Jin, H. Ju, K. Hao, L.-P. Xu, H. Lu and X. Zhang, *Anal. Chem.*, 2012, **84**, 8670.

120. A. P. de Silva and S. Uchiyama, *Nat. Nanotechnol*, 2007, **2**, 399.

121. J. Zhu, T. Li, L. Zhang, S. Dong and E. Wang, *Biomaterials*, 2011, **32**, 7318.

122. A. Lake, S. Shang and D. M. Kolpashchikov, *Angew. Chem., Int. Ed.*, 2010, **49**, 4459.

123. W. Guo, J. Yuan and E. Wang, *Chem. Commun.*, 2011, **47**, 10930.

124. Z. Huang, Y. Tao, F. Pu, J. Ren and X. Qu, *Chem. – Eur. J*, 2012, **18**, 6663.

125. Z. Zhou, Y. Liu and S. Dong, *Chem. Commun.*, 2013, **49**, 3107.

126. C. Eggeling, J. Widengren, R. Rigler and C. A. M. Seidel, *Anal. Chem.*, 1998, **70**, 2651.

127. T. Schmidt, U. Kubitscheck, D. Rohler and U. Nienhaus, *Single Mol.*, 2002, **3**, 327.

128. J. Yu, S. Choi and R. M. Dickson, *Angew. Chem., Int. Ed.*, 2009, **48**, 318.

129. Y. Antoku, J.-i. Hotta, H. Mizuno, R. M. Dickson, J. Hofkens and T. Vosch, *Photochem. Photobiol. Sci.*, 2010, **9**, 716.

130. Z. Sun, Y. Wei, R. Liu, H. Zhu, Y. Cui, Y. Zhao and X. Gao, *Chem. Commun.*, 2011, **47**, 11960.

131. D. Shangguan, Z. Tang, P. Mallikaratchy, Z. Xiao and W. Tan, *ChemBioChem*, 2007, **8**, 603.

132. P. J. Bates, D. A. Laber, D. M. Miller, S. D. Thomas and J. O. Trent, *Exp. Mol. Pathol.*, 2009, **86**, 151.

133. Y. C. Mi, S. D. Thomas, X. H. Xu, L. K. Casson, D. M. Miller and P. J. Bates, *J. Biol. Chem,.*, 2003, **278**, 8572.

134. J. Li, X. Zhong, F. Cheng, J.-R. Zhang, L.-P. Jiang and J.-J. Zhu, *Anal. Chem.*, 2012, **84**, 4140.

135. D. A. Giljohann, D. S. Seferos, P. C. Patel, J. E. Millstone, N. L. Rosi and C. A. Mirkin, *Nano Lett.*, 2007, **7**, 3818.

136. Y. Antoku, J. I. Hotta, H. Mizuno, R. M. Dickson, J. Hofkens and T. Vosch, *Photochem. Photobiol. Sci.*, 2010, **9**, 716.

137. P. L. Xavier, K. Chaudhari, A. Baksi and T. Pradeep, *Nano Rev.*, 2012, **3**, 14767.

138. J. Xie, Y. Zheng and J. Y. Ying, *J. Am. Chem. Soc.*, 2009, **131**, 888.

139. F. Wen, Y. Dong, L. Feng, S. Wang, S. Zhang and X. Zhang, *Anal. Chem.*, 2011, **83**, 1193.

140. H. Wei, Z. Wang, L. Yang, S. Tian, C. Hou and Y. Lu, *Analyst*, 2010, **135**, 1406.

141. P. L. Xavier, K. Chaudhari, P. K. Verma, S. K. Pal and T. Pradeep, *Nanoscale*, 2010, **2**, 2769.

142. R. Zhou, M. Shi, X. Chen, M. Wang and H. Chen, *Chem. – Eur. J*, 2009, **15**, 4944.

143. C. Tyerk and L. Gold, *Science*, 1990, **249**, 505.

CHAPTER 13

Synthesis of Fluorescent Platinum Nanoclusters for Biomedical Imaging

SHIN-ICHI TANAKA AND YASUSHI INOUYE*

Graduate School of Frontier Biosciences, Osaka University, 1-3 Yamadaoka, Suita, Osaka 565-0871, Japan
*Email: ya-inoue@ap.eng.osaka-u.ac.jp

13.1   Introduction

Functional nanomaterials with high biocompatibility, stability, low toxicity and specificity of targeting to desired organs or cells are of interest in nanobiology, biomedical imaging, biosensing and disease diagnostics. Recently, noble metal nanoclusters,[1–43] consisting of a few to roughly a hundred atoms, have attracted considerable attention as the next generation of luminescent nanomaterials because of high photostability, chirality[10,11] and magnetism.[12] Particularly, owing to their fine structure below 1 nm, since free electrons are confined to a dimension smaller than the Fermi wavelength in the conduction band, they exhibit unique size-dependent optical and electronic properties. In addition, low cytotoxicity of noble metal nanoclusters provides great opportunities for application in bioimaging,[3,17–23,35,36] biosensors[3,13–17,31–34] and nanomedicine.[17,21]

Fluorescent gold $(Au)^{1,2,4}$ and silver $(Ag)^{24,35,36}$ nanoclusters have been reported by Dickson *et al.* for the first time. Dickson and co-workers synthesized and stabilized Au and Ag nanoclusters with poly(amidoamine)

RSC Smart Materials No. 7
Functional Nanometer-Sized Clusters of Transition Metals: Synthesis, Properties and Applications
Edited by Wei Chen and Shaowei Chen
© The Royal Society of Chemistry 2014
Published by The Royal Society of Chemistry, www.rsc.org

(PAMAM) dendrimer and oligonucleotide, and successfully demonstrated bioimaging of actin filament labeled with these nanoclusters.[36] Other research groups extended the preparation of fluorescent Au and Ag nanoclusters in peptides,[22,23] proteins,[7,19,21] or polymers.[5,12] For example, Sun and co-workers created red-emitting Au nanoclusters with ferritins,[21] because Au ion can be strongly bound to the imidazole ring of histidine residue at the ferroxidase center of ferritins. The resulting Au nanocluster–ferritin nanoprobe has been successfully applied to both *in vitro* and *in vivo* imaging with nu/nu (nude) female mice.[21]

While, among the noble metals, small platinum (Pt) nanoparticles have already been utilized as a chemical catalyst,[44] it is only recently that fluorescent Pt nanoclusters have been developed by our group,[37,38] Kawasaki *et al.*,[39] and Guével *et al.*[40] Kawasaki *et al.* showed a one-pot synthesis of Pt nanoclusters using *N,N*-dimethylformamide as a capping agent, while Guével *et al.* fabricated Pt nanoclusters from Pt nanoparticles by ligand etching. However, the exact composition and the fluorescent properties of their Pt nanoclusters were not evaluated, because synthesis and isolation still remain a big challenge in Pt nanocluster research.

In our study, we used a PAMAM dendrimer-based template to prepare the Pt nanoclusters. Because this dendrimer has a nanometer-scale size and a uniform structure comprising an external shell and an internal core, which contains tertiary amines that can form coordination bonds with the metallic ions,[45–48] we can regulate the cluster size and can tune photoluminescence wavelength with such a dendrimer. Furthermore, we have achieved the isolation of Pt nanoclusters by ligand-exchange with mercaptoacetic acid (MAA) and by using high performance liquid chromatography (HPLC). Therefore, we found the synthesized Pt nanoclusters were atomically monodispersed and emit a brighter fluorescence than Au and Ag nanoclusters. In this chapter, we describe water-soluble Pt nanoclusters that have low cytotoxicity and introduce their application to bioimaging.

13.2   Experimental

Materials: Fourth-generation (G4-OH) PAMAM dendrimer was purchased from Sigma-Aldrich. H_2PtCl_6, $NaBH_4$, $(NH_4)_2S_2O_8$, trisodium citrate and MAA were purchased from Wako Pure Chemical Industries (Japan). 1-[(3-Dimethylamino)-propyl]-3-ethylcarbodiimide hydrochloride (EDC) and Protein A were purchased from Thermo Fisher Scientific K.K. *N*-Hydroxysulfosuccinimide (Sulfo-NHS) was purchased from Molecular Biosciences Inc. Anti-chemokine receptor (CXCR4) antibody (anti-CXCR4-Ab) was purchased from BioLegend, Inc. HeLa cells and CHO-K1 cells were purchased from DS Pharma Biomedical Co., Ltd.

HPLC: The HPLC system consisted of a 500 µL sample loop, pump (L-2130), UV-Vis absorbance detector (L-2400) and fluorescence detector (L-2485) (Hitachi High-Technologies Corp. Japan). Two TSK gel G2000SW$_{XL}$

size exclusion HPLC columns (double column) and TSK SW$_{XL}$ guard column were purchased from the TOSOH corp. (Japan). HPLC was performed at room temperature. Phosphate buffer (20 mM, pH 7.5) was used as the mobile phase. The flow rate was maintained at 0.5 mL min^{-1} and an injection volume of 100 μL was used for all samples. The detection wavelength for UV absorption was set to 290 nm. Dickson *et al.* have reported the absorption peak of PAMAM (G4-OH) to be around 290 nm.[4,37,38]

Fluorescence lifetime: a femtosecond pulse from a regenerative amplifier (Libra HE, Coherent) was led to an optical parametric amplifier (OPerA Solo, Coherent), which generated 60 fs pulses at 1 kHz repetition rate. The beam was directed to two BBO crystals to generate a third harmonic pulse with a wavelength of 380 nm and 460 nm. The excitation light was softly focused into samples contained in 1 cm cuvettes. The power density of the excitation beam was set to ∼5 μJ cm^{-2}, which was sufficiently low to avoid any saturation effects. Time-resolved spectra were obtained using a photon-counting streak camera (C4780, Hamamatsu Photonics, Japan) through a 25 cm monochromator (250is, Chromex).

Fluorescence spectra: the excitation and fluorescence spectra of Pt$_5$ nanoclusters and PAMAM (G4-OH) were measured with a spectrofluorometer (FP-6200, Jasco Inc. Japan). The excitation wavelength for the fluorescence spectra was set to 380 nm for Pt$_5$ nanoclusters and 375 nm for PAMAM (G4-OH); the emission wavelength for the excitation spectra was set to 470 nm for Pt$_5$ nanoclusters and 450 nm for PAMAM (G4-OH). The excitation–emission matrices spectra of Pt$_8$ nanoclusters were measured with a spectrofluorophotometer (RF-5300PC, SHIMAZDU).

Inductively-coupled plasma mass spectrometry (ICP-MS): ICP-MS on HPLC fractions from Pt nanoclusters sample were measured with Agilent 7500s (Agilent Technologies, Inc. Headquarters).

Furnace atomic absorption spectrometry: Furnace atomic absorption spectrometry (AA-6700F (SHIMAZDU)) was implemented with the 266 nm line.

Electrospray ionization (ESI) mass: ESI mass spectra of Pt nanoclusters were measured using LTQ XL (Thermo Fisher Scientific K.K.). The samples were dissolved in a 50% (v/v) water–methanol system.[30]

Absolute quantum yields (QY): QYs were measured by using a quantum yield measurement system (C10027, Hamamatsu Photonics, Japan). QY is given by $QY = PN_{em} / PN_{ab}$, where PN_{em} and PN_{ab} are the number of emitted and absorbed photons by the fluorescent particles, respectively.

Confocal fluorescence microscopy: a culture dish containing HeLa cells was washed with PBS buffer. 500 μL DMEM were added. Confocal fluorescence imaging was performed with FV1000 (Olympus) using an oil immersion objective lens (60×, N.A. = 1.35), appropriate emission filters and excitation laser.

Cell viability test: culture medium (5.0 μL) containing HeLa cells was mixed with 5.0 μL Trypan Blue (Invitrogen Japan K.K.). 10 μL of the sample were put on a chamber slide set to a Countess Automated Cell Counter (Invitrogen Japan K.K.) to check for cell viability.

Dynamic light scattering (DLS): DLS of Pt$_5$ nanoclusters in water and ZnSe quantum dots[49] in chloroform were measured using a 633 nm He/Ne laser (Zetasizer Nano, Malvern Instruments Ltd).

13.3   Blue-Emitting Platinum Nanoclusters

13.3.1   Preparation of Blue-Emitting Platinum Nanoclusters

PAMAM (G4-OH) (0.5 μmol) was added to 2 mL of millipore water (18.2 MΩ) and then mixed with H$_2$PtCl$_6$ (0.5 M, 6.0 μL). The Pt ions are coordinated with the interior tertiary amines of the PAMAM dendrimers.[45–48] NaBH$_4$ (3.0 μmol) was slowly added to the mixture under continuous stirring (Scheme 13.1). To complete the reaction, the mixture was incubated at room temperature under continuous stirring for two weeks. Large Pt colloidal nanoparticles were removed by ultracentrifugation (Optima MAX-XP Benchtop Ultracentrifuge, Beckman Coulter, Inc.; 100 000 G) for 30 min at 4 °C. The supernatant emitted a strong blue fluorescence under UV light (365 nm) irradiation, indicating Pt nanoclusters.

PAMAM (G4-OH) dendrimers have a sterically-bulky molecular structure and there is a concern that they sterically hinder the biomolecular dynamics in living cells. Furthermore, Bard *et al.* have shown that simple oxidation of PAMAM (G4-OH) with (NH$_4$)$_2$S$_2$O$_8$ can also produce species that emit blue

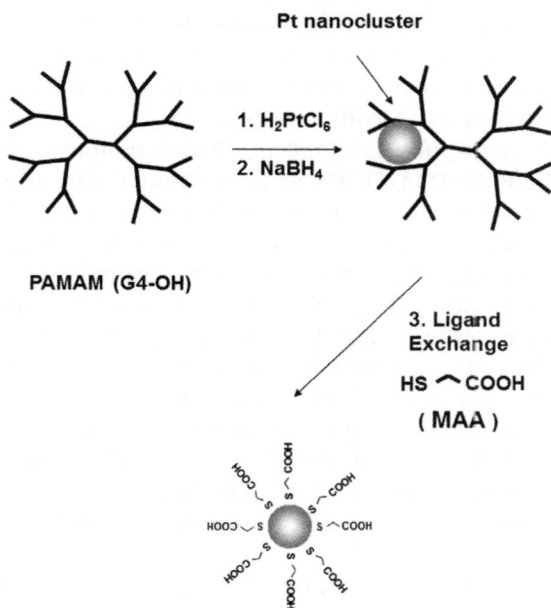

Scheme 13.1   Schematic representation of Pt nanocluster synthesis and ligand exchange with mercaptoacetic acid (MAA).
(Reprinted from ref. 37, with permission from John Wiley & Sons, Inc.)

photoluminescence.[50] Therefore, the Pt nanoclusters needed to be separated from the PAMAM (G4-OH) dendrimers and to be purified from other chemical species using HPLC. To do this, MAA was added to the supernatant to replace PAMAM (G4-OH) as the ligand for the Pt nanoclusters. The reaction mixture was allowed to stand at room temperature for 3 days. The PAMAM (G4-OH) : MAA molar ratio was set to 1 : 20 (Scheme 13.1).

The supernatant with added MAA, which emitted blue fluorescence, was separated into several fractions using HPLC, and the fluorescence spectrum and lifetime of each fraction were measured.[37] The fraction emitting blue fluorescence at 470 nm [Figure 13.1(a)] exhibited a fluorescence lifetime of

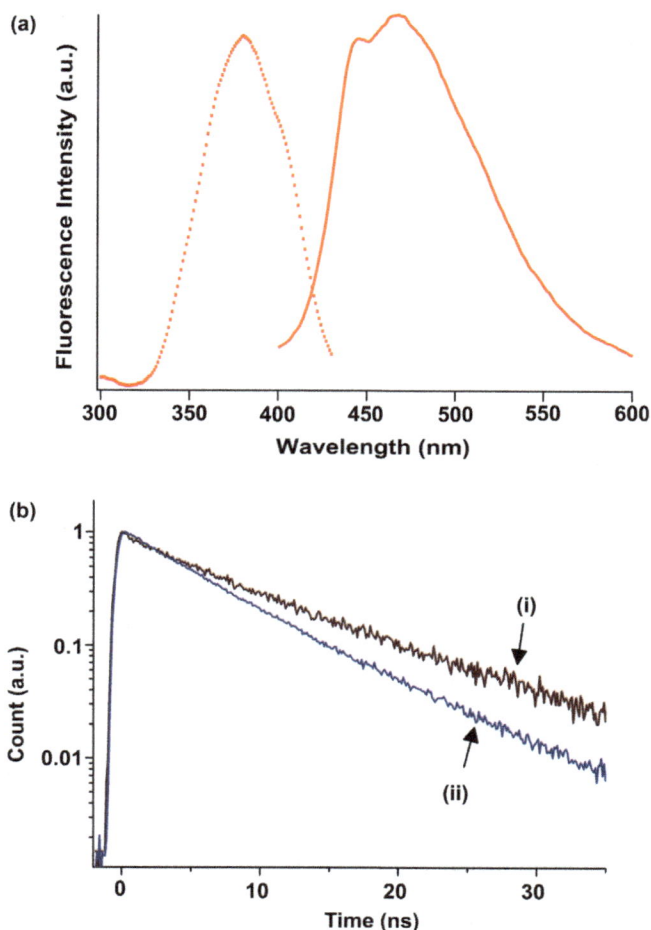

Figure 13.1 (a) Excitation (dashed line) and emission spectra (solid line) of Pt$_5$ nanoclusters in water. (b) Fluorescence lifetime of Pt$_5$ nanoclusters. The fluorescence lifetimes of Pt$_5$ nanoclusters (i) and PAMAM (G4-OH) (ii) were obtained by single exponential fitting (8.8 ± 0.5 ns and 6.3 ± 0.5 ns, respectively).
(Reprinted from ref. 37, with permission from John Wiley & Sons, Inc.)

8.8 ± 0.5 ns while another fraction showed a lifetime of 6.3 ± 0.5 ns [Figure 13.1(b)]. Since a 6.3 ± 0.5 ns lifetime corresponds to deliberately oxidized PAMAM (G4-OH), our results suggest that 8.8 ± 0.5 ns lifetime fractions contained purified Pt nanoclusters.

13.3.2   Characterization

To characterize the Pt nanoclusters, ICP-MS was done for the HPLC fraction and showed blue fluorescence at 470 nm, indicating only Pt (871.64 µg L^{-1}) was present. The ESI data showed a main peak at $m/z = 1712$,[37] which was assigned to $Pt_5L^I_8$ ($L^I = C_2H_4O_2S$) (MW = 1712.3) and revealed that the synthesized nanoclusters were atomically monodispersed Pt_5. The DLS data showed a diameter of 1.3 ± 0.5 nm for Pt_5 nanoclusters, which is much smaller than that of ZeSe quantum dots (7 nm).[49]

Fluorescent probe brightness is crucial for obtaining well-defined images with a high signal-to-noise ratio. We evaluated the absolute QY of the synthesized Pt_5 nanoclusters. The QY for the Pt_5 nanoclusters was found to be 18% in water, while that for another blue photoluminescent, ZnSe quantum dots in chloroform, was 5%.[49] On the other hand, we measured the absolute QY of newly synthesized Au_8 nanoclusters[51] and found that the QY is 3.3%[5] in water. Absolute QY measurements indicate that the fluorescence of Pt_5 nanoclusters is at least 3 times brighter than that of ZnSe quantum dots and Au_8 nanoclusters.

13.3.3   Application to Bioimaging

We investigated the feasibility of our Pt nanoclusters as fluorescent probes for cellular imaging. In this study, we employed human epithelial carcinoma HeLa cells in which chemokine receptors are highly expressed.[52] First, we conjugated the $Pt_5L^I_8$ (57 nM, 400 µL) with Protein A (24 µM, 9.5 µL) by using a EDC (0.1 mM, 4.56 µL)/Sulfo-NHS (0.1 mM, 4.56 µL) coupling reaction. We bound the $Pt_5L^I_8$–(Protein A) to an anti-chemokine receptor antibody (anti-CXCR4-Ab) ((6.67µM, 13.7 µL)) via the Fc moiety of the antibody (Scheme 13.2).[37,38,53,54] The concentration of Pt nanoclusters was evaluated by using fluorescence correlation spectroscopy with Rhodamine 6G as the standard solution.[55]

Figure 13.2(a) shows a confocal fluorescence image of HeLa cells labeled with $Pt_5L^I_8$. Blue-emitting fluorescence (470 nm) was observed on cell membranes where the receptors were exhibited. No fluorescence signal was detected in a control sample labeled without $Pt_5L^I_8$ [Figure 13.2(b)], which rules out any autofluorescence at 405 nm excitation. In order to examine the specific binding of $Pt_5L^I_8$–(Protein A)–(anti-CXCR4-Ab) to the chemokine receptor, chemokine receptor-negative CHO-K1 cells were used as a negative control.[56] In Figure 13.2(c), the fluorescence signal was not observed, indicating CHO-K1 cells were not stained by $Pt_5L^I_8$. These results argue that

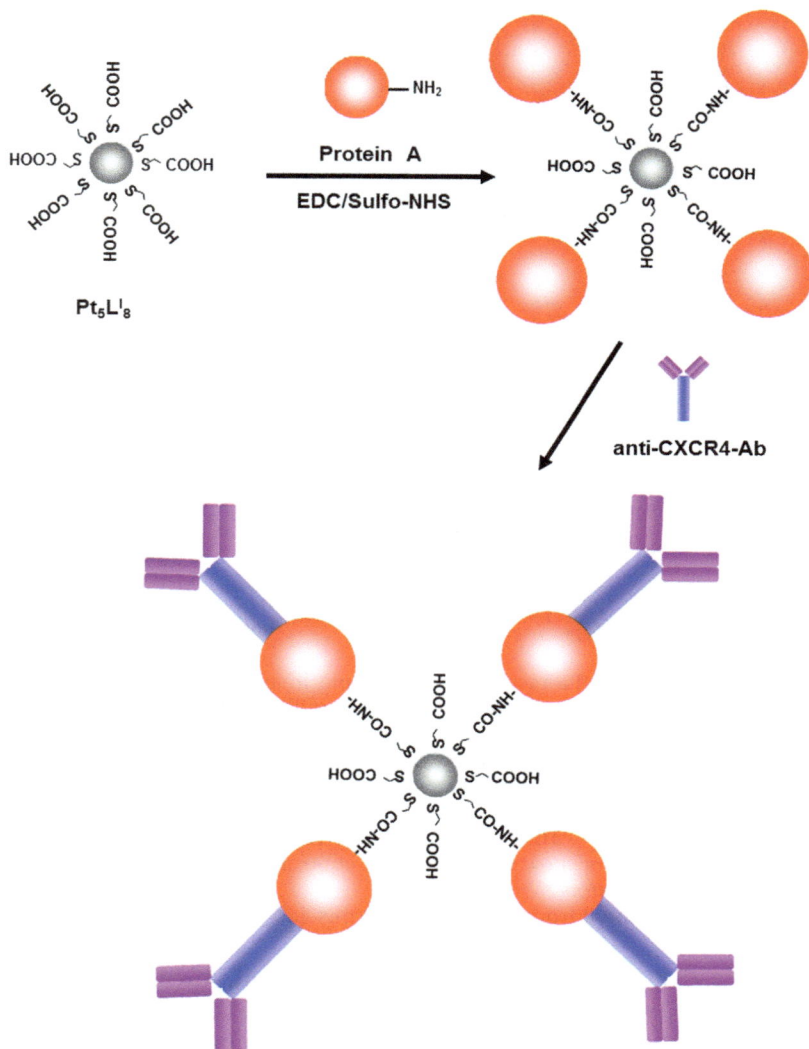

Scheme 13.2 Schematic representation for the preparation of $Pt_5L^I_8$–(Protein A)–(anti-CXCR4-Ab).
(Reprinted from ref. 37, with permission from John Wiley & Sons, Inc.)

$Pt_5L^I_8$–(Protein A)–(anti-CXCR4-Ab) specifically bound to the chemokine receptor.

The cytotoxity of the Pt_5 nanoclusters in HeLa cells was also examined by using a cell counter. Figure 13.2(d) shows the viability of cells labeled with three different concentrations of Pt_5 nanoclusters (1 nM, 10 nM and 100 nM) used in the same manner as that shown in Figure 13.2(a) and incubated at

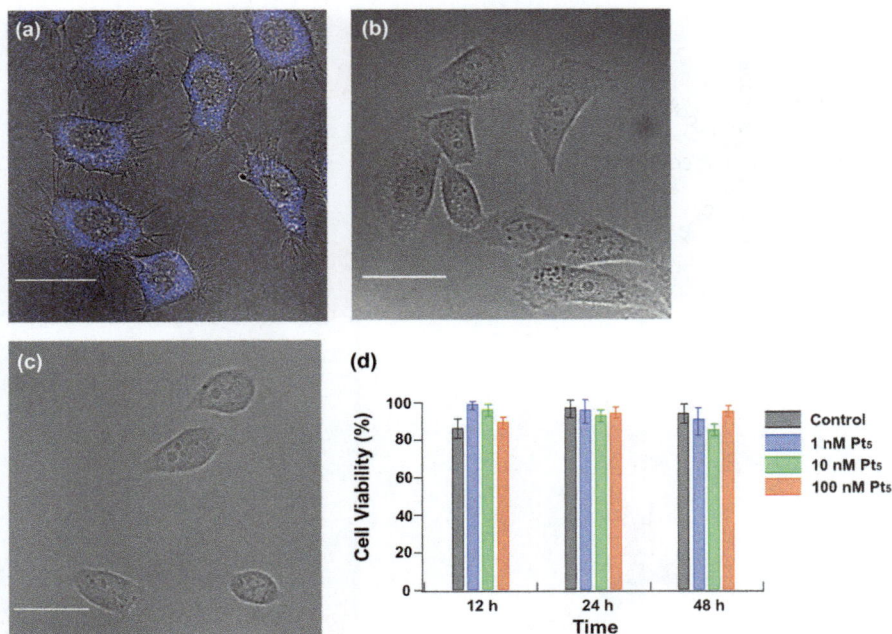

Figure 13.2   Confocal fluorescence microscopic images merged with differential interference contrast (DIC) images of living HeLa cells labeled with (a) and without (b) $Pt_5L^I_8$–(Protein A)–(anti-CXCR4-Ab). (c) Confocal fluorescence microscopic images merged with DIC images of living CHO-K1 cells in the presence of $Pt_5L^I_8$–(Protein A)–(anti-CXCR4-Ab). The scale bars are 20 μm. (d) Cell viability of HeLa cells. Cells were incubated with 1 nM, 10 nM and 100 nM Pt_5 nanoclusters at 37 °C. Black bars show cell viability of control samples in the absence of Pt_5 nanoclusters.
(Reprinted from ref. 37, with permission from John Wiley & Sons, Inc.)

37 °C at three different times (12 h, 24 h and 48 h). After 12 h incubation more than 89% of cells were alive. Even after 48 h incubation, cell viability was still more than 85%. These viabilities were comparable to controls.

13.4   Green-Emitting Platinum Nanoclusters

13.4.1   Preparation of Green-Emitting Platinum Nanoclusters

PAMAM (G4-OH) (0.25 μmol) was added to 5 mL of millipore water (18.2 MΩ) and then mixed with H_2PtCl_6 (0.5 M, 30 μL). The Pt ions are coordinated with the interior tertiary amines of the PAMAM dendrimers.[45–48] The reaction mixture was allowed to stand in the dark at 4 °C to minimize Pt ion oxidization of the PAMAM dendrimers. Since the ligand-to-metal charge transfer (LMCT) UV absorption at 250 nm indicates the complexation of Pt(II) ions with PAMAM (G4-OH),[47,48] we measured the UV-Vis absorbance spectrum of the Pt–PAMAM (G4-OH) solution during incubation by using an UV-2450 spectrophotometer (SHIMAZDU) and quartz cell with 1 cm path length. The background spectrum was subtracted by using an identical cell filled with millipore water (18.2 MΩ).

Figure 13.3(a) illustrates time evolution of absorbance at 250 nm. Although the LMCT band at 250 nm sharply decreased in the first 3 hours, it slowly increased thereafter, reaching its maximum in the first day (24 h). After that, the absorbance gradually decreased again. On the other hand, as shown in Figure 13.3(b), the absorbance of PtCl$_6^{2-}$ at 262 nm[57] and that of PAMAM (G4-OH) at around 200 nm precipitously decreased in the first 3 hours, and slowly changed thereafter. In the first 3 hours, Pt(IV) ions interacted with the external tertiary amines of PAMAM (G4-OH) and were reduced to Pt(II) ions. This reducing reaction corresponds to a decrease in absorbance at both 200 nm and 262 nm. When Pt(II) ions formed coordination bonds

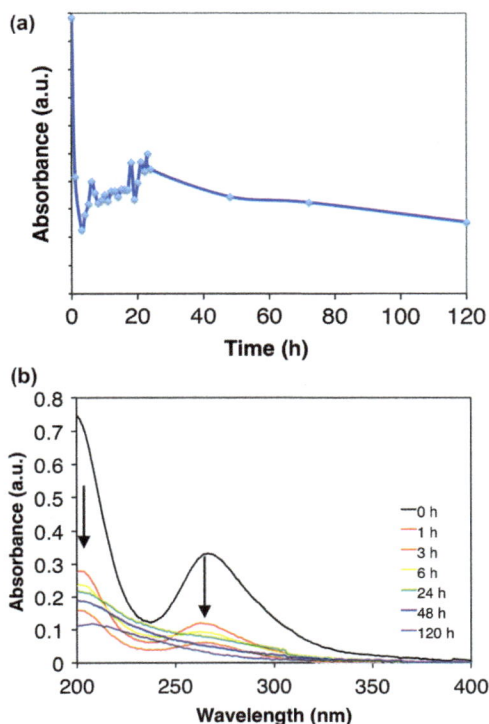

Figure 13.3 (a) Time evolution of absorbance for Pt–PAMAM (G4-OH) complexation at 250 nm. (b) UV-Vis spectra showing the complexation of Pt ions with PAMAM (G4-OH): the absorbance at 250 nm corresponds to the LMCT band. The left and right arrows show the decrease in the absorbance of PAMAM (G4-OH) at around 200 nm and that of PtCl$_6^{2-}$ at 262 nm, respectively.
(Reprinted from ref. 38, with permission from The Optical Society.)

with the internal tertiary amines of PAMAM (G4-OH), the absorbance of the LMCT band at 250 nm increased with incubation time from 3 to 24 hours. Complexation of Pt(II) ions with PAMAM (G4-OH) reached equilibrium after 24 hours. We believe the coordination bonds are broken when Pt(II) ions oxidize PAMAM (G4-OH). This result indicates that the number of Pt ions complexed with PAMAM dendrimers reaches the maximum approximately one day after the reaction started. Therefore, we decided that the reduction reaction started 24 hours after pre-equilibrating the Pt–PAMAM complex. We added a reductant (trisodium citrate; 1 M, 300 μL) to the pre-equilibrated Pt–PAMAM (G4-OH) solution, and incubated this reaction mixture at 90 °C for two weeks under continuous stirring to form nanoclusters. No precipitates were observed after incubation. Then, ultracentrifugation (100 000 G) was performed for 30 min at 4 °C to remove Pt colloidal nanoparticles.

The supernatant to which MAA was added was separated into four fractions by using size-exclusion HPLC [Figure 13.4(a)]. After collecting the fractions, we measured the excitation–emission matrices spectra. The first broad fraction (fraction 1), eluted from 28 min to 38 min during the retention time, showed no fluorescence (data not shown). While the second fraction (fraction 2) had fluorescence at around 420 nm [Figure 13.4(b)], the third fraction (fraction 3) emitted fluorescence at around 420 nm and 520 nm [Figure 13.4(c)]. The fourth fraction (fraction 4) exhibited a single fluorescent component at around 520 nm [Figure 13.4(d)].

13.4.2   Characterization

Furnace atomic absorption spectrometry was implemented for fractions 2 to 4 in Figure 13.4. We found only Pt to be present in fraction 3 (5.74 mg L^{-1}) and 4 (1.64 mg L^{-1}), while we did not detect Pt in fraction 2. ESI mass spectrometry was then performed to determine the molecular weight of the chemical constituents in fraction 4. As shown in Figure 13.5, the main peak was detected at $m/z = 2353.22$, which is assigned to $[Pt_8L^{II}_8 + 3Na + 4H]^-$ ($L^{II} = C_2H_2O_2S$). From the results, we found that the synthesized nanoclusters were composed of eight platinum atoms and that they were monodispersed.[38] The fluorescent peaks (excitation: 460 nm, emission: 520 nm) observed in Figure 13.4(c) and (d) originate from Pt nanoclusters, while the peaks (excitation: 330 nm, emission: 420 nm) observed in Figure 13.4(b) and (c) are attributed to PAMAM (G4-OH), of which the fluorescence was reported in ref. 37. We deduce from these results that fraction 4 contains $Pt_8L^{II}_8$ whereas fraction 3 contains Pt_8 nanoclusters enclosed in PAMAM (G4-OH). Consequently, these results show that the HPLC technique can be used to purify Pt nanoclusters from PAMAM (G4-OH).

We also evaluated the absolute QY and the fluorescent lifetime of these Pt_8 nanoclusters. The purified green fluorescence Pt_8 nanoclusters showed a fluorescence lifetime of 6.5 ± 0.5 ns[38] (Figure 13.6), which were obtained by single exponential fitting. These fluorescence decay curves match well to the single exponential decaying function, which indicates that the synthesized

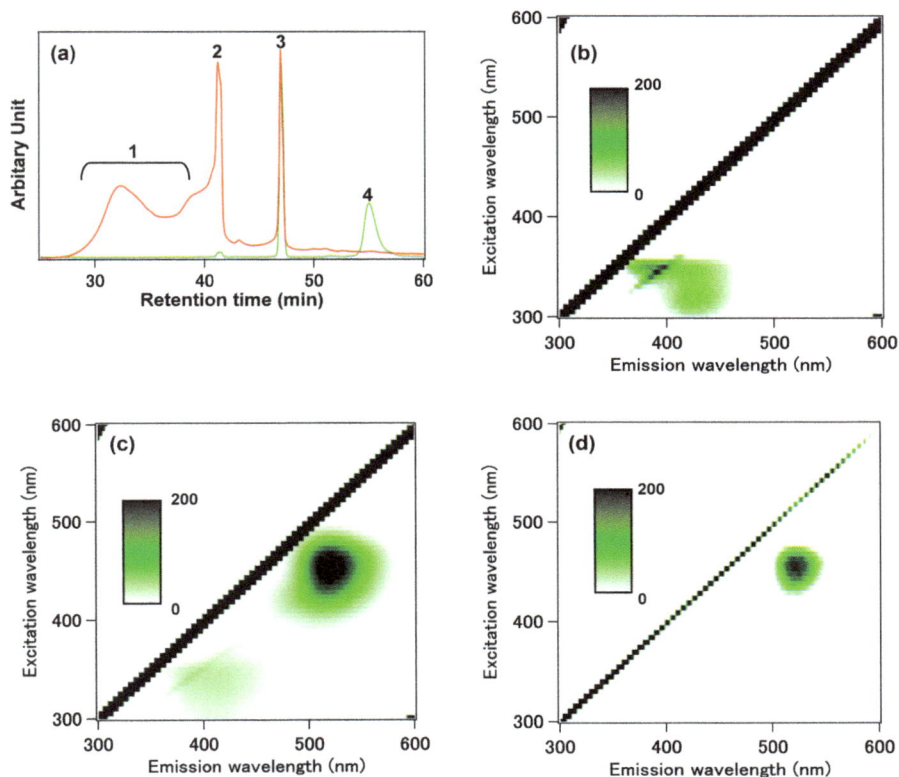

Figure 13.4 (a) Size-exclusion HPLC chromatogram of the supernatant to which mercaptoacetic acid (MAA) was added after centrifugation. HPLC was monitored by UV absorption at 290 nm (red line) and fluorescence at 520 nm (green line). Excitation–emission matrices spectra of fraction 2 (b), 3 (c) and 4 (d).
(Reprinted from ref. 38, with permission from The Optical Society.)

Pt nanoclusters have a single molecular structure. The Pt_8 nanoclusters exhibited a QY of 28% in water, which betters the 18% QY of Pt_5.[37] This value also exceeds those of other green-emitting nanoclusters such as Au (QY = 25%, em = 510 nm)[1,2] and Ag (QY = 16%, em = 520 nm) nanoclusters.[24]

Next, we evaluated the photobleaching characteristic of the Pt_8 nanoclusters by observing decay in fluorescence intensity. The fluorescence intensity decay was measured and fitted by a single exponential function, $F(t) = F_0 \exp(-t/T_d)$ (t: irradiation time, F_0: fluorescence intensity at $t = 0$, T_d: decay time which corresponds to $F(T_d)/F_0 = 1/e$). The decay time of the Pt_8 nanoclusters was 125 min, while that of the organic fluorophore [1,1'-dioctadecyl-3,3,3',3'-tetramethylindocarbocyamine iodide (DiI)], which is well-known as a high resistant fluorophore to photobleaching, was 50 min. We found that the Pt_8 nanoclusters were less subject to photobleaching than DiI.

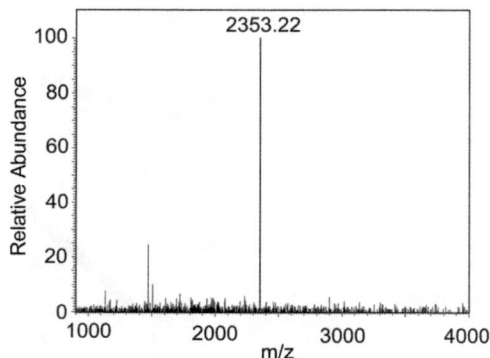

Figure 13.5   ESI mass spectrum of fraction 4. The peak, $m/z = 2353.22$, is assigned
to $[Pt_8L^{II}_8 + 3Na + 4H]^-$ ($L^{II} = C_2H_2O_2S$), and shows Pt nanoclusters
consist of eight platinum atoms.
(Reprinted from ref. 38, with permission from The Optical Society.)

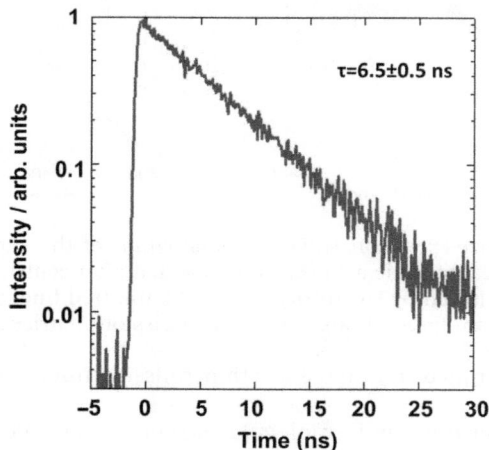

Figure 13.6   Fluorescence lifetime of Pt_8 nanoclusters: the fluorescence lifetime of Pt_8
nanoclusters was obtained by single exponential fitting (6.5 ± 0.5 ns).
(Reprinted from ref. 38, with permission from The Optical Society.)

13.4.3   Application to Bioimaging

We also conjugated $Pt_8L^{II}_8$ with an anti-chemokine receptor (anti-CXCR4-Ab)
antibody in a similar way. Green fluorescence (520 nm) was observed on the
cellular membranes of HeLa cells treated with Pt_8 nanoclusters
[Figure 13.7(a)].[38] Conversely, no fluorescent signal was detected in a control
sample labeled without the Pt_8 nanoclusters [Figure 13.7(b)]. Therefore, the
observed fluorescence in the presence of Pt_8 nanoclusters is not

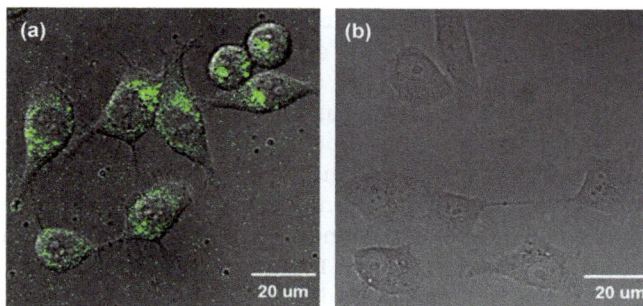

Figure 13.7 Laser confocal fluorescent microscopic image overlaid with DIC images of living HeLa cells labeled with (a) and without (b) $(Pt_8L^{II}_8)$–(Protein A)–(anti-CXCR4-Ab).
(Reprinted from ref. 38, with permission from The Optical Society.)

autofluorescence from the HeLa cell. In addition, we investigated the cytotoxity of Pt_8 nanoclusters in living HeLa cells by using the same cell viability test. These results showed that Pt nanoclusters have very low cytotoxicity, and verify that Pt nanoclusters are relatively harmless fluorescent probes that can be used for the long-term imaging of living cells.[38]

13.5   Conclusion

In conclusion, in our research, we have achieved the development of a synthesis method for atomically monodispersed Pt nanoclusters, the evaluation of fluorescent properties and the determination of their precise composition. In particular, the ligand-exchange with MAA was found to be important for stabilization of Pt nanoclusters, and we could isolate these Pt nanoclusters from other chemical species by using HPLC.

Furthermore, we prepared green-emitting Pt_8 nanoclusters (excitation: 460 nm, emission: 520 nm) that achieved a 28% QY.[38] A key to this method is the pre-equilibration of the Pt ions with PAMAM dendrimers, which promotes the formation of coordination bonds between the Pt ions and the tertiary amine groups of the PAMAM dendrimers, increasing the number of Pt ions trapped in the PAMAM dendrimers. Since the number of tertiary amine groups is increased by using a higher generation of dendrimer, more Pt ions can be coordinated with the higher generation of PAMAM dendrimers. Therefore, synthesis of larger Pt nanoclusters opens the door to realizing longer wavelength photoluminescence with such a higher generation of dendrimer. Using this method, we expect to extend the photoluminescent wavelength of Pt nanoclusters to the NIR region, which is often preferable for *in vivo* imaging experiments. NIR-emission is especially suitable for whole body imaging applications, disease diagnostics and clinical settings, where low background autofluorescence and deep penetration depth are desirable characteristics.

Finally, we successfully labeled chemokine receptors in living HeLa cells with Pt nanoclusters bound to an antibody *via* a conjugated protein to fluorescently image the cells. This suggests that MAA modification can provide an external coating for Pt nanoclusters that is ideal for the attachment of further proteins or other biomolecules using standard conjugation chemistry. We further demonstrate that Pt nanoclusters have significant cell viability for long-term cellular imaging. Since Pt nanoclusters are less cytotoxic as well as smaller than other nanoparticles such as quantum dots[58] and carbon nanotubes,[59] our fluorescent Pt nanoclusters promise to be a useful fluorescent probe for bioimaging and subcellular targeting.

References

1. J. Zheng, C. Zhang and R. M. Dickson, *Phys. Rev. Lett.*, 2004, **93**, 077402.
2. J. Zheng, P. R. Nicovich and R. M. Dickson, *Annu. Rev. Phys. Chem.*, 2007, **58**, 409.
3. L. Shang, S. Dong and G. U. Nienhaus, *Nano Today*, 2011, **6**, 401.
4. J. Zheng, J. T. Petty and R. M. Dickson, *J. Am. Chem. Soc.*, 2003, **125**, 7780.
5. N. Schaeffer, B. Tan, C. Dickinson, M. J. Rosseinsky, A. Laromaine, D. W. McComb, M. M. Stevens, Y. Wang, L. Petit, C. Barentin, D. G. Spiller, A. I. Cooper and R. Lévy, *Chem. Commun.*, 2008, 3986.
6. H. Duan and S. Nie, *J. Am. Chem. Soc.*, 2007, **129**, 2412.
7. J. Xie, Y. Zheng and J. Y. Ying, *J. Am. Chem. Soc.*, 2009, **131**, 888.
8. X. L. Guével, B. Hötzer, G. Jung and M. Schneider, *J. Mater. Chem.*, 2011, **21**, 2974.
9. Y. Bao, C. Zhong, D. M. Vu, J. P. Temirov, R. B. Dyer and J. S. Martinez, *J. Phys. Chem. C*, 2007, **111**, 12194.
10. C. E. Román-Velázquez, C. Noguez and I. L. Garzón, *J. Phys. Chem. B*, 2003, **107**, 12035.
11. C. Gautier and T. Bürgi, *J. Am. Chem. Soc.*, 2008, **130**, 7077.
12. B. S. González, M. J. Rodríguez, C. Blanco, J. Rivas, M. A. López-Quintela and J. M. G. Martinho, *Nano Lett.*, 2010, **10**, 4217.
13. Z. Yuan, M. Peng, Y. He and E. S. Yeung, *Chem. Commun.*, 2011, **47**, 11981.
14. H. Liu, X. Zhang, X. Wu, L. Jiang, C. Burda and J.-J. Zhu, *Chem. Commun.*, 2011, **47**, 4237.
15. J. Xie, Y. Zheng and J. Y. Ying, *Chem. Commun.*, 2010, **46**, 961.
16. Y.-C. Shiang, C.-C. Huang and H.-T. Chang, *Chem. Commun.*, 2009, 3437.
17. C.-A. J. Lin, C.-H. Lee, J.-T. Hsieh, H.-H. Wang, J. K. Li, J.-L. Shen, W.-H. Chan, H.-I. Yeh and W. H. Chang, *J. Med. Biol. Eng.*, 2009, **29**, 276.
18. C. Wang, J. Li, C. Amatore, Y. Chen, H. Jiang and X.-M. Wang, *Angew. Chem., Int. Ed.*, 2011, **50**, 11644.
19. C.-L. Liu, H.-T. Wu, Y.-H. Hsiao, C.-W. Lai, C.-W. Shih, Y.-K. Peng, K.-C. Tang, H.-W. Chang, Y.-C. Chien, J.-K. Hsiao, J.-T. Cheng and P.-T. Chou, *Angew. Chem., Int. Ed.*, 2011, **50**, 7055.

20. S.-Y. Lin, N.-T. Chen, S.-P. Sum, L.-W. Lo and C.-S. Yang, *Chem. Commun.*, 2008, 4762.
21. C. Sun, H. Yang, Y. Yuan, X. Tian, L. Wang, Y. Guo, L. Xu, J. Lei, N. Gao, G. J. Anderson, X.-J. Liang, C. Chen, Y. Zhao and G. Nie, *J. Am. Chem. Soc.*, 2011, **133**, 8617.
22. M. Yu, C. Zhou, J. Liu, J. D. Hankins and J. Zheng, *J. Am. Chem. Soc.*, 2011, **133**, 11014.
23. Y. Wang, Y. Cui, Y. Zhao, R. Liu, Z. Sun, W. Li and X. Gao, *Chem. Commun.*, 2012, **48**, 871.
24. C. I. Richards, S. Choi, J.-C. Hsiang, Y. Antoku, T. Vosch, A. Bongiorno, Y.-L. Tzeng and R. M. Dickson, *J. Am. Chem. Soc.*, 2008, **130**, 5038.
25. I. Díez, M. Pusa, S. Kulmala, H. Jiang, A. Walther, A. S. Goldmann, A. H. E. Müller, O. Ikkala and R. H. A. Ras, *Angew. Chem., Int. Ed.*, 2009, **48**, 2122.
26. W. Guo, J. Yuan, Q. Dong and E. Wang, *J. Am. Chem. Soc.*, 2010, **132**, 932.
27. T. U. B. Rao and T. Pradeep, *Angew. Chem., Int. Ed.*, 2010, **49**, 3925.
28. H. Xu and K. S. Suslick, *Adv. Mater.*, 2010, **22**, 1078.
29. P. R. O'Neill, K. Young, D. Schiffels and D. K. Fygenson, *Nano Lett.*, 2012, **12**, 5464.
30. Z. Wu, E. Lanni, W. Chen, M. E. Bier, D. Ly and R. Jin, *J. Am. Chem. Soc.*, 2009, **131**, 16672.
31. L. Deng, Z. Zhou, J. Li, T. Li and S. Dong, *Chem. Commun.*, 2011, **47**, 11065.
32. W. Guo, J. Yuan and E. Wang, *Chem. Commun.*, 2011, **47**, 10930.
33. J. Sharma, H.-C. Yeh, H. Yoo, J. H. Werner and J. S. Martinez, *Chem. Commun.*, 2011, **47**, 2294.
34. G.-Y. Lan, C.-C. Huang and H.-T. Chang, *Chem. Commun.*, 2010, **46**, 1257.
35. J. Yu, S. A. Patel and R. M. Dickson, *Angew. Chem., Int. Ed.*, 2007, **46**, 2028.
36. J. Yu, S. Choi and R. M. Dickson, *Angew. Chem., Int. Ed.*, 2009, **48**, 318.
37. S.-I. Tanaka, J. Miyazaki, D. K. Tiwari, T. Jin and Y. Inouye, *Angew. Chem., Int. Ed.*, 2011, **50**, 431.
38. S.-I. Tanaka, K. Aoki, A. Muratsugu, H. Ishitobi, T. Jin and Y. Inouye, *Opt. Mater. Express*, 2013, **3**, 157.
39. H. Kawasaki, H. Yamamoto, H. Fujimori, R. Arakawa, M. Inada and Y. Iwasaki, *Chem. Commun.*, 2010, **46**, 3759.
40. X. L. Guével, V. Trouillet, C. Spies, G. Jung and M. Schneider, *J. Phys. Chem. C*, 2012, **116**, 6047.
41. H. Kawasaki, Y. Kosaka, Y. Myoujin, T. Narushima, T. Yonezawa and R. Arakawa, *Chem. Commun.*, 2011, **47**, 7740.
42. W. Wei, Y. Lu, W. Chen and S. Chen, *J. Am. Chem. Soc.*, 2011, **133**, 2060.
43. M. Hyotanishi, Y. Isomura, H. Yamamoto, H. Kawasaki and Y. Obora, *Chem. Commun.*, 2011, **47**, 5750.
44. S. Mostafa, F. Behafarid, J. R. Croy, L. K. Ono, L. Li, J. C. Yang, A. I. Frenkel and B. R. Cuenya, *J. Am. Chem. Soc.*, 2010, **132**, 15714.

45. R. M. Crooks, M. Zhao, L. Sun, V. Chechik and L. K. Yeung, *Acc. Chem. Res.*, 2001, **34**, 181.
46. Y. Niu and R. M. Crooks, *Chem. Mater.*, 2003, **15**, 3463.
47. M. R. Knecht, M. G. Weir, V. S. Myers, W. D. Pyrz, H. Ye, V. Petkov, D. J. Buttrey, A. I. Frenkel and R. M. Crooks, *Chem. Mater.*, 2008, **20**, 5218.
48. Y. Gu, P. Sanders and H. J. Ploehn, *Colloids Surf., A*, 2010, **356**, 10.
49. ZnSe qauntum dots were synthesized by the method described in: H.-S. Chen, B. Lo, J.-Y. Hwang, G.-Y. Chang, C.-M. Chen, S.-J. Tasi and S.-J. J. Wang, *J. Phys. Chem. B*, 2004, **108**, 17119
50. W. I. Lee, Y. Bae and A. J. Bard, *J. Am. Chem. Soc.*, 2004, **126**, 8358.
51. Au_8 nanocluster was synthesized by the method described in ref. 4.
52. A. Müller, B. Homey, H. Soto, N. Ge, D. Catron, M. E. Buchanan, T. McClanahan, E. Murphy, W. Yuan, S. N. Wagner, J. L. Barrera, A. Mohar, E. Verástegui and A. Zlotnik, *Nature*, 2001, **410**, 50.
53. Z. D. Liu, S. F. Chen, C. Z. Huang, S. J. Zhen and Q. G. Liao, *Anal. Chim. Acta*, 2007, **599**, 279.
54. T. Jin, D. K. Tiwari, S.-I. Tanaka, Y. Inouye, K. Yoshizawa and T. M. Watanabe, *Mol. BioSyst.*, 2010, **6**, 2325.
55. T. Jin, F. Fujii, Y. Komai, J. Seki, A. Seiyama and Y. Yoshioka, *Int. J. Mol. Sci.*, 2008, **9**, 2044.
56. A. Amara, O. Lorthioir, A. Valenzuela, A. Magerus, M. Thelen, M. Montes, J.-L. Virelizier, M. Delepierre, F. Baleux, H. Lortat-Jacob and F. Arenzana-Seisdedos, *J. Biol. Chem.*, 1999, **274**, 23916.
57. C. H. Gammons, *Geochim. Cosmochim. Acta*, 1996, **60**, 1683.
58. C. Kirchner, T. Liedl, S. Kudera, T. Pellegrino, A. M. Javier, H. E. Gaub, S. Stölzle, N. Fertig and W. J. Parak, *Nano Lett.*, 2005, **5**, 331.
59. A. Magrez, S. Kasas, V. Salicio, N. Pasquier, J. W. Seo, M. Celio, S. Catsicas, B. Schwaller and L. Forró, *Nano Lett.*, 2006, **6**, 1121.

CHAPTER 14

Janus Nanoparticles by Interfacial Engineering

YANG SONG,[a] XIAOJUN LIU[b] AND SHAOWEI CHEN*[a,b]

[a] Department of Chemistry and Biochemistry, University of California, 1156 High Street, Santa Cruz, California 95064, USA; [b] New Energy Research Center, South China University of Technology, Guangzhou 510006, China
*Email: shaowei@ucsc.edu

14.1   Introduction

The tremendous development in the design and production of isotropic nanomaterials has provided access to nanoparticles with pre-selected composition, shape, size and functionality. These nanostructured materials exhibit diverse applications in optics, magnetics, plasmonics, colloidal chemistry, and biomedicines due to their unique material properties. Janus nanostructures (named after the two-faced Roman god, Janus) represent a new class of functional nanomaterials with asymmetric composition and/or surface structures, and pave a way to new types of materials by a deliberate variation of the composition and functionalities. Different properties at opposite sides enable these particles to mimic the behavior of conventional surfactant molecules, leading to completely new self-assembled structures, with many more possibilities than those found in isotropic particles. Such possibilities rely on a strict control over the balanced forces involved in colloidal stability, so as to form clusters with predefined size, shape and properties.[1]

RSC Smart Materials No. 7
Functional Nanometer-Sized Clusters of Transition Metals: Synthesis, Properties and Applications
Edited by Wei Chen and Shaowei Chen
© The Royal Society of Chemistry 2014
Published by the Royal Society of Chemistry, www.rsc.org

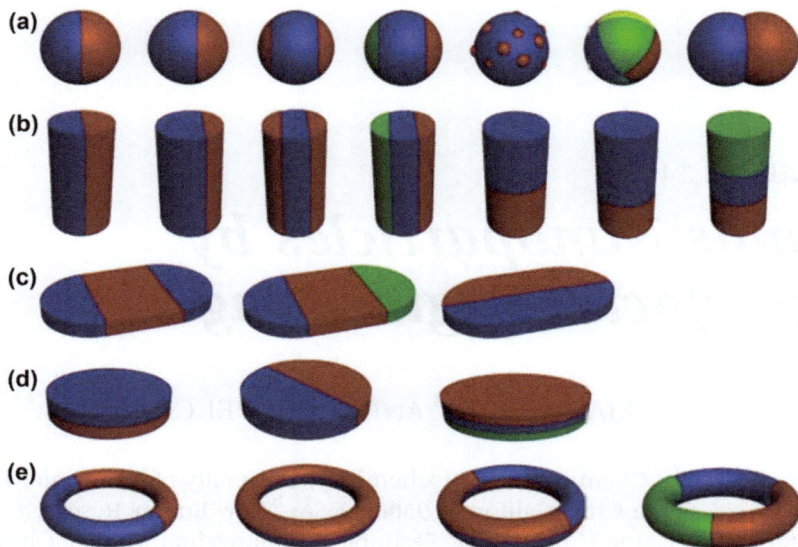

Figure 14.1 Schematic illustration of the (a) zero-, (b) one- and (c, d, e) two-dimensional Janus particles. Different colours represent different components.
Reprinted with permission from ref. 3b. Copyright (2012) The Royal Society of Chemistry.

Depending on their shape, Janus particles can be divided into Janus spheres, rods, cylinders, and sheets (Figure 14.1). Because of their amphiphilic nature like proteins in the shape of "amphiphilic helices", the formation of structures of higher hierarchical orders is expected, leading to novel properties in solution, at interfaces, or in the bulk. Amphiphilic copolymers as a member of Janus structured materials with nonlinear and noncentrosymmetric architectures have been attracting much interest due to their unprecedented properties.[2] In addition, the presence of metallic components in Janus nanostructures can completely change the final properties of the nanostructures and thus expand their potential applications.[3a]

In this chapter, the state of the art in the field of Janus nanostructured materials is summarized with a broad introduction on the general synthetic procedures, characterization methods and properties. We have divided this chapter into polymer-based Janus structures and metallic heterostructures.

14.2 Polymer-Based Janus Structures

14.2.1 Copolymerization

Janus nanostructures are first exemplified by polymeric materials. For instance, the Müller group reported the first experimental results of Janus structured copolymers in 2001, in which cross-linked Janus polymer nanoparticles[4] were prepared by the self-assembly of terpolymers. Their pioneering work took advantage of the wide variety of complex morphologies with a high degree of spatial control[5] that can be obtained spontaneously by the self-organization of terpolymers during film casting, depending on the chemical nature and molecular weights of the different blocks. The first examples consisted of lamellae-sphere phases formed by polystyrene-*b*-polybutadiene-*b*-poly(methyl methacrylate) [PS-*b*-PB-*b*-PMMA] terpolymers with very narrow molecular weight distributions (obtained *via* anionic polymerization), where PS and PMMA formed alternating lamellae, while PB formed small spheres located in between the lamellae.[4a] By changing the molecular weight of the blocks, different morphologies of the three phases were obtained, such as lamellae-cylinders,[6] and more recently, a fully lamellar structure.[6b] The intermediate block was always chosen to be PB, located in the spheres for the lamellae-sphere case, in the cylinders in the case of the lamellae-cylinders, and in one of the three lamellae in the fully lamellar morphology, because PB blocks belonging to different chains located in the same sphere cylinder, or lamella can be cross-linked either *via* vulcanization, or using azobisisobutyronitrile (AIBN) with a polythiol.[7] Upon dissolution of the films by means of a suitable solvent, Janus cross-linked micelles with a spherical, cylindrical or disk-like shape were recovered, where their shape depended upon the morphology of the PB domain, and with two distinct faces consisting of PMMA and PS. This group also reported the preparation of amphiphilic Janus spherical micelles, using the same strategy, but followed by alkaline hydrolysis of the PMMA ester groups, leading to negatively charged hydrophilic polymethacrylic acid groups.[8]

Self-assembly of diblock or triblock copolymers has also been used by other groups for the preparation of Janus nanoparticles. The preparation begins in most cases with the synthesis of the copolymers in a common solvent, and upon a change of the solvent, self-assembly is induced. For instance, Sfika *et al.*[9] prepared poly(2-vinylpyridine)-*b*-poly(methyl methacrylate)-*b*-poly(acrylic acid) [P2VP-*b*-PMMA-*b*-PAA] Janus micelles in water at low pH, with PMMA in the core and the two hydrophilic blocks phase-separated on the surface. Voets *et al.*[10] mixed two different diblock copolymers, poly(2-methylvinylpyridinium iodide)-*b*-poly(ethylene oxide) [P2MVP-*b*-PEO] and poly(acrylic acid)-*b*-poly(acryl amide) [PAA-*b*-PAAm], which self-assembled because of electrostatic interactions between the negatively charged PAA blocks and the positively charged P2MVP blocks, while the PEO and PAAm blocks phase-separated on the surface of the disk-like micelles. Ma *et al.*[11] prepared polymeric micelles by mixing P2VP-*b*-PEO and poly(2-vinylpyridine)-*b*-poly(*N*-isopropylacrylamide) [P2VP-*b*-PNIPAm] diblock copolymers, which self-assembled upon the addition of divalent sulfate ions that physically cross-linked the positively charged P2VP blocks, while the PEO and PNIPAm blocks phase-separated on the surface. Cheng *et al.* prepared polymeric Janus nanoparticles[12] both from PS-*b*-P2VP-*b*-PEO triblock copolymers, with chemical cross-linking of the P2VP blocks *via* 1,4-dibromobutane, and from mixtures of two diblock copolymers, PEO-*b*-PAA and

PAA-*b*-P2VP, followed by non-covalent cross-linking of the PAA blocks using 1,2-propanediamine.[12b] Wurm *et al.*[13] produced PEO-*b*-PPO diblock co-polymers with a ruthenium end-group *via* ROMP and demonstrated their assembly into Janus micelles with the ruthenium groups exposed on the surface. Du and Armes[14] have recently prepared Janus nanoparticles by exploiting the self-assembly of the triblock copolymer poly(ethylene oxide)-*b*-poly(ε-caprolactone)-*b*-poly(2-aminoethyl methacrylate) [PEO-*b*-PCL-*b*-PAMA] in water, driven by the hydrophobic PCL blocks, with PEO and PAMA hemispheres in the corona. The addition of tetraethoxysilane (TEOS) leads to the formation of a silica core thanks to the catalytic activity of the basic PAMA groups, which phase-separated from PCL. Li *et al.*[15] assembled PEO-*b*-P2VP copolymers into Janus micelles *via* protonation–deprotonation in THF. Upon treatment with chloroauric acid, Janus nanoparticles with segregated gold nanocrystals were also prepared.

Zubarev *et al.*[16] prepared 2 nm amphiphilic gold Janus nanoparticles by functionalizing them with a PS-*b*-PEO block copolymer with a binding group for the gold nanoparticle surface sandwiched between the two blocks. The small size of the nanoparticles allowed the binding of only a few copolymer chains on the surface of each nanocrystal, thus leading to a confinement of PEO and PS chains on two separate hemispheres. This led to Janus nano-particles having the capability to self-assemble into well-defined cylindrical micelles in water.

Another type of polymeric Janus structure is cylindrical polymer brushes which refer to a 1D brush possessing densely grafted side chains on a linear polymer main chain. Cylindrical polymer brushes exhibit the unique property of forming extended chain conformations, based on the intramolecular excluded-volume interactions between densely grafted side chains. As shown in Figure 14.2, generally there are three methods for the synthesis of cylin-drical polymer brushes possessing densely grafted side chains covalently bonded to a linear backbone: "grafting through", "grafting onto", and "grafting from".[17] In addition, non-covalent interactions, such as hydrogen bonding, ionic interaction, and coordination bonding, have also been used to graft surfactants onto linear polymer chains to form polymer brush architectures. Schlüter and coworkers[18] have also synthesized amphiphilic copolymer brushes by the polycondensation of Suzuki-type dendronized monomers that form phase-separated hydrophobic/hydrophilic hemi-cylinders with polymerization degree (PD) = 156, due to the incompatibility of hydrophobic PS and hydrophilic PEO side chains. Janus cylinders where the separation plane is parallel to the cylinder axis were reported by Liu *et al.* as shown in Figure 14.3. These particles were obtained in a similar way as the Janus spheres by using a polystyrene-*b*-polybutadiene-*b*-poly(methyl methacrylate) (SBM) block terpolymer forming the lamellae-cylinder morphology. The average diameters of Janus cylinders-SBM-1 and -SBM-2 are 100 ± 10 nm and 70 ± 10 nm, respectively. These diameters are about 2.5 times larger than in the bulk state but 3–4 times smaller than the contour lengths of the PS and PMMA chains.

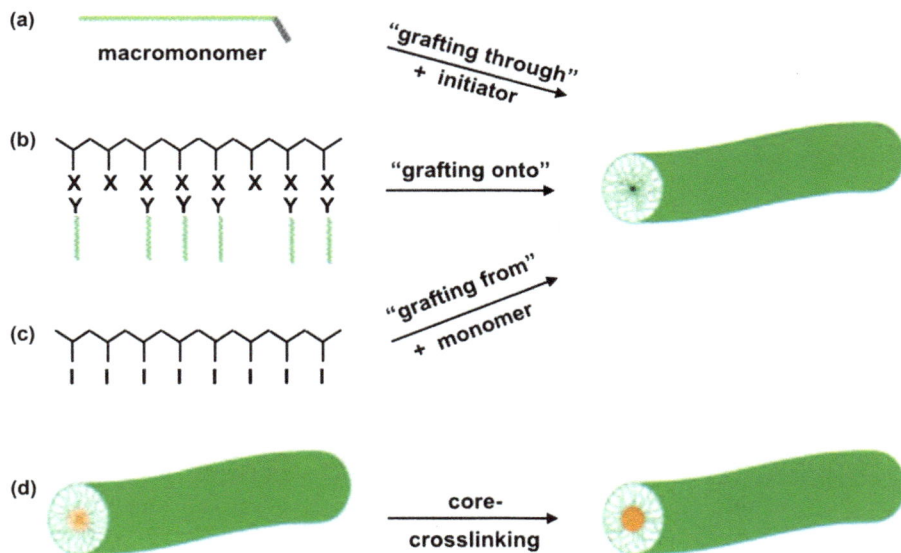

Figure 14.2   Strategies for the synthesis of cylindrical polymer brushes: (a) "grafting through", (b) "grafting onto" (X and Y are functional groups capable of coupling), (c) "grafting from" (I is an initiating group), and (d) core cross-linking of cylindrical micelles (or cross-linking of cylindrical microdomains of microphase-separated block copolymer in bulk).
Reprinted by permission from ref. 17. Copyright (2005) John Wiley and Sons.

14.2.2 Electrospinning

A pioneering study by Roh *et al.*[19] shows how biphasic Janus particles with nanoscale anisotropy can be prepared through electrodynamic jetting of two liquid solutions of immiscible polymers. The shape of the obtained particles was spherical in some cases, and snowman-like in others. Several types of polymer pairs could be used with this method, and particles were created where different dyes were encapsulated in the two polymers to prove their segregation. A schematic of the process and a confocal picture of dye functionalized particles are shown in Figure 14.4. In a typical experiment, polymer solutions of 0.5% poly(acrylic acid) (PAA) and 4.5% poly(acrylamide-coacrylic acid, sodium salts) (PAAm-co-AA) were prepared in double distilled water (ddH$_2$O). To enable monitoring of the biphasic nanocolloids by fluorescence, FITC (fluorescein isocyanate)-conjugated dextran (MW 250 kDa) or rhodamine B-conjugated dextran (MW 70 kDa), or Cascade Blue-conjugated dextran (MW 10 kDa) were added at a final concentration of 0.3%. Following electrified jetting, the biphasic nanocolloids were incubated at 175 °C for 3 h to induce cross-linking by thermal imidization between carboxylic acid and amide groups to form an imide group.

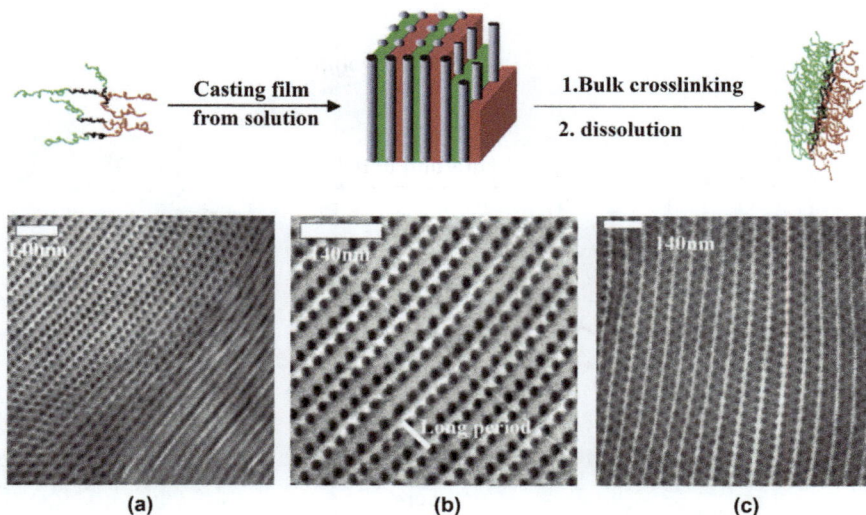

Figure 14.3 (Top) Schematic of the synthesis of Janus cylinders. (Bottom) TEM micrographs of SBM-1 before (a, b) and after (c) cross linking.
Reprinted with permission from ref. 6. Copyright (2003) American Chemical Society.

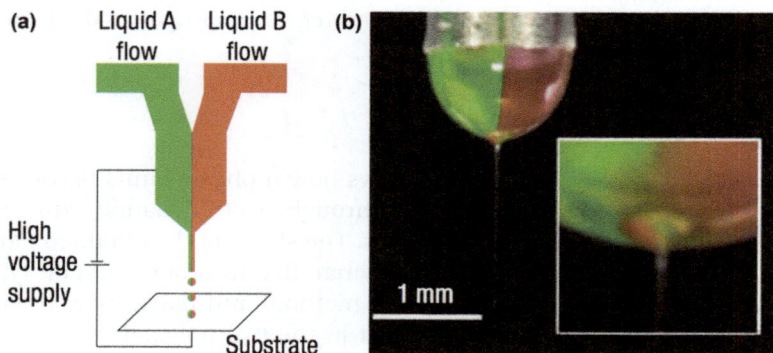

Figure 14.4 (a) A schematic diagram of the experimental setup used for electro-hydrodynamic processing. (b) A digital image of a typical biphasic Taylor cone with jet. PEO $(MW = 600\,000 \text{ g mol}^{-1})$ dissolved in distilled water (2% w/v) was used in both jetting fluids. Each phase was labelled by the addition of 0.5% w/v of a fluorescent dye, that is, (fluorescein isothiocyanate)-conjugated dextran (green) and rhodamine B-conjugated dextran (red). The inset shows a detailed image of the swirl-like jet ejection point. The formation of the vortex was reproducible and did not seem to interfere with the actual synthesis of the biphasic particles.
Reprinted with permission from ref. 19. Copyright (2005) Nature Publishing Group.

The gross morphology of the cells was not greatly affected by the presence of the particles. Slight aggregation occurred at high particle concentrations due to residual interactions of carboxylic acid groups with serum proteins. Note that the biphasic nanocolloids provide simultaneous access to two compartments that can be independently loaded with different therapeutics as well as imaging moieties, such as dyes or magnetic nanomaterials, and at the same time, can be selectively modified to control orientation relative to the cell surface. This property may be exploited for unique imaging capabilities, as manifested in confocal laser scanning microscopy analysis of biphasic nanocolloid association to cells. Furthermore, anisotropic nanoparticles may provide further functions, such as self-assembly into three-dimensional tissue constructs or the potential to mimic extracellular matrices with subcellular organization and order, which will be very difficult to achieve with more conventional, isotropic nanoparticles

Additionally, a model has been built to predict the shape and size of the final particles. The same group prepared[20] Janus sub-micrometer particles *via* electrohydrodynamic co-jetting of two almost identical polymer solutions in water, one containing a dye and the other gold nanoparticles, followed by thermal cross-linking, as well as Janus particles with two different nanocrystals (TiO_2 and Fe_3O_4) segregated in the two hemispheres.[21] In a further study, Janus nanoparticles were prepared with spatially controlled affinity to cells by co-jetting two different solutions of PAAm-*b*-PAA and functionalizing one of them with biotine moieties.[22]

14.2.3 Polymer–Inorganic Heterodimers

Reculusa *et al.*[23] prepared heterodimers of silica and polystyrene particles by hydrophobically modifying 170 nm silica particles with a polymerizable moiety, and then performing seeded polymerization of polystyrene. At low surface density of polymerizable groups heterodimers were formed. Perro *et al.*[24] followed the same recipe to prepare heterodimers, and subsequently functionalized the exposed surfaces of the silica particles with methyl groups. Subsequent addition of surfactants combined with sonication resulted in the separation of the silica from the polystyrene. The previously protected portion of the silica particle surfaces was further functionalized with amino groups, and the addition of negatively charged gold nanoparticles resulted in silica particles decorated with gold crystals bound only on the amino-functionalized side. Along the same line, Ge *et al.*[25] prepared asymmetric nanoparticles, which in special cases are real Janus nanoparticles. The procedure relied on polymerizing, in the presence of styrene, silica-coated magnetic nanoparticles with surfaces functionalized by a polymerizable group. Polymerization gave rise to an asymmetric particle, with the silica-coated magnetic particle eccentrically positioned in the polystyrene matrix, as a result of the contraction of the polystyrene caused by the interfacial tension with silica. The extent of this eccentricity was reduced by the addition of a cross-linker, or enhanced by the addition of more monomers. Under certain conditions, full phase separation occurred, leading to the formation of Janus nanoparticles. Lu *et al.*[26] produced silica–PS heterodimers starting from a miniemulsion of styrene and two silicon precursors (one of them was polymerizable), which were first polymerized to produce a polystyrene core, then treated with ammonia to induce nucleation of a silica nodule connected to the polystyrene. By using an analogous method, but a higher percentage of silicon precursors, Zhang *et al.*[27] obtained Janus PS–silica nanoparticles with one hemisphere made of PS and the other of silica. Mushroom-type particles made of PS and silica have been prepared by Zhang *et al.*,[28] who prepared silica–PS heterodimers with a procedure similar to that introduced by Reculusa *et al.*[23] Subsequently, a selective etching of the silica using HF was used to tune the morphology of the heterodimer and create a fresh and non-functionalized silica surface on the heterodimer. Teo *et al.*[29] prepared magnetite–PS–silica Janus nanoparticles by miniemulsifying magnetite, TEOS, and styrene, and initiating the polymerization *via* ultrasonication, obtaining a phase separation between TEOS and PS. Upon addition of ammonia during the sonication, magnetic PS–silica heterodimers were obtained. Zhang *et al.*[30] synthesized magnetite silica heterodimers using magnetic colloids prepared by using a high temperature hydrolysis reaction in ethylene glycol, followed by addition of ammonia and a cationic surfactant, and finally emulsification by a variable amount of TEOS. The hydrolysis of TEOS led to the formation of heterodimers with a tunable aspect ratio, which increased with the ratio of TEOS to magnetite. Magnetic polystyrene Janus nanoparticles have also been reported by the Elaissari group,[31] who swelled ferrofluid oil nanodroplets with styrene monomers, the polymerization of which induced phase separation of magnetite from the polymer and the creation of biphasic particles. Two other studies using dispersion polymerization techniques to prepare heterodimers are worth mentioning. Ohnuma *et al.*[32] modified conventional precipitation polymerization of polystyrene by adding gold nanocrystals a few min after the beginning of the polymerization process, leading to the production of hybrid particles with only one gold nanocrystal protruding out of the surface of each polystyrene particle. The mechanism of formation was investigated, revealing that the timing of nanocrystal addition was crucial to obtain heterodimers. Similarly, Feyen *et al.*[33] prepared PS-magnetic heterodimers by an analogous procedure, but using hydrophobic magnetite (or maghemite nanocrystals). In a second step, silica was nucleated from the exposed iron oxide nanocrystals, leading to a mushroom-like particle, with a tunable silica lobe size. The removal of the iron oxide core using hydrochloric acid resulted in a hollow structure with a cavity size close to the dimension of the iron oxide nanocrystals.

14.3   Metal Nanocrystals

14.3.1   Solid Masks

Masking is another technique for the preparation of Janus nanoparticles. This means that only one hemisphere (or in general a portion) of their surface is exposed to a secondary reaction environment for asymmetric functionalization, while the rest of the surface is protected. Typically, masking has been achieved by either trapping the nanoparticles at the interface between two immiscible phases, or by depositing them or adsorbing them on a solid surface. Masking is a flexible technique used to prepare Janus nanoparticles, since it is applicable to virtually any type of material, and offers the possibility of modifying the surfaces of nanoparticles with a very wide variety of functional groups. However, masking suffers greatly from scalability as the functionalization of particles can only be achieved after deposition of the particles on flat solid substrates. In addition, the use of dispersed phases of particles and droplets to provide the surfaces upon which nanoparticles and nanocrystals can be trapped is a viable alternative to obtain larger quantities of Janus nanoparticles, but does limit the type of functionalization that can be performed.

The first pioneering work was carried out by Takei and Shimizu,[34] who deposited gold on micrometer-sized latex spheres leading to the production of structurally asymmetrical particles. The technique was also adopted to functionalize sub-micrometer particles by the Whitesides group,[35] who prepared silica particles with partial metal (Au, Pd, Ti, or Pt) coating by dropcasting silica particle suspensions onto a glass surface and then coating with metal particles. Subsequently, the semispherical metallic shells were obtained by etching away the silica part. Meanwhile, the metallic shells could be further functionalized with alkanethiols. The asymmetrical gold-coated silica particles could also be annealed at elevated temperatures (*e.g.*, 700 °C) for a few hours such that the gold layer on a particle surface, which initially formed a hemisphere, beaded up into a microcrystal by a dewetting process.[36] A similar protocol was used to prepare magnetic polystyrene beads with partial gold coating. The polystyrene part was then annealed and pyrolyzed into carbon, creating heterodimers consisting of a magnetic carbon bead and a gold crystal. More recently, nanosized Janus particles were prepared by Valadares *et al.*,[37] using the technique proposed by the Whitesides group, by first depositing a monolayer of silica particles on an inert substrate, followed by deposition of a thick layer of platinum on the exposed surfaces of the particles, which were then annealed at high temperature.

Janus particles have also been prepared based on the protection–deprotection mechanism. For instance, the Velegol group[38] used poly(allylamine hydrochloride) (PAH) coated silica particles with various diameters (1.0–3.0 μm), which were positively charged and electrostatically adsorbed on a glass substrate in order to protect a portion of their surface. Subsequently, negatively charged sulfate latex nanoparticles were added. Janus nanoparticles were formed as the exposed silica particle surfaces were decorated with latex nanoparticles as shown in Figure 14.5. Multiple layers of latex nanoparticles could be achieved by reversing the charge of the Janus nanoparticles through the addition of PAH, followed by the addition of more negatively charged particles. In another study, Sardar and Shumaker-Parry[39] prepared Janus gold nanoparticles by depositing them onto a silanized glass

(a)

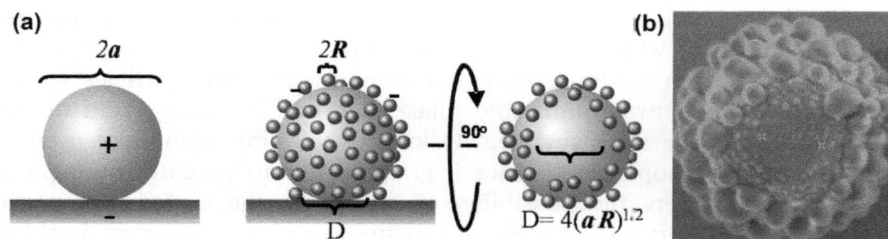

(b)

$$D = 4(aR)^{1.2}$$

Figure 14.5 (a) Schematic of patterning colloidal particles with particle litho-
graphy. The size of the lithographed region is primarily a function
of the core particle radius (a) and the coating radius (R), and simple
geometry shows that the diameter (D) of the lithographed region
should be $D = 4(aR)^{1/2}$. (b) FESEM image of a silica particle with a
lithographed region.
Reprinted with permission from ref. 38. Copyright (2007) American
Chemical Society.

surface with an amino moiety that bonded to the gold nanoparticles by replacing the initial protecting citrate groups; and Janus nanoparticles were
obtained by functionalizing the exposed part of the nanoparticle surface
with 11-mercapto-1-undecanol and removing from the glass surface by
sonication in the presence of 2-mercaptoethylamine to protect the part of the
particle surface that was previously blocked by the glass substrate. A double
stamping procedure has also been employed for Janus particle preparation
as reported by Jiang et al.[40] A trivalent silica particle monolayer was deposited on a hard substrate, followed by functionalization of their free surfaces with a dye using a PDMS stamp. The stamp was used to capture up to
80% of the deposited particles, removing them from the hard substrate. The
other side can be functionalized by a different dye with the same procedure.
Drop-casting a monolayer on a hard substrate is another way to partially
mask nanoparticles. Wu et al.[41] reported the preparation of Janus nanocorals by first producing a monolayer of hexagonal polystyrene particles on a
glass surface by a drop-cast method. Oxygen plasma treatment was then
used to etch the exposed surface of the particles and induced controlled
shrinkage of these surfaces. Finally, gold was deposited on the etched surfaces, leading to the formation of Janus particles.

Janus structured nanoparticles have also been prepared by other methods.
A novel masking technique has been developed by Ho et al.[42] where silica
particles were electrostatically adsorbed on PMMA-co-P4VP fibers *via* electrospinning, and then trapped inside the polymer matrix at increasing
temperatures (Figure 14.6). The depth of the penetration of the particles into
the polymer fibers increased as the temperature approached the glass
transition temperature of the polymer. The exposed surfaces of the particles
were then silanized, and gold colloids were asymmetrically adsorbed on
them. Lin et al. continued the work by creating ternary functionalized particles.[43] After a first treatment identical to that used by Ho et al.,[42] during

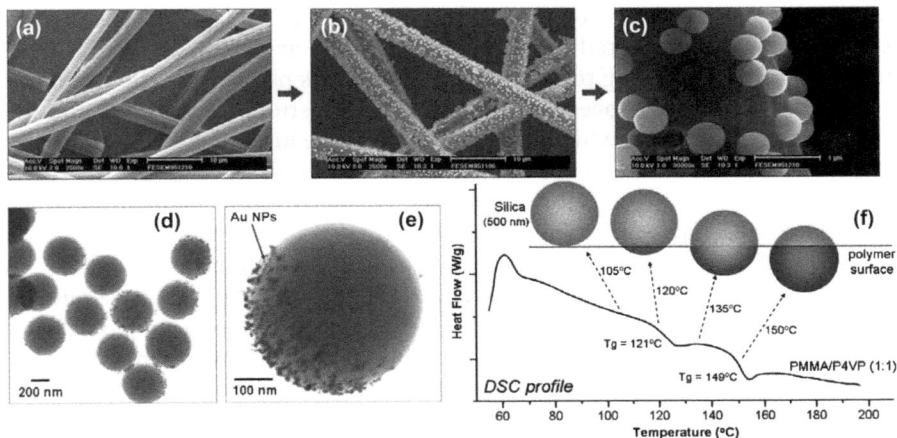

Figure 14.6 SEM image of electrospun fibers (a) before and (b, c) after heat treatment. (d, e) TEM images of gold nanoparticle coated silica Janus particles. (f) Schematic of penetration of the particle in polymer fiber at various temperatures.
Reprinted with permission from ref. 42. Copyright (2008) American Chemical Society.

which particles were trapped in the polymer matrix to a depth of about one third of their size, the penetration of gold-coated silica particles was increased by another third of their size, and finally the exposed surface etched, in order to remove the gold nanocolloids. The resulting particles had only a belt of gold colloids attached to them. A similar approach was used by Skirtach *et al.*[44] who deposited gold nanoparticles on a polymeric surface, and used laser illumination of the nanoparticles to increase the temperature and emboss them into the polymer matrix, to an extent that depended on the intensity of illumination and on the glass transition temperature of the polymeric substrate. McConnell *et al.*[45] instead used amino functionalized silica particles as small as 100 nm and partially embedded them into a polymeric film made of random PS-co-PAA copolymer *via* amidation chemistry. The penetration inside the film was controlled by the reaction time, and the exposed part of the particles was functionalized by depositing gold or Pt nanocolloids.

In comparison to partial masking of particles by hard substrates as mentioned above, trapping the particles at interphase is another effective protocol to limit functionalization on only one hemisphere (or partial surface) of the particles. It has been reported that the self-assembly of mixed ligands could be achieved in different fashions at the surface of metallic nanoparticles,[46] leading in certain cases to the formation of asymmetrically structured Janus particles.[47,48] A recent example is the formation of Au amphiphilic particles.[49] Andala and co-workers used 9.2 nm Au spheres stabilized with dodecylamine (DDA) dispersed in a toluene–water mixture, to which the same amount of hydrophobic dodecanethiol (DDT) and hydrophilic, fully deprotonated mercaptoundecanoic acid (MUA) were simultaneously added, leading to the formation of Janus particles at the interface. The thermodynamic aspects of the process were studied by adding the ligands in different order, which resulted in the same kind of particles. Contact angle measurements of water droplets at functionalized silicon substrates were carried out to confirm the asymmetric distribution of ligands at the nanoparticle surface. A similar approach was used by Sashuk *et al.* to prepare positively charged gold and negatively charged silver Janus particles.[50] Au nanospheres (7.7 nm) were functionalized with *N,N,N*-trimethyl(11-mercaptoundecyl)ammonium chloride (TMA) and 1-undecanethiol (UDT), which are positively charged and neutral, respectively; and Ag nanospheres (6.2 nm) were modified with negatively charged MUA and neutral UDT. When the particles with mixed ligands in appropriate proportion were spread at the water–air interface using a Langmuir–Blodgett (LB) trough, ligands self-assembled at the surface of the metal cores can spontaneously turn into Janus particles due to repulsion between charged and neutral thiolate ligands and adsorb at the air–water interface. Such a monolayer can be compressed to form a close-packed hexagonal structure, and then easily transferred onto a solid substrate with the Langmuir–Blodgett technique.

Other groups have also prepared Janus particles using Langmuir-based techniques. For instance, the Chen group prepared Au Janus nanoparticles by spreading mercapto-capped gold nanoparticles on the water surface of an LB trough. Compression of the monolayer limited the rotation and movement of the nanoparticles. Injection of water-soluble thiol ligands into the water sub-phase led to the formation of asymmetrically modified particles, as shown in Figure 14.7.[51] A modification of this method was also reported which comprised deposition of a monolayer of the mercapto-capped Au nanoparticles on a glass slide using the LB technique, followed by immersion of the glass slide into a solution of a different thiol, where interfacial ligand exchange reactions occurred.[52] The amphiphilic structure was confirmed by contact angle measurements of nanoparticle monolayers deposited onto a glass surface, adhesion force measurements of individual nanoparticles, and NOESY NMR measurements of the spin interactions between protons of neighboring ligands on the nanoparticle surface.[45,51–53]

14.3.2 Soft Masks

Janus particles have also been prepared by partial masking of the particle surface by surface ligands. For instance, the wettability of calcite cubes can be tuned by the size of polystyrene nanoparticles attached to their surfaces.[54] Calcite cubes were prepared by adding calcium nitrate into a sodium carbonate solution under agitation, and the mixture was left at ambient temperature. Then, the purified calcite cubes were lightly silanized by reacting with a limited amount of vinyltriethoxysilane in ethanol. The polystyrene

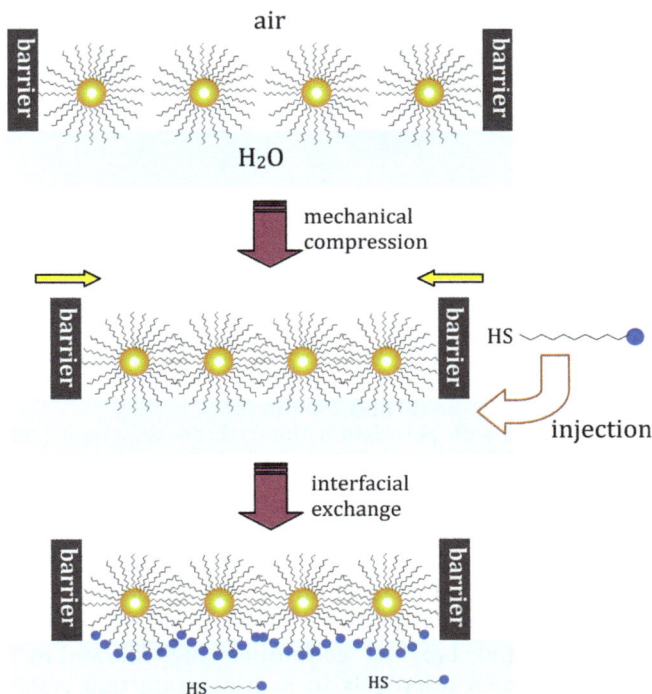

Figure 14.7 Schematic of the preparation of Janus nanoparticles based on the Langmuir technique.
Reprinted with permission from ref. 51. Copyright (2007) John Wiley and Sons.

nanoparticles were discrete and were selectively located at edges and defects where relatively dense vinyltriethoxysilane molecules were coupled.

The Stellacci group demonstrated a unique way towards Janus structured nanoparticles based on the hairy-ball theorem that on a sphere covered in fur, it is not possible to brush all the hairs flat without creating two diametrically opposed singularities, known as poles.[55a–c] In other words, the formation of a two-dimensional ordering is topologically possible only if two "poles" or defects are present at opposite ends of the nanoparticle. Therefore by replacing the thiolate ligands at the poles of gold nanoparticles with longer-chain carboxylic acid-terminated thiol molecules, the nanoparticles might be rendered to behave as divalent molecules that could be bound to other molecules or particles in two distinct directions as molecular building blocks.

In another study, Janus structured Au nanoparticle-conjugated poly(*N*-isopropylacrylamide-co-(acrylic acid)) (pNA) microgels were prepared.[56] The pNA microgels were adsorbed onto oil–water interfaces and formed stable Pickering emulsions, which allowed further introduction of separated functional groups into the microgels if the functionalization occurred on

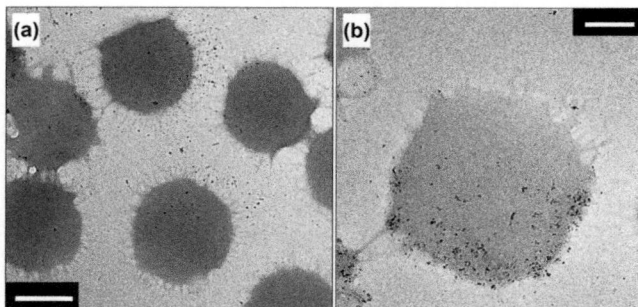

Figure 14.8   Field emission TEM images of Au nanoparticle-conjugated pNA-EDh
microgels on a carbon-coated copper grid dried at 25 °C. The scale
bars are 500 nm (a) and 200 nm (b).
Reprinted with permission from ref. 56. Copyright (2007) American
Chemical Society.

only one side (the side immersed in oil or the water phase) of the microgels.
In this study, one hemisphere of pNA microgel was functionalized by a
simple carbodiimide coupling reaction in the oil phase of the Pickering
emulsions. The asymmetric distribution of the amino groups was confirmed
using the Au nanoparticle labeling technique, which showed in TEM studies
of AuNP-conjugated pNA microgels in a dried state that AuNPs could be
identified only in a particular area of a microgel (Figure 14.8).

14.3.3   Controlled Phase Separation and Growth

Preparation of non-spherical Janus particles has also become an attractive
playground for the design of novel nanoparticle systems. The widely
employed seeded growth method is usually the basis for obtaining non-
spherical Janus particles. Deka *et al.*[57] described the preparation of semi-
conductor-magnetic Janus (CdSe/CdS–Co) nanorods by decomposition of
$Co_2(CO)_8$ under an inert atmosphere at high temperatures to form Janus
structured CdSe@CdS nanorods, which acted as seeds for Co growth. Janus
structured CdSe@CdS nanorods with an aspect ratio of ~5 were prepared by
injecting CdSe nanocrystals into a CdO solution in tri-*n*-octylphosphine
oxide (TOPO) at 350 °C under a nitrogen atmosphere. The growth mechanism was found to involve the nucleation of Co at the edge opposite to
where CdSe seeds were located in the heterodimers, followed by selective
autocatalytic growth. They claimed that the driving force for this directional
growth was due to the semiconductor tips, therefore the hexagonal close-
packed (hcp) structure of Co was obtained in the Janus nanorods rather than
the cubic ε-phase obtained under the same conditions without seeds. It
should be noted that only the matchstick structure was obtained, with no
other intermediate particles that could evolve to the final geometry by
ripening processes. Chakrabortty *et al.* reported the growth of Au particles on

one tip of preformed CdSe@CdS nanorods.[58] Deposition of Au could occur at the tips of the rod or at the location of the CdSe seed by an electrochemical Ostwald ripening process. Under UV excitation, large Au domains may be exclusively deposited at one end of the seeded CdSe@CdS rods because of electron migration to one of the metal tips; and under ambient light conditions, the deposition of Au along the nanorod was strongly influenced by the temperature-dependent, ligand-mediated defect sites. CdS rods were grown on CdSe spherical seeds and then the CdSe@CdS nanoparticles were dispersed in a mixture of toluene and *n*-octadecylphosphonic acid, which provided anchor sites for Au particle growth by the injection of $HAuCl_4$, tetraoctylammonium bromide and dodecylamine in toluene. The growth location of Au particles could be at either one or both tips, or at rod sides that varied with the amount of gold added. This was related to the discrepancy in the reactivity of the different facets. With a sufficiently low amount of Au precursors, growth occurred only at the rod tip that was farther away from the CdSe seed, hence forming Janus structured nanorods. The role of the protecting ligands was studied in a related work by Maynadié *et al.*,[59] in which Co grew over the tips of preformed CdSe nanorods. When the concentration of the ligands increased, a decrease in the number of Co-tipped CdSe rods was observed. When both Co and ligand concentrations were reduced relative to the concentration of CdSe rods, Co particles grew with a very inhomogeneous size distribution. In this case, several Co particles grown on the lateral sides of the CdSe rods were observed. By varying the reaction temperature, they were also able to grow a full Co rod over the CdSe rod seeds.

In addition, masking methods have also been applied for preparation of Janus structured nanorods. An example of asymmetric nanostructures involving Co nanospheres grown on iron oxide tetrapods was reported by Casavola *et al.*[60] Noncoordinating 1-octadecene (ODE) capped iron oxide nanocrystal tetrapods (TPs) acted as foreign seed substrates, onto which heterogeneous nucleation and growth of Co components was selectively accomplished at the tips of iron oxide TPs by thermal decomposition of a $Co_2(CO)_8$ precursor in appropriate proportions. Experimental proofs indicated that the hybridized nanocrystals indeed originated through a temporally limited burst of heterogeneous Co nucleation onto the seeds, which was followed by the progressive enlargement of the initially formed metal nuclei over time, proportionally to the amount of organometallic precursors supplied to the reaction vessel.

SiO_2–Au matchsticks have been synthesized in a similar fashion. This entails the synthesis of silica nanorods followed by alignment along their long axis over a substrate, assisted by an alternating current (AC) electric field.[61] In a third step, titanium and gold layers were evaporated over the rods. The end-functionalized region was protected with a PDMS stamp and a gold etching process was carried out to mitigate the presence of imperfections. After removing the stamp, the matchsticks were released by ultrasonication (Figure 14.9). In another study, He *et al.* prepared lollipop-,

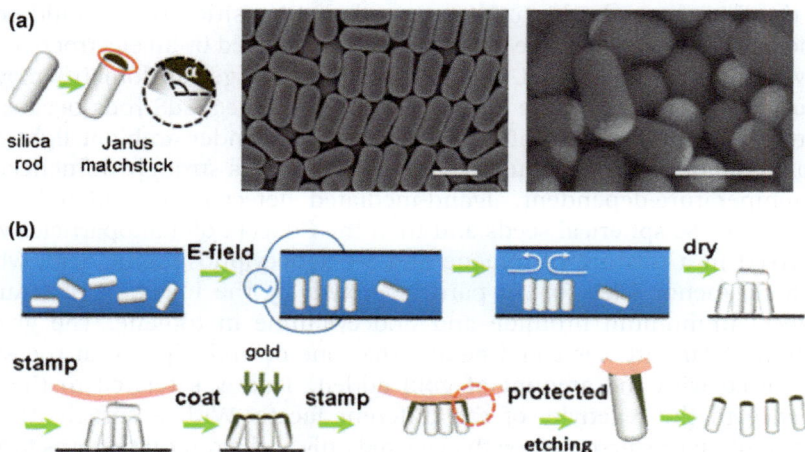

Figure 14.9   (a) Schematic (left) and SEM images of non-functionalized (middle) and gold-tip-coated (right) silica rods ($L/D = 2.1$). The patch angle (α) was approximated assuming a spherical cap. Scale bars are 2 μm. (b) Schematic view of fabrication pathway for transforming silica rods to Janus matchsticks. First, the rods were aligned in an AC electric field and dried (top). Next, misaligned rods were selectively removed using a PDMS stamp (bottom, left). The remaining rods were successively coated with a titanium adhesion layer (2 nm) and a gold layer (25 nm) (bottom, middle). Finally, the rods were picked up using a PDMS stamp, which was used to protect their tips during selective gold etching.
Reprinted with permission from ref. 61. Copyright (2012) American Chemical Society.

dumbbell- and frog egg-like PANI–Au nanoparticles by using interfacial reactions.[62] In brief, organic monomers (*e.g.*, aniline) were dissolved in an organic solvent (*e.g.*, hexane or toluene), while inorganic precursors (*e.g.*, HAuCl$_4$) were dissolved in water at a predetermined concentration. The immiscible solutions were gently placed in a glass vial to form a water–oil interface and incubated at various temperatures overnight. These precursors reacted at the interface for the polymerization of aniline to generate poly-aniline (PANi) NPs and for the reduction of HAuCl$_4$ to form gold nano-particles (AuNPs). The protocol was applied to prepare eccentric–(concentric-Au@Ag)@PANI nanoparticles with gold nanorods by Xing *et al.*[63] where citrate-protected Au colloids were added into an organic phase with aniline and AgNO$_3$ in water phase instead of HAuCl$_4$. Choi *et al.* synthesized Cu$_2$S–In$_2$S$_3$ heterostructured materials with acorn, bottle or larva shapes by the decomposition of copper and indium oleates at different temperatures.[64a] The preparation of Cu–Cu$_3$P Janus nanocrystals were reported by Trizio *et al.*[64b] Cu nanocrystals were prepared by controlling the amounts of CuCl and trioctylphosphine, which evolved into Cu$_3$P nanocrystals through Janus Cu–Cu$_3$P intermediates. X-Ray powder diffraction (XRD), high resolution

TEM (HRTEM) and energy filtered TEM (EFTEM) confirmed the presence of the two components in the structure. Cu crystals were formed by annealing Cu_3P nanocrystals at 400–500 °C, where dephosphorization occurred by phosphorus evaporation. Both kinetic (as the inhomogeneous coverage of surfactants), and thermodynamic aspects (as the presence of epitaxial interfaces between Cu and Cu_3P domains that would minimize the energy) explained the formation of Janus structures instead of the expected centrosymmetric particles. Ag_2S–Ag heterodimers and heterotrimers were prepared by Pang *et al.* by the assembly of Ag_2S and Ag with ethylenediamine.[65]

14.3.4 Bimetallic

Bimetallic nanocrystals are often advantageous over single-component systems for a range of fundamental studies and applications because the associated variations in compositions and spatial distributions provide additional parameters for experimentally maneuvering both the structures and properties. For example, a bimetallic system may exhibit either localized surface plasmon resonance (LSPR) and/or catalytic properties markedly different from those of the individual constituent components because of electronic coupling between the two metals. Up till now, a number of noble metals, such as Au, Ag, Pd, Pt, and Rh have been combined to generate bimetallic nanocrystals with tunable and enhanced properties. With regards to their spatial distributions in a bimetallic system, three patterns have been observed and exploited: (1) homogeneously distributed as in an alloy or intermetallic compound; (2) separated into two concentric layers as in a core–shell structure; and (3) separated into two side-by-side regions as in a dimeric structure. For instance, Janus-like AgCo clusters exhibit unique magneto-optical features that could be used for many applications, ranging from the switching on/off of their catalytic properties depending on the orientation of the nanoparticles to plasmonic applications for cancer treatment and integrated circuits. Parsina and Baletto[66a] reported AgCo Janus-like nanostructures that consist of two subunits: one is a Ag cluster, often delimited only by (111) facets, and the other is an icosahedral Co core covered by a single monolayer of silver. Both computational and experimental results showed that cobalt prefers to bunch together and grow around subsurface sites at any concentration. The cobalt unit tries to adopt an icosahedral arrangement, in agreement with the strong tendency of pure cobalt to form an icosahedral shape at small and intermediate sizes. Interestingly, further deposition of cobalt over a Janus-like system results in well-defined core–shell structures with few defects, depending on the growth conditions, whereas the addition of silver over a perfect core–shell leads to the formation of a Janus-like motif.

Bimetallic nanocrystals have also been prepared by seed-mediated growth, where pre-formed seeds of one metal serve as the sites of nucleation and then growth for another metal. It has been observed that the nucleation and growth mode of the second metal are governed by a number of physical parameters, including the (mis)match of lattice structures and constants, the correlation of surface and interface energies, and the difference in electronegativity between the two metals. For two given metals, these parameters would direct the heterogeneous nucleation and growth on a seed to follow a conformal or site-selective mode, generating a core–shell or hybrid structure. Xia *et al.* recently demonstrated the use of a syringe pump as an effective means of manipulating the precursor concentration and thus selectively enhancing the overgrowth of Rh along the corners and edges rather than side faces of a cubic seed.[66b] A similar fashion of control was applied to extend the capability of this approach to achieve nucleation and growth of Ag on one, three, and six of the equivalent (100) faces on a Pd seed. By simply manipulating the rate at which $AgNO_3$ was added, they found that Pd–Ag bimetallic nanocrystals with three distinct structures could be produced under otherwise similar conditions. When the injection rate was very fast, the Ag atoms around a Pd seed could be maintained at a level sufficiently high to facilitate nucleation on all six faces of a Pd seed towards core–shell bimetallic nanocrystals; dimers of eccentric hybrid bars were obtained with Ag grown on three adjacent faces of a Pd seed when $AgNO_3$ was injected at a moderate rate and the concentration of Ag atoms was only high enough to allow nucleation on half the surface of a Pd seed; and Ag–Pd bimetallic Janus nanocrystals were produced when the amount of $AgNO_3$ added might just be sufficient for nucleation and growth on one of the six faces of a Pd seed with even lower injection rate at the early stage. Significantly, these Pd–Ag bimetallic nanocrystals displayed substantially different LSPR properties as compared with those made of pure Pd, Ag, or their alloys. For the bimetallic Pd–Ag Janus nanocrystals, the major LSPR peak associated with the Ag portion was observed in the range of 420–425 nm, accompanied by a shoulder peak around 350 nm. The peak positions were slightly red-shifted as the deposited Ag evolved from thin plates into bars, reflecting the size sensitivity for the LSPR peak of Ag nanostructures.

In another study, hybrid gold–cobalt (Au–Co) nanorods with asymmetric structure Co–Au were synthesized, as shown in Figure 14.10.[67] The results supported the hypothesis that the stabilizing ligands are coordinated preferentially on the sides of the nanorods and that a fast exchange occurs on the surface of the rods, which involves the stabilizing ligands and also probably triphenylphosphine (PPh_3). Moreover, chemical control of both the gold precursors and the ligands allowed (1) control of the gold growth process (heterogeneous nucleation *versus* galvanic displacement) and (2) control of the location of the gold nanoparticle growth (tip or whole body) through kinetic control of the reaction. These results showed the possibility of selective production of gold-tipped ferromagnetic cobalt nanorods. The physical properties of these new nanoobjects were also explored. For instance, preliminary optical measurements showed in both cases (random deposited gold nanoparticles or gold nanoparticles on the tips) apparent absorptions near 550 nm as well as a very weak one near 720 nm which were attributed to the plasmon absorption, whereas magnetic properties of the

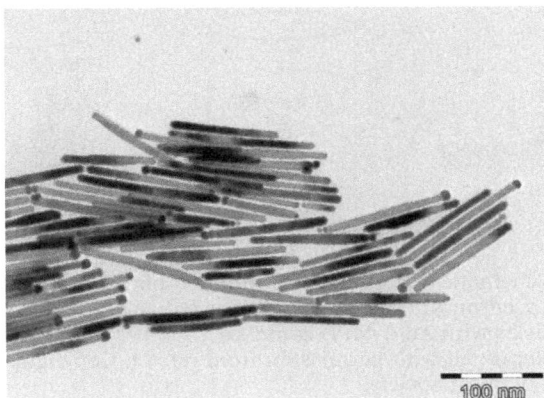

Figure 14.10 TEM image of (a) gold-tipped Co nanorods prepared at 40 °C. [Co] = 1.3×10⁻³ M, [Au] = 1.34×10⁻⁴ M, reaction time 48 h. Reprinted with permission from ref. 67. Copyright (2007) John Wiley and Sons.

new Au–Co nanorods were investigated as the effect of Au presence and location on the magnetic anisotropy of these objects.

Similarly, a green-chemical route has been reported for the growth of Au nanoparticles onto Te nanowires.[68] By controlling the pH of the reaction media, the rate of the redox reaction between Au(III) and the Te nanowires, and thus, the rate of formation of the Au seeds was tuned. In the presence of 10 mM cetyltrimethylammonium bromide (CTAB), Au–Te heterodimers and Au-coated nanowires were obtained at pH 4.0 and 5.0, respectively. Furthermore, photovoltaic studies revealed that the resistance of the Te nanowire-based thin films was controlled by the degree of deposition of the Au nanoparticles. It was suspected that Au–Te and/or Au–Te–Au based thin films hold great potential for use in the fabrication of electronic devices.

Galvanic reactions between metal nanoparticles and a salt of a more noble metal are another way to alter the nanoparticle's constituent metals. Such reactions have, for example, been employed to replace Cu or Ag particles encapsulated in dendrimers by more noble metals (Au, Pt, Pd) and to prepare bi- and trimetallic particles. Galvanic displacement reactions have also been used to prepare the bimetallic sulfides of PbCd and ZnCd,[69] CoPt core–shell bimetallic nanoparticles,[70] AuAg, PdAg, AuPd, and AuCu bimetallic alloy nanoparticles[71] and asymmetric bimetallic nanocrystals such as Pd–Au tadpoles,[72] as well as Au–Ni and Au–(Ni–Pd).[73]

Recently, Song *et al.* took advantage of the Langmuir method as well as galvanic exchange reactions to prepare bimetallic nanosized Ag–Au Janus particles (Figure 14.11).[74] Specifically, a monolayer of 1-hexanethiolate-passivated silver (AgC6) nanoparticles was deposited onto a glass slide surface by the LB method, which was then immersed into a solution of gold(I)-mercaptopropanediol [Au(I)–MPD] complex. Galvanic exchange

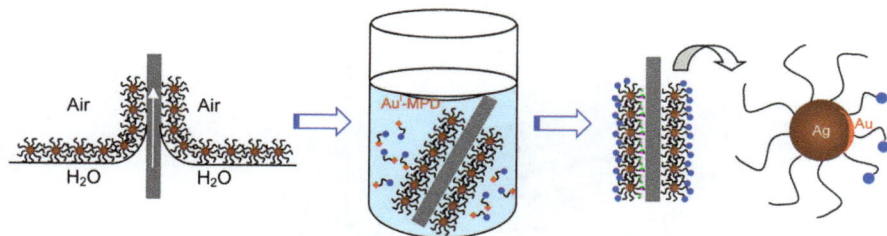

Figure 14.11   Schematic for the preparation of bimetallic Janus nanoparticles
based on interfacial galvanic exchange reactions of AgC6 nanoparticles with Au(ɪ)–MPD complex.
Reprinted with permission from ref. 74. Copyright (2012) American
Chemical Society.

reactions between the AgC6 nanoparticles and Au(ɪ)–MPD led to the displacement of part of the Ag atoms on the nanoparticle cores with Au and
concurrently the replacement of the original hydrophobic hexanethiolates
with the more hydrophilic MPD ligands. Due to the surface ligand inter-
calation between neighboring particles that impeded the accessibility of the
bottom face to the solution ligands the reactions were restricted to the side of
the nanoparticles that was in direct contact with the solution. The asymmetrical chemical structures of the resulting nanoparticles were characterized by a variety of spectroscopic measurements. Interestingly, the bimetallic
Janus nanoparticles exhibited enhanced electrocatalytic activity in oxygen
reduction reactions (ORR), a critical reaction in fuel cell electrochemistry, as
compared with the original monometallic Ag nanoparticles or the bimetallic
nanoparticles that were prepared in a single phase (the so-called bulk-
exchange particles), suggesting that interfacial engineering is key to the
manipulation and optimization of the nanoparticle electronic properties and
hence the interactions with oxygen and ultimately the ORR performance.

Ligand segregation has also been exploited to prepare bimetallic Janus
particles.[75] Chen *et al.* prepared SiO_2–Au–Ag particles by capping 44 nm Au
spheres with 4-mercaptophenylacetic acid and poly(acrylic acid) as the
competitive ligands, followed by silica coating to partially encapsulate the
gold particles, where these Janus particles were used as seeds to reduce silver
nitrate and obtain the final hybrid ternary structure.

14.4   Conclusion

In this chapter a variety of leading synthetic procedures for the preparation
of amphiphilic copolymers and metallic Janus particles are summarized and
highlighted. The primary motivation is to introduce additional structural
variables for the manipulation of the material's properties that may not be
accessible with the structurally symmetrical counterparts. For polymeric
materials, the formation of asymmetrical Janus structures is typically driven
by self-assembly by virtue of the discrepancy of the chemical reactivity of

different polymer blocks, whereas for metal Janus nanoparticles, the asymmetric structures are generally formed by chemical modification/functionalization at an interface. In both cases, the resulting Janus nanostructures exhibit materials properties that are analogous to those of conventional surfactant molecules and may be exploited for novel applications because of the unprecedented control of the arrangements of structural elements.

Acknowledgements

The authors are grateful for financial support from the National Science Foundation and the American Chemical Society – Petroleum Research Funds.

References

1. (a) S. Jiang, Q. Chen, M. Tripathy, E. Luijten, K. S. Schweizer and S. Granick, Janus Particle Synthesis and Assembly, *Adv. Mater.*, 2010, **22**(10), 1060–1071; (b) Q. Chen, S. C. Bae and S. Granick, Directed self-assembly of a colloidal kagome lattice, *Nature*, 2011, **469**(7330), 381–384; (c) Q. Chen, J. K. Whitmer, S. Jiang, S. C. Bae, E. Luijten and S. Granick, Supracolloidal Reaction Kinetics of Janus Spheres, *Science*, 2011, **331**(6014), 199–202.
2. N. Hadjichristidis, M. Pitsikalis, S. Pispas and H. Iatrou, Polymers with Complex Architecture by Living Anionic Polymerization, *Chem. Rev.*, 2001, **101**(12), 3747–3792.
3. (a) D. Rodríguez-Fernández and L. M. Liz-Marzán, Metallic Janus and Patchy Particles, *Part. Part. Syst. Charact.*, 2013, **30**(1), 46–60; (b) S. Yang, F. Guo, B. Kiraly, X. Mao, M. Lu, K. W. Leong and T. J. Huang, Micro-fluidic synthesis of multifunctional Janus particles for biomedical applications, *Lab Chip*, 2012, **12**(12), 2097–2102.
4. (a) R. Erhardt, A. Boker, H. Zettl, H. Kaya, W. Pyckhout-Hintzen, G. Krausch, V. Abetz and A. H. E. Müller, Janus micelles, *Macromolecules*, 2001, **34**(4), 1069–1075; (b) H. Xu, R. Erhardt, V. Abetz, A. H. E. Müller and W. A. Goedel, Janus micelles at the air/water interface, *Langmuir*, 2001, **17**(22), 6787–6793.
5. A. Walther and A. H. E. Müller, Janus particles, *Soft Matter*, 2008, **4**(4), 663–668.
6. (a) Y. F. Liu, V. Abetz and A. H. E. Müller, Janus cylinders, *Macromolecules*, 2003, **36**(21), 7894–7898; (b) A. Walther, M. Drechsler, S. Rosenfeldt, L. Harnau, M. Ballauff, V. Abetz and A. H. E. Müller, Self-Assembly of Janus Cylinders into Hierarchical Superstructures, *J. Am. Chem. Soc.*, 2009, **131**(13), 4720–4728.
7. A. Walther, A. Göldel and A. H. E. Müller, Controlled crosslinking of polybutadiene containing block terpolymer bulk structures: A facile way towards complex and functional nanostructures, *Polymer*, 2008, **49**(15), 3217–3227.

8. R. Erhardt, M. F. Zhang, A. Boker, H. Zettl, C. Abetz, P. Frederik, G. Krausch, V. Abetz and A. H. E. Müller, Amphiphilic Janus micelles with polystyrene and poly(methacrylic acid) hemispheres, *J. Am. Chem. Soc.*, 2003, **125**(11), 3260–3267.

9. V. Sfika, C. Tsitsilianis, A. Kiriy, G. Gorodyska and M. Stamm, pH responsive heteroarm starlike micelles from double hydrophilic ABC terpolymer with ampholitic A and C blocks, *Macromolecules*, 2004, **37**(25), 9551–9560.

10. (a) I. K. Voets, A. de Keizer, P. de Waard, P. M. Frederik, P. H. H. Bomans, H. Schmalz, A. Walther, S. M. King, F. A. M. Leermakers and M. A. C. Stuart, Double-faced micelles from water-soluble polymers, *Angew. Chem., Int. Ed.*, 2006, **45**(40), 6673–6676; (b) I. K. Voets, R. Fokkink, T. Hellweg, S. M. King, P. de Waard, A. de Keizer and M. A. C. Stuart, Spontaneous symmetry breaking: formation of Janus micelles, *Soft Matter*, 2009, **5**(5), 999–1005.

11. R. Ma, B. Wang, Y. Xu, Y. An, W. Zhang, G. Li and L. Shi, Surface phase separation and morphology of stimuli responsive complex micelles, *Macromol. Rapid Commun.*, 2007, **28**(9), 1062–1069.

12. (a) L. Cheng, G. Hou, J. Miao, D. Chen, M. Jiang and L. Zhu, Efficient Synthesis of Unimolecular Polymeric Janus Nanoparticles and Their Unique Self-Assembly Behavior in a Common Solvent, *Macromolecules*, 2008, **41**(21), 8159–8166; (b) L. Cheng, G. Zhang, L. Zhu, D. Chen and M. Jiang, Nanoscale Tubular and Sheetlike Superstructures from Hierarchical Self-Assembly of Polymeric Janus Particles, *Angew. Chem. Int. Ed.*, 2008, **47**(52), 10171–10174.

13. F. Wurm, H. M. Koenig, S. Hilf and A. F. M. Kilbinger, Janus micelles induced by olefin metathesis, *J. Am. Chem. Soc.*, 2008, **130**(18), 5876–5877.

14. J. Du and S. P. Armes, Patchy multi-compartment micelles are formed by direct dissolution of an ABC triblock copolymer in water, *Soft Matter*, 2010, **6**(19), 4851–4857.

15. X. Li, H. Yang, L. Xu, X. Fu, H. Guo and X. Zhang, Janus Micelle Formation Induced by Protonation/Deprotonation of Poly(2-vinylpyridine)-block-Poly(ethylene oxide) Diblock Copolymers, *Macromol. Chem. Phys.*, 2010, **211**(3), 297–302.

16. E. R. Zubarev, J. Xu, A. Sayyad and J. D. Gibson, Amphiphilicity-driven organization of nanoparticles into discrete assemblies, *J. Am. Chem. Soc.*, 2006, **128**(47), 15098–15099.

17. M. Zhang and A. H. E. Müller, Cylindrical polymer brushes, *J. Polym. Sci., Part A: Polym. Chem.*, 2005, **43**(16), 3461–3481.

18. (a) Z. S. Bo, J. P. Rabe and A. D. Schlüter, A poly(para-phenylene) with hydrophobic and hydrophilic dendrons: Prototype of an amphiphilic cylinder with the potential to segregate lengthwise, *Angew. Chem., Int. Ed.*, 1999, **38**(16), 2370–2372; (b) A. D. Schlüter and J. P. Rabe, Dendronized polymers: Synthesis, characterization, assembly at interfaces, and manipulation, *Angew. Chem., Int. Ed.*, 2000, **39**(5), 864–883.

19. K. H. Roh, D. C. Martin and J. Lahann, Biphasic Janus particles with nanoscale anisotropy, *Nat. Mater.*, 2005, **4**(10), 759–763.
20. D. W. Lim, S. Hwang, O. Uzun, F. Stellacci and J. Lahann, Compartmentalization of Gold Nanocrystals in Polymer Microparticles using Electrohydrodynamic Co-Jetting, *Macromol. Rapid Commun.*, 2010, **31**(2), 176–182.
21. S. Hwang, K.-H. Roh, D. W. Lim, G. Wang, C. Uher and J. Lahann, Anisotropic hybrid particles based on electrohydrodynamic co-jetting of nanoparticle suspensions, *Phys. Chem. Chem. Phys.*, 2010, **12**(38), 11894–11899.
22. M. Yoshida, K.-H. Roh, S. Mandal, S. Bhaskar, D. Lim, H. Nandivada, X. Deng and J. Lahann, Structurally Controlled Bio-hybrid Materials Based on Unidirectional Association of Anisotropic Microparticles with Human Endothelial Cells, *Adv. Mater.*, 2009, **21**(48), 4920–4925.
23. S. Reculusa, C. Poncet-Legrand, A. Perro, E. Duguet, E. Bourgeat-Lami, C. Mingotaud and S. Ravaine, Hybrid dissymmetrical colloidal particles, *Chem. Mater.*, 2005, **17**(13), 3338–3344.
24. A. Perro, S. Reculusa, F. Pereira, M. H. Delville, C. Mingotaud, E. Duguet, E. Bourgeat-Lami and S. Ravaine, Towards large amounts of Janus nanoparticles through a protection-deprotection route, *Chem. Commun.*, 2005, **44**, 5542–5543.
25. J. Ge, Y. Hu, T. Zhang and Y. Yin, Superparamagnetic composite colloids with anisotropic structures, *J. Am. Chem. Soc.*, 2007, **129**(29), 8974–8975.
26. W. Lu, M. Chen and L. Wu, One-step synthesis of organic-inorganic hybrid asymmetric dimer particles *via* miniemulsion polymerization and functionalization with silver, *J. Colloid Interface Sci.*, 2008, **328**(1), 98–102.
27. J. Zhang, M. Wu, Q. Wu, J. Yang, N. Liu and Z. Jin, Facile Fabrication of Inorganic/Polymer Janus Microspheres by Miniemulsion Polymerization, *Chem. Lett.*, 2010, **39**(3), 206–207.
28. C. Zhang, B. Liu, C. Tang, J. Liu, X. Qu, J. Li and Z. Yang, Large scale synthesis of Janus submicron sized colloids by wet etching anisotropic ones, *Chem. Commun.*, 2010, **46**(25), 4610–4612.
29. B. M. Teo, S. K. Suh, T. A. Hatton, M. Ashokkumar and F. Grieser, Sonochemical Synthesis of Magnetic Janus Nanoparticles, *Langmuir*, 2011, **27**(1), 30–33.
30. L. Zhang, F. Zhang, W.-F. Dong, J.-F. Song, Q.-S. Huo and H.-B. Sun, Magnetic-mesoporous Janus nanoparticles, *Chem. Commun.*, 2011, **47**(4), 1225–1227.
31. M. M. Rahman, F. Montagne, H. Fessi and A. Elaissari, Anisotropic magnetic microparticles from ferrofluid emulsion, *Soft Matter*, 2011, **7**(4), 1483–1490.
32. A. Ohnuma, E. C. Cho, P. H. Camargo, L. Au, B. Ohtani and Y. Xia, A Facile Synthesis of Asymmetric Hybrid Colloidal Particles, *J. Am. Chem. Soc.*, 2009, **131**(4), 1352–1353.
33. M. Feyen, C. Weidenthaler, F. Schueth and A.-H. Lu, Regioselectively Controlled Synthesis of Colloidal Mushroom Nanostructures and Their Hollow Derivatives, *J. Am. Chem. Soc.*, 2010, **132**(19), 6791–6799.

34. H. Takei and N. Shimizu, Gradient Sensitive Microscopic Probes Prepared by Gold Evaporation and Chemisorption on Latex Spheres, *Langmuir*, 1997, **13**(7), 1865–1868.

35. J. C. Love, B. D. Gates, D. B. Wolfe, K. E. Paul and G. M. Whitesides, Fabrication and wetting properties of metallic half-shells with submicron diameters, *Nano Lett.*, 2002, **2**(8), 891–894.

36. Y. Lu, H. Xiong, X. C. Jiang, Y. N. Xia, M. Prentiss and G. M. Whitesides, Asymmetric dimers can be formed by dewetting half-shells of gold deposited on the surfaces of spherical oxide colloids, *J. Am. Chem. Soc.*, 2003, **125**(42), 12724–12725.

37. L. F. Valadares, Y.-G. Tao, N. S. Zacharia, V. Kitaev, F. Galembeck, R. Kapral and G. A. Ozin, Catalytic Nanomotors: Self-Propelled Sphere Dimers, *Small*, 2010, **6**(4), 565–572.

38. A. M. Yake, C. E. Snyder and D. Velegol, Site-specific functionalization on individual colloids: Size control, stability, and multilayers, *Langmuir*, 2007, **23**(17), 9069–9075.

39. R. Sardar and J. S. Shumaker-Parry, Asymmetrically functionalized gold nanoparticles organized in one-dimensional chains, *Nano Lett.*, 2008, **8**(2), 731–736.

40. S. Jiang and S. Granick, A Simple Method to Produce Trivalent Colloidal Particles, *Langmuir*, 2009, **25**(16), 8915–8918.

41. L. Y. Wu, B. M. Ross, S. Hong and L. P. Lee, Bioinspired Nanocorals with Decoupled Cellular Targeting and Sensing Functionality, *Small*, 2010, **6**(4), 503–507.

42. C.-C. Ho, W.-S. Chen, T.-Y. Shie, J.-N. Lin and C. Kuo, Novel fabrication of Janus particles from the surfaces of electrospun polymer fibers, *Langmuir*, 2008, **24**(11), 5663–5666.

43. C.-C. Lin, C.-W. Liao, Y.-C. Chao and C. Kuo, Fabrication and Characterization of Asymmetric Janus and Ternary Particles, *ACS Appl. Mater. Interfaces*, 2010, **2**(11), 3185–3191.

44. A. G. Skirtach, D. G. Kurth and H. Mohwald, Laser-embossing nanoparticles into a polymeric film, *Appl. Phys. Lett.*, 2009, **94**(9), 093103–093106.

45. M. D. McConnell, M. J. Kraeutler, S. Yang and R. J. Composto, Patchy and Multiregion Janus Particles with Tunable Optical Properties, *Nano Lett.*, 2010, **10**(2), 603–609.

46. (a) A. Centrone, Y. Hu, A. M. Jackson, G. Zerbi and F. Stellacci, Phase separation on mixed-monolayer-protected metal nanoparticles: A study by infrared spectroscopy and scanning tunneling microscopy, *Small*, 2007, 3(5), 814–817; (b) C. Gentilini and L. Pasquato, Morphology of mixed-monolayers protecting metal nanoparticles, *J. Mater. Chem.*, 2010, **20**(8), 1403–1412.

47. (a) J. van Herrikhuyzen, G. Portale, J. C. Gielen, P. C. M. Christianen, N. A. J. M. Sommerdijk, S. C. J. Meskers and A. P. H. J. Schenning, Disk micelles from amphiphilic Janus gold nanoparticles, *Chem. Commun.*, 2008, (6), 697–699; (b) R. T. M. Jakobs, J. van Herrikhuyzen, J. C. Gielen,

P. C. M. Christianen, S. C. J. Meskers and A. P. H. J. Schenning, Self--assembly of amphiphilic gold nanoparticles decorated with a mixed shell of oligo(p-phenylene vinylene)s and ethyleneoxide ligands, *J. Mater. Chem.*, 2008, **18**(29), 3438–3441.

48. H. Kim, R. P. Carney, J. Reguera, Q. K. Ong, X. Liu and F. Stellacci, Synthesis and Characterization of Janus Gold Nanoparticles, *Adv. Mater.*, 2012, **24**(28), 3857–3863.

49. D. M. Andala, S. H. R. Shin, H. Y. Lee and K. J. M. Bishop, Templated Synthesis of Amphiphilic Nanoparticles at the Liquid–Liquid Interface, *ACS Nano*, 2012, **6**(2), 1044–1050.

50. V. Sashuk, R. Holyst, T. Wojciechowski and M. Fialkowski, Close-packed monolayers of charged Janus-type nanoparticles at the air–water inter-face, *J. Colloid Interface Sci.*, 2012, **375**, 180–186.

51. S. Pradhan, L. P. Xu and S. W. Chen, Janus nanoparticles by interfacial engineering, *Adv. Funct. Mater.*, 2007, **17**(14), 2385–2392.

52. S. Pradhan, L. E. Brown, J. P. Konopelski and S. W. Chen, Janus nano-particles: reaction dynamics and NOESY characterization, *J. Nanopart. Res.*, 2009, **11**(8), 1895–1903.

53. L. P. Xu, S. Pradhan and S. W. Chen, Adhesion force studies of Janus nanoparticles, *Langmuir*, 2007, **23**(16), 8544–8548.

54. Y. He, T. Li, X. Yu, S. Zhao, J. Lu and J. He, Tuning the wettability of calcite cubes by varying the sizes of the polystyrene nanoparticles at-tached to their surfaces, *Appl. Surf. Sci.*, 2007, **253**(12), 5320–5324.

55. (a) F. Dmitrii and F. Rosei, Metal nanoparticles: From "Artificial Atoms" to "Artificial Molecules", *Angew. Chem., Int. Ed.*, 2007, **46**(32), 6006–6008; (b) A. M. Jackson, J. W. Myerson and F. Stellacci, Spontaneous assembly of subnanometre-ordered domains in the ligand shell of monolayer-protected nanoparticles, *Nat. Mater.*, 2004, **3**(5), 330–336; (c) A. M. Jackson, Y. Hu, P. Silva and F. Stellacci, From Homoligand- to Mixed-Ligand-Monolayer-Protected Metal Nanoparticles: A Scanning Tunneling Micro-scopy Investigation, *J. Am. Chem. Soc.*, 2006, **128**(34), 11135–11149.

56. D. Suzuki, S. Tsuji and H. Kawaguchi, Janus Microgels Prepared by Surfactant-Free Pickering Emulsion-Based Modification and Their Self-Assembly, *J. Am. Chem. Soc.*, 2007, **129**(26), 8088–8089.

57. S. Deka, A. Falqui, G. Bertoni, C. Sangregorio, G. Poneti, G. Morello, M. De Giorgi, C. Giannini, R. Cingolani, L. Manna and P. D. Cozzoli, Fluorescent asymmetrically cobalt-tipped CdSe@CdS core@shell nanorod heterostructures exhibiting room-temperature ferromagnetic behavior, *J. Am. Chem. Soc.*, 2009, **131**(35), 12817–12828.

58. S. Chakrabortty, J. A. Yang, Y. M. Tan, N. Mishra and Y. Chan, Asym-metric dumbbells from selective deposition of metals on seeded semi-conductor nanorods, *Angew. Chem.*, 2010, **49**(16), 2888–2892.

59. J. Maynadie, A. Salant, A. Falqui, M. Respaud, E. Shaviv, U. Banin, K. Soulantica and B. Chaudret, Cobalt growth on the tips of CdSe nanorods, *Angew. Chem.*, 2009, **48**(10), 1814–1817.

60. M. Casavola, A. Falqui, M. A. Garcia, M. Garcia-Hernandez, C. Giannini, R. Cingolani and P. D. Cozzoli, Exchange-coupled bimagnetic cobalt/iron oxide branched nanocrystal heterostructures, *Nano Lett.*, 2009, **9**(1), 366–376.

61. K. Chaudhary, Q. Chen, J. J. Juarez, S. Granick and J. A. Lewis, Janus Colloidal Matchsticks, *J. Am. Chem. Soc.*, 2012, **134**(31), 12901–12903.

62. J. He, M. T. Perez, P. Zhang, Y. Liu, T. Babu, J. Gong and Z. Nie, A general approach to synthesize asymmetric hybrid nanoparticles by interfacial reactions, *J. Am. Chem. Soc.*, 2012, **134**(8), 3639–3642.

63. S. Xing, Y. Feng, Y. Y. Tay, T. Chen, J. Xu, M. Pan, J. He, H. H. Hng, Q. Yan and H. Chen, Reducing the symmetry of bimetallic Au@Ag nanoparticles by exploiting eccentric polymer shells, *J. Am. Chem. Soc.*, 2010, **132**(28), 9537–9539.

64. (a) S. H. Choi, E. G. Kim and T. Hyeon, One-pot synthesis of copper-indium sulfide nanocrystal heterostructures with acorn, bottle, and larva shapes, *J. Am. Chem. Soc.*, 2006, **128**(8), 2520–2521; (b) L. De Trizio, A. Figuerola, L. Manna, A. Genovese, C. George, R. Brescia, Z. Saghi, R. Simonutti, M. Van Huis and A. Falqui, Size-Tunable, Hexagonal Plate-like Cu3P and Janus-like Cu-Cu3P Nanocrystals, *ACS Nano*, 2012, **6**(1), 32–41.

65. M. Pang, J. Hu and H. C. Zeng, Synthesis, morphological control, and antibacterial properties of hollow/solid Ag_2S/Ag heterodimers, *J. Am. Chem. Soc.*, 2010, **132**(31), 10771–10785.

66. (a) I. Parsina and F. Baletto, Tailoring the Structural Motif of AgCo Nanoalloys: Core/Shell versus Janus-like, *J. Phys. Chem. C*, 2010, **114**(3), 1504–1511; (b) H. Zhang, W. Li, M. Jin, J. Zeng, T. Yu, D. Yang and Y. Xia, Controlling the morphology of rhodium nanocrystals by manipulating the growth kinetics with a syringe pump, *Nano Lett.*, 2011, **11**(2), 898–903.

67. T. Wetz, K. Soulantica, A. Talqui, M. Respaud, E. Snoeck and B. Chaudret, Hybrid Co-Au nanorods: Controlling Au nucleation and location, *Angew. Chem., Int. Ed.*, 2007, **46**(37), 7079–7081.

68. T. P. Vinod, M. Yang, J. Kim and N. A. Kotov, Self-Guided One-Sided Metal Reduction in Te Nanowires Leading to Au-Te Matchsticks, *Langmuir*, 2009, **25**(23), 13545–13550.

69. I. Moriguchi, K. Matsuo, M. Sakai, K. Hanai, Y. Teraoka and S. Kagawa, Synthesis of size-quantized metal sulfides of Pb–Cd and Zn–Cd bi-metallic systems in stearate Langmuir–Blodgett films, *J. Chem. Soc., Faraday Trans.*, 1998, **94**(15), 2199–2204.

70. (a) Y. S. Shon, G. B. Dawson, M. Porter and R. W. Murray, Monolayer-protected bimetal cluster synthesis by core metal galvanic exchange re-action, *Langmuir*, 2002, **18**(10), 3880–3885; (b) J. Zhang, F. H. B. Lima, M. H. Shao, K. Sasaki, J. X. Wang, J. Hanson and R. R. Adzic, Platinum monolayer on nonnoble metal–noble metal core–shell nanoparticle electrocatalysts for O_2 reduction, *J. Phys. Chem. B*, 2005, **109**(48), 22701–22704.

71. C. M. Cobley and Y. Xia, Engineering the properties of metal nanostructures *via* galvanic replacement reactions, *Mater. Sci. Eng., R*, 2010, **70**(3–6), 44–62.

72. P. H. C. Camargo, Y. Xiong, L. Ji, J. M. Zuo and Y. Xia, Facile synthesis of tadpole-like nanostructures consisting of Au heads and Pd tails, *J. Am. Chem. Soc.*, 2007, **129**(50), 15452–15453.

73. L. O. Mair, B. Evans, A. R. Hall, J. Carpenter, A. Shields, K. Ford and M. Millard, and R. Superfine, R., Highly controllable near-surface swimming of magnetic Janus nanorods: application to payload capture and manipulation, *J. Phys. D: Appl. Phys*, 2011, **44**(12), 125001–1.

74. Y. Song, K. Liu and S. Chen, AgAu Bimetallic Janus Nanoparticles and Their Electrocatalytic Activity for Oxygen Reduction in Alkaline Media, *Langmuir*, 2012, **28**(49), 17143–17152.

75. T. Chen, G. Chen, S. X. Xing, T. Wu and H. Y. Chen, Scalable Routes to Janus Au–SiO$_2$ and Ternary Ag–Au–SiO$_2$ Nanoparticles, *Chem. Mater.*, 2010, **22**(13), 3826–3828.

Subject Index

absorption spectroscopy
    and DNA-based metal
        nanoclusters, 363
    and fluorescence, 363
    furnace atomic, 393, 400
    silver nanoparticles
        ligand effects on, 69–70
        size effects on, 69
AC (alternating current), 421
activation of C-H bonds, and Pt
    nanoclusters, 338–340
aerobic oxidation, and metal
    nanoclusters, 282–283
Ag nanoclusters (Ag NCs). *See* silver
    nanoclusters (Ag NCs)
alcohol oxidation, and metal clusters
    heterogeneous catalysis,
        245–247
    homogeneous catalysis,
        228–232
aldehyde hydrogenation, and metal
    clusters, 239–240
    and unsaturated ketone,
        238–239
alkene hydrogenation, and metal
    clusters, 239–240
alkene oxidation, and metal clusters
    heterogeneous catalysis,
        247–249
    homogeneous catalysis,
        232–235
alkylthiols RSH, 8–11
alternating current (AC), 421
anti-chemokine receptor,
    392, 396, 402

applications. *See also* specific types
    bioimaging, Platinum
        nanoclusters
        of blue-emitting, 396–398
        of green-emitting, 402–403
    DNA-based Ag NCs, 371–383
        biological imaging and
            labeling, 381–383
        detection of DNA and
            miRNA, 377–380
        detection of metal ions
            and small
            biomolecules, 371–375
        logic gates, 380–381
        protein analysis, 375–377
    of fluorescent silver
        nanoclusters, 92–96
        in biosensing, 93–96
        in chemical sensing,
            92–93
    of protein protected metal
        clusters, 199–209
        bio-imaging, 202–207
        molecular imaging
            guided delivery, 207–209
        sensing, 199–202
    stabilized/protected metal
        clusters catalysts, 301–306
        other reactions, 306
        oxidations, 301–302
        reductions, 302–306
    supported metal clusters
        catalysts, 312–315
        other reactions, 313–315
        oxidations, 312–313

AQCs. *See* atomic quantum clusters
  (AQCs)
aqueous synthesis, silver
  nanoparticles, 52–55
atomic quantum clusters (AQCs), 25
    and bottom-up approach
      chemical reduction,
        31–33
      electrochemical
        synthesis, 35–36
      microemulsions, 36–37
      microwave irradiation,
        37–38
      photoreduction, 33–34
      sonochemistry, 34–35
      template-assisted
        synthesis, 38–43
    characteristics of, 25–27
    and top-down approach,
      27–30
Au/Ag NPs, thiol etching of, 156–162
    etchants selection, 156–158
      and chain length of
        thiolate ligands,
        157–158
      and size of thiolate
        ligands, 158
    and etching environment,
      158–161
      and kinetic trapping,
        160–161
      and thermodynamic
        selection, 158–160
    and NP precursors, 161–162
      size and size distribution,
        161
      surface properties of, 162
Au(I)/Ag(I)-SR complexes, and
  thiolate-protected Au/Ag NCs.
  See reductive decomposition,
  of Au(I)/Ag(I)-SR complexes
Au nanoclusters (Au NCs). *See also*
  silver nanoclusters (Ag NCs)
    $Au_{25}(SR)_{18}$, 264–266
      and polydisperse thiol
        etching, 147–148

$Au_{38}(SR)_{24}$, 266–268
    and polydisperse thiol
      etching, 148–149
$Au_{102}(SR)_{44}$, 268–270
$Au_{144}(SR)_{60}$, 270–271
    and polydisperse thiol
      etching, 149
  for chiral recognition,
    344–346
  DNA-based metal nanoclusters,
    383–384
  precursors
    non-thiolate-protected,
      150–151
    protecting ligands of,
      152–153
    size distribution of,
      151–152
    thiolate-protected, 150
  stability origin of
    thiolate-protected,
    149–150
Au NCs. *See* Au nanoclusters
  (Au NCs)
Au_{QCs}, and metal clusters
  @BSA, 186
  in Lf, 184–186
  @Lyz, 186–188

bimetallic nanocrystals, 423–426
bioimaging applications
  DNA-templated Ag NCs,
    381–383
  Platinum nanoclusters
    of blue-emitting,
      396–398
    of green-emitting,
      402–403
  protein protected clusters,
    202–207
    in vitro, 202–205
    in vivo, 205–207
biologically important molecules
  detection, 200–202
biomolecules detection,
  DNA-templated Ag NCs, 371–375

biomolecules sensing applications,
   of noble metal clusters,
   200–202
biosensing applications, of
   fluorescent silver nanoclusters,
   93–96
blue-emitting Platinum
   nanoclusters, 394–398
      bioimaging applications of,
      396–398
      characterization of, 396
      preparation of, 394–396
bottom-up approach, and AQCs
   chemical reduction, 31–33
   electrochemical synthesis,
      35–36
   microemulsions, 36–37
   microwave irradiation, 37–38
   photoreduction, 33–34
   sonochemistry, 34–35
   template-assisted synthesis,
      38–43
Brust-Schiffrin two-phase method
   (BSM), 2
   ligand addition, 8–16
      alkylthiols RSH, 8–11
      dialkyl diselenide
         RSe-SeR, 11–14
      dialkyl ditelluride
         RTe-TeR, 14–16
   phase transfer of metal ions,
      5–7
   proton NMR evidence of
      encapsulated water, 2–5
   role of water, 16–19
   for thiolate-protected Au/Ag
      nanoclusters, 133–136
   and zero-valence metal
      nanoparticles formation,
      19–22
      normal reduction
         sequence, 19–20
      reversed reduction
         sequence, 20–22
BSM. *See* Brust-Schiffrin two-phase
   method (BSM)

CAD (coronary artery disease),
   207
calcination, 310
cancer cells sensing applications,
   of noble metal clusters,
   200–202
carbon monoxide oxidation, and
   metal clusters, 243–245
catalysis. *See also* catalysts
   heterogeneous
      (*See* heterogeneous catalysis,
      and metal clusters)
   homogeneous
      (*See* homogeneous catalysis,
      and metal clusters)
   of transition metal
      nanoclusters, 330–340
catalysts. *See also* catalysis
   stabilized/protected metal
      clusters, 294–306
   supported metal clusters,
      306–315
CD. *See* circular dichroism (CD)
CE (collision energies), 64
cell viability test, 393
characterization
   of blue-emitting Platinum
      nanoclusters, 396
   DNA-based metal nanoclusters,
      355–362
      circular dichroism (CD),
         360–362
      extended x-ray absorption
         fine structure (EXAFS)
         technique, 360
      mass spectrometry (MS),
         357–359
      Nuclear Magnetic
         Resonance (NMR)
         spectroscopy,
         359–360
      Transmission Electron
         Microscopy (TEM
         and HRTEM), 357
      UV-Vis spectroscopy,
         356–357

functional transition metal
nanoclusters, 328–330
of green-emitting Platinum
nanoclusters, 400–402
protein protected metal
clusters, 178–179
C-H bonds activation, and
Pt nanoclusters, 338–340
chemical reduction synthesis
and AQCs, 31–33
of fluorescent silver
nanoclusters, 86–92
chemical sensing, of fluorescent
silver nanoclusters, 92–93
chirality properties
Au nanoclusters for, 344–346
of DNA-based metal
nanoclusters, 364–365
circular dichroism (CD), 360–362
classification, metal clusters
catalysts
stabilized/protected, 294
supported, 294
CMC (critical micelle
concentration), 3
collision energies (CE), 64
color tunability, of DNA-templated
Ag NCs, 368
composition-controlled synthesis
stabilized/protected metal
clusters catalysts, 299–300
supported metal clusters
catalysts, 310–312
computer tomography (CT), 204
conductive ink, silver nanoclusters
as, 346–348
confocal fluorescence microscopy, 393
Continuous Set of Gauge
Transformation (CSGT), 329
controlled phase separation, metal
nanocrystals, 420–423
CO oxidation, and metal
nanoclusters, 276–278
copolymerization, Janus
nanoparticles, 408–411
coronary artery disease (CAD), 207

C-rich single-strand DNA (ssDNA),
366–367
critical micelle concentration
(CMC), 3
CROC (cyclic reduction in oxidative
conditions), 123
CSGT (Continuous Set of Gauge
Transformation), 329
CT (computer tomography), 204
Cu nanoclusters, 273–275
CV (Cyclic Voltammometry), 344
cyclic reduction in oxidative
conditions (CROC), 123
Cyclic Voltammometry (CV), 344
cyclization oxidation, and metal
clusters
heterogeneous catalysis, 251
homogeneous catalysis, 235–237
Lewis acidic, 236–237
oxidative coupling, 235–236
cyclohexane oxidation, and metal
clusters, 250–251

dendrimers, 39, 295
density functional theory (DFT), 154,
279, 297, 327, 369
dialkyl diselenide RSe-SeR, 11–14
dialkyl ditelluride RTe-TeR, 14–16
DLS (dynamic light scattering), 394
DNA and metal ions interactions,
354–355
DNA-based metal nanoclusters
Ag NCs, 365–371
applications of, 371–383
color tunability of, 368
C-rich ssDNA and G-rich
ssDNA, 366–367
description, 365–366
gap site-directed
formation of, 370–371
i-motif, G-quadruplex,
duplex and triplex
structures, 369–370
and T-containing
oligonucleotides,
367–368

DNA-based metal nanoclusters
(*continued*)
  Au NCs, 383–384
  characterization of, 355–362
    CD, 360–362
    EXAFS technique, 360
    MS, 357–359
    NMR spectroscopy,
      359–360
    TEM, 357
    UV-Vis spectroscopy,
      356–357
  description of, 352–354
  DNA and metal ions
    interactions, 354–355
  synthesis of, 355
  unique properties of, 362–365
    absorption and
      fluorescence, 363
    chirality properties,
      364–365
    solvatochromic effect,
      363
    two-photon absorption,
      363–364
DNA detection, and DNA-templated
  Ag NCs, 377–380
DNA microarray strategy, 368
DNA-templated Ag NCs, 365–371
  applications of, 371–383
    biological imaging and
      labeling, 381–383
    detection of DNA and
      miRNA, 377–380
    detection of metal ions
      and small
      biomolecules, 371–375
    logic gates, 380–381
    protein analysis,
      375–377
  color tunability of, 368
  C-rich ssDNA and G-rich
    ssDNA, 366–367
  description, 365–366
  gap site-directed formation of,
    370–371

i-motif, G-quadruplex, duplex
  and triplex structures,
  369–370
  and T-containing
    oligonucleotides, 367–368
DNA triple helix, 369
double-strand DNA (dsDNA), 367
duplex structure, and DNA-
  templated Ag NCs, 369–370
dynamic light scattering (DLS), 394

ECL (electrochemiluminescence),
  105
  sensing of, 202
ECSA (electrochemical surface area),
  279
EDC coupling, 203
EFTEM (energy filtered TEM), 423
EGF (epidermal growth factor), 205
eggshell membrane (ESM), 108
electrochemical surface area (ECSA),
  279
electrochemical synthesis, and
  AQCs, 35–36
electrochemiluminescence (ECL),
  105
electronic structure theory,
  and silver nanoparticles,
  66–67
electrospinning, Janus
  nanoparticles, 411–413
electrospray ionization mass
  spectrometry (ESI-MS), 134, 264
energy approach, 228
energy filtered TEM (EFTEM), 423
enhanced permeability and
  retention (EPR), 205
epidermal growth factor (EGF),
  205
epoxidation reactions, and metal
  clusters, 249–250
EPR (enhanced permeability and
  retention), 205
ESI-MS (electrospray ionization mass
  spectrometry), 134, 264
ESM (eggshell membrane), 108

etchants, and Au/Ag NPs thiol
    etching, 156–158
        and kinetic trapping, 160
        thiolate ligands
            chain length of,
                157–158
            size of, 158
etching, thiol. *See* thiol etching
EXAFS (extended x-ray absorption
    fine structure) technique,
    179, 360
experimental Platinum
    nanoclusters, 392–394
extended x-ray absorption fine
    structure (EXAFS) technique, 179,
    360

fluorescence
        and absorption spectroscopy,
            363
        and DNA-based metal
            nanoclusters, 363
        lifetime, 393
        microscopy, confocal, 393
        spectra, 393
fluorescence spectroscopy, silver
    nanoparticles, 70–74
        effect of chemical environment
            on, 73–74
        size effects on, 72–73
fluorescent probe brightness,
    396
fluorescent silver nanoclusters,
    water-soluble. *See* water-soluble
    fluorescent silver nanoclusters
folate receptors (FR), 203
Fourier-transform infrared (FTIR)
    spectroscopy, 230
FR (folate receptors), 203
FTIR (Fourier-transform infrared)
    spectroscopy, 230
functional transition metal
    nanoclusters. *See* transition metal
    nanoclusters
furnace atomic absorption
    spectrometry, 393, 400

gap site-directed formation, of
    DNA-templated Ag NCs,
    370–371
gas phase clusters, and protein
    templates, 209–214
Gauge Independent Atomic Orbital
    (GIAO), 329
Gaussview software, 329
GIAO (Gauge Independent Atomic
    Orbital), 329
G-quadruplex structure, and
    DNA-templated Ag NCs,
    369–370
green-emitting Platinum
    nanoclusters, 398–403
        bioimaging applications,
            402–403
        characterization of,
            400–402
        preparation of, 398–400
G-rich single-strand DNA (ssDNA),
    366–367

heterodimers, polymer-inorganic,
    413–414
heterogeneous catalysis, and metal
    clusters, 243–254
        hydrogenation reactions,
            252–253
        oxidation reactions, 243–252
            alcohol, 245–247
            alkene, 247–249
            carbon monoxide,
                243–245
            cyclization, 251
            cyclohexane, 250–251
            epoxidation, 249–250
            other types of, 251–252
        reduction reactions,
            253–254
highest occupied molecular orbital
    (HOMO), 293, 328–329
high performance liquid
    chromatography (HPLC),
    275, 392
high resolution TEM (HRTEM), 423

homogeneous catalysis, and metal
  clusters, 228–243
    degradation reactions, 241–243
    hydrogenation reactions,
      238–240
        aldehyde and alkene
          hydrogenation,
          239–240
        unsaturated ketone and
          aldehyde
          hydrogenation,
          238–239
    other reactions, 243
    oxidation reactions, 228–237
      alcohol, 228–232
      alkene, 232–235
      cyclization, 235–237
    reduction reactions,
      240–241
HOMO (highest occupied molecular
  orbital), 293, 328–329
    –LUMO energy gap, 66–68,
      72–73, 107, 194, 213, 226,
      261, 271
HPLC (high performance liquid
  chromatography), 275, 392
HRTEM (high resolution TEM), 423
hydrogenation, and metal clusters
  heterogeneous catalysis,
    252–253
  homogeneous catalysis,
    238–240

IBANs (intensely and broadly
  absorbing nanoparticles),
  67, 122
ICP-MS (inductively-coupled plasma
  mass spectrometry), 393
IDE (insulin degrading enzyme),
  204
i-motif structure, and
  DNA-templated Ag NCs, 369–370
impregnation method, supported
  metal clusters, 306–307
inductively-coupled plasma mass
  spectrometry (ICP-MS), 393

in silico
  characterization of transition
    metal nanoclusters,
    328–330
  synthesis
    description, 323–324
    of transition metal
      nanoclusters, 324–328
insulin degrading enzyme (IDE),
  204
intensely and broadly absorbing
  nanoparticles (IBANs), 67, 122
inverse micelle formation, and BSM
  phase transfer of metal ions,
    5–7
  proton NMR evidence of
    encapsulated water, 2–5
in vitro bioimaging applications,
  202–205
in vivo bioimaging applications,
  205–207
Ir nanoclusters, sinter-resistant,
  331–334

Janus nanoparticles
  characteristics of, 407–408
  metal nanocrystals, 414–426
    bimetallic, 423–426
    controlled phase
      separation and growth,
      420–423
    soft masks, 418–420
    solid masks, 414–418
  polymer-based, 408–414
    copolymerization,
      408–411
    electrospinning,
      411–413
    polymer-inorganic
      heterodimers,
      413–414

kinetics, reductive decomposition,
  141–146
  creating uniform reduction
    environment, 145–146

delivering mild reducing
   power, 141–145
   and reducing agents,
      142–143
   using mild reducing
      agents, 143–145
kinetics, thiol etching, 154–155
kinetic trapping, Au/Ag NPs, thiol
   etching of, 160–161

labeling applications. *See*
   bioimaging applications
Langmuir–Blodgett (LB), 418
laser desorption/ionization (LDI), 61
   time of flight mass
      spectroscopies (TOF-MS),
      105
LB (Langmuir–Blodgett), 418
LDI (laser desorption/ionization), 61
   TOF-MS (time of flight mass
      spectroscopies), 105
LDs (lipid droplets), 207
Lewis acidic cyclization oxidation,
   236–237
ligand addition, and BSM, 8–16
   alkylthiols RSH, 8–11
   dialkyl diselenide RSe-SeR,
      11–14
   dialkyl ditelluride RTe-TeR,
      14–16
ligand-protected metal clusters,
   296–299
ligands. *See also* thiolate ligands
   for cluster synthesis, 171–172
   polydisperse Au/Ag NCs, thiol
      etching of
      of Au NC precursors,
         152–153
      thiolating, 153–154
limit of detection (LOD), 200
lipid droplets (LDs), 207
localized surface plasmon resonance
   (LSPR), 423
LOD (limit of detection), 200
logic gates, and DNA-templated Ag
   NCs, 380–381

lowest unoccupied molecular orbital
   (LUMO), 293, 328–329
LSPR (localized surface plasmon
   resonance), 423
luminescent system, and protein
   protected metal clusters, 192–198.
   See also fluorescence
LUMO (lowest unoccupied
   molecular orbital), 293, 328–329
   HOMO, 66–68, 72–73, 107, 194,
   213, 226, 261, 271, 293, 328–329

MALDI. *See* matrix-assisted laser
   desorption/ionization (MALDI)
mass spectrometry (MS)
   DNA-based metal nanoclusters,
      357–359
   MALDI (matrix-assisted laser
      desorption/ionization)-TOF
      (time-of-flight), 105
   and metal clusters, 179–183
      growth in protein
         templates, 184–188
   and silver magic-number
      clusters, 61–65
Materials Studio, 326
matrix-assisted laser desorption/
   ionization (MALDI), 61, 295
   time of flight (TOF)
      mass spectroscopy (MS),
         105
   time-of-flight (TOF), 138
MB (molecular beacon) probe, 380
MD (molecular dynamics)
   simulations, 324–328
metal cluster catalysis/catalysts
   classification, 294
   description, 291–294
   mediated
      heterogeneous catalysis,
         243–254
      homogeneous catalysis,
         228–243
   overview, 226–228
   stabilized/protected, 294–306
   supported, 306–315

metal clusters. *See also* metal
nanoclusters
and catalysts (*See* metal cluster
catalysts/catalysts)
classification, 294
stabilized/protected, 294
supported, 294
general characterization,
178–179
general properties of, 169–171
growth in protein templates,
184–188
Au_{QCs}@BSA, 186
Au_{QCs} in Lf, 184–186
Au_{QCs}@Lyz, 186–188
ligands choice for, 171–172
and mass spectrometry, 179–183
mediated (*See* metal cluster
catalysis/catalysts)
peptide protected, 191–192
protein protected, 173–177
synthesis of, 177–178
metal complex
formation, 5–7
reduction sequence, 8–16
metal ions
detection, DNA-templated Ag
NCs, 371–375
induced quenching, 196–198
interactions with DNA,
354–355
phase transfer of, 5–7
sensing applications, of noble
metal clusters, 199–200
metal nanoclusters. *See also* metal
clusters; specific types
DNA-templated (*See* DNA-
based metal nanoclusters)
overview, 261–262
Platinum (*See* Platinum
nanoclusters)
size-controlled synthesis of,
262–276
Ag, 271–273
$Au_{25}(SR)_{18}$, 264–266
$Au_{38}(SR)_{24}$, 266–268

$Au_{102}(SR)_{44}$, 268–270
$Au_{144}(SR)_{60}$, 270–271
Cu, 273–275
monodisperse
$Au_n(SR)_m$, 271
Pd, 275–276
Pt, 275
size-dependent catalytic
activity of, 276–285
aerobic oxidation,
282–283
CO oxidation, 276–278
ORR, 278–282
other catalytic
applications, 283–285
transition (*See* transition metal
nanoclusters)
metal nanocrystals
Janus nanoparticles, 414–426
bimetallic, 423–426
controlled phase
separation and growth,
420–423
soft masks, 418–420
solid masks, 414–418
microemulsions, and AQCs, 36–37
microwave irradiation synthesis
and AQCs, 37–38
of fluorescent silver
nanoclusters, 85–86
miRNA detection, and DNA-
templated Ag NCs, 377–380
molecular beacon (MB) probe, 380
molecular dynamics (MD)
simulations, 324–328
molecular imaging guided delivery,
207–209
molecular orbitals (MOs), 122
monodisperse $Au_n(SR)_m$ metal
nanoclusters, 271
MOs (molecular orbitals), 122
MS. *See* mass spectrometry (MS)
multicomponent Pt alloy
nanoclusters, 334–338
multiple small proteins, and single
cluster core, 188

near-infrared (NIR) photons, 363
NMR spectroscopy. *See* nuclear magnetic resonance (NMR) spectroscopy
noble metal clusters, in protein templates. *See* protein protected metal clusters
non-aqueous synthesis, silver nanoparticles, 55–57
non-thiolate-protected Au NC precursors, 150–151
normal reduction sequence, 19–20
NP precursors, and etching
    pH-assisted, 159
    size and size distribution of, 161–162
    surface properties of, 162
    temperature-assisted, 159
nuclear magnetic resonance (NMR) spectroscopy
    DNA-based metal nanoclusters, 359–360
    and transition metal nanoclusters, 329
nucleotides, and AQCs, 39–41

one-phase synthesis, 55
one-photon excitation (OPE), 363
on-the-fly calculations, 324–328
OPE (one-photon excitation), 363
optical properties, silver nanoparticles, 65–75
    absorption spectroscopy, 67–70
        ligand effects on, 69–70
        size effects on, 69
    electronic structure theory, 66–67
    fluorescence spectroscopy, 70–74
        effect of chemical environment on, 73–74
        size effects on, 72–73
    spectroscopy challenges, 74–75
optimization calculations, 328
ORR. *See* Oxygen Reduction Reaction (ORR)

oxidation. *See also* reduction reactions
    aerobic, 282–283
    CO, 276–278
    heterogeneous catalysis, 243–252
        alcohol, 245–247
        alkene, 247–249
        carbon monoxide, 243–245
        cyclization, 251
        cyclohexane, 250–251
        epoxidation, 249–250
        other types of, 251–252
    homogeneous catalysis, 228–237
        alcohol, 228–232
        alkene, 232–235
        cyclization, 235–237
    stabilized/protected catalysts, 301–302
    supported metal clusters catalysts, 312–313
oxidative coupling cyclization oxidation, 235–236
oxygen reduction reaction (ORR), 278–282, 426

PAGE. *See* polyacrylamide gel electrophoresis (PAGE)
PDGF-BB (platelet-derived growth factor B-chain homodimer), 376
Pd nanoclusters, 275–276
    for sensing CH_4, 341–344
peptide protected metal clusters, 191–192
peptide-protected silver nanoclusters, 110–113
PES (potential energy surface), 326
PET (photoinduced electron transfer), 118
phase transfer, of metal ions, 5–7
phase transition, structural, 331
pH-assisted etching, of NP precursors, 159

photochemical reduction synthesis, of fluorescent silver nanoclusters, 82–83

photoinduced electron transfer (PET), 118

photoreduction, and AQCs, 33–34

platelet-derived growth factor B-chain homodimer (PDGF-BB), 376

Platinum nanoclusters, 275
  for activation of C-H bonds in CH$_4$, 338–340
  blue-emitting, 394–398
    bioimaging applications of, 396–398
    characterization of, 396
    preparation of, 394–396
  characteristics of, 391–392
  experimental, 392–394
  green-emitting, 398–403
    bioimaging applications, 402–403
    characterization of, 400–402
    preparation of, 398–400
  multicomponent, for O$_2$ reduction, 334–338

polyacrylamide gel electrophoresis (PAGE), 58, 263
  and silver magic-number clusters, 58–61

polydisperse Au/Ag NCs, thiol etching of, 146–156
  facile size-focusing process, 150–151
    non-thiolate-protected Au NC precursors, 150–151
    thiolate-protected Au NC precursors, 150
  ligands
    of Au NC precursors, 152–153
    thiolating, 153–154
  size distribution of Au NC precursors, 151–152

size-focusing process, 146–150
  stability origin of, 149–150
  stable Au$_{25}$(SR)$_{18}$ NCs, 147–148
  stable Au$_{38}$ (SR)$_{24}$ NCs, 148–149
  stable Au144(SR)60 NCs, 149
  versatile, 156

polymer-based Janus nanoparticles, 408–414
  copolymerization, 408–411
  electrospinning, 411–413
  polymer-inorganic heterodimers, 413–414

polymer-inorganic heterodimers, Janus nanoparticles, 413–414

polymers protected silver nanoclusters, 102–107

polymer-stabilized metal clusters, 295–296

potential energy surface (PES), 326

precursors
  NP (*See* NP precursors)
  supported metal clusters, 309–310

probe brightness, fluorescent, 396

protein analysis, and DNA-templated Ag NCs, 375–377

protein protected metal clusters, 173–177
  applications of, 199–209
    bio-imaging, 202–207
    molecular imaging guided delivery, 207–209
    sensing, 199–202
  characterization, 178–179
  conformational changes in, 188–191
  and gas phase clusters, 209–214

and mass spectrometry,
179–183
growth in protein
templates, 184–188
and metal ion induced
quenching, 196–198
origin and properties of
luminescence in, 192–198
synthesis, 177–178
protein-protected silver
nanoclusters, 108–110
proteins
and AQCs, 41–43
single cluster core in, 186–188
templates, and metal clusters,
184–188
proton NMR evidence of
encapsulated water, 2–5
Pt nanoclusters. *See* Platinum
nanoclusters

quenching, metal ion induced,
196–198

radiolytic reduction synthesis, of
fluorescent silver nanoclusters,
81–82
Raman spectroscopy, 329
reducing agents, and reductive
decomposition kinetics
mild usage of, 143–145
reducing power of, 142–143
reduction reactions
and metal clusters
heterogeneous catalysis,
253–254
homogeneous catalysis,
240–241
stabilized/protected
catalysts, 302–306
by NaBH$_4$ in BSM synthesis,
19–22
reduction sequence
of metal complex, 8–16
normal, 19–20
reversed, 20–22

reduction synthesis
Ag nanoclusters
photochemical, 82–83
radiolytic, 81–82
reductive decomposition, of Au(I)/
Ag(I)-SR complexes, 136–146
kinetics of, 141–146
size of, 136–139
structure of, 139–141
RES (retico-endothelial system), 206
retico-endothelial system (RES), 206
reversed reduction sequence, 20–22

SAXS (small-angle X-ray scattering),
120
scanning transmission electron
microscopy (STEM), 331
SC-XRD (single-crystal X-ray
diffraction), 61, 264, 300
SEC (size exclusion
chromatography), 267
sensing applications, of noble metal
clusters, 199–202
biomolecules and cancer cells,
200–202
biologically important
molecules detection,
200–202
ECL detections, 202
metal ions, 199–200
SERS (surface enhanced Raman
scattering), 108
shale gas, 338
short molecule-protected silver
nanoclusters, 113–125
silver magic-number clusters
mass spectrometry, 61–65
optical properties, 65–75
absorption spectroscopy,
67–70
electronic structure
theory, 66–67
fluorescence
spectroscopy, 70–74
spectroscopy challenges,
74–75

silver magic-number clusters
(*continued*)
   PAGE separations, 58–61
   synthesis of, 52–57
      aqueous, 52–55
      non-aqueous, 55–57
      solid-state, 57
silver nanoclusters (Ag NCs),
   271–273. *See also* Au nanoclusters
   (Au NCs)
   characteristics of, 100–101
   chemical reduction for
      preparation of, 86–92
   as conductive ink, 346–348
   C-rich ssDNA, G-rich ssDNA,
      duplex and hairpin
      structures, 366–367
   DNA-based metal nanoclusters,
      365–371
      applications of, 371–383
   microwave-assisted synthesis
      of, 85–86
   reduction synthesis of
      photochemical, 82–83
      radiolytic, 81–82
   short molecule-protected,
      113–125
   sonochemical preparation
      of, 84
   synthesis
      peptide-protected,
         110–113
      polymers protected,
         102–107
      protein-protected,
         108–110
   water-soluble fluorescent
      (*See* water-soluble
      fluorescent silver
      nanoclusters)
silver nanoparticles. *See* silver
   magic-number clusters
single cluster core
   and multiple small proteins,
      188
   in proteins, 186–188

single-crystal X-ray diffraction
   (SC-XRD), 61, 264, 300
sinter-resistant Ir nanoclusters,
   331–334
size-controlled synthesis
   metal nanoclusters, 262–276
      Ag, 271–273
      $Au_{25}(SR)_{18}$, 264–266
      $Au_{38}(SR)_{24}$, 266–268
      $Au_{102}(SR)_{44}$, 268–270
      $Au_{144}(SR)60$, 270–271
      Cu, 273–275
      monodisperse $Au_n(SR)_m$,
         271
      Pd, 275–276
      Pt, 275
stabilized/protected metal
   clusters catalysts, 294–299
      ligand-protected,
         296–299
      polymer-stabilized,
         295–296
supported metal clusters
   catalysts, 306–310
      impregnation method,
         306–307
      precursor method,
         309–310
      soft-landing method,
         307–309
size-dependent catalytic activity
   metal nanoclusters, 276–285
      aerobic oxidation,
         282–283
      CO oxidation, 276–278
      ORR, 278–282
      other catalytic
         applications, 283–285
size exclusion chromatography
   (SEC), 267
small-angle X-ray scattering (SAXS),
   120
soft-landing method, supported
   metal clusters, 307–309
soft masks, metal nanocrystals,
   418–420

solid masks, metal nanocrystals, 414–418

solid-state synthesis, silver nanoparticles, 57

solvatochromic effect, and DNA-based metal nanoclusters, 363

sonochemistry
  and AQCs, 34–35
  and fluorescent silver nanoclusters, 84

spectroscopy, and silver nanoparticles. *See also* specific types
  absorption
    ligand effects on, 69–70
    size effects on, 69
  fluorescence, 70–74
    effect of chemical environment on, 73–74
    size effects on, 72–73
  NMR (*See* Nuclear Magnetic Resonance (NMR) spectroscopy)
  UV-Vis (*See* UV-Vis spectroscopy)

SPR (surface plasmon resonance), 25, 132

SP (Surface Plasmon) band, 344

SSB (SsDNA-binding protein), 376

SsDNA-binding protein (SSB), 376

stability origin, of thiolate-protected Au NCs, 149–150

stabilized/protected metal clusters catalysts, 294–306
  applications, 301–306
    other reactions, 306
    oxidations, 301–302
    reductions, 302–306
  classification, 294
  composition-controlled synthesis, 299–300
  size-controlled synthesis, 294–299
    ligand-protected, 296–299
    polymer-stabilized, 295–296

STEM (scanning transmission electron microscopy), 331

structural phase transition, 331

supported metal clusters catalysts, 306–315
  applications, 312–315
    other reactions, 313–315
    oxidations, 312–313
  classification, 294
  composition-controlled synthesis, 310–312
  size-controlled synthesis, 306–310
    impregnation method, 306–307
    precursor method, 309–310
    soft-landing method, 307–309

surface enhanced Raman scattering (SERS), 108

surface plasmon resonance (SPR), 25, 132

Surface Plasmon (SP) band, 344

synthesis. *See also* specific types
  Brust and Brust-like, 133–136
  BSM (*See* Brust-Schiffrin two-phase method (BSM))
  chemical reduction
    and AQCs, 31–33
    of fluorescent silver nanoclusters, 86–92
  cluster, ligands for, 171–172
  composition-controlled
    stabilized/protected metal clusters catalysts, 299–300
    supported metal clusters catalysts, 310–312
  of DNA-based metal nanoclusters, 355
  electrochemical, and AQCs, 35–36
  protein protected metal clusters, 177–178

synthesis. *See also* specific types
  (*continued*)
    reduction, Ag NCs
      photochemical, 82–83
      radiolytic, 81–82
    in silico of transition metal
      nanoclusters, 324–328
    of silver magic-number
      clusters, 52–57
      aqueous, 52–55
      non-aqueous, 55–57
      solid-state, 57
    silver nanoclusters
      peptide-protected,
        110–113
      polymers protected,
        102–107
      protein-protected,
        108–110
    size-controlled (*See*
      size-controlled synthesis)
    template-assisted, and AQCs,
      38–43
      dendrimers, 39
      nucleotides, 39–41
      proteins, 41–43
    of thiolate-protected Au/Ag
      nanoclusters, 133–136
    of water-soluble fluorescent
      silver nanoclusters, 81–92
      chemical reduction,
        86–92
      microwave irradiation,
        85–86
      photochemical reduction,
        82–83
      radiolytic reduction,
        81–82
      sonochemical
        preparation, 84

T-containing oligonucleotides, and
  DNA-templated Ag NCs, 367–368
TDDFT (time-dependent
  density-functional theory), 105
TEM. *See* transmission electron
  microscopy (TEM)

temperature-assisted etching, of NP
  precursors, 159
template-assisted synthesis, and
  AQCs, 38–43
    dendrimers, 39
    nucleotides, 39–41
    proteins, 41–43
Tf receptor (TfR), 206
thermodynamic selection, and
  Au/Ag NPs thiol etching,
  158–160
    pH-assisted, 159
    temperature-assisted, 159
thiolate ligands. *See also* ligands
    chain length of, 157–158
    size of, 158
    for thiol etching, 153–154
thiolate-protected Au/Ag
  nanoclusters (Au/Ag NCs)
    and Au(I)/Ag(I)-SR complexes,
      136–146
      reductive decomposition
        kinetics, 141–146
      size of, 136–139
      structure of, 139–141
    Brust and Brust-like methods
      for, 133–136
    characteristics, 131–133
    precursors, 150
    stability origin of, 149–150
    synthesis, 133–136
thiol etching
    kinetics, 154–155
    of large Au/Ag NPs, 156–162
    of polydisperse Au/Ag NCs,
      146–156
    thiolate ligands for,
      153–154
time-dependent density-functional
  theory (TDDFT), 105
T5 loop (TTTTT– loop), 383
transition metal nanoclusters
    catalysis of, 330–340
      activation of C-H bonds,
        338–340
      multicomponent Pt alloy
        nanoclusters, 334–338

Pt nanoclusters,
338–340
sinter-resistant Ir
nanoclusters, 331–334
other functionalities of,
340–348
Ag nanoclusters as
conductive ink,
346–348
Au nanoclusters for chiral
recognition, 344–346
Pd nanoclusters for
sensing CH_4, 341–344
in silico
characterization, 328–330
synthesis, 324–328
transmission electron microscopy
(TEM), 295
DNA-based metal nanoclusters,
357
triple helix, DNA, 369
triplex structure, and
DNA-templated Ag NCs, 369–370
TTTTT– loop (T5 loop), 383
two-phase synthesis, 55
two-photon absorption, and
DNA-based metal nanoclusters,
363–364

unsaturated ketone and aldehyde
hydrogenation, and metal
clusters, 238–239
UV-Vis spectroscopy
metal nanoclusters, 267
DNA-based, 356–357
and transition, 329

water role, in BSM synthesis, 16–19
water-soluble fluorescent silver
nanoclusters. *See also* silver
nanoclusters (Ag NCs)
applications of, 92–96
in biosensing, 93–96
in chemical sensing,
92–93
characteristics of, 80–81
synthesis of, 81–92
chemical reduction,
86–92
microwave irradiation,
85–86
photochemical reduction,
82–83
radiolytic reduction,
81–82
sonochemical
preparation, 84

X-ray absorption near-edge
spectroscopy (XANES), 179, 230
X-ray absorption spectroscopy (XAS),
179
X-ray photoelectron spectroscopy
(XPS), 230
X-Ray powder diffraction (XRD), 297,
422

zero-valence metal nanoparticles,
and BSM, 19–22
normal reduction sequence,
19–20
reversed reduction sequence,
20–22